Springer Series on
SIGNALS AND COMMUNICATION TECHNOLOGY

Signals and Communication Technology

Acoustic MIMO Signal Processing
Y. Huang, J. Benesty, J. Chen
ISBN 3-540-37630-5

Algorithmic Information Theory
Mathematics of Digital Information Processing
P. Seibt
ISBN 3-540-33218-9

Distributed Cooperative Laboratories:
Networking, Instrumentation, and Measurements
F. Davoli, S. Palazzo, S. Zappatore (Eds.)
ISBN 0-387-29811-8

The Variational Bayes Method in Signal Processing
V. Šmídl, A. Quinn
ISBN 3-540-28819-8

Topics in Acoustic Echo and Noise Control
Selected Methods for the Cancellation of Acoustical Echoes, the Reduction of Background Noise, and Speech Processing
E. Hänsler, G. Schmidt (Eds.)
ISBN 3-540-33212-X

EM Modeling of Antennas and RF Components for Wireless Communication Systems
F. Gustrau, D. Manteuffel
ISBN 3-540-28614-4

Interactive Video
Algorithms and Technologies
R. Hammoud (Ed.)
ISBN 3-540-33214-6

Continuous-Time Signals
Y. Shmaliy
ISBN 1-4020-4817-3

Voice and Speech Quality Perception
Assessment and Evaluation
U. Jekosch
ISBN 3-540-24095-0

Advanced Man-Machine Interaction
Fundamentals and Implementation
K.-F. Kraiss (Ed.)
ISBN 3-540-30618-8

Orthogonal Frequency Division Multiplexing for Wireless Communications
Y. Li, G.L. Stuber (Eds.)
ISBN 0-387-29095-8

Circuits and Systems Based on Delta Modulation
Linear, Nonlinear and Mixed Mode Processing
D. G. Zrilic
ISBN 3-540-23751-8

Functional Structures in Networks
AMLn – A Language for Model Driven Development of Telecom Systems
T. Muth
ISBN 3-540-22545-5

Radio Wave Propagation for Telecommunication Applications
H. Sizun
ISBN 3-540-40758-8

Electronic Noise and Interfering Signals
Principles and Applications
G. Vasilescu
ISBN 3-540-40741-3

DVB
The Family of International Standards for Digital Video Broadcasting, 2nd ed.
U. Reimers
ISBN 3-540-43545-X

Digital Interactive TV and Metadata
Future Broadcast Multimedia
A. Lugmayr, S. Niiranen, and S. Kalli
ISBN 0-387-20843-7

Adaptive Antenna Arrays
Trends and Applications
S. Chandran (Ed.)
ISBN 3-540-20199-8

Neuro-Fuzzy and Fuzzy-Neural Applications in Telecommunications
P. Stavroulakis (Ed.)
ISBN 3-540-40759-6

SDMA for Multipath Wireless Channels
Limiting Characteristics and Stochastic Models
I.P. Kovalyov
ISBN 3-540-40225-X

Digital Television
A Practical Guide for Engineers
W. Fischer
ISBN 3-540-01155-2

Speech Enhancement
J. Benesty, S. Makino, J. Chen (Eds.)
ISBN 3-540-24039-X

continued after index

DUE
14 NOV 2008
LIBREX-

Yiteng Huang · Jacob Benesty · Jingdong Chen

Acoustic MIMO Signal Processing

With 71 Figures

Yiteng Huang
Lucent Technologies
Bell Laboratories
Mountain Avenue 600-700
Murray Hill, NJ 07974-0636
USA

Jacob Benesty
Université de Quebec
INRS-EMT
de la Gauchetiere Ouest 800
H5A 1K6 Montreal
Canada

Jingdong Chen
Lucent Technologies
Bell Laboratories
Mountain Avenue 600-700
Murray Hill, NJ 07974-0636
USA

ISBN 10 3-540-37630-5 **Springer Berlin Heidelberg New York**
ISBN 13 978-3-540-37630-9 **Springer Berlin Heidelberg New York**

Library of Congress Control Number: 2006931130

Springer is a part of Springer Science+Business Media

springer.com

Printed in Germany

Typesetting: Data conversion by authors.
Final processing by PTP-Berlin Protago-TEX-Production GmbH, Germany
Cover-Design: WMXDesign, Heidelberg
Printed on acid-free paper 62/3141/Yu - 5 4 3 2 1 0

Preface

The book is the crystallization of the synergistic collaboration among the three authors, which can trace back to five years ago when we were all working with Bell Laboratories, the research arm of Lucent Technologies, at Murray Hill, New Jersey. In the last century, Bell Labs was known as a creative haven for innovations. In particular, it is well known and widely accepted that acoustics and speech scientists at Bell Labs made great contributions to the advances of multimedia communications. It's our hope to keep alive this tradition of excellence wherever we are now and share with many readers some interesting and useful ideas that all started in that magical place along with several other great researchers.

The idea of writing a monograph book was floated after we saw the success of a number of books we edited and when we thought that our own research had produced adequate results, which are useful and original. In addition, we believed that the reader would benefit from the fact that various acoustic MIMO (multiple-input multiple output) signal processing problems could be addressed in a more unified framework with more consistent notation in a monograph book than in an edited book. We had a good plan, but the route from word to deed was longer than we initially hoped for. We started writing our first chapter almost a year after signing the contract with Springer. While the writing process was sometimes slow (particularly when the end was approaching), it was in general quite enjoyable: we not only summarized what we knew, but also got new ideas about what we can do in the future. We hope that you will find the reading of this book to be as stimulating as we have found the writing to be.

Finally, we deeply thank Dieter Merkel and Christoph Baumann from Springer (Germany) for their wonderful help in the preparation and publication of this book.

Murray Hill, New Jersey/Montréal, Québec

Yiteng Huang
Jacob Benesty
Jingdong Chen

Contents

1 Introduction 1
1.1 Acoustic MIMO Signal Processing 1
1.2 Organization of the Book 4

Part I Theory

2 Acoustic MIMO Systems 9
2.1 Signal Models 9
2.1.1 SISO Model 9
2.1.2 SIMO Model 11
2.1.3 MISO Model 12
2.1.4 MIMO Model 12
2.2 Characteristics of Acoustic Channels 13
2.2.1 Linearity and Shift-Invariance 14
2.2.2 FIR Representation 14
2.2.3 Time-Varying Channel Impulse Responses 14
2.2.4 Frequency Selectivity 15
2.2.5 Reverberation Time 15
2.2.6 Channel Invertibility and Minimum-Phase Filter 16
2.2.7 Multichannel Diversity and the Common-Zero Problem 18
2.2.8 Sparse Impulse Response 19
2.3 Measurement and Simulation of MIMO Acoustic Systems 21
2.3.1 Direct Measurement of Acoustic Impulse Responses 22
2.3.2 Image Model for Acoustic Impulse Response Simulation 24
2.4 Summary 29

3 **Wiener Filter and Basic Adaptive Algorithms** 31
3.1 Introduction 31
3.2 Wiener Filter 32
3.3 Impulse Response Tail Effect 34
3.4 Condition Number 35
3.4.1 Decomposition of the Correlation Matrix 36
3.4.2 Condition Number with the Frobenius Norm 37
3.4.3 Fast Computation of the Condition Number 39
3.5 Basic Adaptive Algorithms 41
3.5.1 Deterministic Algorithm 41
3.5.2 Stochastic Algorithm 44
3.5.3 Sign Algorithms 46
3.6 MIMO Wiener Filter 48
3.7 Numerical Examples 53
3.8 Summary 56

4 **Sparse Adaptive Filters** 59
4.1 Introduction 59
4.2 Notation and Definitions 60
4.3 The NLMS, PNLMS, and IPNLMS Algorithms 61
4.4 Universal Criterion 64
4.4.1 Linear Update 65
4.4.2 Non-Linear Update 67
4.5 Exponentiated Gradient Algorithms 68
4.5.1 The EG Algorithm for Positive Weights 68
4.5.2 The EG± Algorithm for Positive and Negative Weights 69
4.5.3 The Exponentiated RLS (ERLS) Algorithm 71
4.6 The Lambert W Function Based Gradient Algorithm 72
4.7 Some Important Links Among Algorithms 74
4.7.1 Link Between NLMS and EG± Algorithms 74
4.7.2 Link Between IPNLMS and EG± Algorithms 75
4.7.3 Link Between LWG and EG± Algorithms 77
4.8 Numerical Examples 78
4.9 Summary 83

5 **Frequency-Domain Adaptive Filters** 85
5.1 Introduction 85
5.2 Derivation of SISO FD Adaptive Algorithms 86
5.2.1 Criterion 86
5.2.2 Normal Equations 89
5.2.3 Adaptive Algorithms 91
5.2.4 Convergence Analysis 93
5.3 Approximation and Special Cases 96
5.3.1 Approximation 96
5.3.2 Special Cases 98

5.4 FD Affine Projection Algorithm ... 99
5.5 Generalization to the MISO System Case ... 101
5.6 Numerical Examples ... 104
5.7 Summary ... 106

6 Blind Identification of Acoustic MIMO Systems ... 109
6.1 Introduction ... 109
6.2 Blind SIMO Identification ... 111
6.2.1 Identifiability and Principle ... 111
6.2.2 Constrained Time-Domain Multichannel LMS and Newton Algorithms ... 113
6.2.3 Unconstrained Multichannel LMS Algorithm with Optimal Step-Size Control ... 120
6.2.4 Frequency-Domain Unnormalized and Normalized Multichannel LMS Algorithms ... 122
6.2.5 Adaptive Multichannel Exponentiated Gradient Algorithm ... 135
6.2.6 Numerical Examples ... 141
6.3 Blind MIMO Identification ... 147
6.3.1 Problem Formulation and Background Review ... 148
6.3.2 Memoryless MIMO System with White Inputs ... 151
6.3.3 Memoryless MIMO System with Colored Inputs ... 152
6.3.4 Convolutive MIMO Systems with White Inputs ... 154
6.3.5 Convolutive MIMO Systems with Colored Inputs ... 156
6.3.6 Frequency-Domain Blind Identification of Convolutive MIMO Systems and Permutation Inconsistency ... 157
6.3.7 Convolutive MIMO Systems with White but Quasistationary Inputs ... 158
6.4 Summary ... 160
6.5 Appendix. Blind SIMO Identification: A Derivation Directly from the Covariance Matrices of the System Outputs ... 161

7 Separation and Suppression of Co-Channel and Temporal Interference ... 169
7.1 Introduction ... 169
7.2 Separating Co-Channel and Temporal Interference ... 170
7.2.1 Example: Conversion of a 2×3 MIMO System to Two SIMO Systems ... 170
7.2.2 Generalization to $M \times N$ MIMO Systems with $M > 2$ and $M < N$ ... 174
7.3 Suppressing Temporal Interference ... 177
7.3.1 Direct Inverse (Zero-Forcing) Equalizer ... 178
7.3.2 MMSE Equalizer ... 179
7.3.3 MINT Equalizers ... 179
7.4 Summary ... 182

Part II Applications

8 Acoustic Echo Cancellation and Audio Bridging 185
8.1 Introduction 185
8.2 Network Echo Problem 186
8.3 Single-Channel Acoustic Echo Cancellation 188
8.4 Multichannel Acoustic Echo Cancellation 190
8.4.1 Multi versus Mono 190
8.4.2 Multichannel Identification and the Nonuniqueness Problem 192
8.4.3 Impulse Response Tail Effect 194
8.4.4 Some Different Solutions for Decorrelation 195
8.5 Hybrid Mono/Stereo Acoustic Echo Canceler 199
8.6 Double-Talk Detection 200
8.6.1 Basics 200
8.6.2 Double-Talk Detection Algorithms 202
8.6.3 Performance Evaluation of DTDs 206
8.7 Audio Bridging 206
8.7.1 Principle 206
8.7.2 Interchannel Differences for Synthesizing Stereo Sound . 209
8.7.3 Choice of Interchannel Differences for Stereo AEC 211
8.8 Summary 212

9 Time Delay Estimation and Acoustic Source Localization . . 215
9.1 Time Delay Estimation 215
9.2 Cross-Correlation Method 217
9.3 Magnitude-Difference Method 219
9.4 Maximum Likelihood Method 220
9.5 Generalized Cross-Correlation Method 223
9.6 Adaptive Eigenvalue Decomposition Algorithm 226
9.7 Multichannel Cross-Correlation Algorithm 227
9.7.1 Forward Spatial Linear Prediction 228
9.7.2 Backward Spatial Linear Prediction 230
9.7.3 Spatial Linear Interpolation 231
9.7.4 Time Delay Estimation Using Spatial Linear Prediction 232
9.7.5 Spatial Correlation Matrix and Its Properties 233
9.7.6 Multichannel Cross-Correlation Coefficient 235
9.7.7 Time Delay Estimation Using MCCC 235
9.8 Adaptive Multichannel Time Delay Estimation 236
9.9 Acoustic Source Localization 238
9.10 Measurement Model and Cramèr-Rao Lower Bound 239
9.11 Algorithm Overview 242
9.12 Maximum Likelihood Estimator 243
9.13 Least-Squares Estimators 244

9.13.1 Least-Squares Error Criteria ... 245
9.13.2 Spherical Intersection (SX) Estimator ... 247
9.13.3 Spherical Interpolation (SI) Estimator ... 247
9.13.4 Linear-Correction Least-Squares Estimator ... 248
9.14 Example System Implementation ... 254
9.15 Summary ... 259

10 Speech Enhancement and Noise Reduction ... 261
10.1 Introduction ... 261
10.2 Noise-Reduction and Speech-Distortion Measures ... 263
10.2.1 Noise-Reduction Factor and Noise-Reduction Gain Function ... 264
10.2.2 Speech-Distortion Index and Attenuation Frequency Distortion ... 265
10.2.3 Signal-to-Noise Ratio ... 265
10.2.4 Log-Spectral Distance ... 266
10.2.5 Itakura Distance ... 266
10.2.6 Itakura-Saito Distance ... 268
10.2.7 Mean Opinion Score ... 269
10.3 Single-Channel Noise-Reduction Algorithms: a Brief Overview ... 269
10.4 Time-Domain Wiener Filter ... 270
10.4.1 Estimation of the Clean Speech Samples ... 270
10.4.2 Estimation of the Noise Samples ... 273
10.4.3 Noise Reduction versus Speech Distortion ... 274
10.4.4 *A Priori* SNR versus *a Posteriori* SNR ... 277
10.4.5 Bounds for Noise Reduction and Speech Distortion ... 281
10.4.6 Particular Case: White Gaussian Noise ... 282
10.4.7 A Suboptimal Filter ... 283
10.5 Frequency-Domain Wiener Filter ... 287
10.5.1 Estimation of the Clean Speech Spectrum ... 287
10.5.2 *A Priori* SNR versus *a Posteriori* SNR ... 290
10.6 Noise Reduction Through Spectral Magnitude Restoration ... 292
10.7 Spectral Subtraction ... 293
10.7.1 Estimation of the Spectral Magnitude of the Clean Speech ... 293
10.7.2 Estimation of the Noise Spectrum ... 295
10.7.3 Relationship Between Spectral Subtraction and Wiener Filtering ... 296
10.7.4 Estimation of the Wiener Gain Filter ... 299
10.7.5 Simulations ... 300
10.8 Adaptive Noise Cancellation ... 302
10.8.1 Estimation of the Clean Speech ... 302
10.8.2 Ideal Noise Cancellation Performance ... 304
10.8.3 Signal Cancellation Problem ... 305
10.8.4 Simulations ... 307

10.9 Noise Reduction with a Microphone Array 309
10.9.1 Delay-and-Sum Algorithm 310
10.9.2 Linearly Constrained Algorithms 312
10.10 Summary 317

11 Source Separation and Speech Dereverberation 319
11.1 Cocktail Party Effect 319
11.2 Source Separation 323
11.2.1 Microphone Array Beamforming. 323
11.2.2 Independent Component Analysis and Blind Source Separation 331
11.3 A Synergistic Solution to Source Separation and Speech Dereverberation 341
11.4 Summary 350

References 353

Index 373

List of Tables

2.1 Average reverberation time for acoustic channel impulse responses measured by Härmä in the varechoic chamber as a function of the percentage of open panels. 24
2.2 Image method for simulating an acoustic FIR channel. 29

3.1 Computation of the condition number with the Levinson-Durbin algorithm. 41
3.2 The normalized LMS (NLMS) algorithm. 47
3.3 The MISO NLMS algorithm. 50

4.1 The improved PNLMS algorithm. 64
4.2 The (regularized) affine projection algorithm. 66
4.3 The RLS algorithm. 66
4.4 The EG± algorithm. 71

5.1 The constrained and unconstrained FLMS algorithms. 99
5.2 The constrained and unconstrained MDF algorithms. 100
5.3 The constrained and unconstrained two-channel FLMS algorithms. 103

6.1 The constrained multichannel LMS adaptive algorithm for the blind identification of a SIMO FIR system. 117
6.2 The constrained multichannel Newton adaptive algorithm for the blind identification of a SIMO FIR system. 119
6.3 The variable step-size unconstrained multichannel LMS adaptive algorithm for the blind identification of a SIMO FIR system. 123
6.4 The frequency-domain constrained multichannel LMS (FCMCLMS) adaptive algorithm for the blind identification of a SIMO FIR system. 129

6.5 The frequency-domain normalized multichannel LMS (FNMCLMS) adaptive algorithm for the blind identification of a SIMO FIR system. 136
6.6 The multichannel exponentiated gradient algorithm with positive and negative weights (MCEG$\pm$) for the blind identification of a sparse SIMO FIR system. 140

9.1 The cross-correlation method for time delay estimation. 219
9.2 The magnitude-difference method for time delay estimation. . . . 220
9.3 The maximum likelihood method for time delay estimation. 224
9.4 Commonly used weighting functions in the GCC method. 225
9.5 The generalized cross-correlation method for time delay estimation. 225
9.6 The adaptive eigenvalue decomposition algorithm for time delay estimation. 227
9.7 The multichannel cross-correlation method for time delay estimation. 237
9.8 The multichannel adaptive algorithm for time delay estimation. 238

11.1 Performance of the source separation and speech dereverberation algorithm for extracting a speech signal of interest from a 3×4 acoustic MIMO system in a reverberant environment simulated with data measured in the varechoic chamber at Bell Labs with different panel configurations. 347

1

Introduction

1.1 Acoustic MIMO Signal Processing

When Alexander Graham Bell applied for a patent to protect the epoch-making invention of the telephone in 1876, he probably could not have imagined the kind of communication we have today. Not only are we now able to talk to someone on the other side of the earth (or even in the space) at any time, we also can have a clear conversation with an amazingly short delay for an unbelievably cheap price. It is all thanks to the tremendous advances that we have made in voice telecommunication technology for the last century, from analog to digital signals, from manual to automatic switching, from narrowband to broadband speech, from wireline to wireless terminals, from circuit-switched to packet-switched networks, just to name a few.

In spite of the fascinating developments that have been accomplished, most forms of modern speech telecommunication, however, are still not natural enough to be considered effortless or without distraction, leading to diminished interaction and unsatisfactory productivity. Many companies are still believing that cheap air tickets are far more important than anything else in improving communication among their employees and with their customers. But a desire has never been so strong for less business travels because of a creeping concern with public safety (e.g., fear of terrorist attack) and growing travel expenses due to the recent surge of gas and oil price. Consequently it is no longer a luxury but truly a rational demand to create a lifelike communication mode that makes tele-collaboration and online conferencing more enjoyable and easier to manage without compromising productivity.

By tactically adding video, we may arguably become more efficient at collaborating with others in a multi-party teleconferencing: a quick response and smooth turn-taking can be elicited; subtle emotion and opinion that we convey through body language can be faithfully shared with other participants; and we can easily monitor the involvement of the others. But voice is anyway the most natural form of human communications. Therefore, speech remains one of the focal points of telecommunications and there still exists a strong interest in innovative acoustic and speech techniques that can improve voice communication.

An agreement seems to have been reached recently by researchers in the speech communication society that users of voice communication and speech processing systems will benefit from an acoustic interface that replicates attributes of face-to-face communication, including (a) full-duplex exchange, (b) freedom of movement without body-worn or tethered microphones (i.e., hands free in the broad sense), (c) high-fidelity (i.e., low-distortion), distractionless speech signals captured from a distance, (d) spatial realism of sound rendering. Such an interface will give the involved people the impression of being in the same acoustic environment, which is commonly referred to as "*immersive experience*" in the multimedia communication literature.

Although attractive, the forward-looking vision of the next-generation speech communication systems poses great challenges on acoustic signal processing. It is certainly a difficult problem to render a sound field with spatial richness for multiple sound sources at the listener's ears. But what we are primarily concerned with in this book are the problems on the speech acquisition side.

On this side, we are facing a complicated acoustic environment. In the microphone signals, the speech of interest is possibly mixed with four different kinds of distractions:

- *Noise.* Ambient noise is generally inevitable and is anywhere at all time.
- *Echo.* Acoustic echo occurs due to the coupling between the loudspeakers and the microphones. The existence of echo will make conversations very difficult.
- *Reverberation.* Reverberation is the result of multipath propagation and is introduced by an enclosure. Although reverberation imparts useful information about the size of the surrounding enclosure and can enhance the harmonic effects of music, it causes spectral distortion, which impairs speech intelligibility, and makes directional cues for source localization unclear. In speech communication, reverberation is typically considered as a destructive factor.
- *Interfering signals from other concurrent sound sources.* There are multiple competing sound sources since tele-collaboration potentially involves a number of participants at every conferencing site.

Combating these distractions leads to the developments of diverse acoustic signal processing techniques. They include noise reduction, echo cancellation, speech dereverberation, and source separation, each of which is a rich subject for research.

In addition to speech acquisition with smart control of these interfering distractions, we also want to extract the spatial information associated with the speech sources of interest. On one hand, the knowledge of source positions is a prerequisite of some speech acquisition algorithms like beamforming. On the other hand, without the knowledge, locally reproducing acoustic environments of the far ends is impossible. Therefore acoustic source localization is indispensable and we have more comprehensive requirements for the immersive acoustic interface.

These acoustic signal processing problems, as a matter of fact, are not new. They have been actually investigated for years. But new opportunities emerge as multiple microphones and loudspeakers will be likely employed in the next-generation speech communication systems. It will be interesting to have all these problems be investigated in a unified framework and we attempt to do so in this book.

An immersive acoustic interface by its nature involves multiple sound sources (including loudspeakers) and multiple microphones, fitting well into the multiple-input multiple-output (MIMO) structure. In a MIMO signal model, the channel that connects a sound source to a microphone is described with a finite impulse response (FIR) filter. The spatial information of the sound source and the reverberation effect are all self-contained in the FIR filter. Using such a model to explore the acoustic signal processing problems, we shift our focus from solving the wave equation with proper boundary conditions to identifying the channel impulse responses. Although in many acoustic signal processing problems we do not have to explicitly estimate the channel impulse responses, the challenges of these problems can be easily explained by our incapability of dynamically identifying an acoustic MIMO system. It is our belief that the mother of all challenges in acoustic signal processing is how to accurately estimate the impulse responses of an acoustic MIMO system in real time. So we put a specific emphasis on recent important results of acoustic MIMO identification in this book.

Over the last several years, the MIMO model has been extensively investigated in wireless communications since a multiple-antenna system holds the promise of much higher spectral efficiencies for wireless channels. Although wireless and acoustic channels have many things in common (e.g., time varying, frequency selective), acoustic MIMO systems are substantially different from that of wireless communications. As opposed to communication receivers, the human ear has an extremely wide dynamic range and is much more sensitive to weak tails of the channel impulse responses. As a result, the length of acoustic channel impulse response models is significantly greater than that in wireless communications. Filter lengths of thousands of samples are not uncommon in acoustic MIMO systems while wireless impulse responses consist of usually not more than a few tens of taps. In addition, since communication systems can use a pilot signal for channel identification, such a problem has never been an obstacle to developing practical wireless MIMO communication systems. But, in acoustics, we seek techniques for human talkers. The source signals in acoustic systems are random and their statistics are considerably different from that in wireless communications. The speech signal is neither stationary nor white, and does not form a known set of signal alphabet as in wireless communications, making the identification of an acoustic MIMO system apparently much more challenging.

This book is designed to address these challenges. It is tutorial in nature with a detailed exposition of the state-of-the-art in the field, while open questions and future research directions are also explored. We hope that the result will be a useful text for a large audience, ranging from researchers, Ph.D. students, to practicing engineers, either just approaching the field or already having years of experience.

1.2 Organization of the Book

The main body of this book is organized into two parts. In the first part (Chap. 2 to Chap. 7), the theory of acoustic MIMO signal processing is developed. The second part (Chap. 8 to Chap. 11) focuses on various applications.

Chapter 2 provides a complete overview of acoustic systems. We discuss how to model an acoustic system according to the number of its inputs and outputs, and introduce the acoustic MIMO models in both the time and frequency domains. We also try to characterize acoustic channels. Both single-channel and multichannel properties are discussed. Finally, facilities for directly measuring acoustic impulse responses and the image method for their simulation are explored.

The Wiener filter is a very important tool in signal processing, not only for acoustic applications, but also in many others. Chapter 3 derives the Wiener filter in the context of identification of acoustic MIMO systems when a reference signal is available. Some fundamental differences between the SISO and MISO cases are discussed. Also in this chapter, some basic adaptive algorithms such as the deterministic algorithm, LMS, NLMS, and sign algorithms are derived.

Acoustic impulse responses are usually sparse. However, all classical algorithms such as the NLMS and RLS adaptive filters, do not take this information into account. In Chap. 4, we show how this feature can be utilized to obtain adaptive algorithms with much better initial convergence and tracking abilities than the classical ones. Important sparse adaptive algorithms, such as PNLMS, IPNLMS, and exponentiated gradient algorithms, are explained. It is also shown how all these algorithms are related to each other.

When things come to implementation, it is fundamental to have adaptive filters that are very efficient from an arithmetic complexity point of view. Because frequency-domain algorithms use essentially the fast Fourier transform (FFT) to compute the convolution and to update the coefficients of the filter, they are excellent candidates. Chapter 5 explains how frequency-domain algorithms can be derived rigorously from a recursive least-squares criterion, with a block size independent of the length of the adaptive filter. All classical algorithms (in single-channel and multichannel cases) can be obtained from this approach.

In most cases, the speech inputs to an acoustic MIMO system are unknown and we can use only the observations at the outputs to blindly identify the system. In Chap. 6, we show how blind identification of SIMO and MIMO systems can be performed, with emphasis on second-order-statistics based methods. For SIMO systems, a rich set of adaptive algorithms (MCLMS, MCN, VSS-UMCLMS, FNMCLMS, etc.) are developed. For MIMO systems, we analyze different scenarios and show when the problem can and when it cannot be solved.

One of the challenges in acoustic MIMO signal processing problems lies in the fact that there exist both co-channel and temporal interference in the outputs. Chapter 7 discusses the key issue of separating and suppressing co-channel and temporal interference. We show the conditions of separability for co-channel and temporal interference and illustrate the three approaches to suppressing temporal interference: the direct inverse, MMSE, and MINT methods.

Acoustic signal processing for speech and audio has its debut with acoustic echo cancellation (AEC). It is well-known that in hands-free telephony, the acoustic coupling between the loudspeaker and the microphone generates echoes that can be extremely harmful. So AEC is required. Chapter 8 gives an overview on this important area of research. The multichannel aspect is emphasized since teleconferencing systems of the future will, without a doubt, have multiple loudspeakers and microphones. This chapter also discusses the concept of stereo audio bridging, which we believe will be a big part of the picture of the next-generation voice over IP systems.

Chapter 9 consists of two parts, one on time delay estimation and the other on acoustic source localization. The former deals with the measurement of time difference of arrival (TDOA) between signals received by spatially separated microphones. It addresses a wide variety of techniques ranging from the simple cross-correlation method to the advanced blind system identification based algorithms, with significant attention being paid to combating the reverberation effects. The latter discusses how to employ the TDOA measurements to locate, in the acoustic wavefield, radiating sources that emit signals to microphones. It formulates the problem from an estimation-theory point of view and presents various location estimators, some of which have the potential to achieve the Cramèr-Rao lower bound.

Chapter 10 is devoted to the speech-enhancement/noise-reduction problem, which aims at estimating the speech of interest from its observations corrupted by additive noise. In an acoustic MIMO system, speech enhancement can be achieved by processing the waveform received by a single microphone, but often it is advantageous to use multiple microphones. This chapter covers not only the well-recognized single-channel techniques such as Wiener filtering and spectral subtraction, but also the multichannel techniques such as adaptive noise cancellation and spatio-temporal filtering approaches.

Finally, in Chap. 11, we discuss source separation and speech dereverberation. We study the two difficult problems together in one chapter since they are closely associated with each other in acoustic MIMO systems. We begin with a review of the cocktail party effect and try to shed some light on what implications we can derive for developing more effective source separation algorithms. We compare the beamforming and independent component analysis methods for the source separation problem. We also present a synergistic solution to source separation and speech dereverberation based on blind identification of acoustic MIMO systems.

Part I

Theory

2

Acoustic MIMO Systems

System modeling is fundamental to signal-processing and control theories, and so is to acoustic applications. Creating a mathematical representation of the acoustic environment helps us gain a better understanding of what is going on and enables better visualization of the main elements of an acoustic signal processing problem. Certainly it also forms a basis for discussion of various acoustic problems using the same convenient language of mathematics employed in the rest of this book. For these reasons we will investigate signal models and characteristics of acoustic systems in this chapter.

2.1 Signal Models

For many acoustic signal processing problems and applications, particularly those to be discussed later in this book, the numbers of inputs and outputs of the acoustic system have been found to crucially account for the choice of algorithms and their complexity. Therefore, while there might be many ways to classify types of acoustic systems, we believe that the most important one is in terms of whether there are single or multiple inputs and outputs, which leads to four basic signal models as detailed in the following.

2.1.1 SISO Model

We choose to present the signal models from easy to complex. The first is the single-input single-output (SISO) system, as shown in Fig. 2.1(a). The output signal is given by

$$x(k) = h * s(k) + b(k), \tag{2.1}$$

where h is the channel impulse response, the symbol $*$ denotes the linear convolution operator, $s(k)$ is the source signal at time k, and $b(k)$ is the additive noise at the output. Here we assume the system is linear and shift-invariant, which are routinely used for formulating acoustic signal problems.

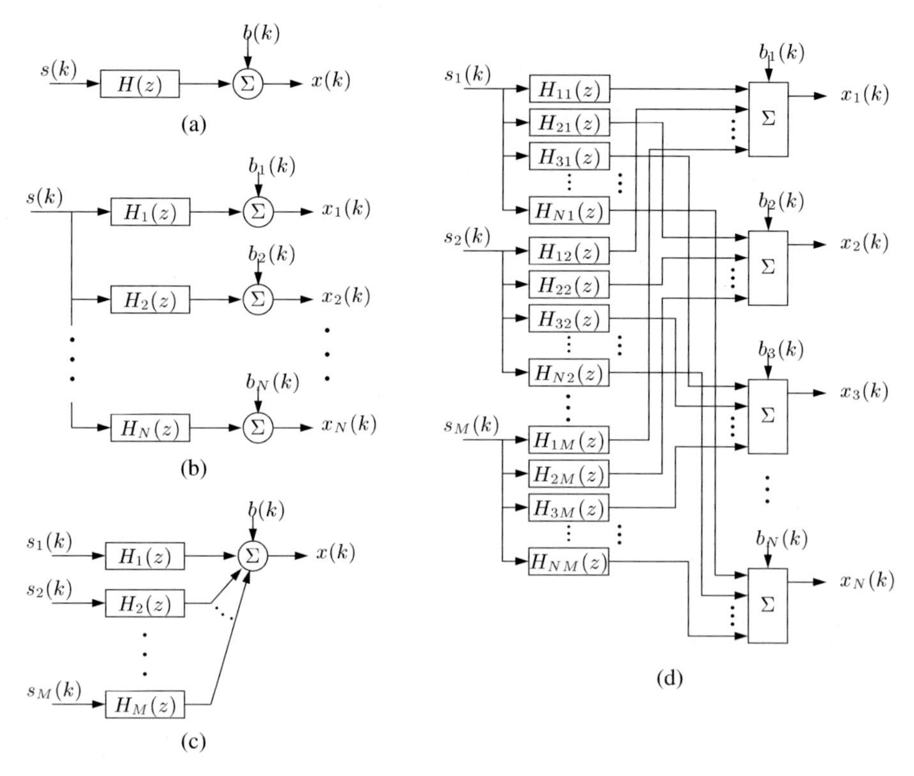

Fig. 2.1. Illustration of four distinct types of acoustic systems. (a) A single-input single-output (SISO) system. (b) A single-input multiple-output (SIMO) system. (c) A multiple-input single-output (MISO) system. (d) A multiple-input multiple-output (MIMO) system.

The channel impulse response is delineated usually with an FIR filter rather than an infinite impulse response (IIR) filter, as explained later in this chapter when characteristics of acoustic channels are discussed. In vector/matrix form, the SISO signal model (2.1) is written as:

$$x(k) = \mathbf{h}^T \mathbf{s}(k) + b(k), \tag{2.2}$$

where

$$\begin{aligned} \mathbf{h} &= \begin{bmatrix} h_0 & h_1 & \cdots & h_{L-1} \end{bmatrix}^T, \\ \mathbf{s}(k) &= \begin{bmatrix} s(k) & s(k-1) & \cdots & s(k-L+1) \end{bmatrix}^T, \end{aligned}$$

$[\cdot]^T$ denotes the transpose of a matrix or a vector, and L is the channel length.

Using the z transform, the SISO signal model (2.2) is described as follows:

$$X(z) = H(z)S(z) + B(z), \tag{2.3}$$

where $X(z)$, $S(z)$, and $B(z)$ are the z-transforms of $x(k)$, $s(k)$, and $b(k)$, respectively, and $H(z) = \sum_{l=0}^{L-1} h_l z^{-l}$.

The SISO model is simple and is probably the most widely used and studied model in communication, signal processing, and control theories.

2.1.2 SIMO Model

The diagram of a single-input multiple-output (SIMO) system is illustrated by Fig. 2.1(b), in which there are N outputs from the same sound source as input and the nth output is expressed as:

$$x_n(k) = \mathbf{h}_n^T \mathbf{s}(k) + b_n(k), \quad n = 1, 2, \cdots, N, \tag{2.4}$$

where $x_n(k)$, $\mathbf{h}_n$, and $b_n(k)$ are defined in a similar way to those in (2.2), and L is the length of the longest channel impulse response in this SIMO system. A more comprehensive expression of the SIMO model is given by

$$\mathbf{x}(k) = \mathbf{H}\mathbf{s}(k) + \mathbf{b}(k), \tag{2.5}$$

where

$$\begin{aligned}
\mathbf{x}(k) &= \begin{bmatrix} x_1(k) & x_2(k) & \cdots & x_N(k) \end{bmatrix}^T, \\
\mathbf{H} &= \begin{bmatrix} h_{1,0} & h_{1,1} & \cdots & h_{1,L-1} \\ h_{2,0} & h_{2,1} & \cdots & h_{2,L-1} \\ \vdots & \vdots & \ddots & \vdots \\ h_{N,0} & h_{N,1} & \cdots & h_{N,L-1} \end{bmatrix}_{N \times L}, \\
\mathbf{b}(k) &= \begin{bmatrix} b_1(k) & b_2(k) & \cdots & b_N(k) \end{bmatrix}^T.
\end{aligned}$$

The SIMO model (2.5) is described in the z-transform domain as:

$$\vec{\mathbf{X}}(z) = \vec{\mathbf{H}}(z)S(z) + \vec{\mathbf{B}}(z), \tag{2.6}$$

where

$$\begin{aligned}
\vec{\mathbf{X}}(z) &= \begin{bmatrix} X_1(z) & X_2(z) & \cdots & X_N(z) \end{bmatrix}^T, \\
\vec{\mathbf{H}}(z) &= \begin{bmatrix} H_1(z) & H_2(z) & \cdots & H_N(z) \end{bmatrix}^T, \\
H_n(z) &= \sum_{l=0}^{L-1} h_{n,l} z^{-l}, \quad n = 1, 2, \cdots, N, \\
\vec{\mathbf{B}}(z) &= \begin{bmatrix} B_1(z) & B_2(z) & \cdots & B_N(z) \end{bmatrix}^T.
\end{aligned}$$

As you may have observed from (2.6), we put an arrow on the top of a bold, uppercase character to represent a vector in the z-transform domain in this book.

2.1.3 MISO Model

In the third type of acoustic systems as drawn in Fig. 2.1(c), we suppose that there are M sources but only one microphone whose signal is then expressed as:

$$\begin{aligned} x(k) &= \sum_{m=1}^{M} \mathbf{h}_m^T \mathbf{s}_m(k) + b(k), \\ &= \mathbf{h}^T \mathbf{s}(k) + b(k), \end{aligned} \quad (2.7)$$

where

$$\begin{aligned} \mathbf{h} &= \begin{bmatrix} \mathbf{h}_1^T & \mathbf{h}_2^T & \cdots & \mathbf{h}_M^T \end{bmatrix}^T, \\ \mathbf{h}_m &= \begin{bmatrix} h_{m,0} & h_{m,1} & \cdots & h_{m,L-1} \end{bmatrix}^T, \\ \mathbf{s}(k) &= \begin{bmatrix} \mathbf{s}_1^T(k) & \mathbf{s}_2^T(k) & \cdots & \mathbf{s}_M^T(k) \end{bmatrix}^T, \\ \mathbf{s}_m(k) &= \begin{bmatrix} s_m(k) & s_m(k-1) & \cdots & s_m(k-L+1) \end{bmatrix}^T. \end{aligned}$$

In the z-transform domain, the MISO model is given by

$$X(z) = \vec{\mathbf{H}}^T(z)\vec{\mathbf{S}}(z) + B(z), \quad (2.8)$$

where

$$\begin{aligned} \vec{\mathbf{H}}(z) &= \begin{bmatrix} H_1(z) & H_2(z) & \cdots & H_M(z) \end{bmatrix}^T, \\ H_m(z) &= \sum_{l=0}^{L-1} h_{m,l} z^{-l}, \quad m = 1, 2, \cdots, M, \\ \vec{\mathbf{S}}(z) &= \begin{bmatrix} S_1(z) & S_2(z) & \cdots & S_M(z) \end{bmatrix}^T. \end{aligned}$$

Note that $\vec{\mathbf{H}}(z)$ defined here is slightly different from that in (2.6). We do not deliberately distinguish them since their dimension can be easily deduced from the context if slight attention is paid.

2.1.4 MIMO Model

Figure 2.1(d) depicts a multiple-input multiple-output (MIMO) system. A MIMO system with M inputs and N outputs is referred to as an $M \times N$ system. At time k, we have

$$\mathbf{x}(k) = \mathbf{H}\mathbf{s}(k) + \mathbf{b}(k), \quad (2.9)$$

where

$$
\begin{aligned}
\mathbf{x}(k) &= \begin{bmatrix} x_1(k) & x_2(k) & \cdots & x_N(k) \end{bmatrix}^T, \\
\mathbf{H} &= \begin{bmatrix} \mathbf{H}_1 & \mathbf{H}_2 & \cdots & \mathbf{H}_M \end{bmatrix}, \\
\mathbf{H}_m &= \begin{bmatrix} h_{1m,0} & h_{1m,1} & \cdots & h_{1m,L-1} \\ h_{2m,0} & h_{2m,1} & \cdots & h_{2m,L-1} \\ \vdots & \vdots & \ddots & \vdots \\ h_{Nm,0} & h_{Nm,1} & \cdots & h_{Nm,L-1} \end{bmatrix}_{N \times L}, \\
& \quad m = 1, 2, \cdots, M, \\
\mathbf{b}(k) &= \begin{bmatrix} b_1(k) & b_2(k) & \cdots & b_N(k) \end{bmatrix}^T,
\end{aligned}
$$

h_{nm} $(n = 1, 2, \cdots, N,\ m = 1, 2, \cdots, M)$ is the impulse response of the channel from input m to output n, and $\mathbf{s}(k)$ is defined similarly to that in (2.7). Again, we have the model presented in the z-transform domain as

$$
\vec{\mathbf{X}}(z) = \mathbf{H}(z)\vec{\mathbf{S}}(z) + \vec{\mathbf{B}}(z), \tag{2.10}
$$

where

$$
\begin{aligned}
\mathbf{H}(z) &= \begin{bmatrix} H_{11}(z) & H_{12}(z) & \cdots & H_{1M}(z) \\ H_{21}(z) & H_{22}(z) & \cdots & H_{2M}(z) \\ \vdots & \vdots & \ddots & \vdots \\ H_{N1}(z) & H_{N2}(z) & \cdots & H_{NM}(z) \end{bmatrix}, \\
H_{nm}(z) &= \sum_{l=0}^{L-1} h_{nm,l} z^{-l}, \quad n = 1, 2, \cdots, N, \ m = 1, 2, \cdots, M.
\end{aligned}
$$

Clearly the MIMO system is the most general model and all other three systems can be treated as special examples of a MIMO system. The difference among these models might look trivial here. But as our study proceeds, the significance of this classification will become clear.

2.2 Characteristics of Acoustic Channels

While the research centered on the MIMO structure in acoustic signal processing systems is relatively quite young, the MIMO model has been extensively investigated in wireless communications in the last decade since a multiple-antenna system holds the promise of much higher spectral efficiencies for wireless channels [291], [120]. However, acoustic channels are substantially different from wireless and other channels. The distinctive characteristics of acoustic channels are what set it apart from the others and are what we need to pay special attention to in order to develop more effective and more efficient algorithms for MIMO acoustic systems. Therefore, we would like to use this section to summarize the characteristics of acoustic channels.

2.2.1 Linearity and Shift-Invariance

We know that not all systems are linear and shift-invariant. But fortunately acoustic channels are *linear shift-invariant* (LSI) systems. Linearity and shift-invariance are the two most important properties for simplifying the analysis and design of discrete-time systems. A linear system ought to satisfy the rules of *homogeneity* and *additivity*, which, taken together, are often referred to as the principle of *superposition*. For a homogeneous system, scaling the input by a constant results in the output being scaled by the same constant. For an additive system, the response of the system to a sum of two signals is the sum of the two responses. A system is deemed to be shift-invariant if a time shift in its input leads to the same shift in its output. With these properties, an LSI system can be easily characterized by its impulse response. Once we know the impulse response, we can predict how the LSI system would respond to any other possible stimuli. This procedure is accomplished by linear convolution.

2.2.2 FIR Representation

The impulse response of an acoustic channel is usually very long, but FIR (finite impulse response) filters are more frequently used than IIR (infinite impulse response) filters in acoustic applications. There are several reasons for the popularity of FIR filters. First and foremost, the stability of FIR filters is easily controlled by ensuring that the filter coefficients are bounded. Second, there is a large number of adaptive algorithms developed for FIR filters and the performance of these algorithms is well understood in terms of convergence, tracking, and stability. Third, FIR filters can model acoustic channels accurately enough to satisfy most design criteria. Finally, FIR filters can be realized efficiently in hardware. Therefore, we will use FIR filters to model acoustic channels throughout this book.

2.2.3 Time-Varying Channel Impulse Responses

As explained above, in order to use only channel impulse response to characterize the input-output relationship of a system, the system needs to be linear and shift-invariant. However, like many other communication channels with different physical medium, acoustic channels are inherently time-varying systems. Sound sources may keep moving and the environment may be disturbed by such events as opening doors or windows from time to time. Even if an acoustic environment is relatively motionless, it can still vary in time with changes of atmospheric conditions. For example, a slight change of half a degree Celsius in the room temperature may result in a significant discrepancy in the transfer function of an acoustic channel inside the room. But this time-varying feature usually does not prevent us from using FIR filters to model acoustic channels since acoustic systems generally change slowly compared to the length of their channel impulse responses. Therefore we can divide time into periods. In each period an acoustic channel is assumed to be stationary and can be modeled with an FIR filter.

2.2.4 Frequency Selectivity

Acoustic waves are pressure disturbances propagating in the air. With spherical radiation and spreading, the inverse-square law rules and the sound level falls off as a function of distance from the sound source. As a rule of thumb, the loss is 6 dB for every doubling of distance. But when acoustic sound propagates over a long distance (usually greater than 100 feet or 30 meter), an excess attenuation of the high-frequency components can often be observed in addition to the normal inverse-square losses, which indicates that the acoustic channel is frequency selective. The level of this high-frequency excess attenuation is highly dependent on the air humidity and other atmospheric conditions.

The inverse-square law governs free-space propagation of sound. But in such enclosures as offices, conference rooms, and cars, acoustic waveforms might be reflected many times by the enclosure surfaces before they reach a microphone. The attenuation due to reflection is generally frequency dependent. But for audio signals, this dependency is not significant (except those surfaces with very absorptive materials), unlike radio-frequency signals in indoor wireless communications. For acoustic channels in these environments, it is the multipath-propagation aspect that leads to frequency-selective characteristics. Frequency-selective fading is viewed in the frequency domain. In the time domain, it is called multipath delay spread and induces sound reverberation analogous to inter-symbol interference (ISI) observed in data communications.

Frequency selectivity is one of the most important properties of acoustic channels. It is the central issue of most, if not all, acoustic signal processing techniques in terms of modeling, estimation, analysis, and equalization, which will be comprehensively discussed in the rest of this book.

2.2.5 Reverberation Time

Room reverberation is usually regarded as destructive since sound in reverberant environments is subject to temporal and spectral smearing, which results in distortion in both the envelope and fine structure of the acoustic signal. If the sound is speech, then speech intelligibility will be impaired. However, room reverberation is not always detrimental. Reverberation helps us to better perceive a listening environment. Although it may not be realized consciously, reverberation is one of many cues used by a listener for sound source localization and orientation in a given space. In addition, reverberation adds "warmth" to sound due to the colorization effect, which is very important to musical quality. The balance between sound clarity and spaciousness is the key to the design of attractive acoustic spaces and audio instruments, while the balance is achieved by controlling the level of reverberation.

The level of reverberation is typically measured by the T_{60} reverberation time, which was introduced by Sabine in [263] and is now a part of the ISO (International Organization for Standardization) reverberation measurement procedure [187]. The T_{60} reverberation time is defined as the length of time that it takes the reverberation to decay 60 dB from the level of the original sound. Logically, many reverberation time estimation algorithms are based on sound decay curves and the most widely used method for measuring the sound decay curves is to employ an excitation signal and record the acoustic channel's response with a microphone. If the excitation signal is an ideal impulse, then the sound decay curve is identical to the impulse response of the acoustic channel. Therefore, we will briefly explain how to estimate the reverberation time with respect to an obtained acoustic impulse response.

The reverberation energy in an acoustic impulse response decays exponentially over time, as shown in Fig. 2.2(a) by an example acoustic impulse response measured in the Bell Labs varechoic chamber. If the energy is measured in dB, then it decays linearly in time as clearly seen in Fig. 2.2(b). The reverberation time is determined from the estimated energy decay rate. By using Schroeder's backward integration method [269], a smoothed envelope for more accurate decay rate estimation can be obtained as illustrated in Fig. 2.2(c). Since the noise floor limits the minimum level of the measured impulse response in Fig. 2.2(c), which leads to errors, we follow the general guideline suggested in [232] to truncate the measured impulse response at the knee where the main decay slope intersects the noise floor. For this particular example, the knee is around 400 ms by observation. Line fitting is then performed with some margin on the both ends. From the slope of the fitting line, we find that the T_{60} reverberation time is about 395 ms.

2.2.6 Channel Invertibility and Minimum-Phase Filter

The invertibility of an acoustic channel is of particular interest in many applications such as speech enhancement and dereverberation. A system is said to be invertible if the input to the system can be uniquely determined by processing the output with a stable filter. In other words, there exists a stable inverse filter that exactly compensates the effect of the invertible system. A stable, causal, rational system requires that its poles be inside the unit circle. Therefore, a stable, causal system has a stable and causal inverse only if both its poles and zeros are inside the unit circle. Such a system is commonly referred to as a *minimum-phase* system. Minimum-phase systems have the properties of minimum phase-lag, minimum group delay, and minimum energy delay. Therefore a more precise terminology is *minimum phase-lag*, *minimum group-delay*, or *minimum energy-delay* system, but none of these is widely used and "minimum phase is historically the established terminology" [243].

Although many systems are minimum phase, room acoustic impulse responses are unfortunately almost never minimum phase [239]. This implies

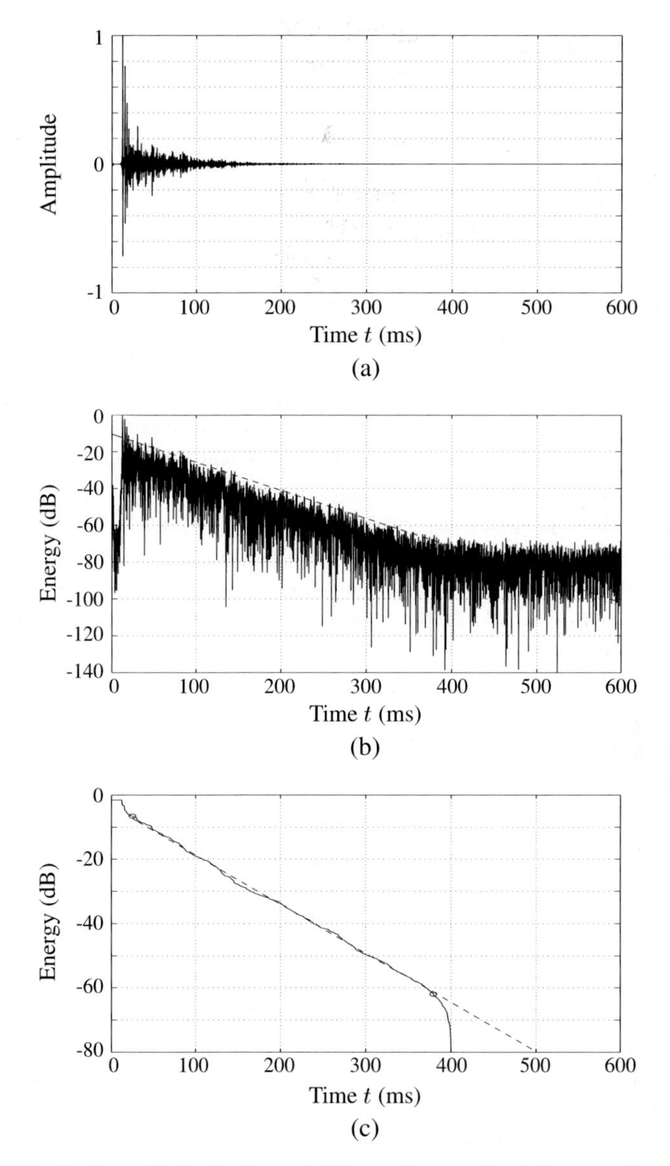

Fig. 2.2. Illustration of the backward integration method for reverberation time estimation. (a) Sample impulse response measured in the varechoic chamber at Bell Labs. (b) Squared impulse response. (c) Backward integration of squared impulse response with truncation time 400 ms (solid) and linear fitting curve (dashed).

that perfect deconvolution of an acoustic channel can be accomplished only with an acausal filter. This in itself may not be a serious problem for off-line processing since we can incorporate an overall time delay in the inverse filter and make it causal. But the delay is usually quite long for acoustic channels and the idea is difficult to implement with real-time systems. Speech dereverberation will be studied later in Chaps. 7 and 11.

2.2.7 Multichannel Diversity and the Common-Zero Problem

One of the most important features of multichannel systems is channel diversity. Diversity is what makes a multichannel system different from its single-channel counterpart. If the channels of a multichannel system are all identical as an extreme example, then the multichannel system is in fact reduced to a single-channel system. Strictly speaking, in the context of multichannel signal processing and wireless communications, channel diversity implies that different channels of a multichannel system would have no modes in common. If the channels are modeled as FIR filters, channel diversity means that their transfer functions share no common zeros, or in other words, they are co-prime polynomials.

A diverse multichannel system is *irreducible*. Otherwise, a multichannel system whose impulse responses share common zeros can be decomposed into two or more sequential subsystems, which have either single or multiple channels. For instance, a SIMO FIR system whose channel transfer functions are not coprime, i.e.,

$$C(z) = \gcd\{H_1(z),\ H_2(z),\ \cdots,\ H_N(z)\} \neq 0, \tag{2.11}$$

where $\gcd\{\cdot\}$ denotes the greatest common divisor of the polynomials involved. Then we have

$$H_n(z) = C(z)H'_n(z), \quad n = 1, 2, \cdots, N, \tag{2.12}$$

and the SIMO system is decomposed into a SISO system followed by a reduced SIMO system, as illustrated in Fig. 2.3. Obviously the order of these two subsystems does not affect the relationship between the input and outputs. The common-zero problem is important for research on SIMO systems because (a) irreducibility is necessary for blind identification of a SIMO system using only second-order statistics, and (b) irreducibility enables us to use the Bezout theorem to determine the inverse of a SIMO system. The former will be addressed in Chap. 6 on blind channel identification and the latter will be clearly explained in Chap. 7 when we investigate how to suppress reverberation.

For an irreducible SIMO FIR system, it can be shown [1] that the Sylvester matrix associated with the system's channel impulse responses has full column rank. The Sylvester matrix of dimension $NL \times (2L-1)$ is defined as follows:

$$\mathbf{H}_{\text{syl}} \triangleq \begin{bmatrix} \mathbf{h}(0) & \mathbf{h}(1) & \cdots & \mathbf{h}(L-1) & \mathbf{0} & \cdots & \mathbf{0} \\ \mathbf{0} & \mathbf{h}(0) & \cdots & \mathbf{h}(L-2) & \mathbf{h}(L-1) & \cdots & \mathbf{0} \\ \vdots & \vdots & \ddots & \ddots & \ddots & \ddots & \vdots \\ \mathbf{0} & \mathbf{0} & \cdots & \mathbf{h}(0) & \mathbf{h}(1) & \cdots & \mathbf{h}(L-1) \end{bmatrix}, \tag{2.13}$$

where

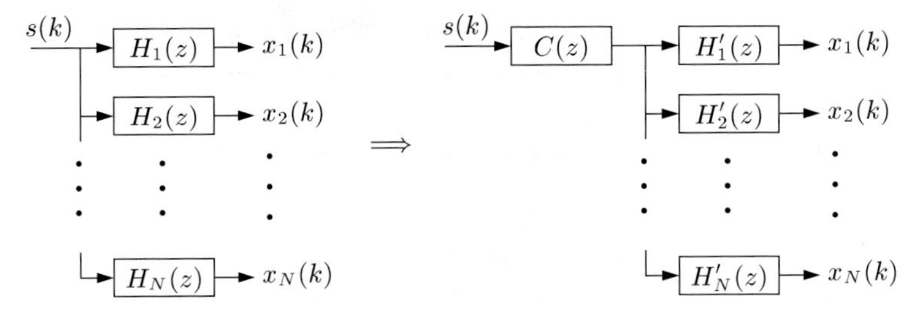

Fig. 2.3. A reducible SIMO FIR system whose channel impulse responses share common zeros.

$$\mathbf{h}(l) \triangleq \left[h_{1,l} \; h_{2,l} \; \cdots \; h_{N,l} \right]^T, \quad l = 0, 1, \cdots, L-1.$$

When the idea of irreducibility is extended from SIMO to MIMO systems, it requires that the z-transform channel matrix $\mathbf{H}(z)$ in (2.10) has full column rank for all $z \in \mathbb{C}$ and $M \leq N$. For an irreducible MIMO FIR system, the transfer functions of the channels that are associated with the same source do not share common zeros. As will be clearly explained in the later relevant chapters, irreducible MIMO systems do not guarantee blind identifiability using only second-order statistics, but promise the existence of stable FIR MIMO equalizers according to the generalized Bezout identity [199].

2.2.8 Sparse Impulse Response

Recently, it has been recognized that most, if not all, acoustic channel impulse responses are sparse and this characteristics can be exploited by adaptive algorithms to improve their performance in terms of initial convergence and tracking. Proportionate normalized LMS (PNLMS) is one of the first adaptive algorithms exploiting sparseness of acoustic channels [102]. In Chap. 4, the PNLMS and another class of exponentiated adaptive algorithms for the identification of sparse acoustic impulse responses will be discussed. In this section, we intend to define a measure to quantitatively, rather than just qualitatively, evaluate the sparseness of an acoustic impulse response.

We believe that a good sparseness measure needs to have the following properties:

- bounded rather than infinite range of definition;
- invariant with a non-zero scaling factor;
- independent of the sorting order of the channel impulse response coefficients.

The first two properties are easy to understand. The third implies that if we exchange two different taps in an acoustic impulse response or we sort the channel impulse response coefficients in a particular, e.g., ascending or descending, order, the sparseness measure will not be any different. This makes sense and it is important since sparseness is all about the dynamic range of the impulse response coefficients and has nothing to do with their order. A variance-like measure is what first comes to mind, but this satisfies only the third requirement, which is not enough.

Here we propose the following measure for assessing sparseness:

$$\xi(\mathbf{h}) \triangleq \frac{L}{L-\sqrt{L}}\left(1-\frac{\|\mathbf{h}\|_1}{\sqrt{L}\|\mathbf{h}\|_2}\right), \tag{2.14}$$

where $L > 1$ is the length of the acoustic channel $\mathbf{h}$, $\|\mathbf{h}\|_1$ and $\|\mathbf{h}\|_2$ are the 1- and 2-norm of $\mathbf{h}$, respectively, i.e.,

$$\|\mathbf{h}\|_1 = \sum_{l=0}^{L-1} |h_l|, \quad \|\mathbf{h}\|_2 = \left[\sum_{l=0}^{L-1} h_l^2\right]^{1/2}.$$

The same definition was proposed in [168].

Before examining the properties of this sparseness measure, we present three symbolic filters:

- The Dirac filter:
$$\mathbf{h}_\mathrm{d} = \begin{bmatrix} 1 & 0 & \cdots & 0 \end{bmatrix}^T. \tag{2.15}$$
- The uniform filter:
$$\mathbf{h}_\mathrm{u} = \begin{bmatrix} 1 & 1 & \cdots & 1 \end{bmatrix}^T. \tag{2.16}$$
- The exponentially decaying filter:
$$\mathbf{h}_\mathrm{e} = \begin{bmatrix} 1 & \exp\left(-\frac{1}{\beta}\right) & \cdots & \exp\left(-\frac{L-1}{\beta}\right) \end{bmatrix}^T, \tag{2.17}$$
where β is a positive number called the decay constant.

The Dirac and uniform filters are actually two particular cases of the exponentially decaying filter:

$$\lim_{\beta\to 0} \mathbf{h}_\mathrm{e} = \mathbf{h}_\mathrm{d}, \tag{2.18}$$

$$\lim_{\beta\to \infty} \mathbf{h}_\mathrm{e} = \mathbf{h}_\mathrm{u}. \tag{2.19}$$

While the Dirac filter is the sparest of all possible impulse responses, the uniform filter is the densest or least sparse one. The filter $\mathbf{h}_\mathrm{e}$ is a good model of acoustic impulse responses where β depends on the reverberation time. For a long reverberation time (large β), $\mathbf{h}_\mathrm{e}$ will decay slowly while for a short reverberation time (small β) $\mathbf{h}_\mathrm{e}$ will decay rapidly.

For the sparseness measure given in (2.14), we have the following properties:

$$\text{(a)} \quad 0 \leq \xi(\mathbf{h}) \leq 1, \tag{2.20}$$

$$\text{(b)} \quad \forall a \neq 0, \quad \xi(a\mathbf{h}) = \xi(\mathbf{h}), \tag{2.21}$$

$$\text{(c)} \quad \xi(\mathbf{h}_\mathrm{d}) = 1, \tag{2.22}$$

$$\text{(d)} \quad \xi(\mathbf{h}_\mathrm{u}) = 0. \tag{2.23}$$

Proof: We only show Property (a); the other three are straightforward. It is easy to see that

$$\|\mathbf{h}\|_2 \leq \|\mathbf{h}\|_1, \tag{2.24}$$

which implies that $\xi(\mathbf{h}) \leq 1$. Moreover, it can be shown using the Cauchy-Schwarz inequality (see [146] for example) that:

$$\|\mathbf{h}\|_1 \leq \sqrt{L}\|\mathbf{h}\|_2. \tag{2.25}$$

As a result, $\xi(\mathbf{h}) \geq 0$.

We see from these properties that the measure is bounded and is not affected by a non-zero scaling factor. Furthermore, the closer the measure is to 1, the sparser is the impulse response. On the contrary, the closer the measure is to 0, the denser or less sparse is the impulse response. Finally, since both the 1-norm and 2-norm of a vector are independent of the order of the vector coefficients, so is the defined sparseness measure.

To further confirm that (2.14) is a good measure of sparseness, Figure 2.4 illustrates what happens for a class of exponentially decaying filters $\mathbf{h}_\mathrm{e}$ of length $L = 256$. Figure 2.4(a) shows $\mathbf{h}_\mathrm{e}$ from $\beta = 1$ to $\beta = 50$ and Fig. 2.4(b) gives the corresponding values of $\xi(\mathbf{h}_\mathrm{e})$. For $\beta = 1$ (very sparse filter), $\xi(\mathbf{h}_\mathrm{e}) \approx 0.97$ and for $\beta = 50$ (quite dense filter), $\xi(\mathbf{h}_\mathrm{e}) \approx 0.4$. From values of the decay constant between 1 and 50, the sparseness measure decreases smoothly.

2.3 Measurement and Simulation of MIMO Acoustic Systems

Channel impulse response in MIMO acoustic systems is currently the nucleus of many acoustic signal processing research. In many if not all acoustic applications, determination of the channel impulse responses is either the primary goal (e.g., acoustic echo cancellation) or the underlying root cause of the difficulties (e.g., speech dereverberation and blind source separation). Therefore, having a collection of acoustic channel impulse responses in different environments and/or under various climate conditions is essential to develop and evaluate effective acoustic signal processing algorithms. In this section, we will describe the facilities that were built to directly measure acoustic channel impulse responses and methods that were developed for their simulation.

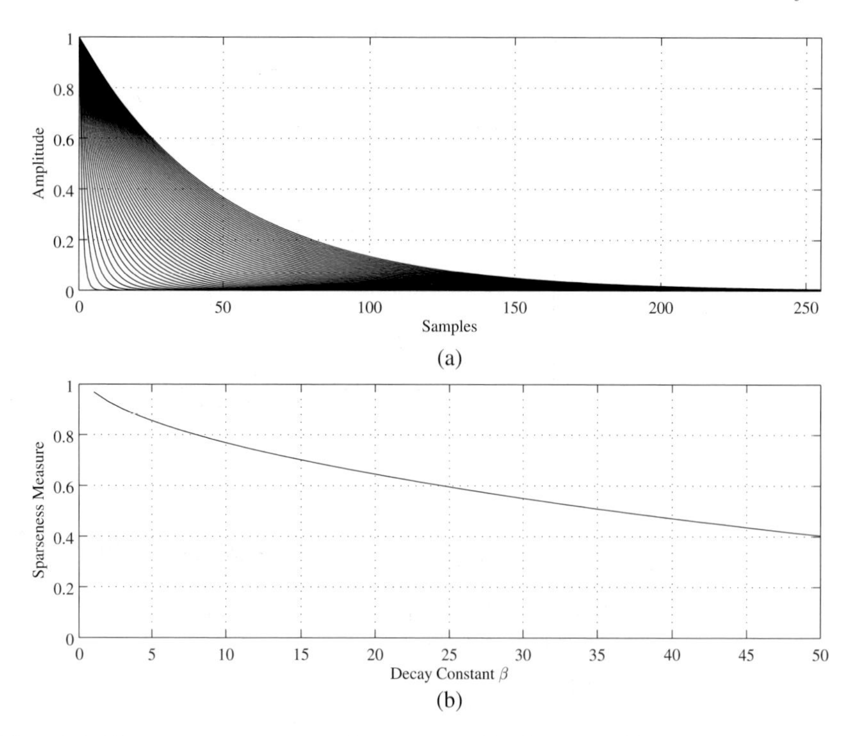

Fig. 2.4. Sparseness measures of exponentially decaying filters with various decay constants. (a) Impulse responses $\mathbf{h}_e$ of length $L = 256$ for values of the decay constant β from 1 to 50. (b) Sparseness measure for $\mathbf{h}_e$ as a function of the decay constant, β.

2.3.1 Direct Measurement of Acoustic Impulse Responses

In order to measure acoustic impulse responses under various conditions, we need a room whose acoustics can be controlled. The varechoic chamber at Bell Labs is a unique facility for such a purpose [312]. The chamber is a rectangular room and measures $x = 6.7$ m wide by $y = 6.1$ m deep by $z = 2.9$ m high. Figure 2.5(a) and (b) show a diagram of the floor plan and a photographic view of the chamber, respectively. For convenience, positions in the floor plan are designated by (x, y) coordinates with reference to the southwest corner and corresponding to meters along the (South, West) walls.

The varechoic chamber is made up of 368 electronically controlled panels on the walls, floor, and ceiling. Each panel consists of two sheets of stainless steel perforated with 0.32 mm holes in a standard staggered (60 degree) pattern with approximately 14% open area and is backed with 10 cm of fiberglass for sound absorption. The rear sheet can be moved relative to the front sheet to seal the holes, allowing the panels to reflect sound, or to expose the holes, allowing the panels to absorb sound. Every panel is individually controlled such that by varying the binary state of each panel in any combination, 2^{368}

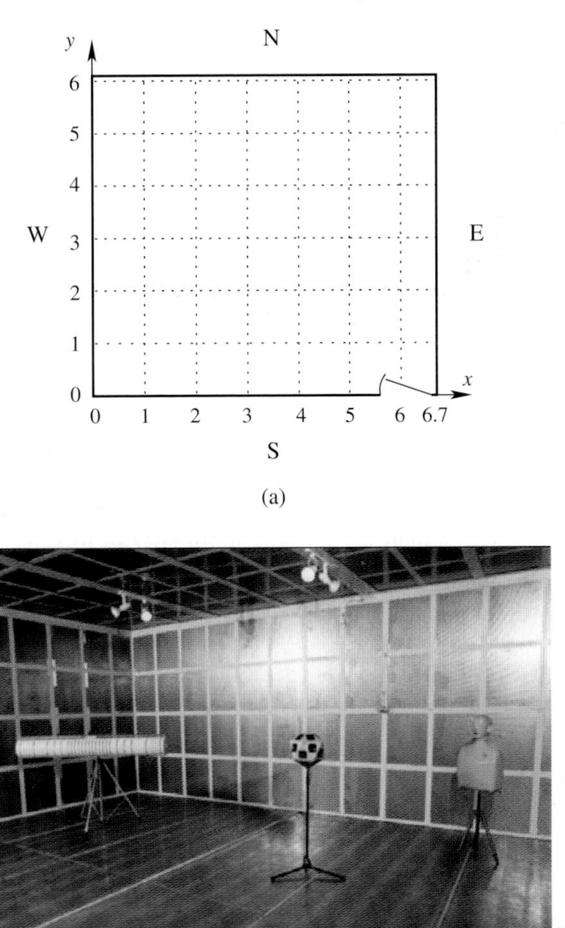

Fig. 2.5. Varechoic chamber at Bell Labs. (a) Floor plan (coordinate values measured in meters). (b) Photographic view of the chamber.

different room characteristics can be simulated. The chamber provides great variability, flexibility, and speed of operation. In addition, the varechoic chamber was build with thick exterior concrete walls providing excellent isolation. There is an independent precision climate control system that allows control of room temperature to 1° Celsius and relative humidity to 1%.

Among many other acoustic measurements made in the varechoic chamber, Härmä's collection [156] has been extensively used in our acoustic MIMO signal processing research. The database consists of impulse response measurements from 31 loudspeaker positions to 22 microphones for four different

Table 2.1. Average reverberation time for acoustic channel impulse responses measured by Härmä in the varechoic chamber as a function of the percentage of open panels.

Percentage Open Panels (%)	Reverberation Time (ms)
0	580
30	380
75	310
89	240

panel configurations. The 22 microphones are omnidirectional Panasonic WM-61a microphones, whose frequency response is flat from 20 Hz to 20 kHz with individual differences in sensitivity within the range of ± 4 dB. The microphones were spaced 10 cm apart and were mounted on an aluminum rod with diameter of 1 cm, which forms an equally-spaced linear array. The array was positioned 1.4 m above the floor, parallel to but 0.5 m away from the north wall, and with an equal distance to the east and west walls. A Cabasse Baltic Murale loudspeaker was used to play a logarithmic sweep excitation signal with 65536 samples at 48 kHz sampling rate. The loudspeaker has a wide effective frequency band ranging from 200 to 20 kHz. The microphone signals were recorded and estimates of channel impulse responses were calculated by spectral division. The four panel configurations are 0%, 30%, 75%, and 89% open. During these measurements the north wall behind the microphone array was covered by a three-inch-thick fiberglass pillow, so that the reverberation time was smaller than reported in [233] for the same panel configuration. (Also, there has been some deterioration over the years due to loss of oil between panels.) The pillow was rectangular of size 3.23 m long and 0.75 m high. Its lower edge was 0.9 m above the floor and its center was 3.565 m from the west wall of the chamber. The average reverberation time measurements for each panel configuration are summarized in Table 2.1. The particular positions of the loudspeaker will be specified when the measurements are used later on for algorithm evaluation.

2.3.2 Image Model for Acoustic Impulse Response Simulation

Direct measurements of acoustic impulse responses are apparently ideal in the sense that they faithfully capture the characteristics of real-world acoustic channels. But the procedure is quite time consuming and the measurement facilities may be prohibitively expensive or may not be readily available. Consequently, a simple analytical model is desired for simulating acoustic impulse responses. The image model proposed by Allen and Berkley [6] is the most widely used method for acoustic simulation although it is restricted to rectangular room geometries.

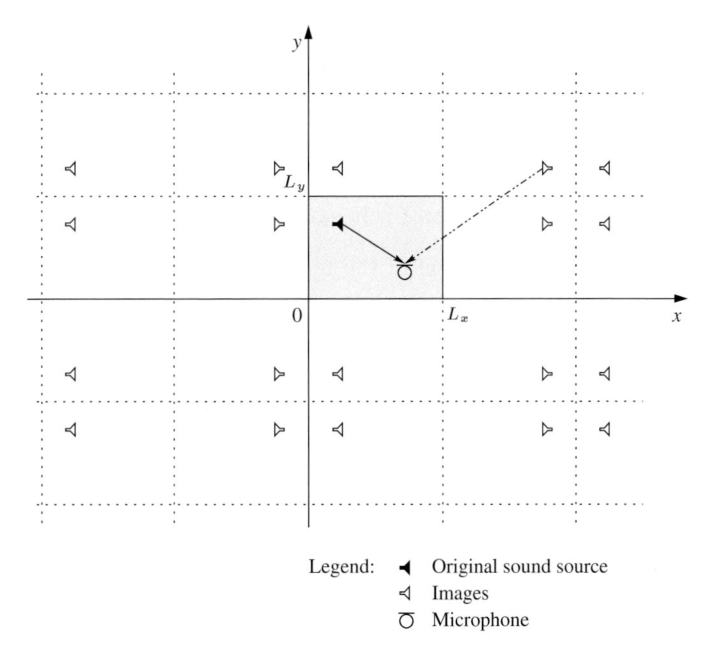

Fig. 2.6. Illustration of the lattice of a sound source and its images in a rectangular room. Only the x-y (floor) plane is shown but the actual modeled space is three dimensional.

Image analysis is based on the fact that the approximate solution of the wave equations for sound reflection from a flat, rigid wall can be interpreted in terms of rays. In other words, sound propagates along straight lines and bounces off walls like billiard balls. If a sound source and a microphone are on the same side of a flat, rigid wall in an infinitely large open space, the pressure at the microphone is the sum from two sound rays (i.e. waves): one from the original sound source and the other from a hypothetical image on the other side of the wall. The amplitude and phase of the image source is identical to those of the original source. If the wall is not perfectly rigid (which is more practical in the real world), the image analysis is altered by reducing the image amplitude by a wall reflection coefficient. In acoustics, the coefficient is usually assumed to be independent of frequency and angle.

For a rectangular room, the images proliferate exactly like a hall of mirrors. Therefore, the key to the image model is to find a way to systematically enumerate all images. We treat a talker as a point source and define the coordinate system with the origin at one corner and the axises along three adjacent boundary lines. As a consequence, the original and image sound sources form a three-dimensional lattice of points, which is illustrated by Fig. 2.6. In this figure, the rectangle with shadow is the real room and the others are image rooms, each containing an image of the original sound source.

Before discussing the image analysis technique, let us first introduce the following notation, which is slightly different from that used in [6]:

L_x, L_y, L_z	Size of the rectangular room in the x, y, and z direction, respectively.
$\mathbf{r}_\mathrm{s} \triangleq [x_\mathrm{s}\ y_\mathrm{s}\ z_\mathrm{s}]^T$	Position of the sound source.
$\mathbf{r}_\mathrm{m} \triangleq [x_\mathrm{m}\ y_\mathrm{m}\ z_\mathrm{m}]^T$	Position of the microphone.
c	Speed of sound in the air.
t_s	Sampling period.
β_{x_0}, β_{x_1}	Reflection coefficients of the walls perpendicular to the x axis, with subscript 0 referring to the wall adjacent to the coordinate origin and subscript 1 the opposing wall.
β_{y_0}, β_{y_1}	Reflection coefficients of the walls perpendicular to the y axis, with subscript 0 referring to the wall adjacent to the coordinate origin and subscript 1 the opposing wall.
β_{z_0}, β_{z_1}	Reflection coefficients of the floor and ceiling, respectively.

As seen from Fig. 2.6, the position of the original or an image source can be written as:

$$\mathbf{r} = \left[\ \pm x_\mathrm{s} + 2q_x L_x, \quad \pm y_\mathrm{s} + 2q_y L_y, \quad \pm z_\mathrm{s} + 2q_z L_z\ \right]^T, \tag{2.26}$$

where q_x, q_y, and q_z are three integers. By using another three integers p_x, p_y, and p_z, we can express (2.26) as follows:

$$\mathbf{r} = \left[\ (-1)^{p_x} x_\mathrm{s} + 2q_x L_x, \quad (-1)^{p_y} y_\mathrm{s} + 2q_y L_y, \quad (-1)^{p_z} z_\mathrm{s} + 2q_z L_z\ \right]^T, \tag{2.27}$$

where $p_x, p_y, p_z \in \{0, 1\}$.

While it is straightforward to calculate the distance that the wave travels from an image source (including the original) to the microphone, it is quite tricky to count how many reflections it experiences before reaching the microphone. Note from Fig. 2.6 that the number of surfaces the path crosses from an image to the microphone equals the number of real reflections. First, we show how many surfaces perpendicular to the x axis the path will cross. Then the same idea can be extended to the y and z directions. Finally the overall number of surfaces the path crosses and the overall loss of wave amplitude due to reflections can be produced.

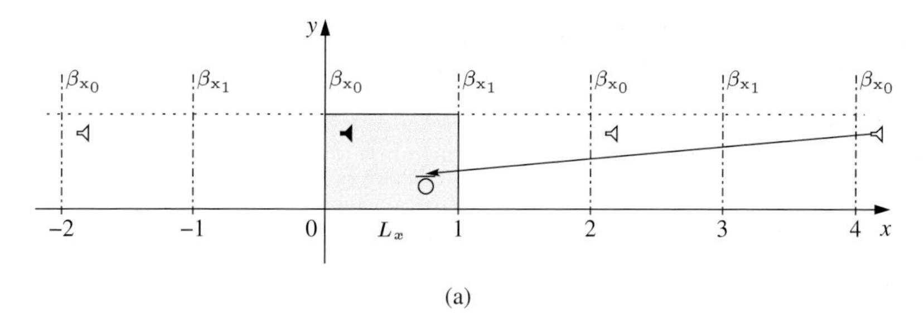

(a)

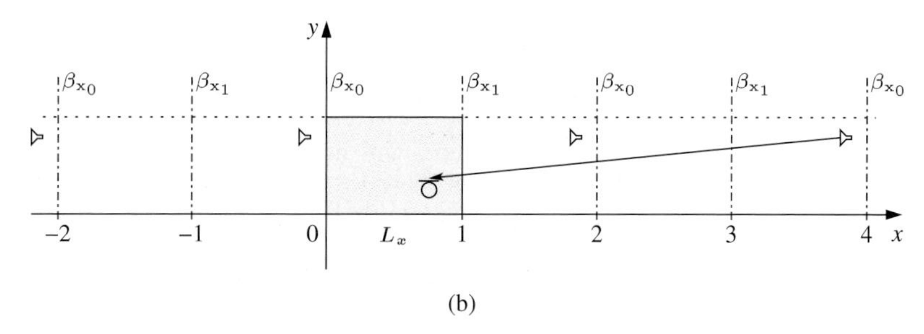

(b)

Fig. 2.7. Illustration of the number of surfaces the path crosses from an image to the microphone in the x direction for (a) $p_x = 0$ and (b) $p_x = 1$. (The legend is the same as that in Fig. 2.6.)

Suppose that an image's position in the x direction is $(-1)^{p_x} x_\mathrm{s} + 2q_x L_x$, which will be simply denoted by (p_x, q_x). The surfaces that are perpendicular to the x direction are indexed according to the number of L_x distances to the coordinate origin, as shown in Fig. 2.7. Then the reflection coefficients corresponding to the surfaces of even and odd indexes are β_{x_0} and β_{x_1}, respectively.

- For $p_x = 0$ as seen from Fig. 2.7(a), the path from the image to the microphone crosses in the x direction the same number ($|q_x|$) of even-indexed and odd-indexed surfaces. Then the reduction in wave amplitude due to the reflections is

$$\beta_\mathrm{x} = \beta_{\mathrm{x}_0}^{|q_x|} \beta_{\mathrm{x}_1}^{|q_x|}. \tag{2.28}$$

- For $p_x = 1$ as seen from Fig. 2.7(b), the number of even-indexed surfaces crossed is different from the number of odd-indexed surfaces crossed by 1. The former is $|q_x - 1|$ and the latter is $|q_x|$. Then we get the reduction as

$$\beta_\mathrm{x} = \beta_{\mathrm{x}_0}^{|q_x - 1|} \beta_{\mathrm{x}_1}^{|q_x|}. \tag{2.29}$$

Combining (2.28) and (2.29) yields the reduction in wave amplitude due to reflections from surfaces perpendicular to the x axis, given by

$$\beta_\mathrm{x} = \beta_{\mathrm{x}_0}^{|q_x - p_x|} \beta_{\mathrm{x}_1}^{|q_x|}, \quad p_x = 0, 1. \tag{2.30}$$

Also applying the same image analysis technique to the y and z directions, we determine the overall reduction in wave amplitude from an image to the microphone in three dimensions as the following:

$$\beta = \beta_{\mathrm{x}_0}^{|q_x - p_x|} \beta_{\mathrm{x}_1}^{|q_x|} \beta_{\mathrm{y}_0}^{|q_y - p_y|} \beta_{\mathrm{y}_1}^{|q_y|} \beta_{\mathrm{z}_0}^{|q_z - p_z|} \beta_{\mathrm{z}_1}^{|q_z|}. \tag{2.31}$$

After taking into account the inverse-square law, the acoustic impulse response is obtained:

$$\begin{aligned} h(t) &= \sum_{p_x, p_y, p_z = 0}^{1} \sum_{q_x, q_y, q_z = -\infty}^{\infty} \beta_{\mathrm{x}_0}^{|q_x - p_x|} \beta_{\mathrm{x}_1}^{|q_x|} \beta_{\mathrm{y}_0}^{|q_y - p_y|} \beta_{\mathrm{y}_1}^{|q_y|} \beta_{\mathrm{z}_0}^{|q_z - p_z|} \beta_{\mathrm{z}_1}^{|q_z|} \\ &\quad \times \frac{\delta\left(t - \|\mathbf{r} - \mathbf{r}_{\mathrm{m}}\|/c\right)}{4\pi \|\mathbf{r} - \mathbf{r}_{\mathrm{m}}\|}, \end{aligned} \tag{2.32}$$

where $\delta(t)$ is the continuous-time Dirac Delta function and $\|\cdot\|$ stands for the Euclidean distance. In discrete time, the acoustic impulse response is simulated as:

$$\begin{aligned} h_l &= \sum_{p_x, p_y, p_z = 0}^{1} \sum_{q_x, q_y, q_z = -\infty}^{\infty} \beta_{\mathrm{x}_0}^{|q_x - p_x|} \beta_{\mathrm{x}_1}^{|q_x|} \beta_{\mathrm{y}_0}^{|q_y - p_y|} \beta_{\mathrm{y}_1}^{|q_y|} \beta_{\mathrm{z}_0}^{|q_z - p_z|} \beta_{\mathrm{z}_1}^{|q_z|} \\ &\quad \times \frac{\delta\left(l - \lfloor \|\mathbf{r} - \mathbf{r}_{\mathrm{m}}\|/c \rceil\right)}{4\pi \|\mathbf{r} - \mathbf{r}_{\mathrm{m}}\|}, \end{aligned} \tag{2.33}$$

where $l \geq 0$, $\delta(k)$ is the discrete-time impulse sequence, and $\lfloor \cdot \rceil$ rounds the parameter to its nearest integer.

In practice, we are more interested in simulating an FIR than an IIR filter. If the amplitude of an acoustic impulse response is negligible after L samples, then we can cut the simulation at that point. For a given L, we can determine the set of images that could contribute to the computed impulse response as follows:

$$\begin{aligned} |q_x| \leq Q_x &= \left\lfloor \frac{L t_{\mathrm{s}} c}{2 L_x} + 1 \right\rceil, \\ |q_y| \leq Q_x &= \left\lfloor \frac{L t_{\mathrm{s}} c}{2 L_y} + 1 \right\rceil, \\ |q_z| \leq Q_x &= \left\lfloor \frac{L t_{\mathrm{s}} c}{2 L_z} + 1 \right\rceil. \end{aligned} \tag{2.34}$$

Then the channel impulse response (2.33) becomes

$$\begin{aligned} h_l &= \sum_{p_x, p_y, p_z = 0}^{1} \sum_{q_x = -Q_x}^{Q_x} \sum_{q_y = -Q_y}^{Q_y} \sum_{q_z = -Q_z}^{Q_z} \beta_{\mathrm{x}_0}^{|q_x - p_x|} \beta_{\mathrm{x}_1}^{|q_x|} \beta_{\mathrm{y}_0}^{|q_y - p_y|} \beta_{\mathrm{y}_1}^{|q_y|} \\ &\quad \times \beta_{\mathrm{z}_0}^{|q_z - p_z|} \beta_{\mathrm{z}_1}^{|q_z|} \frac{\delta\left(l - \lfloor \|\mathbf{r} - \mathbf{r}_{\mathrm{m}}\|/c \rceil\right)}{4\pi \|\mathbf{r} - \mathbf{r}_{\mathrm{m}}\|}, \\ &\quad l = 0, 1, \cdots, L - 1. \end{aligned} \tag{2.35}$$

Table 2.2. Image method for simulating an acoustic FIR channel.

Parameters:	$\mathbf{r}_\mathrm{s} = [x_\mathrm{s},\ y_\mathrm{s},\ z_\mathrm{s}]^T$, $\mathbf{r}_\mathrm{m} = [x_\mathrm{m},\ y_\mathrm{m},\ z_\mathrm{m}]^T$ $L_x,\ L_y,\ L_z,\ \beta_{\mathrm{x}0},\ \beta_{\mathrm{x}1},\ \beta_{\mathrm{y}0},\ \beta_{\mathrm{y}1},\ \beta_{\mathrm{z}0},\ \beta_{\mathrm{z}1}$ $L,\ t_\mathrm{s},\ c$
Initialization:	$\mathbf{h} = \begin{bmatrix} h_0 & h_1 & \cdots & h_{L-1} \end{bmatrix}^T = \mathbf{0}$ $Q_x = \left\lceil \frac{Lt_\mathrm{s}c}{2L_x} + 1 \right\rceil,\quad Q_y = \left\lceil \frac{Lt_\mathrm{s}c}{2L_y} + 1 \right\rceil$ $Q_z = \left\lceil \frac{Lt_\mathrm{s}c}{2L_z} + 1 \right\rceil$
Compute:	For $p_x = 0, 1$, $p_y = 0, 1$, $p_z = 0, 1$ For $q_x = -Q_x, \cdots, Q_x$, $q_y = -Q_y, \cdots, Q_y$, $q_z = -Q_z, \cdots, Q_z$ $\mathbf{r} = \begin{bmatrix} (-1)^{p_x} x_\mathrm{s}, & (-1)^{p_y} y_\mathrm{s}, & (-1)^{p_z} z_\mathrm{s} \end{bmatrix}^T + \begin{bmatrix} 2q_x L_x, & 2q_y L_y, & 2q_z L_z \end{bmatrix}^T$ $\beta = \beta_{\mathrm{x}0}^{\lvert q_x - p_x \rvert} \beta_{\mathrm{x}1}^{\lvert q_x \rvert} \beta_{\mathrm{y}0}^{\lvert q_y - p_y \rvert} \beta_{\mathrm{y}1}^{\lvert q_y \rvert} \beta_{\mathrm{z}0}^{\lvert q_z - p_z \rvert} \beta_{\mathrm{z}1}^{\lvert q_z \rvert}$ $d = \lfloor \lVert \mathbf{r} - \mathbf{r}_\mathrm{m} \rVert / c \rceil$ $h_d = h_d + \frac{\beta}{4\pi \lVert \mathbf{r} - \mathbf{r}_\mathrm{m} \rVert}$ end end
Post-processing:	Apply a high-pass digital filter to $\mathbf{h}$.

Finally, in order to remove the apparently nonphysical behavior of the image model at zero frequency, a high-pass digital filter with cut-off frequency at 0.01 of the sampling frequency is imposed, as suggested in [6]. The image method is summarized in Table 2.2.

2.4 Summary

In this chapter, we described four signal models for acoustic systems according to the number of inputs and outputs. The models were presented in both the time and frequency domains and will be frequently used throughout the rest of this book. The characteristics of acoustic channels were investigated and covered not only conventionally well-recognized features (e.g., linearity and shift-invariance, FIR, time-varying channels, frequency selectivity, invertibility, etc.) but also those characteristics that have not received adequate attention until very recently (like multichannel diversity, the common-zero problem, and sparseness). At the end of this chapter, we discussed facilities for directly measuring acoustic impulse responses and the image method for their simulation.

3

Wiener Filter and Basic Adaptive Algorithms

This chapter and the next two ones deal with the identification of acoustic MIMO systems when a reference signal is available.

3.1 Introduction

In his landmark manuscript on extrapolation, interpolation and smoothing of stationary time series [322], Norbert Wiener was one of the first researchers to treat the filtering problem of estimating a process corrupted by additive noise. The optimum estimate that he derived, required the solution of an integral equation known as the Wiener-Hopf equation [321]. Soon after he published his work, Levinson formulated the same problem in discrete time [213]. Levinson's contribution has had a great impact on the field. Indeed, thanks to him, Wiener's ideas have become more accessible to many engineers. A very nice overview of linear filtering theory and the history of the different discoveries in this area can be found in [198].

In this chapter, we will show that the Wiener theory plays a fundamental role in room acoustics. For example, in many acoustic applications, the impulse response between a loudspeaker and a microphone needs to be identified. Thanks to many (adaptive) algorithms directly derived from the Wiener-Hopf equations, this task is now rather easy.

This chapter is organized as follows. In Sect. 3.2, we derive the optimal Wiener filter for a SISO system. Section 3.3 explains what happens if the length of the modeling filter is shorter than the length of the true impulse response (this case always occurs in practice). It is extremely useful in many acoustic applications to be able to say how the input signal correlation matrix, which appears in the Wiener-Hopf equations, is conditioned. So we dedicate Sect. 3.4 to a detailed discussion on the condition number of this matrix. In Sect. 3.5, we present a collection of basic adaptive filters. We insist on the classical normalized least-mean-square (NLMS) algorithm. Section 3.6 generalizes the Wiener filter to the MIMO system case. While this generalization is straightforward, the optimal solution does not always exist and identification problems may be possible only in some situations. Section 3.7 gives some numerical examples for the identification of SISO and MISO systems with the NLMS algorithm. Finally, at the end of this chapter, we summarize the main results.

3.2 Wiener Filter

In this section, we are interested in the SISO system:

$$x(k) = \mathbf{h}^T \mathbf{s}(k) + b(k), \tag{3.1}$$

where, in our context, $s(k)$ and $x(k)$ are the loudspeaker and microphone signals, respectively. We assume that $x(k)$ and the random noise signal $b(k)$ [independent of $s(k)$] are zero-mean and stationary.

With the Wiener theory, it is possible to identify the impulse response $\mathbf{h}$, given $s(k)$ and $x(k)$. Define the error signal,

$$\begin{aligned} e(k) &= x(k) - \hat{x}(k) \\ &= x(k) - \hat{\mathbf{h}}_\mathrm{f}^T \mathbf{s}_\mathrm{f}(k), \end{aligned} \tag{3.2}$$

where

$$\hat{\mathbf{h}}_\mathrm{f} = \begin{bmatrix} \hat{h}_0 & \hat{h}_1 & \cdots & \hat{h}_{L_\mathrm{f}-1} \end{bmatrix}^T$$

is an estimate of $\mathbf{h}$ of length $L_\mathrm{f} \leq L$ and

$$\mathbf{s}_\mathrm{f}(k) = \begin{bmatrix} s(k) & s(k-1) & \cdots & s(k-L_\mathrm{f}+1) \end{bmatrix}^T.$$

To find the optimal filter, we need to minimize a cost function which is always built around the error signal [eq. (3.2)]. The usual choice for this criterion is the mean-square error (MSE) [160],

$$J\left(\hat{\mathbf{h}}_\mathrm{f}\right) = E\left\{e^2(k)\right\}, \tag{3.3}$$

where $E\{\cdot\}$ denotes mathematical expectation.

The optimal Wiener filter, $\hat{\mathbf{h}}_{\mathrm{f,o}}$, is the one that cancels the gradient of $J\left(\hat{\mathbf{h}}_\mathrm{f}\right)$, i.e.,

$$\frac{\partial J\left(\hat{\mathbf{h}}_\mathrm{f}\right)}{\partial \hat{\mathbf{h}}_\mathrm{f}} = \mathbf{0}_{L_\mathrm{f} \times 1}. \tag{3.4}$$

We have:

$$\frac{\partial J\left(\hat{\mathbf{h}}_{\mathrm{f}}\right)}{\partial \hat{\mathbf{h}}_{\mathrm{f}}} = 2E\left\{e(k)\frac{\partial e(k)}{\partial \hat{\mathbf{h}}_{\mathrm{f}}}\right\}$$
$$= -2E\left\{e(k)\mathbf{s}_{\mathrm{f}}(k)\right\}. \quad (3.5)$$

Therefore, at the optimum, we have:

$$E\left\{e_{\mathrm{o}}(k)\mathbf{s}_{\mathrm{f}}(k)\right\} = \mathbf{0}_{L_{\mathrm{f}}\times 1}, \quad (3.6)$$

where

$$e_{\mathrm{o}}(k) = x(k) - \hat{\mathbf{h}}_{\mathrm{f,o}}^T \mathbf{s}_{\mathrm{f}}(k) \quad (3.7)$$

is the error signal for which $J\left(\hat{\mathbf{h}}_{\mathrm{f}}\right)$ is minimized (i.e., the optimal filter). Expression (3.6) is called the *principle of orthogonality.*

The optimal estimate of $x(k)$ is:

$$\hat{x}_{\mathrm{o}}(k) = \hat{\mathbf{h}}_{\mathrm{f,o}}^T \mathbf{s}_{\mathrm{f}}(k). \quad (3.8)$$

It is then easy to check, with the help of the principle of orthogonality, that we also have:

$$E\left\{e_{\mathrm{o}}(k)\hat{x}_{\mathrm{o}}(k)\right\} = 0. \quad (3.9)$$

The previous expression is called the *corollary to the principle of orthogonality.*

If we substitute (3.7) into (3.6), we find the *Wiener-Hopf equations,*

$$\mathbf{R}_{\mathrm{f}}\hat{\mathbf{h}}_{\mathrm{f,o}} = \mathbf{p}_{\mathrm{f}}, \quad (3.10)$$

where

$$\mathbf{R}_{\mathrm{f}} = E\left\{\mathbf{s}_{\mathrm{f}}(k)\mathbf{s}_{\mathrm{f}}^T(k)\right\}$$

is the correlation matrix of the signal $s(k)$ and

$$\mathbf{p}_{\mathrm{f}} = E\left\{\mathbf{s}_{\mathrm{f}}(k)x(k)\right\}$$

is the cross-correlation vector between $\mathbf{s}_{\mathrm{f}}(k)$ and $x(k)$.

The correlation matrix is symmetric and positive semidefinite. It is also Toeplitz, i.e. a matrix which has constant values along diagonals,

$$\mathbf{R}_{\mathrm{f}} = \begin{bmatrix} r(0) & r(1) & \cdots & r(L_{\mathrm{f}}-1) \\ r(1) & r(0) & \cdots & r(L_{\mathrm{f}}-2) \\ \vdots & \vdots & \ddots & \vdots \\ r(L_{\mathrm{f}}-1) & r(L_{\mathrm{f}}-2) & \cdots & r(0) \end{bmatrix},$$

with $r(l) = E\{s(k)s(k-l)\}$, $l = 0, 1, \cdots, L_{\mathrm{f}} - 1$. In the SISO system case, this matrix is usually positive definite even for quasi-stationary signals like speech; however, it can be very ill conditioned.

Assuming that $\mathbf{R}_\mathrm{f}$ is non-singular, the optimal Wiener filter is:

$$\hat{\mathbf{h}}_{\mathrm{f,o}} = \mathbf{R}_\mathrm{f}^{-1}\mathbf{p}_\mathrm{f}. \tag{3.11}$$

The MSE can be rewritten as:

$$J\left(\hat{\mathbf{h}}_\mathrm{f}\right) = \sigma_x^2 - 2\mathbf{p}_\mathrm{f}^T\hat{\mathbf{h}}_\mathrm{f} + \hat{\mathbf{h}}_\mathrm{f}^T\mathbf{R}_\mathrm{f}\hat{\mathbf{h}}_\mathrm{f}, \tag{3.12}$$

where $\sigma_x^2 = E\{x^2(k)\}$ is the variance of the input signal $x(k)$. The criterion $J\left(\hat{\mathbf{h}}_\mathrm{f}\right)$ is a quadratic function of the filter coefficient vector $\hat{\mathbf{h}}_\mathrm{f}$ and has a single minimum point. This point combines the optimal Wiener filter, as shown above, and a value called the minimum MSE (MMSE), which is obtained by substituting (3.11) in (3.12):

$$\begin{aligned} J_{\mathrm{min}} &= J\left(\hat{\mathbf{h}}_{\mathrm{f,o}}\right) \\ &= \sigma_x^2 - \mathbf{p}_\mathrm{f}^T\mathbf{R}_\mathrm{f}^{-1}\mathbf{p}_\mathrm{f} \\ &= \sigma_x^2 - \sigma_{\hat{x}_\mathrm{o}}^2, \end{aligned} \tag{3.13}$$

where $\sigma_{\hat{x}_\mathrm{o}}^2 = E\{\hat{x}_\mathrm{o}^2(k)\}$ is the variance of the optimal filter output signal $\hat{x}_\mathrm{o}(k)$. This MMSE can be rewritten as:

$$J_{\mathrm{min}} = \sigma_b^2 + \mathbf{h}^T\mathbf{R}\mathbf{h} - \hat{\mathbf{h}}_{\mathrm{f,o}}^T\mathbf{R}_\mathrm{f}\hat{\mathbf{h}}_{\mathrm{f,o}}, \tag{3.14}$$

where $\sigma_b^2 = E\{b^2(k)\}$ is the variance of the noise and $\mathbf{R} = E\{\mathbf{s}(k)\mathbf{s}^T(k)\}$. The value J_{min} is bounded,

$$\sigma_b^2 \leq J_{\mathrm{min}} \leq \sigma_b^2 + \mathbf{h}^T\mathbf{R}\mathbf{h}, \ \forall L_\mathrm{f}. \tag{3.15}$$

We can easily check that for $L_\mathrm{f} = L$, $J_{\mathrm{min}} = \sigma_b^2$, and as L_f decreases compared to L, J_{min} gets closer to its maximum value $\sigma_b^2 + \mathbf{h}^T\mathbf{R}\mathbf{h}$.

We define the normalized MMSE as:

$$\begin{aligned} \tilde{J}_{\mathrm{min}} &= \frac{J_{\mathrm{min}}}{\sigma_x^2} \\ &= 1 - \frac{\sigma_{\hat{x}_\mathrm{o}}^2}{\sigma_x^2}. \end{aligned} \tag{3.16}$$

According to (3.15), the normalized MMSE always satisfies,

$$\frac{\sigma_b^2}{\sigma_x^2} \leq \tilde{J}_{\mathrm{min}} \leq 1. \tag{3.17}$$

3.3 Impulse Response Tail Effect

Room acoustic impulse responses are very long so that the length (L_f) of any FIR modeling filter $\hat{\mathbf{h}}_\mathrm{f}$ will always be shorter than the length (L) of the actual impulse response. Let us split this impulse response into two parts:

$$\mathbf{h} = \begin{bmatrix} \mathbf{h}_{\mathrm{f}} \\ \mathbf{h}_{\mathrm{t}} \end{bmatrix},$$

where $\mathbf{h}_{\mathrm{f}}$ is a vector of size L_{f} and $\mathbf{h}_{\mathrm{t}}$ is the "tail" of the impulse response that is not modeled by $\hat{\mathbf{h}}_{\mathrm{f}}$. Equation (3.1), which represents the SISO system, is now:

$$x(k) = \mathbf{h}_{\mathrm{f}}^T \mathbf{s}_{\mathrm{f}}(k) + \mathbf{h}_{\mathrm{t}}^T \mathbf{s}_{\mathrm{t}}(k - L_{\mathrm{f}}) + b(k), \tag{3.18}$$

where

$$\mathbf{s}_{\mathrm{t}}(k - L_{\mathrm{f}}) = \begin{bmatrix} s(k - L_{\mathrm{f}}) & s(k - L_{\mathrm{f}} - 1) & \cdots & s(k - L + 1) \end{bmatrix}^T.$$

Substituting (3.18) into the cross-correlation vector, we obtain,

$$\begin{aligned} \mathbf{p}_{\mathrm{f}} &= E\left\{\mathbf{s}_{\mathrm{f}}(k)x(k)\right\} \\ &= \mathbf{R}_{\mathrm{f}}\mathbf{h}_{\mathrm{f}} + \mathbf{R}_{\mathrm{t}}\mathbf{h}_{\mathrm{t}}, \end{aligned} \tag{3.19}$$

with $\mathbf{R}_{\mathrm{t}} = E\{\mathbf{s}_{\mathrm{f}}(k)\mathbf{s}_{\mathrm{t}}^T(k - L_{\mathrm{f}})\}$. Finally, plugging the previous expression in the Wiener-Hopf equations (3.11), we get:

$$\hat{\mathbf{h}}_{\mathrm{f,o}} = \mathbf{h}_{\mathrm{f}} + \mathbf{R}_{\mathrm{f}}^{-1}\mathbf{R}_{\mathrm{t}}\mathbf{h}_{\mathrm{t}}. \tag{3.20}$$

It is clear from (3.20) that the underestimation of the length of the impulse response in the Wiener method will introduce a bias (equal to $\mathbf{R}_{\mathrm{f}}^{-1}\mathbf{R}_{\mathrm{t}}\mathbf{h}_{\mathrm{t}}$) in the coefficients of the optimal filter. This bias depends on two things: the energy of the tail impulse response and the correlation of the input signal $s(k)$. If $s(k)$ is white, there is no bias since in this particular case the matrix $\mathbf{R}_{\mathrm{t}}$ is zero. But for highly correlated signals like speech, $\mathbf{R}_{\mathrm{t}}$ may not be negligible and the second term on the right-hand side of (3.20) may therefore be amplified if the energy of the tail is significative. As a consequence, it is important in practice to have a rough idea of the reverberation time of the room, in order to choose an appropriate length for the modeling filter for good identification. As we can see, increasing the length of the filter will improve the accuracy of the solution. On the other hand, the complexity for solving the linear system will increase and the conditioning of $\mathbf{R}_{\mathrm{f}}$ will be worsen. Therefore, depending on the application, a reasonable balance has to be found.

For simplification, in the rest of this chapter, we will assume that $L_{\mathrm{f}} = L$ so that we can drop the subscript "f" in all variables. In this scenario: $\mathbf{R}_{\mathrm{f}} = \mathbf{R}$, $\mathbf{s}_{\mathrm{f}}(k) = \mathbf{s}(k)$, $\hat{\mathbf{h}}_{\mathrm{f,o}} = \hat{\mathbf{h}}_{\mathrm{o}}$, etc.

3.4 Condition Number

The correlation matrix that appears in the Wiener-Hopf equations needs to be inverted to find the optimal filter. If this matrix is ill conditioned and the data is perturbed, the accuracy of the solution will suffer a lot if the linear system is solved directly. One way to improve the accuracy is to regularize the covariance matrix. However, this regularization depends on the condition number: the higher the condition number, the larger the regularization. So it is important to be able to estimate this condition number in an efficient way, in order to use this information for improving the quality of the solution. Many other problems require the knowledge of this condition number for different reasons. For example, the performance of many adaptive algorithms depends on this number. Therefore, it is of great interest to have a detailed discussion on this topic here and to develop a practical algorithm to determine this condition number.

3.4.1 Decomposition of the Correlation Matrix

For a vector of length $L+1$,

$$\mathbf{s}_{L+1}(k) = \begin{bmatrix} s(k) & s(k-1) & \cdots & s(k-L) \end{bmatrix}^T,$$

the covariance matrix of size $(L+1) \times (L+1)$ is:

$$\begin{aligned} \mathbf{R}_{L+1} &= E\left\{\mathbf{s}_{L+1}(k)\mathbf{s}_{L+1}^T(k)\right\} \\ &= \begin{bmatrix} r(0) & \mathbf{r}_L^T \\ \mathbf{r}_L & \mathbf{R}_L \end{bmatrix} \\ &= \begin{bmatrix} \mathbf{R}_L & \mathbf{r}_{\mathrm{b},L} \\ \mathbf{r}_{\mathrm{b},L}^T & r(0) \end{bmatrix}, \end{aligned} \tag{3.21}$$

where $\mathbf{R}_L = E\{\mathbf{s}(k)\mathbf{s}^T(k)\}$ and

$$\begin{aligned} \mathbf{r}_L &= \begin{bmatrix} r(1) & r(2) & \cdots & r(L) \end{bmatrix}^T, \\ \mathbf{r}_{\mathrm{b},L} &= \begin{bmatrix} r(L) & r(L-1) & \cdots & r(1) \end{bmatrix}^T. \end{aligned}$$

By using the Schur complements, it is easy to invert $\mathbf{R}_{L+1}$:

$$\mathbf{R}_{L+1}^{-1} = \begin{bmatrix} \mathbf{R}_L^{-1} + \varrho_L^{-1}\mathbf{b}_L\mathbf{b}_L^T & -\varrho_L^{-1}\mathbf{b}_L \\ -\varrho_L^{-1}\mathbf{b}_L^T & \varrho_L^{-1} \end{bmatrix}, \tag{3.22}$$

where

$$\begin{aligned} \mathbf{b}_L &= \mathbf{R}_L^{-1}\mathbf{r}_{\mathrm{b},L} \\ &= \begin{bmatrix} b_{L,1} & b_{L,2} & \cdots & b_{L,L} \end{bmatrix}^T \end{aligned} \tag{3.23}$$

is the backward predictor of length L,

$$\begin{aligned} \varrho_L &= r(0) - \mathbf{r}_{\mathrm{b},L}^T \mathbf{b}_L \\ &= r(0) - \mathbf{r}_L^T \mathbf{a}_L \end{aligned} \quad (3.24)$$

is the prediction error energy, and $\mathbf{a}_L = \mathbf{J}_L \mathbf{b}_L$ is the forward predictor with $\mathbf{J}_L$ being the co-identity matrix. Equation (3.22) is important and will be used later for a fast computation of the condition number.

3.4.2 Condition Number with the Frobenius Norm

Usually, the condition number is computed by using the 2-norm matrix. However, in the context of Toeplitz matrices, it is more convenient to use the Frobenius norm as explained below and in [47], [42].

To simplify the notation, in this subsection we take $\mathbf{R}_{L+1} = \mathbf{R}$. This matrix is symmetric, positive, and assumed to be non-singular. It can be diagonalized as follows:

$$\mathbf{Q}^T \mathbf{R} \mathbf{Q} = \mathbf{\Lambda}, \quad (3.25)$$

where

$$\mathbf{Q}^T \mathbf{Q} = \mathbf{Q} \mathbf{Q}^T = \mathbf{I}, \quad (3.26)$$

$$\mathbf{\Lambda} = \mathrm{diag}\left\{\lambda_1, \lambda_2, \cdots, \lambda_{L+1}\right\}, \quad (3.27)$$

and $0 < \lambda_1 \leq \lambda_2 \leq \cdots \leq \lambda_{L+1}$. By definition, the square-root of $\mathbf{R}$ is:

$$\mathbf{R}^{1/2} = \mathbf{Q} \mathbf{\Lambda}^{1/2} \mathbf{Q}^T. \quad (3.28)$$

The condition number of a matrix $\mathbf{R}$ is [146]:

$$\chi\left[\mathbf{R}\right] = \|\mathbf{R}\| \left\|\mathbf{R}^{-1}\right\|, \quad (3.29)$$

where $\|\cdot\|$ can be any matrix norm. Note that $\chi\left[\mathbf{R}\right]$ depends on the underlying norm and subscripts will be used to distinguish the different condition numbers.

Consider the Frobenius norm:

$$\|\mathbf{R}\|_{\mathrm{F}} = \left\{\mathrm{tr}\left[\mathbf{R}^T \mathbf{R}\right]\right\}^{1/2}. \quad (3.30)$$

We can easily check that, indeed, $\|\cdot\|_{\mathrm{F}}$ is a matrix norm since for any real matrices $\mathbf{A}$ and $\mathbf{B}$ and a real scalar β, the following three conditions are satisfied:

- $\|\mathbf{A}\|_{\mathrm{F}} \geq 0$ and $\|\mathbf{A}\|_{\mathrm{F}} = 0$ iff $\mathbf{A} = \mathbf{0}_{(L+1)\times(L+1)}$,
- $\|\mathbf{A} + \mathbf{B}\|_{\mathrm{F}} \leq \|\mathbf{A}\|_{\mathrm{F}} + \|\mathbf{B}\|_{\mathrm{F}}$,
- $\|\beta \mathbf{A}\|_{\mathrm{F}} = |\beta| \ \|\mathbf{A}\|_{\mathrm{F}}$.

We have:

$$\begin{aligned}\left\|\mathbf{R}^{1/2}\right\|_{\mathrm{F}} &= \{\mathrm{tr}\,[\mathbf{R}]\}^{1/2} \\ &= \left\{\sum_{l=1}^{L+1}\lambda_l\right\}^{1/2}\end{aligned} \tag{3.31}$$

and

$$\begin{aligned}\left\|\mathbf{R}^{-1/2}\right\|_{\mathrm{F}} &= \left\{\mathrm{tr}\left[\mathbf{R}^{-1}\right]\right\}^{1/2} \\ &= \left\{\sum_{l=1}^{L+1}\frac{1}{\lambda_l}\right\}^{1/2}.\end{aligned} \tag{3.32}$$

Hence, the condition number of $\mathbf{R}^{1/2}$ associated with $\|\cdot\|_{\mathrm{F}}$ is:

$$\chi_{\mathrm{F}}\left[\mathbf{R}^{1/2}\right] = \left\|\mathbf{R}^{1/2}\right\|_{\mathrm{F}}\left\|\mathbf{R}^{-1/2}\right\|_{\mathrm{F}} \geq L+1. \tag{3.33}$$

(The inequality in the previous expression is easy to show by using the Cauchy-Schwartz inequality.) In this section, we choose to work on $\chi_{\mathrm{F}}\left[\mathbf{R}^{1/2}\right]$ (rather than $\chi_{\mathrm{F}}[\mathbf{R}]$), because efficient algorithms can be derived to estimate its value, as will be shown in the next subsection. As far as we know, it does not seem obvious to estimate efficiently $\chi_{\mathrm{F}}[\mathbf{R}]$.

If $\chi\left[\mathbf{R}^{1/2}\right]$ is large, then $\mathbf{R}^{1/2}$ is said to be an ill-conditioned matrix. Note that this is a norm-dependent property. However, according to [146], any two condition numbers $\chi_\alpha\left[\mathbf{R}^{1/2}\right]$ and $\chi_\beta\left[\mathbf{R}^{1/2}\right]$ are equivalent in that constants c_1 and c_2 can be found for which:

$$c_1\chi_\alpha\left[\mathbf{R}^{1/2}\right] \leq \chi_\beta\left[\mathbf{R}^{1/2}\right] \leq c_2\chi_\alpha\left[\mathbf{R}^{1/2}\right]. \tag{3.34}$$

For example, for the 1- and 2-norm matrices and for $\mathbf{R}$, we can show [146]:

$$\frac{1}{(L+1)^2}\chi_2[\mathbf{R}] \leq \frac{1}{L+1}\chi_1[\mathbf{R}] \leq \chi_2[\mathbf{R}]. \tag{3.35}$$

We now show the same principle for the F- and 2-norm matrices and for $\mathbf{R}^{1/2}$. We recall that:

$$\chi_2\left[\mathbf{R}^{1/2}\right] = \sqrt{\frac{\lambda_{L+1}}{\lambda_1}}. \tag{3.36}$$

Since $\mathrm{tr}\left[\mathbf{R}^{-1}\right] \geq 1/\lambda_1$ and $\mathrm{tr}\,[\mathbf{R}] \geq \lambda_{L+1}$, we have:

$$\mathrm{tr}\,[\mathbf{R}]\,\mathrm{tr}\left[\mathbf{R}^{-1}\right] \geq \frac{\mathrm{tr}\,[\mathbf{R}]}{\lambda_1} \geq \frac{\lambda_{L+1}}{\lambda_1}, \tag{3.37}$$

hence,

$$\chi_{\mathrm{F}}\left[\mathbf{R}^{1/2}\right] \geq \chi_2\left[\mathbf{R}^{1/2}\right]. \tag{3.38}$$

Also, since $\mathrm{tr}\left[\mathbf{R}\right] \leq (L+1)\lambda_{L+1}$ and $\mathrm{tr}\left[\mathbf{R}^{-1}\right] \leq (L+1)/\lambda_1$, we obtain:

$$\mathrm{tr}\left[\mathbf{R}\right]\mathrm{tr}\left[\mathbf{R}^{-1}\right] \leq (L+1)\frac{\mathrm{tr}\left[\mathbf{R}\right]}{\lambda_1} \leq (L+1)^2\frac{\lambda_{L+1}}{\lambda_1}, \tag{3.39}$$

thus,

$$\chi_{\mathrm{F}}\left[\mathbf{R}^{1/2}\right] \leq (L+1)\chi_2\left[\mathbf{R}^{1/2}\right]. \tag{3.40}$$

Therefore, we deduce that:

$$\chi_2\left[\mathbf{R}^{1/2}\right] \leq \chi_{\mathrm{F}}\left[\mathbf{R}^{1/2}\right] \leq (L+1)\chi_2\left[\mathbf{R}^{1/2}\right]. \tag{3.41}$$

Moreover, by using the two inequalities,

$$\left[\sum_{l=1}^{L+1}\beta_l\right]^2 \geq \sum_{l=1}^{L+1}\beta_l^2, \tag{3.42}$$

$$\left[\sum_{l=1}^{L+1}\beta_l\right]^2 \leq (L+1)\sum_{l=1}^{L+1}\beta_l^2, \tag{3.43}$$

where $\beta_l > 0,\ \forall l$, it is easy to show that:

$$\frac{1}{L+1}\chi_{\mathrm{F}}^2\left[\mathbf{R}^{1/2}\right] \leq \chi_{\mathrm{F}}\left[\mathbf{R}\right] \leq \chi_{\mathrm{F}}^2\left[\mathbf{R}^{1/2}\right] \leq (L+1)\chi_{\mathrm{F}}\left[\mathbf{R}\right]. \tag{3.44}$$

Note that $\chi_2\left[\mathbf{R}\right] = \chi_2^2\left[\mathbf{R}^{1/2}\right]$ but $\chi_{\mathrm{F}}\left[\mathbf{R}\right] \neq \chi_{\mathrm{F}}^2\left[\mathbf{R}^{1/2}\right]$. According to expressions (3.41) and (3.44), $\chi_{\mathrm{F}}\left[\mathbf{R}^{1/2}\right]$ and $\chi_{\mathrm{F}}^2\left[\mathbf{R}^{1/2}\right]$ are a good measure of the condition number of matrices $\mathbf{R}^{1/2}$ and $\mathbf{R}$, respectively. Basically, there is no difference in the trend of the condition numbers of $\mathbf{R}$ and $\mathbf{R}^{1/2}$. In other words, if $\mathbf{R}^{1/2}$ is ill-conditioned (resp. well-conditioned) so is $\mathbf{R}$. In the next subsection, we will show how to compute $\chi_{\mathrm{F}}^2\left[\mathbf{R}^{1/2}\right]$ by using the Levinson-Durbin algorithm.

3.4.3 Fast Computation of the Condition Number

In this subsection, we need to efficiently compute the two norms $\left\|\mathbf{R}_{L+1}^{1/2}\right\|_{\mathrm{F}}^2$ and $\left\|\mathbf{R}_{L+1}^{-1/2}\right\|_{\mathrm{F}}^2$. The calculation of the first one is straightforward. Indeed:

$$\begin{aligned} \left\| \mathbf{R}_{L+1}^{1/2} \right\|_{\mathrm{F}}^{2} &= \mathrm{tr}\left[\mathbf{R}_{L+1}\right] \\ &= (L+1)r(0). \end{aligned} \tag{3.45}$$

Expression (3.45) requires one multiplication only. Consider the matrix $\mathbf{G}_{L+1} = \mathbf{R}_{L+1}^{-1}$ where its diagonal elements are $g_{L+1,ii}$, $i = 1, 2, \cdots, L+1$. It is clear from (3.22) that the last diagonal component of $\mathbf{G}_{L+1}$ is $g_{L+1,(L+1)(L+1)} = \varrho_L^{-1}$. The Lth diagonal element of $\mathbf{G}_{L+1}$ is $g_{L+1,LL} = \varrho_{L-1}^{-1} + \varrho_L^{-1} b_{L,L}^2$. Continuing the same process, we easily find:

$$g_{L+1,ii} = \varrho_{i-1}^{-1} + \sum_{l=i}^{L} \varrho_l^{-1} b_{l,i}^2, \tag{3.46}$$

with $\varrho_0 = r(0)$. Therefore, from (3.46) we deduce that:

$$\begin{aligned} \left\| \mathbf{R}_{L+1}^{-1/2} \right\|_{\mathrm{F}}^{2} &= \mathrm{tr}\left[\mathbf{G}_{L+1}\right] \\ &= \sum_{l=0}^{L} \varrho_l^{-1} \left[1 + \mathbf{b}_l^T \mathbf{b}_l\right] \\ &= \sum_{l=0}^{L} \varrho_l^{-1} \left[1 + \mathbf{a}_l^T \mathbf{a}_l\right], \end{aligned} \tag{3.47}$$

with $\mathbf{a}_0^T \mathbf{a}_0 = \mathbf{b}_0^T \mathbf{b}_0 = 0$.

Finally, the condition number is:

$$\chi_{\mathrm{F}}^2 \left[\mathbf{R}_{L+1}^{1/2}\right] = (L+1)r(0) \sum_{l=0}^{L} \varrho_l^{-1} \left[1 + \mathbf{a}_l^T \mathbf{a}_l\right]. \tag{3.48}$$

By using the Toeplitz structure, the Levinson-Durbin algorithm solves the linear prediction equation, $\mathbf{a}_L = \mathbf{R}_L^{-1} \mathbf{r}_L$, in $O(L^2)$ operations instead of $O(L^3)$. This algorithm computes all predictors $\mathbf{a}_l$, $l = 1, 2, \cdots, L$, and this is exactly what we need to compute (3.48). Expression (3.48) also shows a very nice link between the condition number and the predictors of all orders. This algorithm, which has roughly the same complexity as the Levinson-Durbin algorithm, is summarized in Table 3.1. Note that a very efficient algorithm was recently proposed by Dias and Leitão [96] to compute $\mathrm{tr}\{\mathbf{T}\mathbf{R}^{-1}\}$ (where $\mathbf{T}$ is a Toeplitz matrix, this form is a much general form than the one used in this section) with the Trench algorithm. Using these techniques here, we can reduce even more the complexity [to $O(L \ln L)$] for the estimation of the overall algorithm.

Table 3.1. Computation of the condition number with the Levinson-Durbin algorithm.

Initialization:	$\varrho_0 = r(0)$
Levinson-Durbin Algorithm:	$k_l = \frac{1}{\varrho_{l-1}}\left[r(l) - \mathbf{a}_{l-1}^T \mathbf{J}_{l-1}\mathbf{r}_{l-1}\right]$
	$\mathbf{a}_l = \begin{bmatrix} \mathbf{a}_{l-1} \\ 0 \end{bmatrix} - k_l \mathbf{J}_l \begin{bmatrix} -1 \\ \mathbf{a}_{l-1} \end{bmatrix}$
	$\varrho_l = \varrho_{l-1}(1 - k_l^2)$
	$l = 1, 2, \cdots, L$
Condition Number:	$\chi_{\mathrm{F}}^2\left[\mathbf{R}_{L+1}^{1/2}\right] = (L+1)r(0)\sum_{l=0}^{L} \varrho_l^{-1}\left[1 + \mathbf{a}_l^T \mathbf{a}_l\right]$

3.5 Basic Adaptive Algorithms

The aim of this section is to give a couple of basic adaptive algorithms that converge to the actual impulse response $\mathbf{h}$ and where the inversion of the correlation matrix $\mathbf{R}$ is avoided.

3.5.1 Deterministic Algorithm

The deterministic or steepest-descent algorithm is actually an iterative algorithm. It is summarized by the simple recursion,

$$\hat{\mathbf{h}}(k) = \hat{\mathbf{h}}(k-1) + \mu\left[\mathbf{p} - \mathbf{R}\hat{\mathbf{h}}(k-1)\right], \quad k = 0, 1, 2, \cdots, \tag{3.49}$$

where μ is a positive constant called the step-size parameter. In this algorithm, $\mathbf{p}$ and $\mathbf{R}$ are supposed to be known. The deterministic algorithm can be reformulated with the error signal:

$$e(k) = x(k) - \hat{\mathbf{h}}^T(k-1)\mathbf{s}(k), \tag{3.50}$$

$$\hat{\mathbf{h}}(k) = \hat{\mathbf{h}}(k-1) + \mu E\{\mathbf{s}(k)e(k)\}. \tag{3.51}$$

Now the important question is: what are the conditions on μ to make the algorithm converge to the true impulse response $\mathbf{h}$? To answer this question, we will examine the *natural modes* of the algorithm [318].

We define the *misalignment vector* as,

$$\mathbf{m}(k) = \mathbf{h} - \hat{\mathbf{h}}(k), \tag{3.52}$$

which is the difference between the true impulse response and the estimated one at time k. The positive quantity $\|\mathbf{m}(k)\|_2^2/\|\mathbf{h}\|_2^2$ is called the *normalized misalignment*. If we substitute (3.1) in the cross-correlation vector, we get,

$$\begin{aligned} \mathbf{p} &= E\{\mathbf{s}(k)x(k)\} \\ &= \mathbf{Rh}. \end{aligned} \tag{3.53}$$

Injecting (3.53) in (3.49) and subtracting $\mathbf{h}$ on both sides of the equation, we obtain:

$$\mathbf{m}(k) = (\mathbf{I} - \mu\mathbf{R})\mathbf{m}(k-1), \tag{3.54}$$

where $\mathbf{I}$ is the identity matrix. Using the eigendecomposition of $\mathbf{R} = \mathbf{Q}\mathbf{\Lambda}\mathbf{Q}^T$ in the previous expression, we get the equivalent form,

$$\mathbf{v}(k) = (\mathbf{I} - \mu\mathbf{\Lambda})\mathbf{v}(k-1), \tag{3.55}$$

where

$$\begin{aligned} \mathbf{v}(k) &= \mathbf{Q}^T\mathbf{m}(k) \\ &= \mathbf{Q}^T\left[\mathbf{h} - \hat{\mathbf{h}}(k)\right]. \end{aligned} \tag{3.56}$$

Thus, for the lth natural mode of the steepest-descent algorithm, we have [160]:

$$v_l(k) = (1 - \mu\lambda_l)v_l(k-1), \; l = 1, 2, \cdots, L, \tag{3.57}$$

or, equivalently,

$$v_l(k) = (1 - \mu\lambda_l)^k v_l(0), \; l = 1, 2, \cdots, L. \tag{3.58}$$

The algorithm converges if,

$$\lim_{k\to\infty} v_l(k) = 0, \; \forall l. \tag{3.59}$$

In this case,

$$\lim_{k\to\infty} \hat{\mathbf{h}}(k) = \mathbf{h}. \tag{3.60}$$

It is straightforward to see from (3.58) that a necessary and sufficient condition for the stability of the deterministic algorithm is that,

$$-1 < 1 - \mu\lambda_l < 1, \; \forall l, \tag{3.61}$$

which implies,

$$0 < \mu < \frac{2}{\lambda_l}, \; \forall l, \tag{3.62}$$

or

$$0 < \mu < \frac{2}{\lambda_{\max}}, \tag{3.63}$$

where $\lambda_{\max}$ is the largest eigenvalue of the correlation matrix $\mathbf{R}$.

Let us evaluate the time needed for each natural mode to converge to a given value. Expression (3.58) gives:

$$\ln \frac{|v_l(k)|}{|v_l(0)|} = k \ln |1 - \mu\lambda_l|, \tag{3.64}$$

hence,

$$k = \frac{1}{\ln |1 - \mu\lambda_l|} \ln \frac{|v_l(k)|}{|v_l(0)|}. \tag{3.65}$$

The *time constant*, τ_l, for the lth natural mode is defined by taking $|v_l(k)|/|v_l(0)| = 1/e$ (where e is the base of the natural logarithm) in (3.65). Therefore,

$$\tau_l = \frac{-1}{\ln |1 - \mu\lambda_l|}. \tag{3.66}$$

We can link the time constant with the condition number of the correlation matrix $\mathbf{R}$. First, let

$$\mu = \frac{\alpha}{\lambda_{\max}}, \tag{3.67}$$

where

$$0 < \alpha < 2, \tag{3.68}$$

to guaranty the convergence of the algorithm. α is called the normalized step-size parameter. Suppose that the smallest eigenvalue is $\lambda_1 = \lambda_{\min}$; in this case,

$$\begin{aligned} \tau_1 &= \frac{-1}{\ln |1 - \alpha\lambda_{\min}/\lambda_{\max}|} \\ &= \frac{-1}{\ln |1 - \alpha/\chi_2[\mathbf{R}]|}, \end{aligned} \tag{3.69}$$

where $\chi_2[\mathbf{R}] = \lambda_{\max}/\lambda_{\min}$. We see that the convergence time of the slowest natural mode depends on the conditioning of $\mathbf{R}$.

From (3.56), we deduce that,

$$\begin{aligned} \mathbf{m}^T(k)\mathbf{m}(k) &= \mathbf{v}^T(k)\mathbf{v}(k) \\ &= \left\| \mathbf{h} - \hat{\mathbf{h}}(k) \right\|_2^2 \\ &= \sum_{l=1}^{L} \lambda_l (1 - \mu\lambda_l)^k v_l(0). \end{aligned} \tag{3.70}$$

This value gives an idea on the global convergence of the filter to the true impulse response. This convergence is clearly governed by the smallest eigenvalues of $\mathbf{R}$.

We now examine the transient behavior of the MSE. Using (3.1), the error signal (3.50) can be rewritten as,

$$\begin{aligned} e(k) &= x(k) - \hat{\mathbf{h}}^T(k-1)\mathbf{s}(k) \\ &= b(k) + \mathbf{m}^T(k-1)\mathbf{s}(k), \end{aligned} \tag{3.71}$$

so that the MSE is:

$$\begin{aligned} J(k) &= E\{e^2(k)\} \\ &= \sigma_b^2 + \mathbf{m}^T(k-1)\mathbf{R}\mathbf{m}(k-1) \\ &= \sigma_b^2 + \mathbf{v}^T(k-1)\mathbf{\Lambda}\mathbf{v}(k-1) \\ &= \sigma_b^2 + \sum_{l=1}^{L} \lambda_l (1-\mu\lambda_l)^{2k-2} v_l^2(0). \end{aligned} \tag{3.72}$$

A plot of $J(k)$ versus k is called the *learning curve*. Note that the MSE decays exponentially. When the algorithm is convergent, we see that,

$$\lim_{k\to\infty} J(k) = \sigma_b^2. \tag{3.73}$$

This value corresponds to the MMSE, $J_{\min}$, obtained with the optimal Wiener filter when $L_{\mathrm{f}} = L$, which is what we assume in this section.

3.5.2 Stochastic Algorithm

The stochastic gradient or least-mean-square (LMS) algorithm, invented by Widrow and Hoff in the late 50s [316], is certainly the most popular algorithm that we can find in the literature of adaptive filters. The popularity of the LMS is probably due to the fact that it is easy to understand, easy to implement, and robust in many respects.

One easy way to derive the stochastic gradient algorithm is by approximating the deterministic algorithm. Indeed, in practice, the two quantities $\mathbf{p} = E\{\mathbf{s}(k)x(k)\}$ and $\mathbf{R} = E\{\mathbf{s}(k)\mathbf{s}^T(k)\}$ are in general not known. If we take their instantaneous estimates:

$$\hat{\mathbf{p}}(k) = \mathbf{s}(k)x(k), \tag{3.74}$$

$$\hat{\mathbf{R}}(k) = \mathbf{s}(k)\mathbf{s}^T(k), \tag{3.75}$$

and replace them in the steepest-descent algorithm [eq. (3.49)], we get:

$$\begin{aligned} \hat{\mathbf{h}}(k) &= \hat{\mathbf{h}}(k-1) + \mu\left[\hat{\mathbf{p}}(k) - \hat{\mathbf{R}}(k)\hat{\mathbf{h}}(k-1)\right] \\ &= \hat{\mathbf{h}}(k-1) + \mu\mathbf{s}(k)\left[x(k) - \mathbf{s}^T(k)\hat{\mathbf{h}}(k-1)\right]. \end{aligned} \tag{3.76}$$

This simple recursion is the LMS algorithm. Contrary to the deterministic algorithm, the LMS weight vector $\hat{\mathbf{h}}(k)$ is now a random vector. The three following equations summarize this algorithm [160],

$$\hat{x}(k) = \mathbf{s}^T(k)\hat{\mathbf{h}}(k-1), \text{ filter output}, \tag{3.77}$$

$$e(k) = x(k) - \hat{x}(k), \text{ error signal}, \tag{3.78}$$

$$\hat{\mathbf{h}}(k) = \hat{\mathbf{h}}(k-1) + \mu\mathbf{s}(k)e(k), \text{ adaptation}, \tag{3.79}$$

which requires $2L$ additions and $2L+1$ multiplications at each iteration.

The stochastic gradient algorithm has been extensively studied and many theoretical results on its performance have been obtained [318], [160], [115]. In particular, we can show the convergence in the mean and mean square (see for example [320]), where under the independence assumption, the condition is remarkably the same as the one obtained for the deterministic algorithm, i.e.,

$$0 < \mu < \frac{2}{\lambda_{\max}}. \tag{3.80}$$

We can show that the asymptotic MSE for the LMS is:

$$\lim_{k\to\infty} J(k) = \sigma_b^2 \left(1 + \frac{\mu}{2} L\sigma_s^2\right), \tag{3.81}$$

where $\sigma_s^2 = E\{s^2(k)\}$ is the variance of the input signal $s(k)$. If we compare (3.81) with the asymptotic MSE for the steepest-descent algorithm [eq. (3.73)], we notice that a positive term,

$$J_{\text{ex}}(\infty) = \frac{\mu}{2} L\sigma_s^2\sigma_b^2, \tag{3.82}$$

is added, called the *excess mean-square error*. This term, of course, have a negative effect on the final MSE and its apparition is due to the approximation discussed at the beginning of this subsection. We can reduce its effect by taking a very small μ. But taking a small step size will increase the convergence time of the LMS. This trade-off between fast convergence and increased MSE is a very well-known fact and is something to consider in any practical implementation.

A simple condition for the stability of LMS is that,

$$|\varepsilon(k)| < |e(k)| \tag{3.83}$$

where

$$\varepsilon(k) = x(k) - \mathbf{s}^T(k)\hat{\mathbf{h}}(k) \tag{3.84}$$

is the *a posteriori* error signal, computed after the filter is updated. This intuitively makes sense since $\varepsilon(k)$ contains more meaningful information than $e(k)$.

This condition is necessary for the LMS to converge to the true impulse response but not sufficient. However, it is very useful to use here and in many other algorithms to find the bounds for the step size μ.

Plugging (3.79) in (3.84) and using the condition (3.83), we find:

$$0 < \mu < \frac{2}{\mathbf{s}^T(k)\mathbf{s}(k)}. \tag{3.85}$$

For L large, $\mathbf{s}^T(k)\mathbf{s}(k) = L\sigma_s^2 = \text{tr}[\mathbf{R}]$. On the other hand, $\text{tr}[\mathbf{R}] = \sum_{l=1}^{L} \lambda_l$ and this implies that $\text{tr}[\mathbf{R}] \geq \lambda_{\max}$. Hence,

$$0 < \mu < \frac{2}{\mathbf{s}^T(k)\mathbf{s}(k)} \leq \frac{2}{\lambda_{\max}}. \tag{3.86}$$

If we now introduce the normalized step size α $(0 < \alpha < 2)$, as we did in the previous subsection, the step size of the LMS will vary with time as follows,

$$\mu(k) = \frac{\alpha}{\mathbf{s}^T(k)\mathbf{s}(k)}, \tag{3.87}$$

and the LMS becomes the normalized LMS (NLMS):

$$\hat{\mathbf{h}}(k) = \hat{\mathbf{h}}(k-1) + \frac{\alpha \mathbf{s}(k)e(k)}{\mathbf{s}^T(k)\mathbf{s}(k)}. \tag{3.88}$$

This algorithm is extremely helpful in practice, especially with non-stationary signals, since $\mu(k)$ can adjust itself at each new iteration. In order to avoid numerical difficulties when the energy of the input signal is small, we regularize the algorithm,

$$\hat{\mathbf{h}}(k) = \hat{\mathbf{h}}(k-1) + \frac{\alpha \mathbf{s}(k)e(k)}{\mathbf{s}^T(k)\mathbf{s}(k) + \delta}, \tag{3.89}$$

where $\delta > 0$ is the regularization factor. Table 3.2 summarizes this very important algorithm. (Note that the definition of $\mu(k)$ in this table is slightly modified in order to include the regularization parameter δ.)

3.5.3 Sign Algorithms

Up to now, the only cost function that we have used is the MSE. What makes this criterion so interesting is that an optimal solution (Wiener) can be easily derived as well as very powerful adaptive algorithms. An alternative to the MSE is the mean absolute error (MAE) [141],

$$\begin{aligned} J_{\text{a}}\left(\hat{\mathbf{h}}\right) &= E\left\{|e(k)|\right\} \\ &= E\left\{\left|x(k) - \hat{\mathbf{h}}^T\mathbf{s}(k)\right|\right\}. \end{aligned} \tag{3.90}$$

Table 3.2. The normalized LMS (NLMS) algorithm.

Initialization:	$\hat{\mathbf{h}}(0) = \mathbf{0}_{L\times 1}$
Parameters:	$0 < \alpha < 2$
	$\delta > 0$
Error:	$e(k) = x(k) - \mathbf{s}^T(k)\hat{\mathbf{h}}(k-1)$
Update:	$\mu(k) = \dfrac{\alpha}{\mathbf{s}^T(k)\mathbf{s}(k) + \delta}$
	$\hat{\mathbf{h}}(k) = \hat{\mathbf{h}}(k-1) + \mu(k)\mathbf{s}(k)e(k)$

The gradient of this cost function is:

$$\frac{\partial J_{\mathrm{a}}\left(\hat{\mathbf{h}}\right)}{\partial \hat{\mathbf{h}}} = -E\left\{\mathbf{s}(k)\mathrm{sgn}[e(k)]\right\}, \tag{3.91}$$

where

$$\mathrm{sgn}[e(k)] = \frac{e(k)}{|e(k)|}. \tag{3.92}$$

From the instantaneous value of the gradient of $J_{\mathrm{a}}\left(\hat{\mathbf{h}}\right)$, we can derive the sign-error adaptive filter:

$$\hat{\mathbf{h}}(k) = \hat{\mathbf{h}}(k-1) + \mu_{\mathrm{a}}\mathbf{s}(k)\mathrm{sgn}[e(k)], \tag{3.93}$$

where μ_{a} is the adaptation step of the algorithm. This algorithm is simplified compared to the LMS since the L multiplications in the update equation are replaced by a sign change of the components of the signal vector $\mathbf{s}(k)$. Using the stability condition, $|\varepsilon(k)| < |e(k)|$, we deduce that:

$$0 < \mu_{\mathrm{a}} < \frac{2|e(k)|}{\mathbf{s}^T(k)\mathbf{s}(k)}. \tag{3.94}$$

Another way to simplify the LMS filter is to replace $s(k)$ with its sign. We get the sign-data algorithm:

$$\hat{\mathbf{h}}(k) = \hat{\mathbf{h}}(k-1) + \mu_{\mathrm{a}}'\mathrm{sgn}[\mathbf{s}(k)]e(k), \tag{3.95}$$

where μ_{a}' is the adaptation step of the algorithm and the stability condition is:

$$0 < \mu_{\mathrm{a}}' < \frac{2}{\mathbf{s}^T(k)\mathrm{sgn}[\mathbf{s}(k)]}. \tag{3.96}$$

Combining the two previous approaches, we derive the sign-sign algorithm:

$$\hat{\mathbf{h}}(k) = \hat{\mathbf{h}}(k-1) + \mu_{\mathrm{a}}'' \mathrm{sgn}[\mathbf{s}(k)]\mathrm{sgn}[e(k)], \tag{3.97}$$

for which the stability condition is:

$$0 < \mu_{\mathrm{a}}'' < \frac{2|e(k)|}{\mathbf{s}^T(k)\mathrm{sgn}[\mathbf{s}(k)]}. \tag{3.98}$$

The algorithms derived in this subsection are very simple to implement and can be very useful in some applications. However, their convergence rate is usually lower than the LMS and their excess MSE is higher [17], [81], [50].

3.6 MIMO Wiener Filter

In this section, we consider a MIMO system with M inputs and N outputs (see Chap. 2 for more details):

$$\mathbf{x}(k) = \mathbf{H}\mathbf{s}(k) + \mathbf{b}(k), \tag{3.99}$$

where

$$\begin{aligned} x_n(k) &= \sum_{m=1}^{M} \mathbf{h}_{nm}^T \mathbf{s}_m(k) + b_n(k) \\ &= \mathbf{h}_{n:}^T \mathbf{s}(k) + b_n(k), \ n = 1, 2, \cdots, N, \end{aligned} \tag{3.100}$$

and

$$\mathbf{H} = \begin{bmatrix} \mathbf{h}_{11}^T & \mathbf{h}_{12}^T & \cdots & \mathbf{h}_{1M}^T \\ \mathbf{h}_{21}^T & \mathbf{h}_{22}^T & \cdots & \mathbf{h}_{2M}^T \\ \vdots & \vdots & \ddots & \vdots \\ \mathbf{h}_{N1}^T & \mathbf{h}_{N2}^T & \cdots & \mathbf{h}_{NM}^T \end{bmatrix}_{N \times ML} = \begin{bmatrix} \mathbf{h}_{1:}^T \\ \mathbf{h}_{2:}^T \\ \vdots \\ \mathbf{h}_{N:}^T \end{bmatrix}. \tag{3.101}$$

We define the error signal at time k at the nth output as:

$$\begin{aligned} e_n(k) &= x_n(k) - \hat{x}_n(k) \\ &= x_n(k) - \hat{\mathbf{h}}_{n:}^T \mathbf{s}(k), \ n = 1, 2, \cdots, N, \end{aligned} \tag{3.102}$$

where $\hat{\mathbf{h}}_{n:}$ is an estimate of $\mathbf{h}_{n:}$. It is more convenient to define an error signal vector for all outputs:

$$\begin{aligned} \mathbf{e}(k) &= \mathbf{x}(k) - \hat{\mathbf{x}}(k) \\ &= \mathbf{x}(k) - \hat{\mathbf{H}}\mathbf{s}(k), \end{aligned} \tag{3.103}$$

where $\hat{\mathbf{H}}$ is an estimate of $\mathbf{H}$ and

$$\mathbf{e}(k) = \begin{bmatrix} e_1(k) & e_2(k) & \cdots & e_N(k) \end{bmatrix}^T.$$

Having written the error signal, we now define the MIMO MSE with respect to the modeling filters as:

$$\begin{aligned} J\left(\hat{\mathbf{H}}\right) &= E\left\{\mathbf{e}^T(k)\mathbf{e}(k)\right\} \\ &= \sum_{n=1}^{N} E\left\{e_n^2(k)\right\} \\ &= \sum_{n=1}^{N} J_n\left(\hat{\mathbf{h}}_{n:}\right). \end{aligned} \tag{3.104}$$

The minimization of (3.104) leads to the MIMO Wiener-Hopf equations:

$$\mathbf{R}_{ss}\hat{\mathbf{H}}_{\mathrm{o}}^T = \mathbf{P}_{sx}, \tag{3.105}$$

where

$$\begin{aligned} \mathbf{R}_{ss} &= E\{\mathbf{s}(k)\mathbf{s}^T(k)\} \\ &= \begin{bmatrix} \mathbf{R}_{s_1s_1} & \mathbf{R}_{s_1s_2} & \cdots & \mathbf{R}_{s_1s_M} \\ \mathbf{R}_{s_2s_1} & \mathbf{R}_{s_2s_2} & \cdots & \mathbf{R}_{s_2s_M} \\ \vdots & \vdots & \ddots & \vdots \\ \mathbf{R}_{s_Ms_1} & \mathbf{R}_{s_Ms_2} & \cdots & \mathbf{R}_{s_Ms_M} \end{bmatrix} \end{aligned} \tag{3.106}$$

is the input signal covariance matrix (which has a block Toeplitz structure) with $\mathbf{R}_{s_ms_i} = E\{\mathbf{s}_m(k)\mathbf{s}_i^T(k)\}$, and

$$\begin{aligned} \mathbf{P}_{sx} &= E\{\mathbf{s}(k)\mathbf{x}^T(k)\} \\ &= \begin{bmatrix} \mathbf{p}_{sx_1} & \mathbf{p}_{sx_2} & \cdots & \mathbf{p}_{sx_N} \end{bmatrix} \end{aligned} \tag{3.107}$$

is the cross-correlation matrix between the input and output signals, with $\mathbf{p}_{sx_n} = E\{\mathbf{s}(k)x_n(k)\}$.

It can be easily seen that the MIMO Wiener-hopf equations [eq. (3.105)] can be decomposed in N independent MISO Wiener-Hopf equations,

$$\mathbf{R}_{ss}\hat{\mathbf{h}}_{n:,\mathrm{o}} = \mathbf{p}_{sx_n}, \quad n = 1, 2, \cdots, N, \tag{3.108}$$

each one corresponding to an output signal of the system. In other words, minimizing $J\left(\hat{\mathbf{H}}\right)$ or minimizing each $J_n\left(\hat{\mathbf{h}}_{n:}\right)$ independently gives exactly the same results from an identification point of view. This observation is very important from a practical point of view when adaptive algorithms need to be designed. Indeed, any MIMO adaptive filter is simplified to N MISO adaptive filters. As an example, we give the MISO NLMS algorithm in Table 3.3. We deduce from this discussion that, obviously, the identification of a SIMO system is equivalent to the identification of N independent SISO systems. As a result, with a reference signal, the identification of any acoustic system simplifies to the identification of SISO or MISO systems.

Table 3.3. The MISO NLMS algorithm.

Initialization:	$\hat{\mathbf{h}}_{n:}(0) = \mathbf{0}_{ML\times 1}$
Parameters:	$0 < \alpha < 2$
	$\delta > 0$
Error:	$e_n(k) = x_n(k) - \mathbf{s}^T(k)\hat{\mathbf{h}}_{n:}(k-1)$
Update:	$\mu(k) = \dfrac{\alpha}{\mathbf{s}^T(k)\mathbf{s}(k) + \delta}$
	$\hat{\mathbf{h}}_{n:}(k) = \hat{\mathbf{h}}_{n:}(k-1) + \mu(k)\mathbf{s}(k)e_n(k)$

Conditioning of the Covariance Matrix

The best possible case for the identification of a MISO system is when the input signals $s_m(k)$, $m = 1, 2, \cdots, M$, are uncorrelated. In this scenario, we have:

$$\mathbf{R}_{s_m s_i} = \mathbf{0}_{L\times L}, \ \forall m, i = 1, 2, \cdots, M, \ m \neq i, \tag{3.109}$$

and the input signal covariance matrix $\mathbf{R}_{ss}$ is block-diagonal. Therefore, if $\mathbf{R}_{s_m s_m}$, $m = 1, 2, \cdots, M$, are non-singular and well conditioned, the impulse responses of the MISO system are easy to estimate. This case, however, does not often occur in practice so it has very little interest.

The worst possible case, from an identification point of view, is when the signals $s_m(k)$ are generated from a unique source $s_\mathrm{s}(k)$, i.e.,

$$s_m(k) = \mathbf{g}_m^T s_\mathrm{s}(k), \ m = 1, 2, \cdots, M, \tag{3.110}$$

where

$$\mathbf{g}_m = \begin{bmatrix} g_{m,0} & g_{m,1} & \cdots & g_{m,L-1} \end{bmatrix}^T$$

is the impulse response between the source $s_\mathrm{s}(k)$ and the signal $s_m(k)$. In this scenario, it can be shown that matrix $\mathbf{R}_{ss}$ is rank-deficient by, at least, $(M-2)L+1$. As a matter of fact, from (3.110), the input signal vector $\mathbf{s}_\mathrm{s}(k)$ can be written as

$$\begin{aligned} \mathbf{s}(k) &= \begin{bmatrix} \mathbf{s}_1^T(k) & \mathbf{s}_2^T(k) & \cdots & \mathbf{s}_M^T(k) \end{bmatrix}^T \\ &= \mathbf{G}\mathbf{s}_\mathrm{s}(k) \end{aligned}$$

$$= \begin{bmatrix} g_{1,0} & g_{1,1} & \cdots & g_{1,L-1} & 0 & 0 & \cdots & 0 \\ 0 & g_{1,0} & g_{1,1} & \cdots & g_{1,L-1} & 0 & \cdots & 0 \\ \vdots & \vdots & \vdots & \vdots & \vdots & \vdots & \ddots & \vdots \\ 0 & \cdots & 0 & g_{1,0} & g_{1,1} & \cdots & & g_{1,L-1} \\ g_{2,0} & g_{2,1} & \cdots & g_{2,L-1} & 0 & 0 & \cdots & 0 \\ 0 & g_{2,0} & g_{2,1} & \cdots & g_{2,L-1} & 0 & \cdots & 0 \\ \vdots & \vdots & \vdots & \vdots & \vdots & \vdots & \ddots & \vdots \\ 0 & \cdots & 0 & g_{2,0} & g_{2,1} & \cdots & & g_{2,L-1} \\ \vdots & \vdots & \vdots & \vdots & \vdots & \vdots & & \vdots \\ g_{M,0} & g_{M,1} & \cdots & g_{M,L-1} & 0 & 0 & \cdots & 0 \\ 0 & g_{M,0} & g_{M,1} & \cdots & g_{M,L-1} & 0 & \cdots & 0 \\ \vdots & \vdots & \vdots & \vdots & \vdots & \vdots & \ddots & \vdots \\ 0 & \cdots & 0 & g_{M,0} & g_{M,1} & \cdots & & g_{M,L-1} \end{bmatrix} \begin{bmatrix} s_{\mathrm{s}}(k) \\ s_{\mathrm{s}}(k-1) \\ s_{\mathrm{s}}(k-2) \\ \vdots \\ s_{\mathrm{s}}(k-2L+3) \\ s_{\mathrm{s}}(k-2L+2) \end{bmatrix},$$

where $\mathbf{G}$ is an $ML \times (2L-1)$ matrix, containing the impulse responses g_m and is assumed to be full column rank, and $\mathbf{s}_{\mathrm{s}}(k)$ is a $(2L-1) \times 1$ vector. We then have:

$$\begin{aligned} \mathbf{R}_{ss} &= E\left\{\mathbf{s}(k)\mathbf{s}^T(k)\right\} \\ &= \mathbf{G}E\left\{\mathbf{s}_{\mathrm{s}}(k)\mathbf{s}_{\mathrm{s}}^T(k)\right\}\mathbf{G}^T \\ &= \mathbf{G}\mathbf{R}_{s_{\mathrm{s}}s_{\mathrm{s}}}\mathbf{G}^T, \end{aligned} \tag{3.111}$$

where $\mathbf{R}_{s_{\mathrm{s}}s_{\mathrm{s}}}$ is the source signal covariance matrix of size $(2L-1) \times (2L-1)$, assumed to be full rank. We immediately see from (3.111), that:

$$\mathrm{Rank}\left[\mathbf{R}_{ss}\right] = \min\left\{ \mathrm{Rank}\left[\mathbf{G}\right], \ \mathrm{Rank}\left[\mathbf{R}_{s_{\mathrm{s}}s_{\mathrm{s}}}\right] \right\} = 2L-1, \tag{3.112}$$

and

$$\begin{aligned} \mathrm{Null}[\mathbf{R}_{ss}] &= ML - \mathrm{Rank}\left[\mathbf{R}_{ss}\right] \\ &= (M-2)L+1, \end{aligned} \tag{3.113}$$

where Null[ ] and Rank[ ] denote the dimension of the null space and the rank of a matrix, respectively.

From the above analysis, one can see that a MISO system is rank deficient if its inputs are the filtered version of the same source signal. Thus, the Wiener-Hopf equations do not have a unique solution.

In most practical situations, the signals $s_m(k),\ m = 1, 2, \cdots, M$, are somehow related. If they are highly coherent, adaptive algorithms will be very slow to converge to the true solution and in some situations, they will converge to a solution that is far from the desired one.

We are now going to show in the particular case of a MISO system with two inputs how a high coherence between these signals affects the condition number of the covariance matrix:

$$\mathbf{R}_{ss} = \begin{bmatrix} \mathbf{R}_{s_1 s_1} & \mathbf{R}_{s_1 s_2} \\ \mathbf{R}_{s_2 s_1} & \mathbf{R}_{s_2 s_2} \end{bmatrix}.$$

For $L \rightarrow \infty$, a Toeplitz matrix is asymptotically equivalent to a circulant matrix if its elements are absolutely summable [148], which is the case in most acoustic applications. In this situation, we can decompose

$$\mathbf{R}_{s_m s_i} = \mathbf{F}^{-1} \underline{\boldsymbol{R}}_{s_m s_i} \mathbf{F}, \ m, i = 1, 2, \tag{3.114}$$

where $\mathbf{F}$ is the Fourier matrix and the diagonal matrix

$$\underline{\boldsymbol{R}}_{s_m s_i} = \mathrm{diag}\left\{ \underline{R}_{s_m s_i}(0), \underline{R}_{s_m s_i}(1), \cdots, \underline{R}_{s_m s_i}(L-1) \right\} \tag{3.115}$$

contains elements corresponding to the L frequency bins which are formed from the discrete Fourier transform (DFT) of the first column of $\mathbf{R}_{s_m s_i}$. Letting $r_{s_m s_i}(l)$ be the auto- and cross-correlation for $m = i$ and $m \neq i$ respectively, we see that the spectral content between two signals is related to the correlation function by

$$\underline{R}_{s_m s_i}(f) = \sum_{l=-\infty}^{\infty} r_{s_m s_i}(l) e^{-j2\pi f l}, \ f = 0, 1, \cdots, L-1. \tag{3.116}$$

Using (3.114), $\mathbf{R}_{ss}$ can be expressed in terms of its spectra as:

$$\begin{aligned} \mathbf{R}_{ss} &= \mathbf{F}_{\mathrm{d}}^{-1} \underline{\boldsymbol{R}}_{ss} \mathbf{F}_{\mathrm{d}} \\ &= \begin{bmatrix} \mathbf{F}^{-1} & \mathbf{0}_{L\times L} \\ \mathbf{0}_{L\times L} & \mathbf{F}^{-1} \end{bmatrix} \begin{bmatrix} \underline{\boldsymbol{R}}_{s_1 s_1} & \underline{\boldsymbol{R}}_{s_1 s_2} \\ \underline{\boldsymbol{R}}_{s_2 s_1} & \underline{\boldsymbol{R}}_{s_2 s_2} \end{bmatrix} \begin{bmatrix} \mathbf{F} & \mathbf{0}_{L\times L} \\ \mathbf{0}_{L\times L} & \mathbf{F} \end{bmatrix}. \end{aligned} \tag{3.117}$$

To compute the condition number $\chi_{\mathrm{F}}^2 \left[\mathbf{R}_{ss}^{1/2} \right]$ (see Sect. 3.4), we need to compute $\mathrm{tr}\left[\mathbf{R}_{ss}\right]$ and $\mathrm{tr}\left[\mathbf{R}_{ss}^{-1}\right]$. The first trace is easy to compute. Indeed, using (3.117), we easily find:

$$\begin{aligned} \mathrm{tr}\left[\mathbf{R}_{ss}\right] &= \mathrm{tr}\left[\mathbf{F}_{\mathrm{d}}^{-1} \underline{\boldsymbol{R}}_{ss} \mathbf{F}_{\mathrm{d}}\right] = \mathrm{tr}\left[\underline{\boldsymbol{R}}_{ss}\right] \\ &= \sum_{l=0}^{L-1} \left[\underline{R}_{s_1 s_1}(l) + \underline{R}_{s_2 s_2}(l) \right]. \end{aligned} \tag{3.118}$$

For the second trace, we have:

$$\mathrm{tr}\left[\mathbf{R}_{ss}^{-1}\right] = \mathrm{tr}\left[\mathbf{F}_{\mathrm{d}}^{-1} \underline{\boldsymbol{R}}_{ss}^{-1} \mathbf{F}_{\mathrm{d}}\right] = \mathrm{tr}\left[\underline{\boldsymbol{R}}_{ss}^{-1}\right]. \tag{3.119}$$

Furthermore, it is easy to show that:

$$\underset{\Rightarrow}{\boldsymbol{R}}_{ss}^{-1} = \begin{bmatrix} \underset{\Rightarrow}{\boldsymbol{R}}_1^{-1} & \mathbf{0}_{L\times L} \\ \mathbf{0}_{L\times L} & \underset{\Rightarrow}{\boldsymbol{R}}_2^{-1} \end{bmatrix} \begin{bmatrix} \mathbf{I}_{L\times L} & -\underset{\Rightarrow}{\boldsymbol{R}}_{s_1s_2}\underset{\Rightarrow}{\boldsymbol{R}}_{s_2s_2}^{-1} \\ -\underset{\Rightarrow}{\boldsymbol{R}}_{s_2s_1}\underset{\Rightarrow}{\boldsymbol{R}}_{s_1s_1}^{-1} & \mathbf{I}_{L\times L} \end{bmatrix}, \tag{3.120}$$

where

$$\underset{\Rightarrow}{\boldsymbol{R}}_1 = \left[\mathbf{I}_{L\times L} - \underset{\Rightarrow}{\boldsymbol{R}}_{s_1s_2}^2\left(\underset{\Rightarrow}{\boldsymbol{R}}_{s_1s_1}^{-1}\underset{\Rightarrow}{\boldsymbol{R}}_{s_2s_2}^{-1}\right)\right]\underset{\Rightarrow}{\boldsymbol{R}}_{s_1s_1}, \tag{3.121}$$

$$\underset{\Rightarrow}{\boldsymbol{R}}_2 = \left[\mathbf{I}_{L\times L} - \underset{\Rightarrow}{\boldsymbol{R}}_{s_1s_2}^2\left(\underset{\Rightarrow}{\boldsymbol{R}}_{s_1s_1}^{-1}\underset{\Rightarrow}{\boldsymbol{R}}_{s_2s_2}^{-1}\right)\right]\underset{\Rightarrow}{\boldsymbol{R}}_{s_2s_2}. \tag{3.122}$$

Hence,

$$\begin{aligned} \mathrm{tr}\left[\mathbf{R}_{ss}^{-1}\right] &= \mathrm{tr}\left[\underset{\Rightarrow}{\boldsymbol{R}}_1^{-1} + \underset{\Rightarrow}{\boldsymbol{R}}_2^{-1}\right] \\ &= \sum_{l=0}^{L-1}\left[1 - |\gamma(l)|^2\right]^{-1}\left[\underset{\Rightarrow}{R}_{s_1s_1}^{-1}(l) + \underset{\Rightarrow}{R}_{s_2s_2}^{-1}(l)\right], \end{aligned} \tag{3.123}$$

where

$$|\gamma(f)|^2 = \frac{|\underset{\Rightarrow}{R}_{s_1s_2}(f)|^2}{\underset{\Rightarrow}{R}_{s_1s_1}(f)\underset{\Rightarrow}{R}_{s_2s_2}(f)}, \quad f = 0, 1, \cdots, L-1, \tag{3.124}$$

is the squared interchannel coherence function of the fth frequency bin.

We finally obtain the relationship between the interchannel coherence and the condition number based on the Frobenius norm:

$$\begin{aligned} \chi_{\mathrm{F}}^2\left[\mathbf{R}_{ss}^{1/2}\right] &= \left\{\sum_{l=0}^{L-1}\left[\underset{\Rightarrow}{R}_{s_1s_1}(l) + \underset{\Rightarrow}{R}_{s_2s_2}(l)\right]\right\} \\ &\times \left\{\sum_{l=0}^{L-1}\left[1 - |\gamma(l)|^2\right]^{-1}\left[\underset{\Rightarrow}{R}_{s_1s_1}^{-1}(l) + \underset{\Rightarrow}{R}_{s_2s_2}^{-1}(l)\right]\right\}. \end{aligned} \tag{3.125}$$

It is now evident from the previous expression that $\chi_{\mathrm{F}}^2\left[\mathbf{R}_{ss}^{1/2}\right]$ increases with the squared interchannel coherence function, hence degrading the condition of $\mathbf{R}_{ss}$; as $\gamma \to 1$, $\chi_{\mathrm{F}}^2\left[\mathbf{R}_{ss}^{1/2}\right] \to \infty$ and the identification of the system is more and more difficult, if not impossible.

3.7 Numerical Examples

In this section, a number of numerical experiments will be presented to better understand how the classical NLMS algorithm identifies acoustic impulse responses in SISO and 2-channel MISO systems.

For the performance measure, we will use the normalized misalignment (in dB),

$$20\log_{10}\frac{\left\|\mathbf{h} - \hat{\mathbf{h}}(k)\right\|_2}{\|\mathbf{h}\|_2}. \tag{3.126}$$

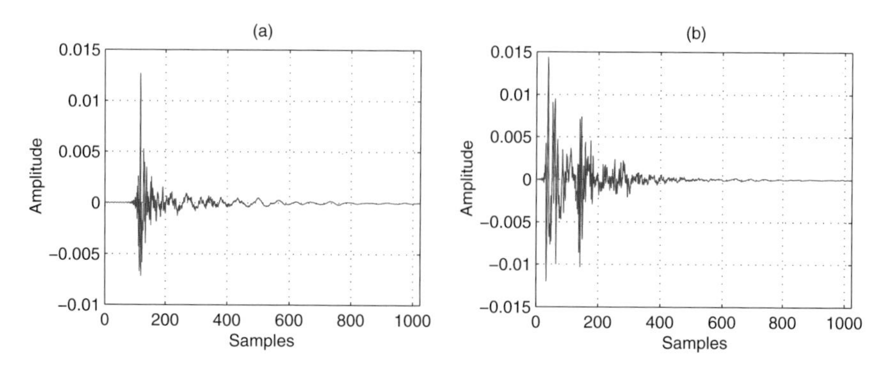

Fig. 3.1. Acoustic impulse responses used in simulations.

The two room acoustic impulse responses, $\mathbf{h}$, to be identified are shown in Fig. 3.1. The impulse response of Fig. 3.1(a) is used for a SISO system while the two of them are used for a 2-channel MISO system. They are both of length $L = 1024$. The sampling rate is 8 kHz and a white Gaussian noise signal with 30 dB SNR (signal-to-noise ratio) is added to the output $x(k)$. The regularization parameter was always chosen to be $\delta = 10\sigma_s^2$.

We first consider a SISO system whose input $s(k)$ is a zero-mean white Gaussian signal. We would like to verify the compromise between initial convergence rate and final MSE and the effect of underestimating the length of the impulse response on the behavior of the NLMS algorithm.

Figure 3.2 shows the convergence of the normalized misalignment of the NLMS algorithm for different values of the normalized step-size parameter α. For $\alpha \leq 1$, we see that decreasing α improves the final MSE, as expected, but the rate of convergence slows down. For $\alpha = 1.5$, both the convergence rate and final MSE are worse than for $\alpha = 1$. This suggests that it is very risky to choose $\alpha > 1$ in practice.

Figure 3.3 presents the convergence of the normalized misalignment of the NLMS algorithm for different lengths of the adaptive filter $\hat{\mathbf{h}}(k)$ with $\alpha = 0.5$. We see clearly from this figure that taking a filter length L_{f} much smaller than the length of the impulse response L increases considerably the final MSE, even for a white input signal. On the other hand, the initial convergence rate is improved.

Finally, to conclude this section, we are going to consider a 2-channel MISO system and verify how a high coherence between these input signals affects the convergence of the NLMS algorithm. We generate $s_1(k)$ and $s_2(k)$ as follows,

$$s_1(k) = w(k) + nl\frac{w(k) + |w(k)|}{2}, \tag{3.127}$$

$$s_2(k) = w(k) + nl\frac{w(k) - |w(k)|}{2}, \tag{3.128}$$

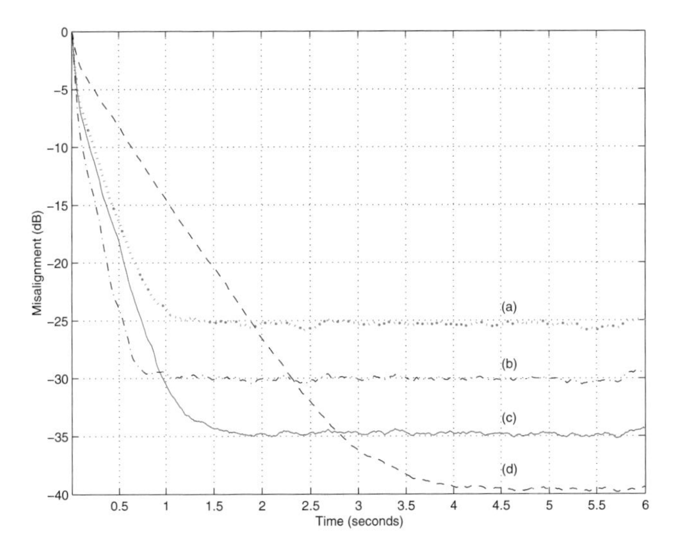

Fig. 3.2. Convergence of the normalized misalignment of the SISO NLMS algorithm for different values of the normalized step-size parameter. (a) $\alpha = 1.5$, (b) $\alpha = 1$, (c) $\alpha = 0.5$, and (d) $\alpha = 0.2$. The length of the adaptive filter is $L = 1024$, the same as the length of the impulse response.

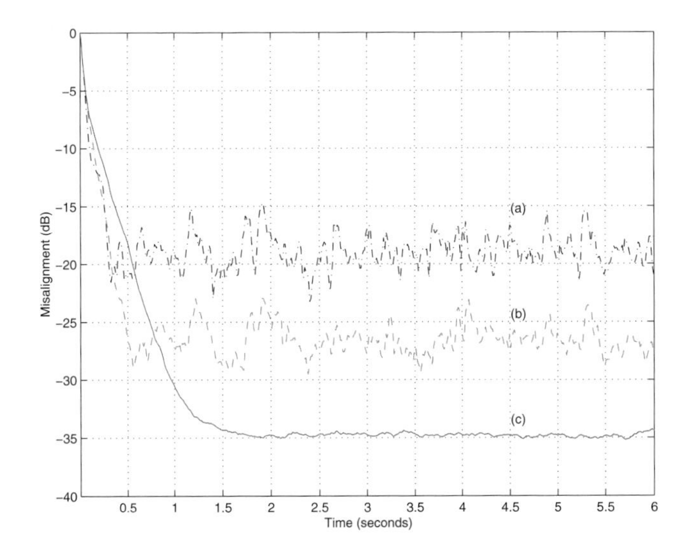

Fig. 3.3. Convergence of the normalized misalignment of the SISO NLMS algorithm for different lengths of the adaptive filter. (a) $L_{\mathrm{f}} = 256$, (b) $L_{\mathrm{f}} = 512$, and (c) $L_{\mathrm{f}} = 1024$. The normalized step-size parameter is $\alpha = 0.5$.

where $w(k)$ is a zero-mean white Gaussian random process and nl is the amount of non-linearity (positive and negative half-wave rectifiers) added to

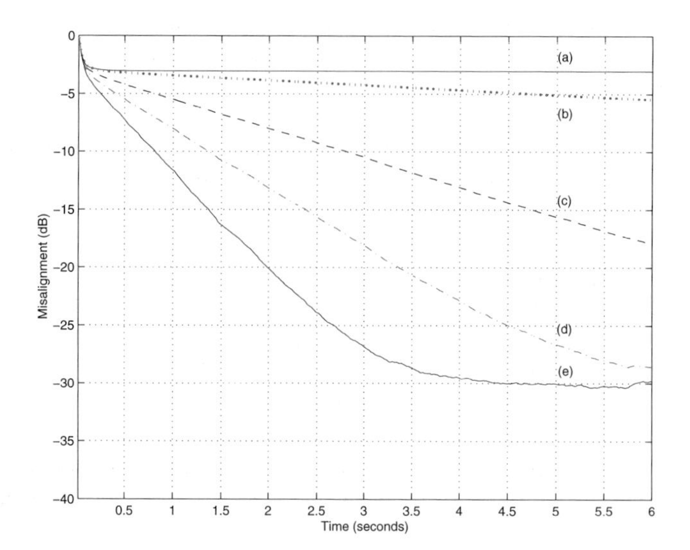

Fig. 3.4. Convergence of the normalized misalignment of the 2-channel MISO NLMS algorithm for different values of the non-linearity nl. (a) $nl = 0$, (b) $nl = 0.3$, (c) $nl = 1$, (d) $nl = 2$, and (e) $nl = 5$. Here $L_{\mathrm{f}} = L$ and $\alpha = 1$.

the signals. In all our simulations, the mean of the processes $s_1(k)$ and $s_2(k)$ is removed. For $nl = 0$, the two input signals $s_1(k)$ and $s_2(k)$ are perfectly coherent (so that the magnitude squared coherence is equal to 1 at all frequencies) and the MISO NLMS algorithm will not be able to identify the two impulse responses. As nl increases, the coherence decreases and better identification is expected.

Figure 3.4 shows the convergence of the normalized misalignment of the 2-channel MISO NLMS algorithm for different values of the non-linearity nl with $L_{\mathrm{f}} = L$ and $\alpha = 1$. For $nl = 0$, the NLMS is stalled at the value -3 dB and the solution obtained is far from the desired one. For $nl = 0.3$, the coherence is still pretty high (close to one), the algorithm seems indicating that it will converge but the time of convergence to a satisfactory value of misalignment seems extremely long. As nl increases, we notice that the identification is better and better and for $nl = 5$, the MISO NLMS algorithm is able to attain, in a reasonable amount of time, the final misalignment.

3.8 Summary

We first derived the optimal Wiener filter for a SISO system and showed that this filter can be a very good approximation of the desired acoustic impulse response if its length covers well enough the reverberation of the room.

We discussed in details the condition number of the input signal correlation matrix. This matrix appears explicitly in the Wiener-Hopf equations and implicitly in all adaptive filters. A high condition number will perturb the accuracy of the solution of the Wiener-Hopf equations and will slow down the rate of convergence of most adaptive algorithms. A fast, efficient algorithm to compute the conditional number was also developed.

We also discussed several important adaptive filters. In particular, the NLMS algorithm, which is extremely popular and useful in practice, was derived.

We generalized the Wiener principle to the MIMO system case. We showed that the MIMO Wiener-Hopf equations can be decomposed in N independent MISO Wiener-Hopf equations. As a result, adaptive filters for SISO and MISO systems, with a reference signal, cover all possible cases. A deep analysis on the conditioning of the input signal covariance matrix was given, showing that identification is not always obvious and depends on the interchannel coherence between the input signals.

Finally, we presented a number of numerical experiments showing how the NLMS algorithm converges in some typical identification situations for SISO and MISO systems.

4

Sparse Adaptive Filters

The aim of the previous chapter was to introduce the first and most fundamental tools that allow the estimation of acoustic impulse responses. In this chapter, we propose to derive more sophisticated SISO adaptive algorithms that better suit the acoustic environment. Generalization of these algorithms to MISO systems is straightforward, so it will not be discussed here.

4.1 Introduction

An impulse response that is sparse has a small percentage of its components with a significant magnitude while the rest are zero or small. Another definition could be the following: an impulse response is sparse if a large fraction of its energy is concentrated in a small fraction of its duration. A mathematical definition of sparseness is given in Chap. 2.

Acoustic impulse responses are usually sparse. Therefore, the applications where they are encountered are numerous: acoustic echo cancellation, feedback cancellation in hearing aids, blind identification of acoustic impulse responses for time delay estimation, source localization, dereverberation, etc.

Classical and most used algorithms such as the normalized least-mean-square (NLMS) [320] or recursive least-squares (RLS) [160] do not take into account whether the impulse responses they try to identify are sparse or not. Intuitively, however, it seems possible to improve the performance of the NLMS algorithm, for example, if the target is sparse.

Perhaps Duttweiler was one of the first persons who exploited this intuition algorithmically in the context of network echo cancellation involving a hybrid transformer in conjunction with variable network delay and where impulse responses are clearly sparse. The so-called proportionate NLMS (PNLMS) algorithm was then introduced [102]. This new algorithm converges and tracks much faster than the NLMS algorithm when the impulse response that we need to identify is sparse. PNLMS and other more sophisticated versions such as improved PNLMS (IPNLMS) [39] are success stories since they are now used in many products.

Recently, another variant of the LMS algorithm, called the exponentiated gradient algorithm with positive and negative weights (EG± algorithm), was proposed by Kivinen and Warmuth in the context of computational learning theory [204]. This new algorithm also converges much faster than the LMS algorithm when the target is sparse. The EG± algorithm has the nice feature that its update rule takes advantage of the sparseness of the impulse response to speed up its initial convergence and to improve its tracking abilities compared to LMS. In [164], a general expression of the mean-square error (MSE) is derived for the EG± algorithm showing that, indeed, for sparse impulse responses, the EG± algorithm, like PNLMS, converges more quickly than the LMS for a given asymptotic MSE. Even though the EG± and PNLMS algorithms may look very different, clearly they must be linked somehow. It is quite remarkable that two equivalent algorithms, as it will be shown later, were proposed in two completely different contexts.

There are two fundamental ways to update the coefficients of an adaptive filter $\hat{\mathbf{h}}(k)$. The linear update (see Chap. 3 for some examples and more will be given later):

$$\hat{\mathbf{h}}(k) = \mathbf{M}(k)\hat{\mathbf{h}}(k-1) + \mathbf{v}(k), \tag{4.1}$$

where $\mathbf{M}(k)$ and $\mathbf{v}(k)$ are respectively a matrix and a vector independent of $\hat{\mathbf{h}}(k-1)$, and the non-linear update:

$$\hat{\mathbf{h}}(k) = \mathbf{M}\left[\hat{\mathbf{h}}(k-1)\right]\hat{\mathbf{h}}(k-1) + \mathbf{v}(k), \tag{4.2}$$

where this time, as indicated, $\mathbf{M}\left[\hat{\mathbf{h}}(k-1)\right]$ depends on $\hat{\mathbf{h}}(k-1)$. All classical algorithms such as NLMS and RLS can be deduced from (4.1) and sparse adaptive filters can be derived from (4.2). This view gives already an answer to the important question: how can this *a priori* information (sparseness) be taken into account to improve convergence and tracking of adaptive algorithms? The study of this question is the main objective of this chapter.

4.2 Notation and Definitions

In derivations and descriptions, the following notation is used:

$$\begin{aligned}
s(k) &= \text{input signal}, \\
x(k) &= \mathbf{h}^T\mathbf{s}(k) + b(k), \text{ output signal of the SISO system}, \\
\mathbf{s}(k) &= \begin{bmatrix} s(k) & s(k-1) & \cdots & s(k-L+1) \end{bmatrix}^T, \text{ excitation vector}, \\
\mathbf{h} &= \begin{bmatrix} h_0 & h_1 & \cdots & h_{L-1} \end{bmatrix}^T, \text{ impulse response of the SISO system}, \\
\hat{\mathbf{h}}(k) &= \begin{bmatrix} \hat{h}_0(k) & \hat{h}_1(k) & \cdots & \hat{h}_{L-1}(k) \end{bmatrix}^T, \text{ estimated impulse response}.
\end{aligned}$$

Here L is the length of the adaptive filter and k is the time index.

We now give some important definitions that will be used in the rest of this chapter:

$$\begin{aligned}
e(k) &= x(k) - \hat{x}(k) \\
&= x(k) - \mathbf{s}^T(k)\hat{\mathbf{h}}(k-1), \ \textit{a priori} \text{ error signal}, && (4.3) \\
\varepsilon(k) &= x(k) - \mathbf{s}^T(k)\hat{\mathbf{h}}(k), \ \textit{a posteriori} \text{ error signal}, && (4.4) \\
e_{\mathrm{n}}(k) &= \left[\mathbf{s}^T(k)\mathbf{G}(k-1)\mathbf{s}(k)\right]^{-1/2} e(k), && (4.5) \\
&\quad \text{normalized } \textit{a priori} \text{ error signal}, \\
\varepsilon_{\mathrm{n}}(k) &= \left[\mathbf{s}^T(k)\mathbf{G}(k-1)\mathbf{s}(k)\right]^{-1/2} \varepsilon(k), && (4.6) \\
&\quad \text{normalized } \textit{a posteriori} \text{ error signal}, \\
\mathbf{e}(k) &= \mathbf{x}(k) - \mathbf{S}^T(k)\hat{\mathbf{h}}(k-1), \ \textit{a priori} \text{ error signal vector}, && (4.7) \\
\boldsymbol{\varepsilon}(k) &= \mathbf{x}(k) - \mathbf{S}^T(k)\hat{\mathbf{h}}(k), \ \textit{a posteriori} \text{ error signal vector}, && (4.8) \\
\mathbf{e}_{\mathrm{n}}(k) &= \left[\mathbf{S}^T(k)\mathbf{G}(k-1)\mathbf{S}(k)\right]^{-1/2} \mathbf{e}(k), && (4.9) \\
&\quad \text{normalized } \textit{a priori} \text{ error signal vector}, \\
\boldsymbol{\varepsilon}_{\mathrm{n}}(k) &= \left[\mathbf{S}^T(k)\mathbf{G}(k-1)\mathbf{S}(k)\right]^{-1/2} \boldsymbol{\varepsilon}(k), && (4.10) \\
&\quad \text{normalized } \textit{a posteriori} \text{ error signal vector},
\end{aligned}$$

where

$$\mathbf{x}(k) = \begin{bmatrix} x(k) & x(k-1) & \cdots & x(k-P+1) \end{bmatrix}^T$$

is a vector containing the P more recent samples of the output signal $x(k)$,

$$\mathbf{S}(k) = \begin{bmatrix} \mathbf{s}(k) & \mathbf{s}(k-1) & \cdots & \mathbf{s}(k-P+1) \end{bmatrix}$$

is an $L \times P$ matrix of the input signal samples $s(k)$, and

$$\mathbf{G}(k-1) = \mathrm{diag}\left\{ g_0(k-1) \ g_1(k-1) \ \cdots \ g_{L-1}(k-1) \right\} \tag{4.11}$$

is an $L \times L$ diagonal matrix, where $g_l(k-1) > 0$, $\forall k, l$. This matrix is context dependent but is usually a function of $\hat{\mathbf{h}}(k-1)$.

4.3 The NLMS, PNLMS, and IPNLMS Algorithms

In this section, we briefly recall the normalized least-mean-square (NLMS) algorithm, and explain the proportionate NLMS (PNLMS) and improved PNLMS (IPNLMS) algorithms. Even though NLMS and IPNLMS may seem coming from the same family of adaptive filters, this is not really the case and the similarity between the two is quite deceiving.

The role of an adaptive filter, $\hat{\mathbf{h}}(k)$, is to estimate the true impulse response, $\mathbf{h}$, at each iteration time, k, when new samples, $s(k)$ and $x(k)$, are available. Depending on the algorithm used for this task, convergence, tracking, complexity, robustness to noise, etc, can be very different. One of the most popular adaptive filters in signal processing applications is the NLMS (see the previous chapter) [320], [160], due to its simplicity and robustness. But its convergence and tracking are slow in general, especially for long impulse responses. In many situations where an adaptive algorithm is required, convergence and tracking are critical for a good performance of the entire system. While in the NLMS, the adaptation step is the same for all components of the filter, in the PNLMS [102], an adaptive individual step size is assigned to each filter coefficient. The step sizes are calculated from the last estimate of the filter coefficients in such a way that a larger coefficient receives a larger increment, thus increasing the convergence rate of that coefficient. This has the effect that active coefficients are adjusted faster than non-active coefficients (i.e. small or zero coefficients). Hence, PNLMS converges much faster than NLMS for sparse impulse responses. Unfortunately, PNLMS behaves much worse than NLMS when the impulse response is not sparse. This problem is due to the fact that the proportionate update is not very well refined. In [39], an IPNLMS was proposed where the adaptive individual step size has a better balance between the fixed step size of NLMS and the large amount of proportionality in PNLMS. As a result, IPNLMS always converges and tracks better than NLMS and PNLMS, no matter how sparse the impulse response is.

The error signal and the coefficient update equation of the three previously discussed algorithms can be written as:

$$e(k) = x(k) - \hat{\mathbf{h}}^T(k-1)\mathbf{s}(k), \tag{4.12}$$

$$\hat{\mathbf{h}}(k) = \hat{\mathbf{h}}(k-1) + \frac{\alpha\mathbf{G}(k-1)\mathbf{s}(k)e(k)}{\mathbf{s}^T(k)\mathbf{G}(k-1)\mathbf{s}(k)+\delta}, \tag{4.13}$$

where

$$\mathbf{G}(k-1) = \operatorname{diag}\left\{\begin{array}{cccc} g_0(k-1) & g_1(k-1) & \cdots & g_{L-1}(k-1)\end{array}\right\} \tag{4.14}$$

is a diagonal matrix that adjusts the step sizes of the individual taps of the filter, α $(0 < \alpha < 1)$ is the overall step-size factor, and δ is the regularization parameter.

The NLMS algorithm is obtained by taking:

$$\mathbf{G}(k) = \mathbf{I}, \tag{4.15}$$

$$\delta = \delta_{\mathrm{NLMS}} = \mathrm{cst}\cdot\sigma_s^2, \tag{4.16}$$

where $\mathbf{I}$, σ_s^2, and cst are the identity matrix, the power of the signal $s(k)$, and a small positive constant, respectively.

In the PNLMS, the diagonal elements of $\mathbf{G}(k) = \mathbf{G}_{\mathrm{p}}(k)$ are calculated as follows [102]:

$$\gamma_{\mathrm{p},l}(k) = \max\left\{\rho \max\left[\delta_{\mathrm{p}}, \left|\hat{h}_0(k)\right|, \cdots, \left|\hat{h}_{L-1}(k)\right|\right], \left|\hat{h}_l(k)\right|\right\}, \quad (4.17)$$

$$g_{\mathrm{p},l}(k) = \frac{\gamma_{\mathrm{p},l}(k)}{\|\boldsymbol{\gamma}_{\mathrm{p}}(k)\|_1}, \quad 0 \leq l \leq L-1, \quad (4.18)$$

where

$$\boldsymbol{\gamma}_{\mathrm{p}}(k) = \left[\, \gamma_{\mathrm{p},0}(k) \ \gamma_{\mathrm{p},1}(k) \ \cdots \ \gamma_{\mathrm{p},L-1}(k) \,\right]^T.$$

Parameters δ_{p} and ρ are positive numbers with typical values $\delta_{\mathrm{p}} = 0.01$, $\rho = 0.01$. The first term in (4.17), ρ, prevents $\hat{h}_l(k)$ from stalling when its magnitude is much smaller than the magnitude of the largest coefficient and δ_{p} regularizes the updating when all coefficients are zero at initialization. For the regularization parameter, we usually choose:

$$\delta_{\mathrm{PNLMS}} = \delta_{\mathrm{NLMS}}/L. \quad (4.19)$$

For the IPNLMS algorithm, the diagonal matrix, $\mathbf{G}(k) = \mathbf{G}_{\mathrm{ip}}(k)$, is computed in a more elegant way [39]:

$$\gamma_{\mathrm{ip},l}(k) = (1-\beta)\frac{\left\|\hat{\mathbf{h}}(k)\right\|_1}{L} + (1+\beta)\left|\hat{h}_l(k)\right|, \quad (4.20)$$

$$\begin{aligned} g_{\mathrm{ip},l}(k) &= \frac{\gamma_{\mathrm{ip},l}(k)}{\|\boldsymbol{\gamma}_{\mathrm{ip}}(k)\|_1} \\ &= \frac{1-\beta}{2L} + (1+\beta)\frac{\left|\hat{h}_l(k)\right|}{2\left\|\hat{\mathbf{h}}(k)\right\|_1}, \quad 0 \leq l \leq L-1, \end{aligned} \quad (4.21)$$

where β $(-1 \leq \beta < 1)$ is a parameter that controls the amount of proportionality in the IPNLMS. For $\beta = -1$, it can be easily checked that the IPNLMS and NLMS algorithms are identical. For β close to 1, IPNLMS behaves like PNLMS. In practice, a good choice for β is -0.5 or 0. With this choice and in simulations, IPNLMS always performs better than NLMS and PNLMS. As for the regularization parameter, it should be taken as:

$$\delta_{\mathrm{IPNLMS}} = \frac{1-\beta}{2L}\delta_{\mathrm{NLMS}}. \quad (4.22)$$

The IPNLMS algorithm is summarized in Table 4.1.

Before finishing this section, it is worth mentioning another variant of PNLMS, called PNLMS++ [137]. In this algorithm, the adaptation of the filter coefficients alternates between NLMS and PNLMS; as a result, PNLMS++ seems a little bit less sensitive to the assumption of a sparse impulse response than PNLMS.

Now, a natural question arises: is it possible to find an optimization criterion that includes the sparseness information?

Table 4.1. The improved PNLMS algorithm.

Initialization:	$\hat{h}_l(0) = 0,\ l = 0, 1, \cdots, L-1$
Parameters:	$-1 \leq \beta < 1$
	$0 < \alpha < 1$
	$\delta_{\text{IPNLMS}} = \text{cst} \cdot \sigma_s^2 \frac{1-\beta}{2L}$
	$\epsilon > 0$, very small number to avoid division by zero
Error:	$e(k) = x(k) - \hat{\mathbf{h}}^T(k-1)\mathbf{s}(k)$
Update:	$g_{\text{ip},l}(k-1) = \frac{1-\beta}{2L} + (1+\beta)\frac{\left\lvert \hat{h}_l(k-1)\right\rvert}{2\left\lVert \hat{\mathbf{h}}(k-1)\right\rVert_1 + \epsilon}$
	$\mu(k) = \frac{\alpha}{\sum_{j=0}^{L-1} s^2(k-j) g_{\text{ip},j}(k-1) + \delta_{\text{IPNLMS}}}$
	$\hat{h}_l(k) = \hat{h}_l(k-1) + \mu(k) g_{\text{ip},l}(k-1) s(k-l) e(k)$
	$l = 0, 1, \cdots, L-1$

4.4 Universal Criterion

In this section, we show how to derive different classes of adaptive filters. As explained in [204], a reasonable adaptive algorithm must find a good balance between its needs to be conservative (retain the information it has acquired in preceding iterations) and corrective (make sure that with new information, the accuracy of the solution is increased). For that, we give a universal criterion that is the sum of two terms: one of them is a distance between the old and new weight vectors (and depending on how we define this distance, we obtain different update rules) and the other one depends on the *a posteriori* error signal. Therefore, according to this principle, one easy way to find adaptive filters that adjust the new weight vector, $\hat{\mathbf{h}}(k)$, from the old one, $\hat{\mathbf{h}}(k-1)$, is to minimize the following cost function:

$$J(k) = d\left[\hat{\mathbf{h}}(k), \hat{\mathbf{h}}(k-1)\right] + \boldsymbol{\varepsilon}_{\text{n}}^T(k)\boldsymbol{\varepsilon}_{\text{n}}(k), \tag{4.23}$$

where $d\left[\hat{\mathbf{h}}(k), \hat{\mathbf{h}}(k-1)\right]$ is some measure of distance from the old to the new weight vectors. Differentiating $J(k)$ with respect to $\hat{\mathbf{h}}(k)$ and setting the resulting vector to zero, we can see that any adaptive algorithm has the form:

$$\begin{aligned} &2\mathbf{P}_s(k)\left[\hat{\mathbf{h}}(k) - \hat{\mathbf{h}}(k-1)\right] + \frac{\partial d\left[\hat{\mathbf{h}}(k), \hat{\mathbf{h}}(k-1)\right]}{\partial \hat{\mathbf{h}}(k)} \\ &= 2\mathbf{S}(k)\left[\mathbf{S}^T(k)\mathbf{G}(k-1)\mathbf{S}(k)\right]^{-1}\mathbf{e}(k), \end{aligned} \tag{4.24}$$

where

$$\mathbf{P}_s(k) = \mathbf{S}(k) \left[\mathbf{S}^T(k)\mathbf{G}(k-1)\mathbf{S}(k)\right]^{-1} \mathbf{S}^T(k) \tag{4.25}$$

is a projection matrix for $\mathbf{G}(k-1) = \mathbf{I}$. It has the two properties:

$$\begin{aligned} \mathbf{P}_s(k)\mathbf{G}(k-1)\mathbf{S}(k) &= \mathbf{S}(k), & (4.26)\\ \mathbf{P}_s(k)\mathbf{G}(k-1)\mathbf{P}_s(k) &= \mathbf{P}_s(k). & (4.27) \end{aligned}$$

Clearly, the choice of the distance $d\left[\hat{\mathbf{h}}(k), \hat{\mathbf{h}}(k-1)\right]$ is significant. Depending on how we choose it, we may have a linear or non-linear update equation with respect to the weight vector.

4.4.1 Linear Update

In this important category of adaptive filters, we take $\mathbf{G}(k-1) = \mathbf{I}$ and we choose for the distance:

$$d\left[\hat{\mathbf{h}}(k), \hat{\mathbf{h}}(k-1)\right] = \left[\hat{\mathbf{h}}(k) - \hat{\mathbf{h}}(k-1)\right]^T \mathbf{Q}_s(k) \left[\hat{\mathbf{h}}(k) - \hat{\mathbf{h}}(k-1)\right], \tag{4.28}$$

where the symmetric matrix $\mathbf{Q}_s(k)$ is positive definite and depends on the input signal $s(n)$ only. Using (4.28) in (4.24), we obtain the update equation:

$$\hat{\mathbf{h}}(k) = \hat{\mathbf{h}}(k-1) + \left[\mathbf{P}_s(k) + \mathbf{Q}_s(k)\right]^{-1} \mathbf{S}(k) \left[\mathbf{S}^T(k)\mathbf{S}(k)\right]^{-1} \mathbf{e}(k). \tag{4.29}$$

There are two important things to pay attention to. First, if we replace $\mathbf{e}(k)$ by its value in (4.29), we can see that this equation is updated linearly with respect to the estimation filter $\hat{\mathbf{h}}(k-1)$. Second, the choice of the matrix $\mathbf{Q}_s(k)$ will lead to well-known algorithms and even to new ones.

Let's take:

$$\mathbf{Q}_s(k) = \alpha^{-1}\mathbf{I} - \mathbf{P}_s(k). \tag{4.30}$$

It can be easily checked that $\mathbf{Q}_s(k)$ is positive definite if $0 < \alpha < 1$. Replacing (4.30) in (4.29), we get the affine projection algorithm (APA) [245]:

$$\hat{\mathbf{h}}(k) = \hat{\mathbf{h}}(k-1) + \alpha\mathbf{S}(k) \left[\mathbf{S}^T(k)\mathbf{S}(k)\right]^{-1} \mathbf{e}(k). \tag{4.31}$$

For the particular case $P = 1$, we obviously have the (non-regularized) NLMS algorithm [160]. Table 4.2 shows a practical way to implement the (regularized) APA. The complexity of this algorithm is $2LP + O(P^2)$ multiples per sample period. Even for a small P, this complexity is quite high. Fortunately, there are many ways to reduce this complexity to $2L + O(P)$, see for example [136], [138], [290], [153].

Consider a recursive estimation of the input signal correlation matrix:

$$\mathbf{R}(k) = \lambda\mathbf{R}(k-1) + \mathbf{s}(k)\mathbf{s}^T(k), \tag{4.32}$$

Table 4.2. The (regularized) affine projection algorithm.

Initialization:	$\hat{\mathbf{h}}(0) = \mathbf{0}_{L\times 1}$
Parameters:	$0 < \alpha < 1$
	$\delta_{\mathrm{APA}} = \mathrm{cst}\cdot\sigma_s^2$
	$\mathbf{s}_P(k) = \left[\begin{array}{cccc} s(k) & s(k-1) & \cdots & s(k-P+1)\end{array}\right]^T$
Error:	$\mathbf{e}(k) = \mathbf{x}(k) - \mathbf{S}^T(k)\hat{\mathbf{h}}(k-1)$
Update:	$\mathbf{S}^T(k)\mathbf{S}(k) = \mathbf{S}^T(k-1)\mathbf{S}(k-1) - \mathbf{s}_P(k-L)\mathbf{s}_P^T(k-L) + \mathbf{s}_P(k)\mathbf{s}_P^T(k)$
	$\hat{\mathbf{h}}(k) = \hat{\mathbf{h}}(k-1) + \alpha\mathbf{S}(k)\left[\mathbf{S}^T(k)\mathbf{S}(k) + \delta_{\mathrm{APA}}\mathbf{I}\right]^{-1}\mathbf{e}(k)$

Table 4.3. The RLS algorithm.

Initialization:	$\hat{\mathbf{h}}(0) = \mathbf{0}_{L\times 1}$
	$\mathbf{R}^{-1}(0) = \left(\mathrm{cst}\cdot\sigma_s^2\right)^{-1}\mathbf{I}$
Parameter:	$0 < \lambda < 1$, forgetting factor
Error:	$e(k) = x(k) - \hat{\mathbf{h}}^T(k-1)\mathbf{s}(k)$
Update:	$\mathbf{k}'_{\mathrm{g}}(k) = \mathbf{R}^{-1}(k-1)\mathbf{s}(k)$, *a priori* Kalman gain
	$\mathbf{k}_{\mathrm{g}}(k) = \dfrac{\mathbf{k}'_{\mathrm{g}}(k)}{\lambda + \mathbf{s}^T(k)\mathbf{k}'_{\mathrm{g}}(k)}$, *a posteriori* Kalman gain
	$\hat{\mathbf{h}}(k) = \hat{\mathbf{h}}(k-1) + \mathbf{k}_{\mathrm{g}}(k)e(k)$
	$\mathbf{R}^{-1}(k) = \lambda^{-1}\mathbf{R}^{-1}(k-1) - \lambda^{-1}\mathbf{k}_{\mathrm{g}}(k)\mathbf{k}'^T_{\mathrm{g}}(k)$

where λ $(0 < \lambda < 1)$ is an exponential forgetting factor. With $P = 1$ and plugging

$$\mathbf{Q}_s(k) = \frac{\mathbf{R}(k)}{\mathbf{s}^T(k)\mathbf{s}(k)} - \mathbf{P}_s(k) \tag{4.33}$$

in (4.29), we obtain the recursive least-squares (RLS) algorithm [160]:

$$\hat{\mathbf{h}}(k) = \hat{\mathbf{h}}(k-1) + \mathbf{R}^{-1}(k)\mathbf{s}(k)e(k). \tag{4.34}$$

A practical way to update the coefficients of this adaptive filter is presented in Table 4.3. The complexity of the RLS is proportional to L^2 but fast versions based on linear prediction techniques exist with an arithmetic complexity proportional to L [17].

Thus, so far in this section, we have seen how to deduce the three most classical adaptive filters that can be found in the literature: NLMS, APA, and RLS.

4.4.2 Non-Linear Update

In this second category of adaptive filters, $\mathbf{G}(k-1)$ is now a function of the filter $\hat{\mathbf{h}}(k-1)$ and the distance is changed accordingly:

$$d\left[\hat{\mathbf{h}}(k), \hat{\mathbf{h}}(k-1)\right] = \left[\hat{\mathbf{h}}(k) - \hat{\mathbf{h}}(k-1)\right]^T \mathbf{Q}_s\left[\mathbf{G}(k-1)\right]\left[\hat{\mathbf{h}}(k) - \hat{\mathbf{h}}(k-1)\right], \tag{4.35}$$

where now the symmetric positive-definite matrix $\mathbf{Q}_s\left[\mathbf{G}(k-1)\right]$ is not only a function of the input signal $s(k)$ but also of $\mathbf{G}(k-1)$ [and indirectly of $\hat{\mathbf{h}}(k-1)$] as well. Minimizing (4.23) with (4.35), we obtain a general form of the adaptive algorithm:

$$\begin{aligned} \hat{\mathbf{h}}(k) &= \hat{\mathbf{h}}(k-1) + \left\{\mathbf{P}_s(k) + \mathbf{Q}_s\left[\mathbf{G}(k-1)\right]\right\}^{-1}\mathbf{S}(k) \\ &\quad \cdot \left[\mathbf{S}^T(k)\mathbf{G}(k-1)\mathbf{S}(k)\right]^{-1}\mathbf{e}(k), \end{aligned} \tag{4.36}$$

where $\mathbf{P}_s(k)$ is defined in (4.25). The main difference between expressions (4.29) and (4.36) is that the former one is linearly updated with respect to the estimation filter $\hat{\mathbf{h}}(k-1)$ while the latter one is not since $\mathbf{G}(k-1)$ is a function of $\hat{\mathbf{h}}(k-1)$.

With the distance defined in (4.35), the parameter space is a curved manifold (non Euclidean). Such a space is a Riemannian space. The $L \times L$ positive-definite matrix $\mathbf{Q}_s\left[\mathbf{G}(k-1)\right]$ is called the *Riemannian metric tensor* and it depends in general on $\hat{\mathbf{h}}(k-1)$. The Riemannian metric tensor characterizes the intrinsic curvature of a particular manifold in L-dimensional space.

Taking $P = 1$ and

$$\mathbf{Q}_s(k) = \alpha^{-1}\mathbf{G}^{-1}(k-1) - \mathbf{P}_s(k), \tag{4.37}$$

we get the natural gradient (NG) algorithm proposed by Amari [8]:

$$\hat{\mathbf{h}}(k) = \hat{\mathbf{h}}(k-1) + \frac{\alpha\mathbf{G}(k-1)\mathbf{s}(k)e(k)}{\mathbf{s}^T(k)\mathbf{G}(k-1)\mathbf{s}(k)}. \tag{4.38}$$

Depending on the choice of $\mathbf{G}(k-1)$, we may obtain PNLMS [102], IPNLMS [39], or other proportionate versions of NLMS [140], [220], [217].

Now with $P > 1$ and with the same definition of $\mathbf{Q}_s(k)$ as in (4.37), we have the natural APA (NAPA):

$$\hat{\mathbf{h}}(k) = \hat{\mathbf{h}}(k-1) + \alpha\mathbf{G}(k-1)\mathbf{S}(k)\left[\mathbf{S}^T(k)\mathbf{G}(k-1)\mathbf{S}(k)\right]^{-1}\mathbf{e}(k). \tag{4.39}$$

Again, the choice of $\mathbf{G}(k-1)$ leads to different interesting algorithms such as proportionate APA (PAPA) [127] or improved PAPA (IPAPA) [167].

Following the same philosophy, we may derive the natural RLS (NRLS) algorithm:

$$\hat{\mathbf{h}}(k) = \hat{\mathbf{h}}(k-1) + \mathbf{G}^{1/2}(k-1)\mathbf{R}_{\hat{h}}^{-1}(k)\mathbf{G}^{1/2}(k-1)\mathbf{s}(k)e(k), \quad (4.40)$$

where

$$\mathbf{R}_{\hat{h}}(k) = \lambda \mathbf{R}_{\hat{h}}(k-1) + \left[\mathbf{G}^{1/2}(k-1)\mathbf{s}(k)\right]\left[\mathbf{G}^{1/2}(k-1)\mathbf{s}(k)\right]^T \quad (4.41)$$

is an estimate of the input signal correlation matrix.

To summarize this subsection, we can say that this relatively new non-linear framework for the update of the coefficients of the filter is very promising since it seems to fit very well the identification of sparse impulse responses: by taking this information into account, the adjustment of the coefficients of the estimated filter is done in a non-uniform manner (e.g., components with large magnitude have a larger step size than components with small magnitude) and the performance of the adaptive filter can be greatly improved. In the next section, we present another class of algorithms having the same feature.

4.5 Exponentiated Gradient Algorithms

The exponentiated gradient (EG) algorithms were first proposed by Kivinen and Warmuth in the context of computational learning theory [204]. These algorithms are highly non-linear and can be easily derived from the criterion explained in Sect. 4.4, by simply using for the distance $d_{\mathrm{re}}\left[\hat{\mathbf{h}}(k), \hat{\mathbf{h}}(k-1)\right]$, the *relative entropy* also known as *Kullback-Leibler divergence*. Since this divergence is not really a distance, it has to be handled with care.

4.5.1 The EG Algorithm for Positive Weights

In this subsection, we assume that the components of the impulse response that we try to identify are all positive, in order that the relative entropy is meaningful.

Taking $P = 1$ and $\mathbf{G}(k-1) = \mathbf{I}$, the criterion (4.23) simplifies to:

$$\begin{aligned} J(k) &= d_{\mathrm{re}}\left[\hat{\mathbf{h}}(k), \hat{\mathbf{h}}(k-1)\right] + \varepsilon_{\mathrm{n}}^2(k) \\ &= d_{\mathrm{re}}\left[\hat{\mathbf{h}}(k), \hat{\mathbf{h}}(k-1)\right] + \left[\mathbf{s}^T(k)\mathbf{s}(k)\right]^{-1}\varepsilon^2(k), \end{aligned} \quad (4.42)$$

where now

$$d_{\mathrm{re}}\left[\hat{\mathbf{h}}(k), \hat{\mathbf{h}}(k-1)\right] = \eta^{-1}\sum_{l=0}^{L-1}\hat{h}_l(k)\ln\frac{\hat{h}_l(k)}{\hat{h}_l(k-1)}, \quad (4.43)$$

with $\eta > 0$. With this formalism, $\hat{\mathbf{h}}(k)$ and $\hat{\mathbf{h}}(k-1)$ are *probability vectors*, which means that their components are nonnegative and $\left\|\hat{\mathbf{h}}(k)\right\|_1 =$

$\left\|\hat{\mathbf{h}}(k-1)\right\|_1 = u > 0$, where u is a scaling factor. Therefore, we minimize $J(k)$ with the constraint that $\sum_l \hat{h}_l(k) = 1$ (i.e. we take here $u = 1$). This optimization leads to:

$$\eta^{-1}\left[\ln\frac{\hat{h}_l(k)}{\hat{h}_l(k-1)} + 1\right] - 2s(k-l)\left[\mathbf{s}^T(k)\mathbf{s}(k)\right]^{-1}\varepsilon(k) + \kappa = 0, \quad (4.44)$$
$$l = 0, 1, \cdots, L-1,$$

where κ is a Lagrange multiplier. Equation (4.44) is highly non-linear so that solving it is very difficult if not impossible. However, if the new weight vector $\hat{\mathbf{h}}(k)$ is close to the old weight vector $\hat{\mathbf{h}}(k-1)$, replacing the *a posteriori* error signal, $\varepsilon(k)$, in (4.44) with the *a priori* error signal, $e(k)$, is a reasonable approximation and the equation

$$\eta^{-1}\left[\ln\frac{\hat{h}_l(k)}{\hat{h}_l(k-1)} + 1\right] - 2s(k-l)\left[\mathbf{s}^T(k)\mathbf{s}(k)\right]^{-1}e(k) + \kappa = 0, \quad (4.45)$$
$$l = 0, 1, \cdots, L-1,$$

is much easier to solve. We then deduce the EG algorithm [204]:

$$\hat{h}_l(k) = \frac{\hat{h}_l(k-1)r_l(k)}{\sum_{j=0}^{L-1}\hat{h}_j(k-1)r_j(k)}, \ l = 0, 1, \cdots, L-1, \quad (4.46)$$

where

$$r_l(k) = \exp\left[\eta(k)s(k-l)e(k)\right], \quad (4.47)$$

with $\eta(k) = 2\eta\left[\mathbf{s}^T(k)\mathbf{s}(k)\right]^{-1}$. The algorithm is initialized with: $h_l(0) = c > 0, \ \forall l$.

4.5.2 The EG± Algorithm for Positive and Negative Weights

The EG algorithm is designed to work for positive weights only, due to the nature of the relative entropy definition. However, there is a simple way to generalize the idea to both positive and negative weights. Indeed, we can always find two vectors $\hat{\mathbf{h}}^+(k)$ and $\hat{\mathbf{h}}^-(k)$ with positive coefficients, in such a way that the vector

$$\hat{\mathbf{h}}(k) = \hat{\mathbf{h}}^+(k) - \hat{\mathbf{h}}^-(k) \quad (4.48)$$

can have positive and negative components. In this case, the *a priori* and *a posteriori* error signals can be written as:

$$e(k) = x(k) - \left[\hat{\mathbf{h}}^+(k-1) - \hat{\mathbf{h}}^-(k-1)\right]^T \mathbf{s}(k), \quad (4.49)$$

$$\varepsilon(k) = x(k) - \left[\hat{\mathbf{h}}^+(k) - \hat{\mathbf{h}}^-(k)\right]^T \mathbf{s}(k), \quad (4.50)$$

and the criterion (4.42) will change to:

$$\begin{aligned} J^{\pm}(k) &= d_{\text{re}}\left[\hat{\mathbf{h}}^{+}(k), \hat{\mathbf{h}}^{+}(k-1)\right] + d_{\text{re}}\left[\hat{\mathbf{h}}^{-}(k), \hat{\mathbf{h}}^{-}(k-1)\right] \\ &\quad + \frac{1}{u}\left[\mathbf{s}^T(k)\mathbf{s}(k)\right]^{-1} \varepsilon^2(k), \end{aligned} \tag{4.51}$$

where u is a positive scaling constant. Using the Kullback-Leibler divergence plus the constraint $\sum_l [\hat{h}_l^+(k) + \hat{h}_l^-(k)] = u$ and the same approximation as for the EG, the minimization of (4.51) gives:

$$\eta^{-1}\left[\ln\frac{\hat{h}_l^+(k)}{\hat{h}_l^+(k-1)} + 1\right] - \frac{2}{u}s(k-l)\left[\mathbf{s}^T(k)\mathbf{s}(k)\right]^{-1} e(k) + \kappa = 0, \tag{4.52}$$

$$\eta^{-1}\left[\ln\frac{\hat{h}_l^-(k)}{\hat{h}_l^-(k-1)} + 1\right] + \frac{2}{u}s(k-l)\left[\mathbf{s}^T(k)\mathbf{s}(k)\right]^{-1} e(k) + \kappa = 0, \tag{4.53}$$

$$l = 0, 1, \cdots, L-1,$$

where κ is a Lagrange multiplier. From the two previous equations, we easily find the EG± algorithm [204]:

$$\hat{h}_l^+(k) = u\frac{\hat{h}_l^+(k-1)r_l^+(k)}{\sum_{j=0}^{L-1}\left[\hat{h}_j^+(k-1)r_j^+(k) + \hat{h}_j^-(k-1)r_j^-(k)\right]}, \tag{4.54}$$

$$\hat{h}_l^-(k) = u\frac{\hat{h}_l^-(k-1)r_l^-(k)}{\sum_{j=0}^{L-1}\left[\hat{h}_j^+(k-1)r_j^+(k) + \hat{h}_j^-(k-1)r_j^-(k)\right]}, \tag{4.55}$$

where

$$r_l^+(k) = \exp\left[\frac{\eta(k)}{u}s(k-l)e(k)\right], \tag{4.56}$$

$$\begin{aligned} r_l^-(k) &= \exp\left[-\frac{\eta(k)}{u}s(k-l)e(k)\right] \\ &= \frac{1}{r_l^+(k)}, \end{aligned} \tag{4.57}$$

with $\eta(k) = 2\eta\left[\mathbf{s}^T(k)\mathbf{s}(k)\right]^{-1}$. We can check that we always have $\left\|\hat{\mathbf{h}}^+(k)\right\|_1 + \left\|\hat{\mathbf{h}}^-(k)\right\|_1 = u$. This algorithm is summarized in Table 4.4.

The fact that,

$$u = \left\|\hat{\mathbf{h}}^+(k)\right\|_1 + \left\|\hat{\mathbf{h}}^-(k)\right\|_1 \geq \left\|\hat{\mathbf{h}}^+(k) - \hat{\mathbf{h}}^-(k)\right\|_1 = \left\|\hat{\mathbf{h}}(k)\right\|_1, \tag{4.58}$$

suggests that the constant u has to be chosen such that $u \geq \|\mathbf{h}\|_1$ in order that $\hat{\mathbf{h}}(k)$ converges to $\mathbf{h}$. If we take $u < \|\mathbf{h}\|_1$, the algorithm will introduce a bias in the coefficients of the filter.

Table 4.4. The EG± algorithm.

Initialization:	$\hat{h}_l^+(0) = \hat{h}_l^-(0) = c > 0,\ l = 0, 1, \cdots, L-1$
Parameters:	$u \geq \|\mathbf{h}\|_1$
	$0 < \alpha < 1$
	$\delta_{\text{EG}} = \text{cst} \cdot \sigma_s^2$
Error:	$e(k) = x(k) - \left[\hat{\mathbf{h}}^+(k-1) - \hat{\mathbf{h}}^-(k-1)\right]^T \mathbf{s}(k)$
Update:	$\mu(k) = \dfrac{\alpha}{\mathbf{s}^T(k)\mathbf{s}(k) + \delta_{\text{EG}}}$
	$r_l^+(k) = \exp\left[L\dfrac{\mu(k)}{u} s(k-l)e(k)\right]$
	$r_l^-(k) = \dfrac{1}{r_l^+(k)}$
	$\hat{h}_l^+(k) = u\dfrac{\hat{h}_l^+(k-1)r_l^+(k)}{\sum_{j=0}^{L-1}\left[\hat{h}_j^+(k-1)r_j^+(k) + \hat{h}_j^-(k-1)r_j^-(k)\right]}$
	$\hat{h}_l^-(k) = u\dfrac{\hat{h}_l^-(k-1)r_l^-(k)}{\sum_{j=0}^{L-1}\left[\hat{h}_j^+(k-1)r_j^+(k) + \hat{h}_j^-(k-1)r_j^-(k)\right]}$
	$l = 0, 1, \cdots, L-1$

The motivation for the EG± (and EG) algorithm can be developed by taking the logarithmic of (4.54) and (4.55). This shows that the logarithmic weights use almost the same update as the NLMS algorithm. Alternatively, this can be interpreted as exponentiating the update, hence the name EG±. This has the effect of assigning larger relative updates to larger weights, thereby deemphasizing the effect of smaller weights. This is qualitatively similar to the PNLMS algorithm which makes the update *proportional* to the size of the weight. This type of behavior is desirable for sparse impulse responses where small weights do not contribute significantly to the *mean* solution but introduce an undesirable noise-like *variance*.

4.5.3 The Exponentiated RLS (ERLS) Algorithm

The RLS algorithm is optimal from a convergence point of view since its convergence does not depend on the condition number of the input signal covariance matrix. It is well known that with ill-conditioned signals (like speech) this condition number can be very large and algorithms like LMS suffer from slow convergence [160]. Thus, it is interesting to compare the RLS algorithm to the other algorithms when the impulse response to identify is sparse. The update equation (4.34) of the RLS algorithm can be rewritten as:

$$\hat{h}_l(k) = \hat{h}_l(k-1) + k_{\text{g},l}(k)e(k),\ 0 \leq l \leq L-1, \tag{4.59}$$

where

$$\begin{aligned}\mathbf{k}_{\mathrm{g}}(k) &= \begin{bmatrix} k_{\mathrm{g},0}(k) & k_{\mathrm{g},1}(k) & \cdots & k_{\mathrm{g},L-1}(k) \end{bmatrix}^T \\ &= \mathbf{R}^{-1}(k)\mathbf{s}(k) \end{aligned} \quad (4.60)$$

is the Kalman gain. A fast RLS (FRLS) can be derived by using the *a priori* Kalman gain $\mathbf{k}'_{\mathrm{g}}(k) = \mathbf{R}^{-1}(k-1)\mathbf{s}(k)$ and the forward and backward predictors. This *a priori* Kalman gain can be computed recursively with only $5L$ multiplications [160], [17].

Following the same approach as for the EG± algorithm, we deduce the exponentiated RLS (ERLS) algorithm [41]:

$$e(k) = x(k) - \left[\hat{\mathbf{h}}^+(k-1) - \hat{\mathbf{h}}^-(k-1)\right]^T \mathbf{s}(k), \quad (4.61)$$

$$\hat{h}_l^+(k) = u \frac{\hat{h}_l^+(k-1)r_l^+(k)}{\sum_{j=0}^{L-1}\left[\hat{h}_j^+(k-1)r_j^+(k) + \hat{h}_j^-(k-1)r_j^-(k)\right]}, \quad (4.62)$$

$$\hat{h}_l^-(k) = u \frac{\hat{h}_l^-(k-1)r_l^-(k)}{\sum_{j=0}^{L-1}\left[\hat{h}_j^+(k-1)r_j^+(k) + \hat{h}_j^-(k-1)r_j^-(k)\right]}, \quad (4.63)$$

where now:

$$\begin{aligned} r_l^+(k) &= \exp\left[\frac{k_{\mathrm{g},l}(k)}{u} e(k)\right] \\ &= \frac{1}{r_l^-(k)}. \end{aligned} \quad (4.64)$$

Obviously, a fast ERLS (FERLS) can be easily derived since the Kalmain gain in (4.64) is the same as the one used in the FRLS. Simulations presented later show that there is not much difference between the FRLS and FERLS for initial convergence, but for tracking, FERLS can be much better than FRLS. Hence, the FERLS algorithm may be of some interest.

4.6 The Lambert W Function Based Gradient Algorithm

As was shown in the previous section, the EG algorithm is derived from the relative entropy which is not a symmetric distance, e.g. $d_{\mathrm{re}}\left[\hat{\mathbf{h}}(k), \hat{\mathbf{h}}(k-1)\right] \neq d_{\mathrm{re}}\left[\hat{\mathbf{h}}(k-1), \hat{\mathbf{h}}(k)\right]$. Moreover, the constraint $\sum_l \hat{h}_l(k-1) = \sum_l \hat{h}_l(k) = 1$ needs to be added in the minimization process to ensure that $d_{\mathrm{re}}\left[\hat{\mathbf{h}}(k), \hat{\mathbf{h}}(k-1)\right] \geq 0$.

Consider the following symmetric distance:

$$d_{\mathrm{lw}}\left[\hat{\mathbf{h}}(k), \hat{\mathbf{h}}(k-1)\right] = \eta'^{-1} \sum_{l=0}^{L-1} \left[\hat{h}_l(k) - \hat{h}_l(k-1)\right]\left[\ln \hat{h}_l(k) - \ln \hat{h}_l(k-1)\right]$$

$$
\begin{aligned}
&= \eta'^{-1} \sum_{l=0}^{L-1} \hat{h}_l(k) \ln \frac{\hat{h}_l(k)}{\hat{h}_l(k-1)} \\
&\quad + \eta'^{-1} \sum_{l=0}^{L-1} \hat{h}_l(k-1) \ln \frac{\hat{h}_l(k-1)}{\hat{h}_l(k)} \\
&= d_{\mathrm{re}} \left[\hat{\mathbf{h}}(k), \hat{\mathbf{h}}(k-1) \right] + d_{\mathrm{re}} \left[\hat{\mathbf{h}}(k-1), \hat{\mathbf{h}}(k) \right]. \qquad (4.65)
\end{aligned}
$$

It is easy to see that $d_{\mathrm{lw}} \left[\hat{\mathbf{h}}(k), \hat{\mathbf{h}}(k-1) \right] \geq 0$ as long as the components $\hat{h}_l(k)$ and $\hat{h}_l(k-1)$ are nonnegative. This means that a criterion using $d_{\mathrm{lw}} \left[\hat{\mathbf{h}}(k), \hat{\mathbf{h}}(k-1) \right]$ does not need to include any constraint, which is not the case if $d_{\mathrm{re}} \left[\hat{\mathbf{h}}(k), \hat{\mathbf{h}}(k-1) \right]$ is used.

If we now take the general case (positive and negative components), we seek to minimize:

$$
\begin{aligned}
J_{\mathrm{lw}}(k) &= d_{\mathrm{lw}}[\hat{\mathbf{h}}^+(k), \hat{\mathbf{h}}^+(k-1)] + d_{\mathrm{lw}}[\hat{\mathbf{h}}^-(k), \hat{\mathbf{h}}^-(k-1)] \\
&\quad + \left[\mathbf{s}^T(k)\mathbf{s}(k) \right]^{-1} \varepsilon^2(n). \qquad (4.66)
\end{aligned}
$$

This minimization with respect to $\hat{\mathbf{h}}^+(k)$ and $\hat{\mathbf{h}}^-(k)$ (then approximating the *a posteriori* error with the *a priori* error) leads to the two equations:

$$
1 - \eta'(k)s(k-l)e(k) = \ln \frac{\hat{h}_l^+(k-1)}{\hat{h}_l^+(k)} + \frac{\hat{h}_l^+(k-1)}{\hat{h}_l^+(k)}, \qquad (4.67)
$$

$$
1 + \eta'(k)s(k-l)e(k) = \ln \frac{\hat{h}_l^-(k-1)}{\hat{h}_l^-(k)} + \frac{\hat{h}_l^-(k-1)}{\hat{h}_l^-(k)}, \qquad (4.68)
$$

$$
l = 0, 1, \cdots, L-1,
$$

where $\eta'(k) = 2\eta' \left[\mathbf{s}^T(k)\mathbf{s}(k) \right]^{-1}$. Exponentiating the two previous equations, we find what we call the Lambert W function based gradient (LWG) algorithm:

$$
\exp\left[1 - \eta'(k)s(k-l)e(k)\right] = w_l^+(k) \exp w_l^+(k), \qquad (4.69)
$$

$$
\exp\left[1 + \eta'(k)s(k-l)e(k)\right] = w_l^-(k) \exp w_l^-(k), \qquad (4.70)
$$

where

$$
w_l^+(k) = \frac{\hat{h}_l^+(k-1)}{\hat{h}_l^+(k)}, \qquad (4.71)
$$

$$
w_l^-(k) = \frac{\hat{h}_l^-(k-1)}{\hat{h}_l^-(k)}. \qquad (4.72)
$$

The Lambert W function is defined to be the multivalued inverse of the function $w \exp w$ [86]. Since $\exp\left[1 \pm \eta'(k)s(k-l)e(k)\right] > 0$, there is a unique value for $w_l^+(k)$ and $w_l^-(k)$.

Obviously, the complexity of the LWG algorithm is quite high since it requires, at each iteration and for each component of the filters, to find the solution of the nonlinear equation: $z = w \exp w$. Iterative algorithms exist for that and MATLAB has a function called "lambertw" to find w.

4.7 Some Important Links Among Algorithms

Non-linear algorithms like EG are not easy to analyze and even when it is possible, very often the information we can get from a tedious analysis is not that much helpful in understanding their behavior. It is sometimes more useful to link a new adaptive filter to a well-studied one such as NLMS, in order to be able to deduce its limitations and potentials.

4.7.1 Link Between NLMS and EG± Algorithms

If we initialize $\hat{h}_l(0) = 0,\ l = 0, 1, \cdots, L-1$, in the (non-regularized) NLMS algorithm, we can easily see that:

$$\begin{aligned} \hat{\mathbf{h}}(k) &= \sum_{i=0}^{k-1} \mu(i+1)\mathbf{s}(i+1)e(i+1) \\ &= \alpha \sum_{i=0}^{k-1} \frac{\mathbf{s}(i+1)e(i+1)}{\mathbf{s}^T(i+1)\mathbf{s}(i+1)}, \end{aligned} \tag{4.73}$$

where $\mu(i+1) = \alpha \left[\mathbf{s}^T(i+1)\mathbf{s}(i+1)\right]^{-1}$.

If we start the adaptation of the EG± algorithm with $\hat{h}_l^+(0) = \hat{h}_l^-(0) = c > 0,\ l = 0, 1, \cdots, L-1$, we can show that (4.54) and (4.55) are equivalent to [45]:

$$\hat{h}_l^+(k) = u \frac{t_l^+(k)}{\sum_{j=0}^{L-1}\left[t_j^+(k) + t_j^-(k)\right]}, \tag{4.74}$$

$$\hat{h}_l^-(k) = u \frac{t_l^-(k)}{\sum_{j=0}^{L-1}\left[t_j^+(k) + t_j^-(k)\right]}, \tag{4.75}$$

where

$$t_l^+(k) = \exp\left[\frac{1}{u}\sum_{i=0}^{k-1} \eta(i+1)s(i+1-l)e(i+1)\right], \tag{4.76}$$

$$\begin{aligned} t_l^-(k) &= \exp\left[-\frac{1}{u}\sum_{i=0}^{k-1} \eta(i+1)s(i+1-l)e(i+1)\right] \\ &= \frac{1}{t_l^+(k)}, \end{aligned} \tag{4.77}$$

and $\eta(i+1) = 2\eta \left[\mathbf{s}^T(i+1)\mathbf{s}(i+1)\right]^{-1}$. Clearly, the convergence of the algorithm does not depend on the initialization parameter c (as long it is positive and nonzero). Now

$$\begin{aligned}\hat{h}_l(k) &= \hat{h}_l^+(k) - \hat{h}_l^-(k) \\ &= u\frac{t_l^+(k) - t_l^-(k)}{\sum_{j=0}^{L-1}[t_j^+(k) + t_j^-(k)]} \\ &= u\frac{\sinh\left[\frac{1}{u}\sum_{i=0}^{k-1}\eta(i+1)s(i+1-l)e(i+1)\right]}{\sum_{j=0}^{L-1}\cosh\left[\frac{1}{u}\sum_{i=0}^{k-1}\eta(i+1)s(i+1-j)e(i+1)\right]}. \end{aligned} \quad (4.78)$$

Note that the sinh function has the effect of exponentiating the update, as previously commented.

For u large enough and using the approximations $\sinh(a) \approx a$ and $\cosh(a) \approx 1$ when $|a| \ll 1$, (4.78) becomes:

$$\hat{h}_l(k) = \frac{2\eta}{L}\sum_{i=0}^{k-1}\frac{s(i+1-l)e(i+1)}{\mathbf{s}^T(i+1)\mathbf{s}(i+1)}, \quad 0 \le l \le L-1. \quad (4.79)$$

Comparing (4.73) and (4.79), we understand that, by taking $\eta = L\alpha/2$ and for u large enough, the NLMS and EG± algorithms have the same performance. Obviously, the choice of u is critical in practice: if we take $u < \|\mathbf{h}\|_1$, the EG± will introduce a bias in the coefficients of the filter, and if $u \gg \|\mathbf{h}\|_1$, the EG± will behave like NLMS.

4.7.2 Link Between IPNLMS and EG± Algorithms

PNLMS and IPNLMS algorithms were developed for use in network echo cancelers [38]. In comparison to the NLMS algorithm, they have very fast initial convergence and tracking when the echo path is sparse. As previously mentioned, the idea behind these "proportionate" algorithms is to update each coefficient of the filter independently of the others by adjusting the adaptation step size in proportion to the estimated filter coefficient.

How are the IPNLMS and EG± algorithms specifically related? In the rest of this subsection, we show that the IPNLMS is in fact an approximation of the EG±.

If we suppose that $\hat{\mathbf{h}}^+(k)$ [resp. $\hat{\mathbf{h}}^-(k)$] is close to $\hat{\mathbf{h}}^+(k-1)$ [resp. $\hat{\mathbf{h}}^-(k-1)$], which is usually the case in all adaptive algorithms (especially for a small step size), the two distances $d_{\text{re}}\left[\hat{\mathbf{h}}^+(k), \hat{\mathbf{h}}^+(k-1)\right]$ and $d_{\text{re}}\left[\hat{\mathbf{h}}^-(k), \hat{\mathbf{h}}^-(k-1)\right]$ in criterion (4.51) can be approximated as follows:

$$d_{\text{re}}\left[\hat{\mathbf{h}}^+(k), \hat{\mathbf{h}}^+(k-1)\right] = \eta^{-1}\sum_{l=0}^{L-1}\hat{h}_l^+(k)\ln\frac{\hat{h}_l^+(k)}{\hat{h}_l^+(k-1)}$$

$$\approx \eta^{-1}\sum_{l=0}^{L-1}\hat{h}_l^+(k)\left[\frac{\hat{h}_l^+(k)}{\hat{h}_l^+(k-1)} - 1\right], \quad (4.80)$$

$$d_{\text{re}}[\hat{\mathbf{h}}^-(k), \hat{\mathbf{h}}^-(k-1)] = \eta^{-1}\sum_{l=0}^{L-1}\hat{h}_l^-(k)\ln\frac{\hat{h}_l^-(k)}{\hat{h}_l^-(k-1)}$$

$$\approx \eta^{-1}\sum_{l=0}^{L-1}\hat{h}_l^-(k)\left[\frac{\hat{h}_l^-(k)}{\hat{h}_l^-(k-1)} - 1\right]. \quad (4.81)$$

Using (4.80) and (4.81) plus the constraint $\sum_l \left[\hat{h}_l^+(k) + \hat{h}_l^-(k)\right] = u$ and the same approximation as for the EG±, the minimization of (4.51) gives the approximated EG± algorithm:

$$\hat{h}_l^+(k) = \hat{h}_l^+(k-1)\left[1 + \frac{\eta(k)}{2u}s(k-l)e(k) - \frac{\eta(k)}{2u^2}\hat{x}(k)e(k)\right], \quad (4.82)$$

$$\hat{h}_l^-(k) = \hat{h}_l^-(k-1)\left[1 - \frac{\eta(k)}{2u}s(k-l)e(k) - \frac{\eta(k)}{2u^2}\hat{x}(k)e(k)\right], \quad (4.83)$$

so that:

$$\begin{aligned}\hat{h}_l(k) &= \hat{h}_l^+(k) - \hat{h}_l^-(k) \\ &= \hat{h}_l(k-1) + \frac{\eta(k)\left[\hat{h}_l^+(k-1) + \hat{h}_l^-(k-1)\right]}{2u}s(k-l)e(k) \\ &\quad - \frac{\eta(k)}{2u^2}\hat{h}_l(k-1)\hat{x}(k)e(k). \end{aligned} \quad (4.84)$$

Neglecting the last term of the right-hand side of (4.84), we get:

$$\hat{h}_l(k) = \hat{h}_l(k-1) + \frac{\eta(k)}{2}\frac{\hat{h}_l^+(k-1) + \hat{h}_l^-(k-1)}{\left\|\hat{\mathbf{h}}^+(k-1)\right\|_1 + \left\|\hat{\mathbf{h}}^-(k-1)\right\|_1}x(n-l)e(n). \quad (4.85)$$

If the true impulse response $\mathbf{h}$ is sparse, it can be shown that if we choose $u = \|\mathbf{h}\|_1$, the (positive) vector $\hat{\mathbf{h}}^+(k-1) + \hat{\mathbf{h}}^-(k-1)$ is also sparse after convergence. This means that the elements

$$\frac{\hat{h}_l^+(k-1) + \hat{h}_l^-(k-1)}{\left\|\hat{\mathbf{h}}^+(k-1)\right\|_1 + \left\|\hat{\mathbf{h}}^-(k-1)\right\|_1}$$

in (4.85) play exactly the same role as the elements $g_{\text{ip},l}(k)$ in the IPNLMS algorithm in the particular case where $\beta = 1$ (PNLMS algorithm). As a result, we can expect the two algorithms (IPNLMS and EG±) to have similar performance. On the other hand, if $u \gg \|\mathbf{h}\|_1$, it can be shown that

$\hat{h}_l^+(k-1) + \hat{h}_l^-(k-1) \approx u/L$, $\forall l$. In this case, the EG± algorithm will behave like IPNLMS with $\beta = -1$ (NLMS algorithm). Thus, the parameter β in IPNLMS operates like the parameter u in EG±. However, the advantage of IPNLMS is that no *a priori* information of the system impulse response is required in order to have a better convergence rate than the NLMS algorithm. Another clear advantage of IPNLMS is that it is much less complex to implement than EG±. We conclude that IPNLMS is a good approximation of EG± and is more useful in practice. Note also that the approximated EG± algorithm (4.85) belongs to the family of natural gradient algorithms [140], [220].

4.7.3 Link Between LWG and EG± Algorithms

As we have already mentioned, the complexity of the LWG is high, so this algorithm is not normally suitable for practical applications. However, it is important to understand how it is related to other algorithms.

If we make the usual assumption that $\hat{\mathbf{h}}^+(k)$ [resp. $\hat{\mathbf{h}}^-(k)$] is close to $\hat{\mathbf{h}}^+(k-1)$ [resp. $\hat{\mathbf{h}}^-(k-1)$], the LWG algorithm can be approximated as follows:

$$\hat{h}_l^+(k) = \frac{\hat{h}_l^+(k-1)}{1 - \frac{\eta'(k)}{2} s(k-l)e(k)}, \quad (4.86)$$

$$\hat{h}_l^-(k) = \frac{\hat{h}_l^-(k-1)}{1 + \frac{\eta'(k)}{2} s(k-l)e(k)}, \quad (4.87)$$

$$l = 0, 1, \cdots, L-1.$$

For $|a| \ll 1$, we have:

$$\frac{1}{1-a} \approx 1 + a, \quad (4.88)$$

$$\frac{1}{1+a} \approx 1 - a. \quad (4.89)$$

Using these approximations in (4.86) and (4.87), we obtain the approximated LWG algorithm:

$$\hat{h}_l^+(k) = \hat{h}_l^+(k-1)\left[1 + \frac{\eta'(k)}{2} s(k-l)e(k)\right], \quad (4.90)$$

$$\hat{h}_l^-(k) = \hat{h}_l^-(k-1)\left[1 - \frac{\eta'(k)}{2} s(k-l)e(k)\right], \quad (4.91)$$

$$l = 0, 1, \cdots, L-1,$$

which is equivalent to the approximated EG± algorithm. Therefore, we can expect that in practice, the EG± and LWG algorithms will perform in a very similar way.

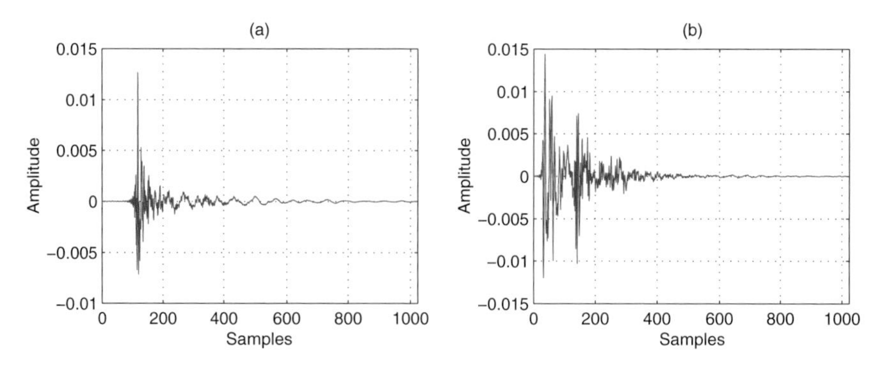

Fig. 4.1. Acoustic impulse responses used in simulations.

4.8 Numerical Examples

The objective of this section is to show, by way of simulations, how some of the algorithms presented in this chapter work in typical conditions of room acoustic impulse response identification. Comparison among the different algorithms is another important aspect we emphasize here. The aim is to give a representatives set of simulation scenarios that are relevant in this context.

The two room acoustic impulse responses, $\mathbf{h}$, to be identified are shown in Fig. 4.1. Using the mathematical definition of sparseness given in Chap. 2, we find that the impulse response of Fig. 4.1(a) is more sparse $[\xi(\mathbf{h}) \approx 0.69]$ than the one of Fig. 4.1(b) $[\xi(\mathbf{h}) \approx 0.65]$. They are both of length $L = 1024$ and the same length is used for all the adaptive filters $\hat{\mathbf{h}}(k)$. The sampling rate is 8 kHz and a white Gaussian noise signal with 30 dB SNR (signal-to-noise ratio) is added to the output $x(k)$. The input signal $s(k)$ is either a white Gaussian signal or a speech signal. The parameter settings chosen (unless stated otherwise) for all the simulations are:

- $\hat{h}_l(0) = 0, \;\; \hat{h}_l^+(0) = \hat{h}_l^-(0) = 1, \;\; l = 0, 1, \cdots, L-1,$
- $\alpha = 0.3, \;\; \delta = 10\sigma_s^2,$
- $\beta = -0.5, \;\; \epsilon = 0.001,$
- $\rho = 0.01, \;\; \delta_{\mathrm{p}} = 0.01,$
- $\lambda = 1 - 1/(3L),$
- $\delta_{\mathrm{NLMS}} = \delta_{\mathrm{EG}} = \delta, \;\; \delta_{\mathrm{PNLMS}} = \delta/L, \;\; \delta_{\mathrm{IPNLMS}} = (1-\beta)\delta/(2L).$

Figures 4.2–4.9 show the convergence of the normalized misalignment (in dB),

$$20 \log_{10} \frac{\left\| \mathbf{h} - \hat{\mathbf{h}}(k) \right\|_2}{\left\| \mathbf{h} \right\|_2}, \tag{4.92}$$

for all the algorithms.

The only simulation that was done with a speech source as excitation signal is shown in Fig. 4.9; all the others were done with a white Gaussian signal.

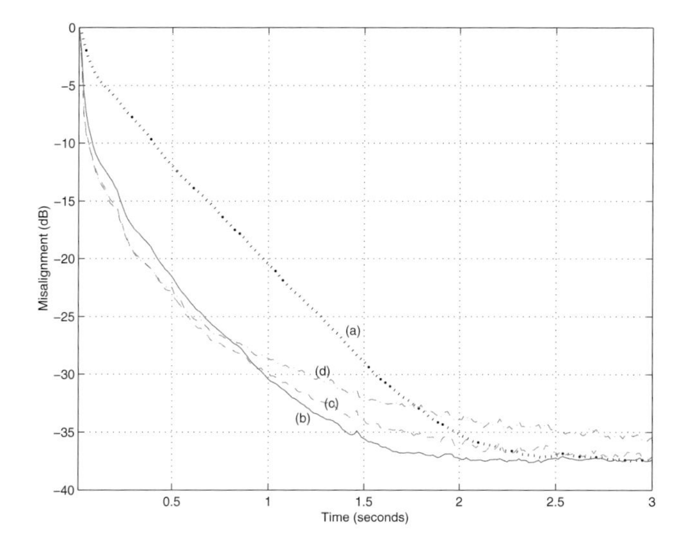

Fig. 4.2. Misalignment of the IPNLMS algorithm for different values of β with a white Gaussian noise as input signal, impulse response of Fig. 4.1(a), and using the true coefficients in $\mathbf{G}(k)$. (a) $\beta = -1$ (equivalent to NLMS), (b) $\beta = -0.5$, (c) $\beta = 0$, and (d) $\beta = 0.9$.

Impulse response of Fig. 4.1(a) was used everywhere except for Fig. 4.6, where impulse response of Fig. 4.1(b) was used.

Figures 4.2 shows how IPNLMS behaves with different values of β. In this unrealistic simulation, we used for the diagonal matrix $\mathbf{G}(k)$, the true values of the coefficients $\mathbf{h}$ instead of the estimated ones. This may seem like what we can do best with the "proportionate" idea. First, we see that when β approaches 1, the algorithm degrades and a good value seems to be $\beta = -0.5$. Second, comparing Fig. 4.2 with Fig. 4.3 where this time the real IPNLMS is evaluated [with the estimated coefficients in $\mathbf{G}(k)$], we see that the difference is not that significant, although better when $\mathbf{G}(k)$ is known a priori. This observation is very important because it shows, in a very simple manner, the limits of natural gradient algorithms in general.

Figure 4.4 presents the misalignment of the EG$\pm$ algorithm for different values of u. As expected, for a large u, this algorithm coincides with NLMS and for $u = 0.5\|\mathbf{h}\|_1$, it introduces a bias in the coefficients of the filter and, as a result, the EG$\pm$ is much worse than NLMS. Clearly, the EG$\pm$ algorithm is not very interesting from a practical point of view since it requires some *a priori* knowledge that we can not have.

Figure 4.5 compares the initial convergence of four algorithms (NLMS, PNLMS, IPNLMS, and EG$\pm$) with the impulse response of Fig. 4.1(a). We see on this figure that the PNLMS, IPNLMS, and EG$\pm$ (with $u = 2\|\mathbf{h}\|_1$) algorithms converge much faster than NLMS. We also see that IPNLMS and

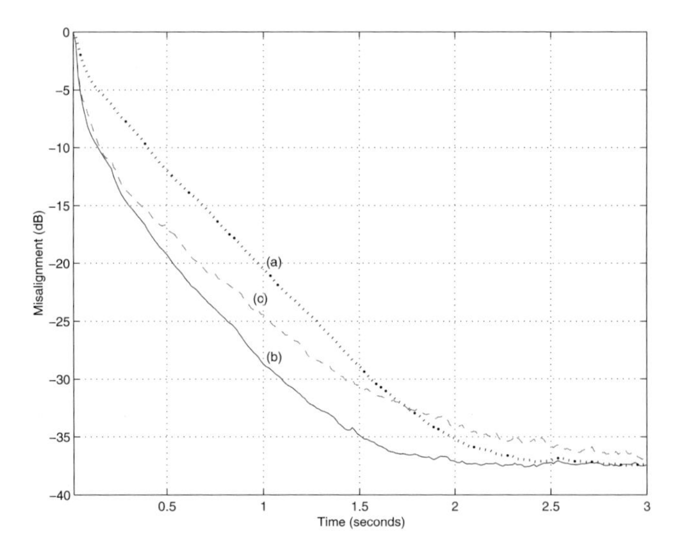

Fig. 4.3. Misalignment of the IPNLMS algorithm for different values of β with a white Gaussian noise as input signal and impulse response of Fig. 4.1(a). (a) $\beta = -1$ (equivalent to NLMS), (b) $\beta = -0.5$, and (c) $\beta = 0.5$ (close to PNLMS).

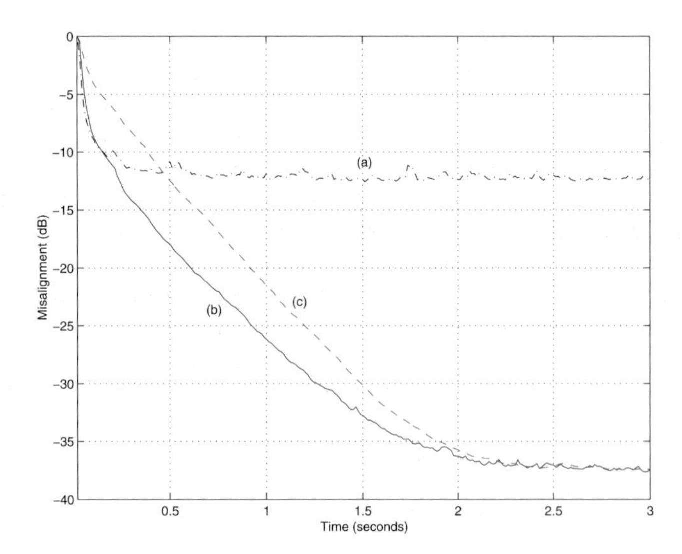

Fig. 4.4. Misalignment of the EG$\pm$ algorithm for different values of u with a white Gaussian noise as input signal and impulse response of Fig. 4.1(a). (a) $u = 0.5\|\mathbf{h}\|_1$, (b) $u = 2\|\mathbf{h}\|_1$, and (c) $u = 50\|\mathbf{h}\|_1$ (equivalent to NLMS).

EG$\pm$ are very close to each other, confirming that these two algorithms are related.

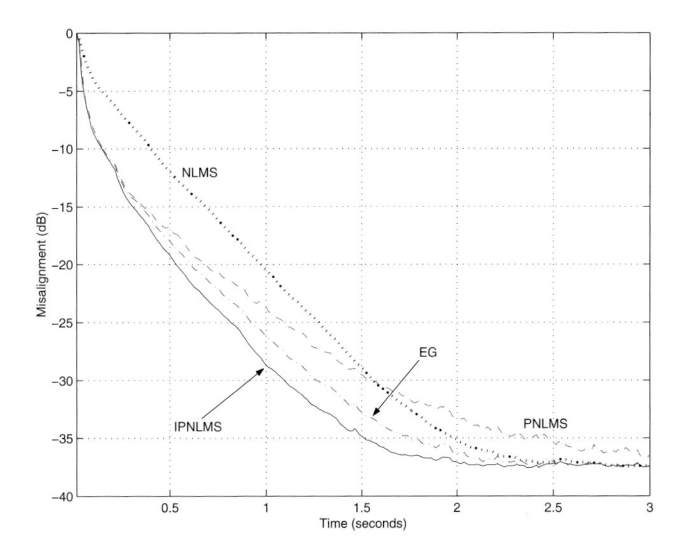

Fig. 4.5. Misalignment of the NLMS (dotted line), PNLMS (dashed line), IPNLMS (solid line), and EG± (dash-dot line) algorithms with white Gaussian noise as input signal and impulse response of Fig. 4.1(a).

In Fig. 4.6, we compare again the initial convergence of the same four algorithms but with the impulse response of Fig. 4.1(b). While IPNLMS and EG± still perform much better than NLMS, PNLMS starts degrading very significantly after 1.2 seconds. This confirms that PNLMS is not very well optimized when the impulse response is not strongly sparse.

Tracking is another very important issue in adaptive algorithms. In applications like room acoustics, it is essential that an adaptive filter tracks fast since impulse responses are not very stationary. Figures 4.7 and 4.8 compare the algorithms in a tracking situation when after 3 seconds the sparse impulse response of Fig. 4.1(a) is shifted to the right by 12 samples. The other conditions of Fig. 4.7 are the same as that in Fig. 4.5. According to this simulation, the PNLMS, IPNLMS, and EG± algorithms track much better than the NLMS algorithm. In Fig. 4.8, the FRLS algorithm is compared to the FERLS algorithm with $u = 6\|\mathbf{h}\|_1$: while the initial convergence of the two algorithms is almost the same, the FERLS tracks faster than the FRLS. It is also worth noticing that IPNLMS tracks better than the FRLS and FERLS algorithms.

In Fig. 4.9, the initial convergence and tracking of the NLMS, PNLMS, IPNLMS, and EG± algorithms are compared with a speech source as input signal and impulse response of Fig. 4.1(a). Here, we changed the adaption step to $\alpha = 0.5$. We notice the same trend as with a white Gaussian noise as input signal. Virtually, IPNLMS and EG± give almost the same results and they

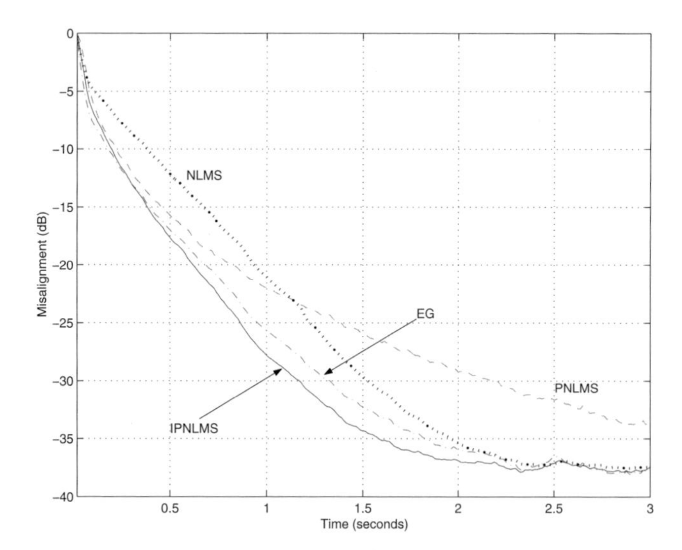

Fig. 4.6. Misalignment of the NLMS (dotted line), PNLMS (dashed line), IPNLMS (solid line), and EG± (dash-dot line) algorithms with white Gaussian noise as input signal and impulse response of Fig. 4.1(b).

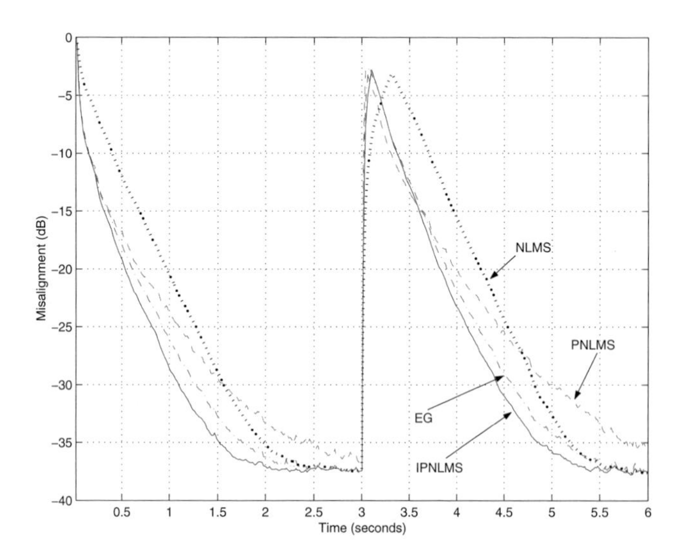

Fig. 4.7. Misalignment during impulse response change. The impulse response changes at time 3 seconds. Other conditions same as in Fig. 4.5.

are slightly better than PNLMS; all three of them are better than NLMS in terms of initial convergence and tracking.

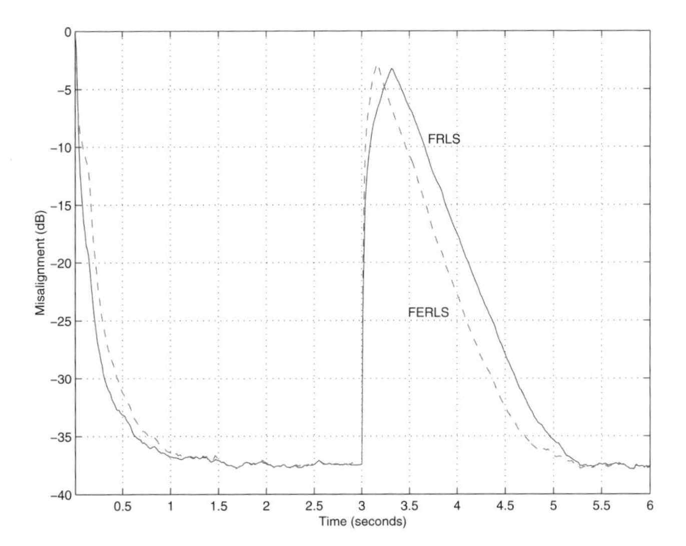

Fig. 4.8. Misalignment, during impulse response change, of the FRLS (solid line) and FERLS (dotted line) algorithms with white Gaussian noise as input signal, impulse response of Fig. 4.1(a), and $u = 6\|\mathbf{h}\|_1$ for the FERLS algorithm.

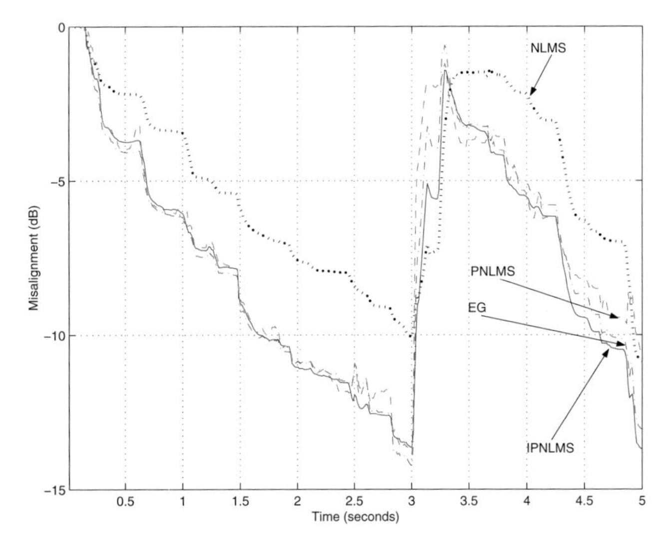

Fig. 4.9. Misalignment of the NLMS (dotted line), PNLMS (dashed line), IPNLMS (solid line), and EG± (dash-dot line) algorithms with a speech source as input signal and impulse response of Fig. 4.1(a). The impulse response changes at time 3 seconds.

4.9 Summary

Throughout this chapter, we have shown how to use *a priori* information on sparseness in the design of adaptive algorithms in order to make them perform better (in terms of initial convergence and tracking) than classical adaptive algorithms. We have first proposed a universal criterion from which any adaptive filter can be derived. It was clearly shown that a non-linear update with respect to the filter weights is advantageous. We have studied and compared in particular the IPNLMS and EG$\pm$, which are the two most important algorithms with non-linear update. The IPNLMS algorithm was introduced in the context of network echo cancellation where there is a strong need to improve convergence rate and tracking. It was known for a long time that unknown echo paths in the network are most of the time sparse and there are many different intuitions on how one should take advantage of that. Kivinen and Warmuth [204] derived the EG$\pm$ algorithm in the context of computational learning theory. We have shown here that a good approximation of the EG$\pm$ leads to the IPNLMS. As a result, the two algorithms have very similar performance in all the simulations we have investigated. We have also shown some links between the EG$\pm$ and NLMS algorithms, so that with appropriate choice of some parameters, the two algorithms can be identical. We have also derived an algorithm called LWG by simply doubling the Kullback-Leibler divergence to have a symmetric distance. In fact, the LWG and EG$\pm$ are almost equivalent.

Finally, all the ideas presented here can be generalized to blind identification of multichannel systems with sparse channels. Some possibilities are presented in [41], [46], and in Chap. 6.

5

Frequency-Domain Adaptive Filters

Frequency-domain (FD) adaptive algorithms are extremely useful in practice because of their great efficiency from an arithmetic complexity point of view. This chapter derives SISO and MISO FD adaptive filters.

5.1 Introduction

In acoustic MIMO systems, we often have to deal with very long filters (sometimes several thousand taps), time-varying environments, and non-stationary, auto-correlated, and highly coherent signals. Therefore, needs for efficient and well-behaved adaptive filters are very strong.

Generally, we distinguish two classes of adaptive algorithms. One class includes filters that are updated in the time domain, sample-by-sample in general, like the classical LMS and RLS algorithms (see the two previous chapters). The other class contains filters that are updated in the frequency domain, block-by-block in general, using the fast Fourier transform (FFT) as an intermediary step. As a result of this block processing, the arithmetic complexity of the algorithms in the latter category is significantly reduced compared to time-domain adaptive algorithms. Use of the FFT is appropriate to the Toeplitz structure, which results from the time-shift properties of the filter input signal. Consequently, deriving a frequency-domain (FD) adaptive algorithm is just a matter of rewriting the time-domain error criterion in a way that Toeplitz and circulant matrices are explicitly shown.

Since its first introduction by Dentino *et al.* [94], adaptive filtering in the frequency domain has progressed rapidly, and different sophisticated algorithms have since been proposed. Ferrara [114] was the first to elaborate an efficient frequency-domain adaptive filter algorithm (FLMS) that converges to the optimal (Wiener) solution. Mansour and Gray [218] derived an even more efficient algorithm, the *unconstrained* FLMS (UFLMS), using only three FFT operations per block instead of five for the FLMS, with comparable performance [211]. However, a handicap with these structures is the delay introduced between input and output. Indeed, this delay is equal to the length of the adaptive filter L, which can be considerable for some applications like acoustic echo cancellation where the number of taps can be easily a thousand or more. A new structure called *multidelay filter* (MDF), using the classical overlap save (OLS) method, was proposed in [284], [19] and generalized in [20], [21] where the block length L_b was made independent of the filter length L; L_b can be chosen as small as desired, so as to limit the block delay. Although from a complexity point of view, the optimal choice is $L_\mathrm{b} = L$, using smaller block sizes ($L_\mathrm{b} < L$) in order to reduce the delay is still more efficient than time-domain algorithms. A more general scheme based on weighted overlap and add (WOLA) methods, the *generalized multidelay filter* (GMDFo), was proposed in [236], [253], where o is the overlapping factor. The settings $o > 1$ appear to be very useful in the context of adaptive filtering, since the filter coefficients can be adapted more frequently (every L_b/o samples instead of every L_b samples in the standard OLS scheme). Thus, this structure introduces one more degree of freedom, but the complexity is increased by roughly a factor of o. Taking the block size in the MDF as large as the delay permits, will increase the convergence rate of the algorithm, while taking the overlapping factor greater than 1 will increase the tracking ability of the algorithm.

In [38], [56], [59], for the first time, frequency-domain adaptive algorithms were derived rigorously from a recursive least-squares criterion, with a block size independent of the length of the adaptive filter. As a result, we better understand these algorithms as well as the choice of the parameters that govern them. This chapter gives an overview of this important class of adaptive filters.

5.2 Derivation of SISO FD Adaptive Algorithms

The first part of this section shows how to write a block recursive least-squares criterion, for a SISO system, in the frequency domain, with a block size independent of the length of the adaptive filter. When the criterion is rigorously defined, the adaptive algorithm follows immediately.

5.2.1 Criterion

In the context of SISO system identification, the error signal at time k between the system and model filter outputs is given by

$$e(k) = x(k) - \hat{x}(k), \tag{5.1}$$

where

$$\hat{x}(k) = \hat{\mathbf{h}}^T \mathbf{s}(k) \tag{5.2}$$

is an estimate of the output signal, $x(k) = \mathbf{h}^T \mathbf{s}(k) + b(k)$, and $\hat{\mathbf{h}}$ is the model filter of length L. The noise $b(k)$ is assumed to be white here.

We now define the block error signal (of length $L_\mathrm{b} \leq L$). For that, we assume that L is an integer multiple of L_b, i.e., $L = PL_\mathrm{b}$. We have:

$$\mathbf{e}(t) = \mathbf{x}(t) - \hat{\mathbf{x}}(t), \tag{5.3}$$

where t is the block time index, and

$$\begin{aligned}
\mathbf{e}(t) &= \begin{bmatrix} e(tL_\mathrm{b}) & \cdots & e(tL_\mathrm{b} + L_\mathrm{b} - 1) \end{bmatrix}^T, \\
\mathbf{x}(t) &= \begin{bmatrix} x(tL_\mathrm{b}) & \cdots & x(tL_\mathrm{b} + L_\mathrm{b} - 1) \end{bmatrix}^T \\
&= \mathbf{S}^T(t)\mathbf{h} + \mathbf{b}(t), \\
\mathbf{s}(tL_\mathrm{b} + i) &= \begin{bmatrix} s(tL_\mathrm{b} + i) & \cdots & s(tL_\mathrm{b} + i - L + 1) \end{bmatrix}^T \\
\mathbf{S}(t) &= \begin{bmatrix} \mathbf{s}(tL_\mathrm{b}) & \cdots & \mathbf{s}(tL_\mathrm{b} + L_\mathrm{b} - 1) \end{bmatrix}, \\
\mathbf{b}(t) &= \begin{bmatrix} b(tL_\mathrm{b}) & \cdots & b(tL_\mathrm{b} + L_\mathrm{b} - 1) \end{bmatrix}^T, \\
\hat{\mathbf{x}}(t) &= \begin{bmatrix} \hat{x}(tL_\mathrm{b}) & \cdots & \hat{x}(tL_\mathrm{b} + L_\mathrm{b} - 1) \end{bmatrix}^T \\
&= \mathbf{S}^T(t)\hat{\mathbf{h}}.
\end{aligned}$$

It can easily be checked that $\mathbf{S}(t)$ is a Toeplitz matrix of size $(L \times L_\mathrm{b})$.

We can show that for $P = L/L_\mathrm{b}$,

$$\begin{aligned}
\hat{\mathbf{x}}(t) &= \mathbf{S}^T(t)\hat{\mathbf{h}} \\
&= \sum_{p=0}^{P-1} \mathbf{T}_s(t-p)\hat{\mathbf{h}}_p,
\end{aligned} \tag{5.4}$$

where

$$\mathbf{T}_s(t-p) = \begin{bmatrix} s(tL_\mathrm{b} - pL_\mathrm{b}) & \cdots & s(tL_\mathrm{b} - pL_\mathrm{b} - L_\mathrm{b} + 1) \\ s(tL_\mathrm{b} - pL_\mathrm{b} + 1) & \ddots & \vdots \\ \vdots & \ddots & \vdots \\ s(tL_\mathrm{b} - pL_\mathrm{b} + L_\mathrm{b} - 1) & \cdots & s(tL_\mathrm{b} - pL_\mathrm{b}) \end{bmatrix}$$

is an $(L_\mathrm{b} \times L_\mathrm{b})$ Toeplitz matrix and

$$\hat{\mathbf{h}}_p = \begin{bmatrix} \hat{h}_{pL_\mathrm{b}} & \hat{h}_{pL_\mathrm{b}+1} & \cdots & \hat{h}_{pL_\mathrm{b}+L_\mathrm{b}-1} \end{bmatrix}^T, \; p = 0, 1, \cdots, P-1,$$

are the sub-filters of $\hat{\mathbf{h}}$. In (5.4), the filter $\hat{\mathbf{h}}$ (of length L) is partitioned into P sub-filters $\hat{\mathbf{h}}_p$ of length L_b and the rectangular matrix $\mathbf{S}^T$ [of size $(L_\mathrm{b} \times L)$] is decomposed to P square sub-matrices of size $(L_\mathrm{b} \times L_\mathrm{b})$ [327].

It is well known that a Toeplitz matrix $\mathbf{T}_s$ can be transformed, by doubling its size, to a circulant matrix,

$$\mathbf{C}_s = \begin{bmatrix} \mathbf{T}'_s & \mathbf{T}_s \\ \mathbf{T}_s & \mathbf{T}'_s \end{bmatrix},$$

where $\mathbf{T}'_s$ is also a Toeplitz matrix obtained by shifting rows of $\mathbf{T}_s$; the diagonal of $\mathbf{C}_s$ is arbitrary, but it is natural and customary to set it equal to the first sample in the previous block. Using circulant matrices, the block error signal can be re-written equivalently:

$$\mathbf{e}(t) = \mathbf{x}(t) - \mathbf{W}^{01}_{L_\mathrm{b} \times 2L_\mathrm{b}} \hat{\mathbf{x}}'_{2L_\mathrm{b}}(t), \tag{5.5}$$

where

$$\mathbf{W}^{01}_{L_\mathrm{b} \times 2L_\mathrm{b}} = \begin{bmatrix} \mathbf{0}_{L_\mathrm{b} \times L_\mathrm{b}} & \mathbf{I}_{L_\mathrm{b} \times L_\mathrm{b}} \end{bmatrix},$$

$$\hat{\mathbf{x}}'_{2L_\mathrm{b}}(t) = \sum_{p=0}^{P-1} \mathbf{C}_s(t-p) \mathbf{W}^{10}_{2L_\mathrm{b} \times L_\mathrm{b}} \hat{\mathbf{h}}_k, \tag{5.6}$$

$$\mathbf{C}_s(t-p) = \begin{bmatrix} \mathbf{T}'_s(t-p) & \mathbf{T}_s(t-p) \\ \mathbf{T}_s(t-p) & \mathbf{T}'_s(t-p) \end{bmatrix},$$

$$\mathbf{T}'_s(t-p) = \begin{bmatrix} s(tL_\mathrm{b} - pL_\mathrm{b} + L_\mathrm{b}) & \cdots & s(tL_\mathrm{b} - pL_\mathrm{b} + 1) \\ s(tL_\mathrm{b} - pL_\mathrm{b} - L_\mathrm{b} + 1) & \ddots & \vdots \\ \vdots & \ddots & \vdots \\ s(tL_\mathrm{b} - pL_\mathrm{b} - 1) & \cdots & s(tL_\mathrm{b} - pL_\mathrm{b} + L_\mathrm{b}) \end{bmatrix},$$

and

$$\mathbf{W}^{10}_{2L_\mathrm{b} \times L_\mathrm{b}} = \begin{bmatrix} \mathbf{I}_{L_\mathrm{b} \times L_\mathrm{b}} \\ \mathbf{0}_{L_\mathrm{b} \times L_\mathrm{b}} \end{bmatrix}.$$

It is also well known that a circulant matrix is easily decomposed as follows: $\mathbf{C}_s = \mathbf{F}^{-1}_{2L_\mathrm{b} \times 2L_\mathrm{b}} \underline{\boldsymbol{D}}_s \mathbf{F}_{2L_\mathrm{b} \times 2L_\mathrm{b}}$, where $\mathbf{F}_{2L_\mathrm{b} \times 2L_\mathrm{b}}$ is the Fourier matrix [of size $(2L_\mathrm{b} \times 2L_\mathrm{b})$] and $\underline{\boldsymbol{D}}_s$ is a diagonal matrix whose elements are the discrete Fourier transform of the first column of $\mathbf{C}_s$. In this chapter only, we assume that all Fourier matrices are normalized, such that they are unitary, i.e. $\mathbf{F}^{-1} = \mathbf{F}^H$, where ^{H} denotes conjugate transpose. If we multiply (5.5) by $\mathbf{F}_{L_\mathrm{b} \times L_\mathrm{b}}$ [Fourier matrix of size $(L_\mathrm{b} \times L_\mathrm{b})$], we get the error signal in the frequency domain:

$$
\begin{aligned}
\underset{\rightarrow}{\boldsymbol{e}}(t) &= \underset{\rightarrow}{\boldsymbol{x}}(t) - \underset{\Rightarrow}{\boldsymbol{W}}^{01}_{L_\mathrm{b}\times 2L_\mathrm{b}} \underset{\rightarrow}{\hat{\boldsymbol{x}}}'_{2L_\mathrm{b}}(t) \\
&= \underset{\rightarrow}{\boldsymbol{x}}(t) - \underset{\Rightarrow}{\boldsymbol{W}}^{01}_{L_\mathrm{b}\times 2L_\mathrm{b}} \sum_{p=0}^{P-1} \underset{\Rightarrow}{\boldsymbol{D}}_s(t-p)\, \underset{\Rightarrow}{\boldsymbol{W}}^{10}_{2L_\mathrm{b}\times L_\mathrm{b}} \underset{\rightarrow}{\hat{\boldsymbol{h}}}_p \\
&= \underset{\rightarrow}{\boldsymbol{x}}(t) - \underset{\Rightarrow}{\boldsymbol{W}}^{01}_{L_\mathrm{b}\times 2L_\mathrm{b}} \sum_{p=0}^{P-1} \underset{\Rightarrow}{\boldsymbol{U}}_s(t-p) \underset{\rightarrow}{\hat{\boldsymbol{h}}}_p \\
&= \underset{\rightarrow}{\boldsymbol{x}}(t) - \underset{\Rightarrow}{\boldsymbol{W}}^{01}_{L_\mathrm{b}\times 2L_\mathrm{b}} \underset{\Rightarrow}{\boldsymbol{U}}_{\mathrm{b},s}(t) \underset{\rightarrow}{\hat{\boldsymbol{h}}},
\end{aligned}
\tag{5.7}
$$

where

$$
\begin{aligned}
\underset{\rightarrow}{\boldsymbol{e}}(t) &= \mathbf{F}_{L_\mathrm{b}\times L_\mathrm{b}} \mathbf{e}(t), \\
\underset{\rightarrow}{\boldsymbol{x}}(t) &= \mathbf{F}_{L_\mathrm{b}\times L_\mathrm{b}} \mathbf{x}(t) \\
&= \underset{\Rightarrow}{\boldsymbol{W}}^{01}_{L_\mathrm{b}\times 2L_\mathrm{b}} \underset{\Rightarrow}{\boldsymbol{U}}_{\mathrm{b},s}(t) \underset{\rightarrow}{\boldsymbol{h}} + \underset{\rightarrow}{\boldsymbol{b}}(t), \\
\underset{\rightarrow}{\boldsymbol{b}}(t) &= \mathbf{F}_{L_\mathrm{b}\times L_\mathrm{b}} \mathbf{b}(t), \\
\underset{\Rightarrow}{\boldsymbol{W}}^{01}_{L_\mathrm{b}\times 2L_\mathrm{b}} &= \mathbf{F}_{L_\mathrm{b}\times L_\mathrm{b}} \mathbf{W}^{01}_{L_\mathrm{b}\times 2L_\mathrm{b}} \mathbf{F}^{-1}_{2L_\mathrm{b}\times 2L_\mathrm{b}}, \\
\underset{\rightarrow}{\hat{\boldsymbol{x}}}'_{2L_\mathrm{b}}(t) &= \mathbf{F}_{2L_\mathrm{b}\times 2L_\mathrm{b}} \hat{\mathbf{x}}'_{2L_\mathrm{b}}(t), \\
\underset{\Rightarrow}{\boldsymbol{W}}^{10}_{2L_\mathrm{b}\times L_\mathrm{b}} &= \mathbf{F}_{2L_\mathrm{b}\times 2L_\mathrm{b}} \mathbf{W}^{10}_{2L_\mathrm{b}\times L_\mathrm{b}} \mathbf{F}^{-1}_{L_\mathrm{b}\times L_\mathrm{b}}, \\
\underset{\rightarrow}{\hat{\boldsymbol{h}}}_p &= \mathbf{F}_{L_\mathrm{b}\times L_\mathrm{b}} \hat{\mathbf{h}}_p, \\
\underset{\rightarrow}{\boldsymbol{h}}_p &= \mathbf{F}_{L_\mathrm{b}\times L_\mathrm{b}} \mathbf{h}_p, \\
\underset{\Rightarrow}{\boldsymbol{D}}_s(t-p) &= \mathbf{F}_{2L_\mathrm{b}\times 2L_\mathrm{b}} \mathbf{C}_s(t-p) \mathbf{F}^{-1}_{2L_\mathrm{b}\times 2L_\mathrm{b}}, \\
\underset{\Rightarrow}{\boldsymbol{U}}_s(t-p) &= \underset{\Rightarrow}{\boldsymbol{D}}_s(t-p)\, \underset{\Rightarrow}{\boldsymbol{W}}^{10}_{2L_\mathrm{b}\times L_\mathrm{b}}, \\
\underset{\Rightarrow}{\boldsymbol{U}}_{\mathrm{b},s}(t) &= \begin{bmatrix} \underset{\Rightarrow}{\boldsymbol{U}}_s(t) & \underset{\Rightarrow}{\boldsymbol{U}}_s(t-1) & \cdots & \underset{\Rightarrow}{\boldsymbol{U}}_s(t-P+1) \end{bmatrix}, \\
\underset{\rightarrow}{\hat{\boldsymbol{h}}} &= \begin{bmatrix} \underset{\rightarrow}{\hat{\boldsymbol{h}}}_0^T & \underset{\rightarrow}{\hat{\boldsymbol{h}}}_1^T & \cdots & \underset{\rightarrow}{\hat{\boldsymbol{h}}}_{P-1}^T \end{bmatrix}^T, \\
\underset{\rightarrow}{\boldsymbol{h}} &= \begin{bmatrix} \underset{\rightarrow}{\boldsymbol{h}}_0^T & \underset{\rightarrow}{\boldsymbol{h}}_1^T & \cdots & \underset{\rightarrow}{\boldsymbol{h}}_{P-1}^T \end{bmatrix}^T.
\end{aligned}
$$

Having derived a frequency-domain error signal, we now define a frequency-domain criterion which is similar to the one proposed in [33], [34]:

$$
J_\mathrm{f}(t) = (1-\lambda_\mathrm{f}) \sum_{i=0}^{t} \lambda_\mathrm{f}^{t-i} \underset{\rightarrow}{\boldsymbol{e}}^H(i) \underset{\rightarrow}{\boldsymbol{e}}(i),
\tag{5.8}
$$

where λ_f ($0 < \lambda_\mathrm{f} < 1$) is an exponential forgetting factor. The main advantage of using (5.8) is to take advantage of the FFT in order to have low complexity adaptive filters.

5.2.2 Normal Equations

Let ∇ be the gradient operator (with respect to $\underset{\rightarrow}{\hat{\boldsymbol{h}}}$). Applying the operator ∇ to the cost function J_f, we obtain the complex gradient vector [53]:

$$\begin{aligned}\nabla J_{\mathrm{f}}(t) &= \frac{\partial J_{\mathrm{f}}(t)}{\partial \hat{\underline{\boldsymbol{h}}}(t)} \\ &= -(1-\lambda_{\mathrm{f}})\sum_{i=0}^{t}\lambda_{\mathrm{f}}^{t-i}\,\underline{\boldsymbol{U}}^{T}_{\mathrm{b},s}(i)\left(\underline{\boldsymbol{W}}^{01}_{L_{\mathrm{b}}\times 2L_{\mathrm{b}}}\right)^{T}\underline{\boldsymbol{x}}^{*}(i) \qquad (5.9)\\ &\quad + (1-\lambda_{\mathrm{f}})\left[\sum_{i=0}^{t}\lambda_{\mathrm{f}}^{t-i}\,\underline{\boldsymbol{U}}^{T}_{\mathrm{b},s}(i)\left(\underline{\boldsymbol{W}}^{01}_{2L_{\mathrm{b}}\times 2L_{\mathrm{b}}}\right)^{T}\underline{\boldsymbol{U}}^{*}_{\mathrm{b},s}(i)\right]\hat{\underline{\boldsymbol{h}}}^{*}(t),\end{aligned}$$

where * denotes complex conjugate,

$$\begin{aligned}\underline{\boldsymbol{W}}^{01}_{2L_{\mathrm{b}}\times 2L_{\mathrm{b}}} &= \left(\underline{\boldsymbol{W}}^{01}_{L_{\mathrm{b}}\times 2L_{\mathrm{b}}}\right)^{H}\underline{\boldsymbol{W}}^{01}_{L_{\mathrm{b}}\times 2L_{\mathrm{b}}} \\ &= \mathbf{F}_{2L_{\mathrm{b}}\times 2L_{\mathrm{b}}}\mathbf{W}^{01}_{2L_{\mathrm{b}}\times 2L_{\mathrm{b}}}\mathbf{F}^{-1}_{2L_{\mathrm{b}}\times 2L_{\mathrm{b}}},\end{aligned}$$

and

$$\mathbf{W}^{01}_{2L_{\mathrm{b}}\times 2L_{\mathrm{b}}} = \begin{bmatrix}\mathbf{0}_{L_{\mathrm{b}}\times L_{\mathrm{b}}} & \mathbf{0}_{L_{\mathrm{b}}\times L_{\mathrm{b}}}\\ \mathbf{0}_{L_{\mathrm{b}}\times L_{\mathrm{b}}} & \mathbf{I}_{L_{\mathrm{b}}\times L_{\mathrm{b}}}\end{bmatrix}.$$

By setting the gradient of the cost function equal to zero, taking the conjugate, noting that $\left(\underline{\boldsymbol{W}}^{01}_{2L_{\mathrm{b}}\times 2L_{\mathrm{b}}}\right)^{H} = \underline{\boldsymbol{W}}^{01}_{2L_{\mathrm{b}}\times 2L_{\mathrm{b}}}$, and defining

$$\begin{aligned}\underline{\boldsymbol{x}}_{2L_{\mathrm{b}}}(t) &= \left(\underline{\boldsymbol{W}}^{01}_{L_{\mathrm{b}}\times 2L_{\mathrm{b}}}\right)^{H}\underline{\boldsymbol{x}}(t) \\ &= \mathbf{F}_{2L_{\mathrm{b}}\times 2L_{\mathrm{b}}}\begin{bmatrix}\mathbf{0}_{L_{\mathrm{b}}\times 1}\\ \mathbf{x}(t)\end{bmatrix},\end{aligned}$$

we obtain the so-called *normal* equations:

$$\underline{\boldsymbol{R}}(t)\hat{\underline{\boldsymbol{h}}}(t) = \underline{\boldsymbol{p}}(t), \qquad (5.10)$$

where

$$\begin{aligned}\underline{\boldsymbol{R}}(t) &= (1-\lambda_{\mathrm{f}})\sum_{i=0}^{t}\lambda_{\mathrm{f}}^{t-i}\,\underline{\boldsymbol{U}}^{H}_{\mathrm{b},s}(i)\,\underline{\boldsymbol{W}}^{01}_{2L_{\mathrm{b}}\times 2L_{\mathrm{b}}}\,\underline{\boldsymbol{U}}_{\mathrm{b},s}(i) \\ &= \lambda_{\mathrm{f}}\underline{\boldsymbol{R}}(t-1) + (1-\lambda_{\mathrm{f}})\,\underline{\boldsymbol{U}}^{H}_{\mathrm{b},s}(t)\,\underline{\boldsymbol{W}}^{01}_{2L_{\mathrm{b}}\times 2L_{\mathrm{b}}}\,\underline{\boldsymbol{U}}_{\mathrm{b},s}(t) \qquad (5.11)\end{aligned}$$

and

$$\begin{aligned}\underline{\boldsymbol{p}}(t) &= (1-\lambda_{\mathrm{f}})\sum_{i=0}^{t}\lambda_{\mathrm{f}}^{t-i}\,\underline{\boldsymbol{U}}^{H}_{\mathrm{b},s}(i)\underline{\boldsymbol{x}}_{2L_{\mathrm{b}}}(i) \\ &= \lambda_{\mathrm{f}}\underline{\boldsymbol{p}}(t-1) + (1-\lambda_{\mathrm{f}})\,\underline{\boldsymbol{U}}^{H}_{\mathrm{b},s}(t)\underline{\boldsymbol{x}}_{2L_{\mathrm{b}}}(t) \\ &= \lambda_{\mathrm{f}}\underline{\boldsymbol{p}}(t-1) + (1-\lambda_{\mathrm{f}})\,\underline{\boldsymbol{U}}^{H}_{\mathrm{b},s}(t)\left(\underline{\boldsymbol{W}}^{01}_{L_{\mathrm{b}}\times 2L_{\mathrm{b}}}\right)^{H}\underline{\boldsymbol{x}}(t). \qquad (5.12)\end{aligned}$$

If the input signal is well-conditioned (non-zero energy at all frequencies), matrix $\underline{\boldsymbol{R}}(t)$ is nonsingular. In this case, the normal equations have a unique solution which is the optimal Wiener solution. For $L_{\mathrm{b}} = 1$, (5.10) is strictly equivalent to the normal equations obtained by minimizing the classical recursive least-squares error criterion [17].

5.2.3 Adaptive Algorithms

There are many different ways to write the adaptive algorithm. In any case, it is derived directly from the normal equations. Enforcing these normal equations at block time indices t and $t-1$, and using (5.11) and (5.12), we easily derive an exact adaptive algorithm:

$$\begin{aligned} \underline{e}(t) &= \underline{x}(t) - \underline{\mathbf{W}}^{01}_{L_b \times 2L_b} \underline{\mathbf{U}}_{b,s}(t) \hat{\underline{h}}(t-1), && (5.13)\\ \hat{\underline{h}}(t) &= \hat{\underline{h}}(t-1) + (1-\lambda_f) \underline{\mathbf{R}}^{-1}(t) \underline{\mathbf{U}}^H_{b,s}(t) (\underline{\mathbf{W}}^{01}_{L_b \times 2L_b})^H \underline{e}(t), && (5.14) \end{aligned}$$

or, multiplying (5.13) by $(\underline{\mathbf{W}}^{01}_{L_b \times 2L_b})^H$,

$$\begin{aligned} \underline{e}_{2L_b}(t) &= \underline{x}_{2L_b}(t) - \underline{\mathbf{W}}^{01}_{2L_b \times 2L_b} \underline{\mathbf{U}}_{b,s}(t) \hat{\underline{h}}(t-1), && (5.15)\\ \hat{\underline{h}}(t) &= \hat{\underline{h}}(t-1) + (1-\lambda_f) \underline{\mathbf{R}}^{-1}(t) \underline{\mathbf{U}}^H_{b,s}(t) \underline{e}_{2L_b}(t), && (5.16) \end{aligned}$$

where we have defined

$$\begin{aligned} \underline{e}_{2L_b}(t) &= \mathbf{F}_{2L_b \times 2L_b} \begin{bmatrix} \mathbf{0}_{L_b \times 1} \\ \mathbf{e}(t) \end{bmatrix} \\ &= (\underline{\mathbf{W}}^{01}_{L_b \times 2L_b})^H \underline{e}(t). \end{aligned}$$

Not only (5.13) implies (5.15) but these two expressions are in fact equivalent. Note that, for $L_b = 1$, this algorithm is exactly the RLS algorithm. Define the following matrix:

$$\underline{\mathbf{W}}^{10}_{2L \times L} = \text{diag}\left[\underline{\mathbf{W}}^{10}_{2L_b \times L_b} \cdots \underline{\mathbf{W}}^{10}_{2L_b \times L_b} \right].$$

If we multiply (5.16) by $\underline{\mathbf{W}}^{10}_{2L \times L}$ and observe that $\underline{\mathbf{U}}_{b,s}(t) = \underline{\mathbf{D}}_{b,s}(t) \underline{\mathbf{W}}^{10}_{2L \times L}$, where

$$\underline{\mathbf{D}}_{b,s}(t) = \left[\underline{\mathbf{D}}_s(t) \;\; \underline{\mathbf{D}}_s(t-1) \;\; \cdots \;\; \underline{\mathbf{D}}_s(t-P+1) \right],$$

we obtain the algorithm:

$$\begin{aligned} \underline{\mathbf{R}}(t) &= \lambda_f \underline{\mathbf{R}}(t-1) && (5.17)\\ &\quad + (1-\lambda_f)(\underline{\mathbf{W}}^{10}_{2L \times L})^H \underline{\mathbf{D}}^H_{b,s}(t) \underline{\mathbf{W}}^{01}_{2L_b \times 2L_b} \underline{\mathbf{D}}_{b,s}(t) \underline{\mathbf{W}}^{10}_{2L \times L}, \\ \underline{e}_{2L_b}(t) &= \underline{x}_{2L_b}(t) - \underline{\mathbf{W}}^{01}_{2L_b \times 2L_b} \underline{\mathbf{D}}_{b,s}(t) \hat{\underline{h}}_{2L}(t-1), && (5.18)\\ \hat{\underline{h}}_{2L}(t) &= \hat{\underline{h}}_{2L}(t-1) \\ &\quad + (1-\lambda_f) \underline{\mathbf{W}}^{10}_{2L \times L} \underline{\mathbf{R}}^{-1}(t) (\underline{\mathbf{W}}^{10}_{2L \times L})^H \underline{\mathbf{D}}^H_{b,s}(t) \underline{e}_{2L_b}(t), && (5.19) \end{aligned}$$

where

$$\begin{aligned} \hat{\underline{h}}_{2L}(t) &= \underline{\mathbf{W}}^{10}_{2L \times L} \hat{\underline{h}}(t) \\ &= \left[\hat{\underline{h}}^T_{2L_b,0}(t) \;\; \hat{\underline{h}}^T_{2L_b,1}(t) \;\; \cdots \;\; \hat{\underline{h}}^T_{2L_b,P-1}(t) \right]^T, \\ \hat{\underline{h}}_{2L_b,p}(t) &= \mathbf{F}_{2L_b \times 2L_b} \begin{bmatrix} \hat{\mathbf{h}}_p(t) \\ \mathbf{0}_{L_b \times 1} \end{bmatrix}. \end{aligned}$$

The rank of the matrix $\underset{\Rightarrow}{\boldsymbol{W}}{}^{10}_{2L\times L}$ is equal to L. Since we have L unknowns to identify, in principle (5.19) is equivalent to (5.16). Indeed, if we multiply (5.19) by $(\underset{\Rightarrow}{\boldsymbol{W}}{}^{10}_{2L\times L})^H$, we obtain exactly (5.16) since $(\underset{\Rightarrow}{\boldsymbol{W}}{}^{10}_{2L\times L})^H \underset{\Rightarrow}{\boldsymbol{W}}{}^{10}_{2L\times L} = \mathbf{I}_{L\times L}$. It is interesting to see how naturally we have ended up using blocks of length $2L_{\mathrm{b}}$ (especially for the error signal) even though we have used an error criterion with blocks of length L_{b}. We can do even better than that and rewrite the algorithm exclusively using FFTs of size $2L_{\mathrm{b}}$. This formulation will be, by far, the most interesting because an explicit link is made with existing frequency-domain algorithms. For convenience, let us first define the $(2L \times 2L)$ matrix:

$$
\begin{aligned}
\underset{\Rightarrow}{\boldsymbol{R}}_{\mathrm{d}}(t) &= (1-\lambda_{\mathrm{f}}) \sum_{i=0}^{t} \lambda_{\mathrm{f}}^{t-i} \underset{\Rightarrow}{\boldsymbol{D}}{}^H_{\mathrm{b},s}(i)\, \underset{\Rightarrow}{\boldsymbol{W}}{}^{01}_{2L_{\mathrm{b}}\times 2L_{\mathrm{b}}} \underset{\Rightarrow}{\boldsymbol{D}}_{\mathrm{b},s}(i) \\
&= \lambda_{\mathrm{f}} \underset{\Rightarrow}{\boldsymbol{R}}_{\mathrm{d}}(t-1) + (1-\lambda_{\mathrm{f}}) \underset{\Rightarrow}{\boldsymbol{D}}{}^H_{\mathrm{b},s}(t)\, \underset{\Rightarrow}{\boldsymbol{W}}{}^{01}_{2L_{\mathrm{b}}\times 2L_{\mathrm{b}}} \underset{\Rightarrow}{\boldsymbol{D}}_{\mathrm{b},s}(t). \qquad (5.20)
\end{aligned}
$$

The relationship with $\underset{\Rightarrow}{\boldsymbol{R}}$ is immediate:

$$
\underset{\Rightarrow}{\boldsymbol{R}}(t) = (\underset{\Rightarrow}{\boldsymbol{W}}{}^{10}_{2L\times L})^H \underset{\Rightarrow}{\boldsymbol{R}}_{\mathrm{d}}(t)\, \underset{\Rightarrow}{\boldsymbol{W}}{}^{10}_{2L\times L}. \qquad (5.21)
$$

Also, define:

$$
\begin{aligned}
\underset{\Rightarrow}{\boldsymbol{W}}{}^{10}_{2L\times 2L} &= \underset{\Rightarrow}{\boldsymbol{W}}{}^{10}_{2L\times L} (\underset{\Rightarrow}{\boldsymbol{W}}{}^{10}_{2L\times L})^H \\
&= \mathrm{diag}\left[\underset{\Rightarrow}{\boldsymbol{W}}{}^{10}_{2L_{\mathrm{b}}\times 2L_{\mathrm{b}}} \cdots \underset{\Rightarrow}{\boldsymbol{W}}{}^{10}_{2L_{\mathrm{b}}\times 2L_{\mathrm{b}}} \right],
\end{aligned}
$$

where

$$
\begin{aligned}
\underset{\Rightarrow}{\boldsymbol{W}}{}^{10}_{2L_{\mathrm{b}}\times 2L_{\mathrm{b}}} &= \underset{\Rightarrow}{\boldsymbol{W}}{}^{10}_{2L_{\mathrm{b}}\times L_{\mathrm{b}}} (\underset{\Rightarrow}{\boldsymbol{W}}{}^{10}_{2L_{\mathrm{b}}\times L_{\mathrm{b}}})^H \\
&= \mathbf{F}_{2L_{\mathrm{b}}\times 2L_{\mathrm{b}}} \mathbf{W}^{10}_{2L_{\mathrm{b}}\times 2L_{\mathrm{b}}} \mathbf{F}^{-1}_{2L_{\mathrm{b}}\times 2L_{\mathrm{b}}},
\end{aligned}
$$

and

$$
\mathbf{W}^{10}_{2L_{\mathrm{b}}\times 2L_{\mathrm{b}}} = \begin{bmatrix} \mathbf{I}_{L_{\mathrm{b}}\times L_{\mathrm{b}}} & \mathbf{0}_{L_{\mathrm{b}}\times L_{\mathrm{b}}} \\ \mathbf{0}_{L_{\mathrm{b}}\times L_{\mathrm{b}}} & \mathbf{0}_{L_{\mathrm{b}}\times L_{\mathrm{b}}} \end{bmatrix}.
$$

We have an interesting relation between the inverse of the two matrices $\underset{\Rightarrow}{\boldsymbol{R}}$ and $\underset{\Rightarrow}{\boldsymbol{R}}_{\mathrm{d}}$:

$$
\underset{\Rightarrow}{\boldsymbol{W}}{}^{10}_{2L\times 2L} \underset{\Rightarrow}{\boldsymbol{R}}{}^{-1}_{\mathrm{d}}(t) = \underset{\Rightarrow}{\boldsymbol{W}}{}^{10}_{2L\times L} \underset{\Rightarrow}{\boldsymbol{R}}{}^{-1}(t) (\underset{\Rightarrow}{\boldsymbol{W}}{}^{10}_{2L\times L})^H. \qquad (5.22)
$$

This can be checked by post-multiplying both sides of (5.22) by $\underset{\Rightarrow}{\boldsymbol{R}}_{\mathrm{d}}(t)\, \underset{\Rightarrow}{\boldsymbol{W}}{}^{10}_{2L\times L}$ and noting that $\underset{\Rightarrow}{\boldsymbol{W}}{}^{10}_{2L\times 2L} \underset{\Rightarrow}{\boldsymbol{W}}{}^{10}_{2L\times L} = \underset{\Rightarrow}{\boldsymbol{W}}{}^{10}_{2L\times L}$.

Using (5.22), the adaptive algorithm can now be written in a much simpler and more convenient way:

$$
\begin{aligned}
\underset{\Rightarrow}{\boldsymbol{R}}_{\mathrm{d}}(t) &= \lambda_{\mathrm{f}} \underset{\Rightarrow}{\boldsymbol{R}}_{\mathrm{d}}(t-1) + (1-\lambda_{\mathrm{f}}) \underset{\Rightarrow}{\boldsymbol{D}}{}^H_{\mathrm{b},s}(t)\, \underset{\Rightarrow}{\boldsymbol{W}}{}^{01}_{2L_{\mathrm{b}}\times 2L_{\mathrm{b}}} \underset{\Rightarrow}{\boldsymbol{D}}_{\mathrm{b},s}(t), \qquad (5.23) \\
\underset{\rightarrow}{\boldsymbol{e}}_{2L_{\mathrm{b}}}(t) &= \underset{\rightarrow}{\boldsymbol{x}}_{2L_{\mathrm{b}}}(t) - \underset{\Rightarrow}{\boldsymbol{W}}{}^{01}_{2L_{\mathrm{b}}\times 2L_{\mathrm{b}}} \underset{\Rightarrow}{\boldsymbol{D}}_{\mathrm{b},s}(t) \underset{\rightarrow}{\hat{\boldsymbol{h}}}_{2L}(t-1), \qquad (5.24) \\
\underset{\rightarrow}{\hat{\boldsymbol{h}}}_{2L}(t) &= \underset{\rightarrow}{\hat{\boldsymbol{h}}}_{2L}(t-1) + (1-\lambda_{\mathrm{f}})\, \underset{\Rightarrow}{\boldsymbol{W}}{}^{10}_{2L\times 2L} \underset{\Rightarrow}{\boldsymbol{R}}{}^{-1}_{\mathrm{d}}(t) \underset{\Rightarrow}{\boldsymbol{D}}{}^H_{\mathrm{b},s}(t) \underset{\rightarrow}{\boldsymbol{e}}_{2L_{\mathrm{b}}}(t). \qquad (5.25)
\end{aligned}
$$

5.2.4 Convergence Analysis

In this section, we analyze the convergence of the algorithm for stationary signals $s(k)$ and $x(k)$ based on (5.13) and (5.14).

Due to the assumed stationarity of the filter input signal, we obtain, after taking the expected value of (5.11):

$$E\left\{\underset{\Rightarrow}{\boldsymbol{R}}(t)\right\} = (1-\lambda_{\mathrm{f}})\sum_{i=0}^{t}\lambda_{\mathrm{f}}^{t-i}\underset{\Rightarrow}{\boldsymbol{R}}_{\mathrm{e}}, \tag{5.26}$$

where

$$\underset{\Rightarrow}{\boldsymbol{R}}_{\mathrm{e}} = E\left\{\underset{\Rightarrow}{\boldsymbol{U}}_{\mathrm{b},s}^{H}(t)\,\underset{\Rightarrow}{\boldsymbol{W}}_{2L_{\mathrm{b}}\times 2L_{\mathrm{b}}}^{01}\,\underset{\Rightarrow}{\boldsymbol{U}}_{\mathrm{b},s}(t)\right\} \tag{5.27}$$

denotes the time-independent ensemble average correlation matrix. Noting that in (5.26) we have a sum of a finite geometric series, it can be simplified to

$$E\left\{\underset{\Rightarrow}{\boldsymbol{R}}(t)\right\} = \left(1-\lambda_{\mathrm{f}}^{t+1}\right)\underset{\Rightarrow}{\boldsymbol{R}}_{\mathrm{e}}. \tag{5.28}$$

For a single realization of the stochastic process $\underset{\Rightarrow}{\boldsymbol{R}}(t)$, we assume that

$$\underset{\Rightarrow}{\boldsymbol{R}}(t) \approx \left(1-\lambda_{\mathrm{f}}^{t+1}\right)\underset{\Rightarrow}{\boldsymbol{R}}_{\mathrm{e}}, \tag{5.29}$$

and for the steady state we see with $0 < \lambda_{\mathrm{f}} < 1$ that

$$\underset{\Rightarrow}{\boldsymbol{R}}(t) \approx \underset{\Rightarrow}{\boldsymbol{R}}_{\mathrm{e}} \tag{5.30}$$

for large t.

Convergence in Mean

By noting that $\left(\underset{\Rightarrow}{\boldsymbol{W}}_{L_{\mathrm{b}}\times 2L_{\mathrm{b}}}^{01}\right)^{H}\underset{\Rightarrow}{\boldsymbol{W}}_{L_{\mathrm{b}}\times 2L_{\mathrm{b}}}^{01} = \underset{\Rightarrow}{\boldsymbol{W}}_{2L_{\mathrm{b}}\times 2L_{\mathrm{b}}}^{01}$, (5.14) can be written as:

$$\begin{aligned}\underset{\rightarrow}{\boldsymbol{h}} - \hat{\underset{\rightarrow}{\boldsymbol{h}}}(t) &= \underset{\rightarrow}{\boldsymbol{h}} - \hat{\underset{\rightarrow}{\boldsymbol{h}}}(t-1)\\ &\quad - (1-\lambda_{\mathrm{f}})\underset{\Rightarrow}{\boldsymbol{R}}^{-1}(t)\,\underset{\Rightarrow}{\boldsymbol{U}}_{\mathrm{b},s}^{H}(t)\,\underset{\Rightarrow}{\boldsymbol{W}}_{2L_{\mathrm{b}}\times 2L_{\mathrm{b}}}^{01}\,\underset{\Rightarrow}{\boldsymbol{U}}_{\mathrm{b},s}(t)\left[\underset{\rightarrow}{\boldsymbol{h}} - \hat{\underset{\rightarrow}{\boldsymbol{h}}}(t-1)\right]\\ &\quad - (1-\lambda_{\mathrm{f}})\underset{\Rightarrow}{\boldsymbol{R}}^{-1}(t)\,\underset{\Rightarrow}{\boldsymbol{U}}_{\mathrm{b},s}^{H}(t)\underset{\rightarrow}{\boldsymbol{b}}(t).\end{aligned} \tag{5.31}$$

Taking mathematical expectation of expression (5.31), using the independence theory, and (5.27) together with (5.30), we deduce for large t that,

$$\begin{aligned}E\{\underset{\rightarrow}{\boldsymbol{m}}(t)\} &= \lambda_{\mathrm{f}}E\{\underset{\rightarrow}{\boldsymbol{m}}(t-1)\}\\ &= \lambda_{\mathrm{f}}^{t}E\{\underset{\rightarrow}{\boldsymbol{m}}(0)\},\end{aligned} \tag{5.32}$$

where $\underset{\rightarrow}{\boldsymbol{m}}(t) = \underset{\rightarrow}{\boldsymbol{h}} - \hat{\underset{\rightarrow}{\boldsymbol{h}}}(t)$ is the misalignment vector in the frequency domain. Equation (5.32) says that the convergence rate of the algorithm is governed by λ_f. Most importantly, the rate of convergence is completely independent of the input statistics. Finally, we have:

$$\lim_{t\to\infty} E\{\underline{\boldsymbol{m}}(t)\} = \mathbf{0}_{L\times 1} \Rightarrow \lim_{t\to\infty} E\left\{\hat{\mathbf{h}}(t)\right\} = \mathbf{h}. \tag{5.33}$$

Now, suppose that λ is the forgetting factor of a sample-by-sample time-domain adaptive algorithm. To have the same effective window length for the two algorithms (sample-by-sample and block-by-block), we should choose $\lambda_\mathrm{f} = \lambda^{L_\mathrm{b}}$. For example, the usual choice for the RLS algorithm is $\lambda = 1 - 1/(3L)$. In this case, a good choice for the frequency-domain algorithm is $\lambda_\mathrm{f} = [1 - 1/(3L)]^{L_\mathrm{b}}$.

Convergence in Mean Square

The convergence of the algorithm in the mean is not sufficient for convergence to the minimum mean-square error as it only assures a bias-free estimate $\hat{\underline{\boldsymbol{h}}}(t)$. An analysis of the algorithm in the mean square is required. The algorithm converges in the mean square if:

$$\lim_{t\to\infty} J'_\mathrm{f}(t) = \text{Constant}, \tag{5.34}$$

where

$$J'_\mathrm{f}(t) = \frac{1}{L_\mathrm{b}} E\left\{\underline{\boldsymbol{e}}^H(t)\underline{\boldsymbol{e}}(t)\right\}. \tag{5.35}$$

From (5.13), the error signal $\underline{\boldsymbol{e}}(t)$ can be written in terms of $\underline{\boldsymbol{m}}(t)$ as:

$$\underline{\boldsymbol{e}}(t) = \underline{\boldsymbol{W}}^{01}_{L_\mathrm{b}\times 2L_\mathrm{b}} \underline{\boldsymbol{U}}_{\mathrm{b},s}(t)\underline{\boldsymbol{m}}(t-1) + \underline{\boldsymbol{b}}(t). \tag{5.36}$$

Expression (5.35) becomes:

$$J'_\mathrm{f}(t) = \frac{1}{L_\mathrm{b}} J_\mathrm{ex}(t) + \sigma_b^2, \tag{5.37}$$

where

$$J_\mathrm{ex}(t) = E\left\{\underline{\boldsymbol{m}}^H(t-1)\,\underline{\boldsymbol{U}}^H_{\mathrm{b},s}(t)\,\underline{\boldsymbol{W}}^{01}_{2L_\mathrm{b}\times 2L_\mathrm{b}}\,\underline{\boldsymbol{U}}_{\mathrm{b},s}(t)\underline{\boldsymbol{m}}(t-1)\right\} \tag{5.38}$$

is the excess mean-square error and σ_b^2 is the variance of the noise signal $b(k)$. Furthermore:

$$\begin{aligned} J_\mathrm{ex}(t) &= E\left\{\mathrm{tr}\left[\underline{\boldsymbol{m}}^H(t-1)\,\underline{\boldsymbol{U}}^H_{\mathrm{b},s}(t)\,\underline{\boldsymbol{W}}^{01}_{2L_\mathrm{b}\times 2L_\mathrm{b}}\,\underline{\boldsymbol{U}}_{\mathrm{b},s}(t)\underline{\boldsymbol{m}}(t-1)\right]\right\} \\ &= E\left\{\mathrm{tr}\left[\underline{\boldsymbol{U}}^H_{\mathrm{b},s}(t)\,\underline{\boldsymbol{W}}^{01}_{2L_\mathrm{b}\times 2L_\mathrm{b}}\,\underline{\boldsymbol{U}}_{\mathrm{b},s}(t)\underline{\boldsymbol{m}}(t-1)\underline{\boldsymbol{m}}^H(t-1)\right]\right\} \\ &= \mathrm{tr}\left[E\left\{\underline{\boldsymbol{U}}^H_{\mathrm{b},s}(t)\,\underline{\boldsymbol{W}}^{01}_{2L_\mathrm{b}\times 2L_\mathrm{b}}\,\underline{\boldsymbol{U}}_{\mathrm{b},s}(t)\underline{\boldsymbol{m}}(t-1)\underline{\boldsymbol{m}}^H(t-1)\right\}\right]. \end{aligned} \tag{5.39}$$

Invoking the independence assumption and using (5.30), we may reduce this expectation to

$$J_{\mathrm{ex}}(t) \approx \mathrm{tr}\left[\underline{\underline{\boldsymbol{R}}}_{\mathrm{e}}\underline{\underline{\boldsymbol{M}}}(t-1)\right], \tag{5.40}$$

where

$$\underline{\underline{\boldsymbol{M}}}(t) = E\left\{\underline{\boldsymbol{m}}(t)\underline{\boldsymbol{m}}^{H}(t)\right\} \tag{5.41}$$

is the misalignment correlation matrix in the frequency domain.

Using the normal equations (5.10) and (5.12), it can be easily shown that:

$$\underline{\boldsymbol{m}}(t) = -(1-\lambda_{\mathrm{f}})\underline{\underline{\boldsymbol{R}}}^{-1}(t)\sum_{i=0}^{t}\lambda_{\mathrm{f}}^{t-i}\underline{\underline{\boldsymbol{U}}}_{\mathrm{b},s}^{H}(i)\left(\underline{\underline{\boldsymbol{W}}}_{L_{\mathrm{b}}\times 2L_{\mathrm{b}}}^{01}\right)^{H}\underline{\boldsymbol{b}}(i). \tag{5.42}$$

Since we have supposed that $\underline{\underline{\boldsymbol{R}}}_{\mathrm{e}} \approx \underline{\underline{\boldsymbol{R}}}(t)$, (5.42) can be written in a recursive way:

$$\begin{aligned} \underline{\boldsymbol{m}}(t) &\approx -(1-\lambda_{\mathrm{f}})\sum_{i=0}^{t}\lambda_{\mathrm{f}}^{t-i}\underline{\underline{\boldsymbol{R}}}_{\mathrm{e}}^{-1}\underline{\underline{\boldsymbol{U}}}_{\mathrm{b},s}^{H}(i)\left(\underline{\underline{\boldsymbol{W}}}_{L_{\mathrm{b}}\times 2L_{\mathrm{b}}}^{01}\right)^{H}\underline{\boldsymbol{b}}(i) \\ &\approx \lambda_{\mathrm{f}}\underline{\boldsymbol{m}}(t-1) - (1-\lambda_{\mathrm{f}})\underline{\underline{\boldsymbol{R}}}^{-1}(t)\underline{\underline{\boldsymbol{U}}}_{\mathrm{b},s}^{H}(t)\left(\underline{\underline{\boldsymbol{W}}}_{L_{\mathrm{b}}\times 2L_{\mathrm{b}}}^{01}\right)^{H}\underline{\boldsymbol{b}}(t). \end{aligned} \tag{5.43}$$

From (5.43) and (5.30), we may rewrite (5.41) as:

$$\underline{\underline{\boldsymbol{M}}}(t) \approx \frac{1-\lambda_{\mathrm{f}}}{2}\sigma_b^2\underline{\underline{\boldsymbol{R}}}_{\mathrm{e}}^{-1}, \tag{5.44}$$

where we have used the following approximations: $\underline{\underline{\boldsymbol{M}}}(t-1) \approx \underline{\underline{\boldsymbol{M}}}(t)$ (for t large) and $1-\lambda_{\mathrm{f}}^2 \approx 2(1-\lambda_{\mathrm{f}})$ (for λ_{f} close to unity). Finally:

$$J_{\mathrm{f}}'(t) \approx \left[(1-\lambda_{\mathrm{f}})\frac{L}{2L_{\mathrm{b}}} + 1\right]\sigma_b^2 \tag{5.45}$$

for t large, so that the mean-square error converges to a constant value. Moreover, the convergence of the algorithm in the mean-square error is independent of the eigenvalues of the ensemble-averaged matrix $\underline{\underline{\boldsymbol{R}}}_{\mathrm{e}}$.

The value

$$J_{\mathrm{mis}}(t) = E\left\{\underline{\boldsymbol{m}}^{H}(t)\underline{\boldsymbol{m}}(t)\right\} \tag{5.46}$$

gives an idea of the convergence of the misalignment. Using (5.44), we deduce that

$$\begin{aligned} J_{\mathrm{mis}}(t) &= \mathrm{tr}[\underline{\underline{\boldsymbol{M}}}(t)] \\ &= \frac{1-\lambda_{\mathrm{f}}}{2}\sigma_b^2\mathrm{tr}\left[\underline{\underline{\boldsymbol{R}}}_{\mathrm{e}}^{-1}\right] \\ &= \frac{1-\lambda_{\mathrm{f}}}{2}\sigma_b^2\sum_{l=0}^{L-1}\frac{1}{\lambda_{\mathrm{e},l}}, \end{aligned} \tag{5.47}$$

where the $\lambda_{e,l}$ are the eigenvalues of the ensemble-averaged matrix $\underline{\boldsymbol{R}}_{\mathrm{e}}$. It is important to note the difference between the convergence of the mean-square error and the misalignment. While the mean-square error does not depend on the eigenvalues of $\underline{\boldsymbol{R}}_{\mathrm{e}}$, the misalignment is magnified by the inverse of the smallest eigenvalue $\lambda_{\mathrm{e,min}}$ of $\underline{\boldsymbol{R}}_{\mathrm{e}}$. The situation is worsened when the variance of the noise (σ_b^2) is large. So in practice, at some frequencies (where the signal is poorly excited) we may have a very large misalignment. In order to avoid this problem and to keep the misalignment low, the adaptive algorithm should be regularized by adding a constant value to the diagonal of $\underline{\boldsymbol{R}}(t)$. This process is called *regularization*. A dynamical regularization strategy is proposed in [59].

5.3 Approximation and Special Cases

We start this section by giving a very useful approximation of the algorithm derived in Sect. 5.2. We then find some examples of classical efficient algorithms. This list is not exhaustive and many other algorithms can also be deduced.

5.3.1 Approximation

Frequency-domain adaptive filters were first introduced to reduce the arithmetic complexity of the LMS algorithm [114]. Unfortunately, the matrix $\underline{\boldsymbol{R}}_{\mathrm{d}}$ is generally not diagonal, so its inversion has a high complexity and the algorithm may not be very useful in practice. Since $\underline{\boldsymbol{R}}_{\mathrm{d}}$ is composed of P^2 sub-matrices,

$$\underline{\boldsymbol{R}}_{\mathrm{d}}(t) = \lambda_{\mathrm{f}} \underline{\boldsymbol{R}}_{\mathrm{d}}(t-1) + (1-\lambda_{\mathrm{f}}) \underline{\boldsymbol{D}}_{\mathrm{b},s}^{H}(t)\, \underline{\boldsymbol{W}}_{2L_{\mathrm{b}}\times 2L_{\mathrm{b}}}^{01} \underline{\boldsymbol{D}}_{\mathrm{b},s}(t), \qquad (5.48)$$

it is desirable that each of those sub-matrices be a diagonal matrix. In the next paragraph, we will argue that $\underline{\boldsymbol{W}}_{2L_{\mathrm{b}}\times 2L_{\mathrm{b}}}^{01}$ can be well approximated by $\mathbf{I}_{2L_{\mathrm{b}}\times 2L_{\mathrm{b}}}/2$, where $\mathbf{I}_{2L_{\mathrm{b}}\times 2L_{\mathrm{b}}}$ is the $2L_{\mathrm{b}} \times 2L_{\mathrm{b}}$ identity matrix; we then obtain the following approximate algorithm (absorbing the factor of 2 in the step size):

$$\begin{aligned}
\underline{\boldsymbol{R}}'(t) &= \lambda_{\mathrm{f}} \underline{\boldsymbol{R}}'(t-1) + (1-\lambda_{\mathrm{f}}) \underline{\boldsymbol{D}}_{\mathrm{b},s}^{H}(t) \underline{\boldsymbol{D}}_{\mathrm{b},s}(t), && (5.49)\\
\underline{\boldsymbol{e}}_{2L_{\mathrm{b}}}(t) &= \underline{\boldsymbol{x}}_{2L_{\mathrm{b}}}(t) - \underline{\boldsymbol{W}}_{2L_{\mathrm{b}}\times 2L_{\mathrm{b}}}^{01} \underline{\boldsymbol{D}}_{\mathrm{b},s}(t) \underline{\hat{\boldsymbol{h}}}_{2L}(t-1), && (5.50)\\
\underline{\hat{\boldsymbol{h}}}_{2L}(t) &= \underline{\hat{\boldsymbol{h}}}_{2L}(t-1) \\
&\quad + \alpha_{\mathrm{f}}(1-\lambda_{\mathrm{f}})\, \underline{\boldsymbol{W}}_{2L\times 2L}^{10} \underline{\boldsymbol{R}}'^{-1}(t) \underline{\boldsymbol{D}}_{\mathrm{b},s}^{H}(t) \underline{\boldsymbol{e}}_{2L_{\mathrm{b}}}(t), && (5.51)
\end{aligned}$$

where each sub-matrix of $\underline{\boldsymbol{R}}'$ is now a diagonal matrix (as a result of this approximation, the algorithm is more attractive from a complexity point of view) and $\alpha_{\mathrm{f}} \leq 2$ is a positive number. A fast version of this algorithm for $L_{\mathrm{b}} < L$ was derived in [57]. Note that the imprecision introduced by the approximation in (5.49) will only affect the convergence rate. Obviously, we cannot permit the same kind of approximation in (5.50), because that would result in approximating a linear convolution by a circular one, which of course can have a disastrous impact in our identification problem. Now the question is the following: is the above approximation justified?

Let's examine the structure of the matrix $\underset{\Rightarrow}{\mathbf{W}}{}^{01}_{2L_\mathrm{b}\times 2L_\mathrm{b}}$. Since $\mathbf{F}_{2L_\mathrm{b}\times 2L_\mathrm{b}} = \mathbf{F}^T_{2L_\mathrm{b}\times 2L_\mathrm{b}}$ and $\mathbf{F}^{-1}_{2L_\mathrm{b}\times 2L_\mathrm{b}} = \mathbf{F}^H_{2L_\mathrm{b}\times 2L_\mathrm{b}}$, we have: $(\underset{\Rightarrow}{\mathbf{W}}{}^{01}_{2L_\mathrm{b}\times 2L_\mathrm{b}})^* = \mathbf{F}^{-1}_{2L_\mathrm{b}\times 2L_\mathrm{b}}\mathbf{W}^{01}_{2L_\mathrm{b}\times 2L_\mathrm{b}}\mathbf{F}_{2L_\mathrm{b}\times 2L_\mathrm{b}}$, and since $\mathbf{W}^{01}_{2L_\mathrm{b}\times 2L_\mathrm{b}}$ is a diagonal matrix, $(\underset{\Rightarrow}{\mathbf{W}}{}^{01}_{2L_\mathrm{b}\times 2L_\mathrm{b}})^*$ is a circulant matrix. Therefore, inverse transforming the diagonal of $\mathbf{W}^{01}_{2L_\mathrm{b}\times 2L_\mathrm{b}}$ gives the first column of $(\underset{\Rightarrow}{\mathbf{W}}{}^{01}_{2L_\mathrm{b}\times 2L_\mathrm{b}})^*$ (up to a scaling factor),

$$\begin{aligned}\mathbf{w}^* &= \begin{bmatrix} w_0^* & w_1^* & \cdots & w_{2L_\mathrm{b}-1}^* \end{bmatrix}^T \\ &= \frac{1}{\sqrt{2L_\mathrm{b}}}\mathbf{F}^{-1}_{2L_\mathrm{b}\times 2L_\mathrm{b}}\,[0 \ \cdots \ 0 \ 1 \ \cdots \ 1]^T .\end{aligned}$$

The elements of vector $\mathbf{w}$ can be written explicitly as:

$$\begin{aligned} w_i &= \frac{1}{2L_\mathrm{b}}\sum_{l=L_\mathrm{b}}^{2L_\mathrm{b}-1}\exp\left(-j\frac{2\pi i l}{2L_\mathrm{b}}\right) \\ &= \frac{(-1)^i}{2L_\mathrm{b}}\sum_{l=0}^{L_\mathrm{b}-1}\exp(-j\pi i l/L_\mathrm{b}), \end{aligned} \tag{5.52}$$

where $j^2 = -1$. Since w_i is the sum of a geometric progression, we have:

$$\begin{aligned} w_i &= \begin{cases} 0.5\ , & i = 0 \\ \dfrac{(-1)^i}{2L_\mathrm{b}}\dfrac{1-\exp(-j\pi i)}{1-\exp(-j\pi i/L_\mathrm{b})}\ , & i \neq 0 \end{cases} \\ &= \begin{cases} 0.5\ , & i = 0 \\ 0\ , & i \text{ even}\ , \\ \dfrac{-1}{2L_\mathrm{b}}\left[1 - j\cot\left(\dfrac{\pi i}{2L_\mathrm{b}}\right)\right]\ , & i \text{ odd} \end{cases} \end{aligned} \tag{5.53}$$

where $L_\mathrm{b} - 1$ elements of vector $\mathbf{w}$ are equal to zero. Moreover, since $(\underset{\Rightarrow}{\mathbf{W}}{}^{01}_{2L_\mathrm{b}\times 2L_\mathrm{b}})^H\underset{\Rightarrow}{\mathbf{W}}{}^{01}_{2L_\mathrm{b}\times 2L_\mathrm{b}} = \underset{\Rightarrow}{\mathbf{W}}{}^{01}_{2L_\mathrm{b}\times 2L_\mathrm{b}}$, then $\mathbf{w}^H\mathbf{w} = w_0 = 0.5$ and we have

$$\mathbf{w}^H\mathbf{w} - w_0^2 = \sum_{l=1}^{2L_\mathrm{b}-1}|w_l|^2 = 2\sum_{l=1}^{L_\mathrm{b}-1}|w_l|^2 = \frac{1}{4}. \tag{5.54}$$

We can see from (5.54) that the first element of vector $\mathbf{w}$, i.e. w_0, is dominant, in a mean-square sense, and from (5.53) that the absolute values of the L_b first elements of $\mathbf{w}$ decrease rapidly to zero as i increases. Because of the conjugate symmetry, i.e., $|w_i| = |w_{2L_b-i}|$ for $i = 1, 2, \cdots, L_b - 1$, the last few elements of $\mathbf{w}$ are not negligible, but this affects only the first and last columns of $\underset{\Rrightarrow}{\boldsymbol{W}}^{01}_{2L_b \times 2L_b}$ since this matrix is circulant with $\mathbf{w}$ as its first column. All other columns have those non-negligible elements wrapped around in such a way that they are concentrated around the main diagonal. To summarize, we can say that for L_b large, only the very first (few) off-diagonals of $\underset{\Rrightarrow}{\boldsymbol{W}}^{01}_{2L_b \times 2L_b}$ will be non-negligible while the others can be completely neglected. We also neglect the influence of the two isolated peaks $|w_{2L_b-1}| = |w_1| < w_0$ on the lower left corner and the upper right corner, respectively. Thus, approximating $\underset{\Rrightarrow}{\boldsymbol{W}}^{01}_{2L_b \times 2L_b}$ by a diagonal matrix, i.e. $\underset{\Rrightarrow}{\boldsymbol{W}}^{01}_{2L_b \times 2L_b} \approx w_0 \mathbf{I}_{2L_b \times 2L_b} = \mathbf{I}_{2L_b \times 2L_b}/2$, is reasonable, and in this case we will have $\alpha_f \approx 1/w_0 = 2$ for an optimal convergence rate. In the rest, we suppose that $0 < \alpha_f \leq 2$.

5.3.2 Special Cases

For $L_b = L$, $\underset{\Rrightarrow}{\boldsymbol{D}}_{b,s}(t) = \underset{\Rrightarrow}{\boldsymbol{D}}_s(t)$ and $\underset{\Rrightarrow}{\boldsymbol{R}}'(t)$ is a diagonal matrix. In this case, the constrained FLMS [114] follows immediately from (5.49)–(5.51). This algorithm requires the computation of 5 FFTs of length $2L$ per block. By approximating $\underset{\Rrightarrow}{\boldsymbol{W}}^{10}_{2L \times 2L}$ in (5.51) by $\mathbf{I}_{2L \times 2L}/2$, we obtain the unconstrained FLMS (UFLMS) algorithm [218] which requires only 3 FFTs per block. Many simulations (in the case $L_b = L$) show that the two algorithms have virtually the same performance. Table 5.1 summarizes these two algorithms.

For $L_b < L$, $\underset{\Rrightarrow}{\boldsymbol{R}}_d(t)$ in (5.48) consists of P^2 sub-matrices that can be approximated as shown above. It is interesting that for $L_b = 1$, the algorithm is strictly equivalent to the RLS algorithm in the time domain. After the approximation, we obtain a new algorithm that we call the *extended multidelay filter* (EMDF) for $1 < L_b < L$ that takes the auto-correlations among blocks into account [57]. Finally, the classical multidelay filter (MDF) [284] is obtained by further approximating $\underset{\Rrightarrow}{\boldsymbol{R}}'(t)$ in (5.49) as follows:

$$\underset{\Rrightarrow}{\boldsymbol{R}}'(t) \approx \text{diag}\left[\underset{\Rrightarrow}{\boldsymbol{R}}_{\text{MDF}}(t) \ \cdots \ \underset{\Rrightarrow}{\boldsymbol{R}}_{\text{MDF}}(t) \right], \tag{5.55}$$

where

$$\underset{\Rrightarrow}{\boldsymbol{R}}_{\text{MDF}}(t) = \lambda_f \underset{\Rrightarrow}{\boldsymbol{R}}_{\text{MDF}}(t-1) + (1-\lambda_f) \underset{\Rrightarrow}{\boldsymbol{D}}^*_s(t) \underset{\Rrightarrow}{\boldsymbol{D}}_s(t) \tag{5.56}$$

is a $(2L_b \times 2L_b)$ diagonal matrix. Table 5.2 gives the constrained and unconstrained versions of this algorithm. The constrained MDF requires the computation of $3 + 2P$ FFTs of size $2L_b$ per block, while the unconstrained requires only 3. Obviously, the unconstrained algorithm is much more efficient when the block size L_b is much smaller than the length L of the adaptive filter, but the performance (compared to the case $L_b = L$) is significantly degraded.

Table 5.1. The constrained and unconstrained FLMS algorithms.

Initialization:	$\underline{\hat{\boldsymbol{h}}}_{2L}(0) = \mathbf{0}$
	$\underline{\underline{\boldsymbol{R}}}'(0) = \sigma_s^2 \mathbf{I}_{2L\times 2L}$
Parameters:	$0 < \alpha_{\mathrm{f}} \leq 2$
	$\delta > 0$
	$\lambda_{\mathrm{f}} = [1 - 1/(3L)]^L$
	$\underline{\underline{\boldsymbol{W}}}^{01}_{2L\times 2L} = \mathbf{F}_{2L\times 2L} \mathbf{W}^{01}_{2L\times 2L} \mathbf{F}^{-1}_{2L\times 2L}$
	$\mathbf{W}^{01}_{2L\times 2L} = \begin{bmatrix} \mathbf{0}_{L\times L} & \mathbf{0}_{L\times L} \\ \mathbf{0}_{L\times L} & \mathbf{I}_{L\times L} \end{bmatrix}$
	$\underline{\underline{\boldsymbol{W}}}^{10}_{2L\times 2L} = \mathbf{F}_{2L\times 2L} \mathbf{W}^{10}_{2L\times 2L} \mathbf{F}^{-1}_{2L\times 2L}$
	$\mathbf{W}^{10}_{2L\times 2L} = \begin{bmatrix} \mathbf{I}_{L\times L} & \mathbf{0}_{L\times L} \\ \mathbf{0}_{L\times L} & \mathbf{0}_{L\times L} \end{bmatrix}$
	$\underline{\underline{\boldsymbol{W}}} = \underline{\underline{\boldsymbol{W}}}^{10}_{2L\times 2L}$, constrained algorithm
	$\underline{\underline{\boldsymbol{W}}} = \mathbf{I}_{2L\times 2L}/2$, unconstrained algorithm
Error:	$\underline{\boldsymbol{e}}_{2L}(t) = \underline{\boldsymbol{x}}_{2L}(t) - \underline{\underline{\boldsymbol{W}}}^{01}_{2L\times 2L} \underline{\underline{\boldsymbol{D}}}_s(t) \underline{\hat{\boldsymbol{h}}}_{2L}(t-1)$
Update:	$\underline{\underline{\boldsymbol{R}}}'(t) = \lambda_{\mathrm{f}} \underline{\underline{\boldsymbol{R}}}'(t-1) + (1-\lambda_{\mathrm{f}}) \underline{\underline{\boldsymbol{D}}}_s^*(t) \underline{\underline{\boldsymbol{D}}}_s(t)$
	$\underline{\hat{\boldsymbol{h}}}_{2L}(t) = \underline{\hat{\boldsymbol{h}}}_{2L}(t-1)$
	$\quad + \alpha_{\mathrm{f}}(1-\lambda_{\mathrm{f}}) \underline{\underline{\boldsymbol{W}}} \left[\underline{\underline{\boldsymbol{R}}}'(t) + \delta \mathbf{I}_{2L\times 2L}\right]^{-1} \underline{\underline{\boldsymbol{D}}}_s^*(t) \underline{\boldsymbol{e}}_{2L}(t)$

5.4 FD Affine Projection Algorithm

The affine projection algorithm (APA) [245] has become very popular in the last decade, especially after fast versions of this algorithm were invented [139], [290]. Its popularity comes from the fact that, for speech signals, it has almost the same performance as the FRLS, with a reduction in the number of operations of roughly a factor of 3. In this section, we derive a frequency-domain APA. A simple and elegant way to do this is to search for an algorithm of the stochastic gradient type cancelling L_{b} *a posteriori* errors [228], [231], [22]. This requirement results in an underdetermined set of linear equations of which the minimum-norm solution is chosen.

By definition, the set of L_{b} *a priori* errors $\underline{\boldsymbol{e}}(t)$ and L_{b} *a posteriori* errors $\underline{\boldsymbol{e}}_{\mathrm{a}}(t)$ are:

$$\underline{\boldsymbol{e}}(t) = \underline{\boldsymbol{x}}(t) - \underline{\underline{\boldsymbol{W}}}^{01}_{L_{\mathrm{b}}\times 2L_{\mathrm{b}}} \underline{\underline{\boldsymbol{D}}}_{\mathrm{b},s}(t) \, \underline{\underline{\boldsymbol{W}}}^{10}_{2L\times L} \underline{\hat{\boldsymbol{h}}}(t-1), \tag{5.57}$$

$$\underline{\boldsymbol{e}}_{\mathrm{a}}(t) = \underline{\boldsymbol{x}}(t) - \underline{\underline{\boldsymbol{W}}}^{01}_{L_{\mathrm{b}}\times 2L_{\mathrm{b}}} \underline{\underline{\boldsymbol{D}}}_{\mathrm{b},s}(t) \, \underline{\underline{\boldsymbol{W}}}^{10}_{2L\times L} \underline{\hat{\boldsymbol{h}}}(t). \tag{5.58}$$

Using these two equations plus the requirement that $\underline{\boldsymbol{e}}_{\mathrm{a}}(t) = \mathbf{0}_{L_{\mathrm{b}}\times 1}$, we obtain:

$$\underline{\underline{\boldsymbol{W}}}^{01}_{L_{\mathrm{b}}\times 2L_{\mathrm{b}}} \underline{\underline{\boldsymbol{D}}}_{\mathrm{b},s}(t) \, \underline{\underline{\boldsymbol{W}}}^{10}_{2L\times L} \left[\underline{\hat{\boldsymbol{h}}}(t) - \underline{\hat{\boldsymbol{h}}}(t-1)\right] = \underline{\boldsymbol{e}}(t). \tag{5.59}$$

Table 5.2. The constrained and unconstrained MDF algorithms.

Initialization: $\underline{\hat{\boldsymbol{h}}}_{2L_\mathrm{b},p}(0) = \mathbf{0}, \; p = 0, 1, \cdots, P-1$

$\underline{\boldsymbol{R}}_\mathrm{MDF}(0) = \sigma_s^2 \mathbf{I}_{2L_\mathrm{b} \times 2L_\mathrm{b}}$

Parameters: $0 < \alpha_\mathrm{f} \leq 2$

$\delta > 0$

$\lambda_\mathrm{f} = [1 - 1/(3L)]^{L_\mathrm{b}}$

$\underline{\boldsymbol{W}}^{01}_{2L_\mathrm{b} \times 2L_\mathrm{b}} = \mathbf{F}_{2L_\mathrm{b} \times 2L_\mathrm{b}} \mathbf{W}^{01}_{2L_\mathrm{b} \times 2L_\mathrm{b}} \mathbf{F}^{-1}_{2L_\mathrm{b} \times 2L_\mathrm{b}}$

$\mathbf{W}^{01}_{2L_\mathrm{b} \times 2L_\mathrm{b}} = \begin{bmatrix} \mathbf{0}_{L_\mathrm{b} \times L_\mathrm{b}} & \mathbf{0}_{L_\mathrm{b} \times L_\mathrm{b}} \\ \mathbf{0}_{L_\mathrm{b} \times L_\mathrm{b}} & \mathbf{I}_{L_\mathrm{b} \times L_\mathrm{b}} \end{bmatrix}$

$\underline{\boldsymbol{W}}^{10}_{2L_\mathrm{b} \times 2L_\mathrm{b}} = \mathbf{F}_{2L_\mathrm{b} \times 2L_\mathrm{b}} \mathbf{W}^{10}_{2L_\mathrm{b} \times 2L_\mathrm{b}} \mathbf{F}^{-1}_{2L_\mathrm{b} \times 2L_\mathrm{b}}$

$\mathbf{W}^{10}_{2L_\mathrm{b} \times 2L_\mathrm{b}} = \begin{bmatrix} \mathbf{I}_{L_\mathrm{b} \times L_\mathrm{b}} & \mathbf{0}_{L_\mathrm{b} \times L_\mathrm{b}} \\ \mathbf{0}_{L_\mathrm{b} \times L_\mathrm{b}} & \mathbf{0}_{L_\mathrm{b} \times L_\mathrm{b}} \end{bmatrix}$

$\underline{\boldsymbol{W}} = \underline{\boldsymbol{W}}^{10}_{2L_\mathrm{b} \times 2L_\mathrm{b}}$, constrained algorithm

$\underline{\boldsymbol{W}} = \mathbf{I}_{2L_\mathrm{b} \times 2L_\mathrm{b}}/2$, unconstrained algorithm

Error: $\underline{\boldsymbol{e}}_{2L_\mathrm{b}}(t) = \underline{\boldsymbol{x}}_{2L_\mathrm{b}}(t) - \underline{\boldsymbol{W}}^{01}_{2L_\mathrm{b} \times 2L_\mathrm{b}} \sum_{p=0}^{P-1} \underline{\boldsymbol{D}}_s(t-p) \underline{\hat{\boldsymbol{h}}}_{2L_\mathrm{b},p}(t-1)$

Update: $\underline{\boldsymbol{R}}_\mathrm{MDF}(t) = \lambda_\mathrm{f} \underline{\boldsymbol{R}}_\mathrm{MDF}(t-1) + (1-\lambda_\mathrm{f}) \underline{\boldsymbol{D}}_s^*(t) \underline{\boldsymbol{D}}_s(t)$

$\underline{\hat{\boldsymbol{h}}}_{2L_\mathrm{b},p}(t) = \underline{\hat{\boldsymbol{h}}}_{2L_\mathrm{b},p}(t-1)$

$\quad + \alpha_\mathrm{f}(1-\lambda_\mathrm{f}) \underline{\boldsymbol{W}} \underline{\boldsymbol{D}}_s^*(t-p) \left[\underline{\boldsymbol{R}}_\mathrm{MDF}(t) + \delta \mathbf{I}_{2L_\mathrm{b} \times 2L_\mathrm{b}} \right]^{-1} \underline{\boldsymbol{e}}_{2L_\mathrm{b}}(t)$

$p = 0, 1, \cdots, P-1$

Equation (5.59) (L_b equations in L unknowns, $L_\mathrm{b} \leq L$) is an underdetermined set of linear equations. Hence, it has an infinite number of solutions, out of which the minimum-norm solution is chosen. This results in:

$$\begin{aligned} \underline{\hat{\boldsymbol{h}}}(t) &= \underline{\hat{\boldsymbol{h}}}(t-1) + (\underline{\boldsymbol{W}}^{10}_{2L \times L})^H \underline{\boldsymbol{D}}^H_{\mathrm{b},s}(t) (\underline{\boldsymbol{W}}^{01}_{L_\mathrm{b} \times 2L_\mathrm{b}})^H \\ &\times \left[\underline{\boldsymbol{W}}^{01}_{L_\mathrm{b} \times 2L_\mathrm{b}} \underline{\boldsymbol{D}}_{\mathrm{b},s}(t) \underline{\boldsymbol{W}}^{10}_{2L \times 2L} \underline{\boldsymbol{D}}^H_{\mathrm{b},s}(t) (\underline{\boldsymbol{W}}^{01}_{L_\mathrm{b} \times 2L_\mathrm{b}})^H \right]^{-1} \underline{\boldsymbol{e}}(t). \end{aligned} \tag{5.60}$$

Multiplying (5.60) by $\underline{\boldsymbol{W}}^{10}_{2L \times L}$, we obtain:

$$\begin{aligned} \underline{\hat{\boldsymbol{h}}}_{2L}(t) &= \underline{\hat{\boldsymbol{h}}}_{2L}(t-1) + \underline{\boldsymbol{W}}^{10}_{2L \times 2L} \underline{\boldsymbol{D}}^H_{\mathrm{b},s}(t) (\underline{\boldsymbol{W}}^{01}_{L_\mathrm{b} \times 2L_\mathrm{b}})^H \\ &\times \left[\underline{\boldsymbol{W}}^{01}_{L_\mathrm{b} \times 2L_\mathrm{b}} \underline{\boldsymbol{D}}_{\mathrm{b},s}(t) \underline{\boldsymbol{W}}^{10}_{2L \times 2L} \underline{\boldsymbol{D}}^H_{\mathrm{b},s}(t) (\underline{\boldsymbol{W}}^{01}_{L_\mathrm{b} \times 2L_\mathrm{b}})^H \right]^{-1} \underline{\boldsymbol{e}}(t), \end{aligned} \tag{5.61}$$

or, equivalently, each sub-filter is updated as follows:

$$\begin{aligned} \underline{\hat{\boldsymbol{h}}}_{2L_\mathrm{b},p}(t) &= \underline{\hat{\boldsymbol{h}}}_{2L_\mathrm{b},p}(t-1) + \underline{\boldsymbol{W}}^{10}_{2L_\mathrm{b} \times 2L_\mathrm{b}} \underline{\boldsymbol{D}}^*_s(t-p) (\underline{\boldsymbol{W}}^{01}_{L_\mathrm{b} \times 2L_\mathrm{b}})^H \\ &\times \left[\underline{\boldsymbol{W}}^{01}_{L_\mathrm{b} \times 2L_\mathrm{b}} \underline{\boldsymbol{D}}_{\mathrm{b},s}(t) \underline{\boldsymbol{W}}^{10}_{2L \times 2L} \underline{\boldsymbol{D}}^H_{\mathrm{b},s}(t) (\underline{\boldsymbol{W}}^{01}_{L_\mathrm{b} \times 2L_\mathrm{b}})^H \right]^{-1} \underline{\boldsymbol{e}}(t), \\ p &= 0, 1, \cdots, P-1. \end{aligned} \tag{5.62}$$

Furthermore, it can be easily checked that:

$$
\begin{aligned}
&(\underset{\Rightarrow}{\boldsymbol{W}}{}^{01}_{L_\mathrm{b}\times 2L_\mathrm{b}})^H \left[\underset{\Rightarrow}{\boldsymbol{W}}{}^{01}_{L_\mathrm{b}\times 2L_\mathrm{b}} \underset{\Rightarrow}{\boldsymbol{D}}_{\mathrm{b},s}(t)\, \underset{\Rightarrow}{\boldsymbol{W}}{}^{10}_{2L\times 2L} \underset{\Rightarrow}{\boldsymbol{D}}{}^H_{\mathrm{b},s}(t) (\underset{\Rightarrow}{\boldsymbol{W}}{}^{01}_{L_\mathrm{b}\times 2L_\mathrm{b}})^H\right]^{-1} \\
&= \left[\underset{\Rightarrow}{\boldsymbol{D}}_{\mathrm{b},s}(t)\, \underset{\Rightarrow}{\boldsymbol{W}}{}^{10}_{2L\times 2L} \underset{\Rightarrow}{\boldsymbol{D}}{}^H_{\mathrm{b},s}(t)\right]^{-1} (\underset{\Rightarrow}{\boldsymbol{W}}{}^{01}_{L_\mathrm{b}\times 2L_\mathrm{b}})^H,
\end{aligned} \tag{5.63}
$$

so that (5.62) is simplified to

$$
\begin{aligned}
\underset{\rightarrow}{\hat{\boldsymbol{h}}}_{2L_\mathrm{b},p}(t) &= \underset{\rightarrow}{\hat{\boldsymbol{h}}}_{2L_\mathrm{b},p}(t-1) + \underset{\Rightarrow}{\boldsymbol{W}}{}^{10}_{2L_\mathrm{b}\times 2L_\mathrm{b}} \underset{\Rightarrow}{\boldsymbol{D}}{}^*_s(t-p) \\
&\quad \times \left[\underset{\Rightarrow}{\boldsymbol{D}}_{\mathrm{b},s}(t)\, \underset{\Rightarrow}{\boldsymbol{W}}{}^{10}_{2L\times 2L} \underset{\Rightarrow}{\boldsymbol{D}}{}^H_{\mathrm{b},s}(t)\right]^{-1} \underset{\rightarrow}{\boldsymbol{e}}_{2L_\mathrm{b}}(t), \\
p &= 0, 1, \cdots, P-1,
\end{aligned} \tag{5.64}
$$

which is also an MDF-type algorithm except that the cross-power spectrum matrix is estimated with a rectangular window instead of an exponential window as in Sect. 5.3.2.

5.5 Generalization to the MISO System Case

The generalization to the multichannel case is rather straightforward. Therefore, in this section we only highlight some important steps and state the algorithms. For convenience, we will use the same notation as previously employed. Let M be the number of channels. Our definition of MISO is that we have a system with M input signals s_m, $m = 1, 2, \cdots, M$, and one output signal x. By definition, the error signal at time k is:

$$
e(k) = x(k) - \sum_{m=1}^{M} \hat{\mathbf{h}}_m^T \mathbf{s}_m(k), \tag{5.65}
$$

where $\hat{\mathbf{h}}_m$ is the estimated impulse response of the mth channel,

$$
\hat{\mathbf{h}}_m = \begin{bmatrix} \hat{h}_{m,0} & \hat{h}_{m,1} & \cdots & \hat{h}_{m,L-1} \end{bmatrix}^T
$$

and

$$
\mathbf{s}_m(k) = \begin{bmatrix} s_m(k) & s_m(k-1) & \cdots & s_m(k-L+1) \end{bmatrix}^T.
$$

Now the block error signal is defined as:

$$
\mathbf{e}(t) = \mathbf{x}(t) - \sum_{m=1}^{M} \mathbf{S}_m^T(t) \hat{\mathbf{h}}_m, \tag{5.66}
$$

where $\mathbf{e}$ and $\mathbf{x}$ are vectors of e and x respectively, and all matrices $\mathbf{S}_m$ are Toeplitz of size $(L \times L_\mathrm{b})$. In the frequency domain, we have:

$$\begin{aligned}
\underline{e}(t) &= \underline{x}(t) - \underline{\underline{W}}^{01}_{L_\mathrm{b}\times 2L_\mathrm{b}} \sum_{m=1}^{M}\sum_{p=0}^{P-1} \underline{\underline{D}}_{s_m}(t-p)\,\underline{\underline{W}}^{10}_{2L_\mathrm{b}\times L_\mathrm{b}}\hat{\underline{h}}_{m,p} \\
&= \underline{x}(t) - \underline{\underline{W}}^{01}_{L_\mathrm{b}\times 2L_\mathrm{b}} \sum_{m=1}^{M}\sum_{p=0}^{P-1} \underline{\underline{U}}_{s_m}(t-p)\hat{\underline{h}}_{m,p} \\
&= \underline{x}(t) - \underline{\underline{W}}^{01}_{L_\mathrm{b}\times 2L_\mathrm{b}} \sum_{m=1}^{M} \underline{\underline{U}}_{\mathrm{b},s_m}(t)\hat{\underline{h}}_{m} \\
&= \underline{x}(t) - \underline{\underline{W}}^{01}_{L_\mathrm{b}\times 2L_\mathrm{b}} \underline{\underline{U}}_{\mathrm{b},s}(t)\hat{\underline{h}}, \qquad (5.67)
\end{aligned}$$

where

$$\begin{aligned}
\hat{\mathbf{h}}_{m,p} &= \left[\, \hat{h}_{m,pL_\mathrm{b}} \;\; \hat{h}_{m,pL_\mathrm{b}+1} \;\; \cdots \;\; \hat{h}_{m,pL_\mathrm{b}+L_\mathrm{b}-1} \,\right]^T, \\
\hat{\underline{h}}_{m,p} &= \mathbf{F}_{L_\mathrm{b}\times L_\mathrm{b}}\hat{\mathbf{h}}_{m,p}, \\
\hat{\underline{h}}_{m} &= \left[\, \hat{\underline{h}}^T_{m,0} \;\; \hat{\underline{h}}^T_{m,1} \;\; \cdots \;\; \hat{\underline{h}}^T_{m,P-1} \,\right]^T, \\
\hat{\underline{h}} &= \left[\, \hat{\underline{h}}^T_{1} \;\; \hat{\underline{h}}^T_{2} \;\; \cdots \;\; \hat{\underline{h}}^T_{M} \,\right]^T, \\
\underline{\underline{D}}_{s_m}(t-p) &= \mathbf{F}_{2L_\mathrm{b}\times 2L_\mathrm{b}}\mathbf{C}_{s_m}(t-p)\mathbf{F}^{-1}_{2L_\mathrm{b}\times 2L_\mathrm{b}}, \\
\underline{\underline{U}}_{s_m}(t-p) &= \underline{\underline{D}}_{s_m}(t-p)\,\underline{\underline{W}}^{10}_{2L_\mathrm{b}\times L_\mathrm{b}}, \\
\underline{\underline{U}}_{\mathrm{b},s_m}(t) &= \left[\, \underline{\underline{U}}_{s_m}(t) \;\; \underline{\underline{U}}_{s_m}(t-1) \;\; \cdots \;\; \underline{\underline{U}}_{s_m}(t-P+1) \,\right], \\
\underline{\underline{U}}_{\mathrm{b},s}(t) &= \left[\, \underline{\underline{U}}_{\mathrm{b},s_1}(t) \;\; \underline{\underline{U}}_{\mathrm{b},s_2}(t) \;\; \cdots \;\; \underline{\underline{U}}_{\mathrm{b},s_M}(t) \,\right], \\
\underline{\underline{D}}_{\mathrm{b},s_m}(t) &= \left[\, \underline{\underline{D}}_{s_m}(t) \;\; \underline{\underline{D}}_{s_m}(t-1) \;\; \cdots \;\; \underline{\underline{D}}_{s_m}(t-P+1) \,\right], \\
\underline{\underline{D}}_{\mathrm{b},s}(t) &= \left[\, \underline{\underline{D}}_{\mathrm{b},s_1}(t) \;\; \underline{\underline{D}}_{\mathrm{b},s_2}(t) \;\; \cdots \;\; \underline{\underline{D}}_{\mathrm{b},s_M}(t) \,\right], \\
\underline{\underline{W}}^{10}_{2ML\times ML} &= \mathrm{diag}\left[\, \underline{\underline{W}}^{10}_{2L_\mathrm{b}\times L_\mathrm{b}} \;\; \cdots \;\; \underline{\underline{W}}^{10}_{2L_\mathrm{b}\times L_\mathrm{b}} \,\right], \\
\underline{\underline{U}}_{\mathrm{b},s}(t) &= \underline{\underline{D}}_{\mathrm{b},s}(t)\,\underline{\underline{W}}^{10}_{2ML\times ML}.
\end{aligned}$$

The size of the matrix $\underline{\underline{U}}_{\mathrm{b},s}$ is $(2L_\mathrm{b} \times ML)$ and the length of $\hat{\underline{h}}$ is ML.

Minimizing the criterion defined in (5.8), we obtain the normal equations for the multichannel case:

$$\underline{\underline{R}}(t)\hat{\underline{h}}(t) = \underline{p}(t), \qquad (5.68)$$

where

$$\begin{aligned}
\underline{\underline{R}}(t) &= \lambda_\mathrm{f}\underline{\underline{R}}(t-1) \qquad (5.69) \\
&\quad + (1-\lambda_\mathrm{f})(\underline{\underline{W}}^{10}_{2ML\times ML})^H \underline{\underline{D}}^H_{\mathrm{b},s}(t)\,\underline{\underline{W}}^{01}_{2L_\mathrm{b}\times 2L_\mathrm{b}}\underline{\underline{D}}_{\mathrm{b},s}(t)\,\underline{\underline{W}}^{10}_{2ML\times ML}
\end{aligned}$$

is an $(ML \times ML)$ matrix and

Table 5.3. The constrained and unconstrained two-channel FLMS algorithms.

$$\begin{aligned}
\text{Initialization:}\quad & \underrightarrow{\hat{\boldsymbol{h}}}_{2L,m}(0) = \mathbf{0},\ m = 1,2 \\
& \underset{\Rightarrow}{\boldsymbol{R}}'_{s_m s_m}(0) = \sigma^2_{s_m}\mathbf{I}_{2L\times 2L},\ m = 1,2 \\
\text{Parameters:}\quad & 0 < \alpha_{\mathrm{f}} \leq 2 \\
& \delta_1,\ \delta_2 > 0 \\
& \lambda_{\mathrm{f}} = [1 - 1/(3L)]^L \\
& \underset{\Rightarrow}{\boldsymbol{W}}^{01}_{2L\times 2L} = \mathbf{F}_{2L\times 2L}\mathbf{W}^{01}_{2L\times 2L}\mathbf{F}^{-1}_{2L\times 2L} \\
& \mathbf{W}^{01}_{2L\times 2L} = \begin{bmatrix} \mathbf{0}_{L\times L} & \mathbf{0}_{L\times L} \\ \mathbf{0}_{L\times L} & \mathbf{I}_{L\times L} \end{bmatrix} \\
& \underset{\Rightarrow}{\boldsymbol{W}}^{10}_{2L\times 2L} = \mathbf{F}_{2L\times 2L}\mathbf{W}^{10}_{2L\times 2L}\mathbf{F}^{-1}_{2L\times 2L} \\
& \mathbf{W}^{10}_{2L\times 2L} = \begin{bmatrix} \mathbf{I}_{L\times L} & \mathbf{0}_{L\times L} \\ \mathbf{0}_{L\times L} & \mathbf{0}_{L\times L} \end{bmatrix} \\
& \underset{\Rightarrow}{\boldsymbol{W}} = \underset{\Rightarrow}{\boldsymbol{W}}^{10}_{2L\times 2L},\ \text{constrained algorithm} \\
& \underset{\Rightarrow}{\boldsymbol{W}} = \mathbf{I}_{2L\times 2L}/2,\ \text{unconstrained algorithm} \\
\text{Error:}\quad & \underrightarrow{\boldsymbol{e}}_{2L}(t) = \underrightarrow{\boldsymbol{x}}_{2L}(t) \\
& \quad - \underset{\Rightarrow}{\boldsymbol{W}}^{01}_{2L\times 2L}\left[\underset{\Rightarrow}{\boldsymbol{D}}_{s_1}(t)\underrightarrow{\hat{\boldsymbol{h}}}_{2L,1}(t-1) + \underset{\Rightarrow}{\boldsymbol{D}}_{s_2}(t)\underrightarrow{\hat{\boldsymbol{h}}}_{2L,2}(t-1)\right] \\
\text{Update:}\quad & \underset{\Rightarrow}{\boldsymbol{R}}'_{s_m s_i}(t) = \lambda_{\mathrm{f}}\underset{\Rightarrow}{\boldsymbol{R}}'_{s_m s_i}(t-1) + (1-\lambda_{\mathrm{f}})\underset{\Rightarrow}{\boldsymbol{D}}^*_{s_m}(t)\underset{\Rightarrow}{\boldsymbol{D}}_{s_i}(t),\ m,i = 1,2 \\
& \underset{\Rightarrow}{\tilde{\boldsymbol{R}}}'_{s_m s_m}(t) = \underset{\Rightarrow}{\boldsymbol{R}}'_{s_m s_m}(t) + \delta_m\mathbf{I}_{2L\times 2L},\ m = 1,2 \\
& \underset{\Rightarrow}{\boldsymbol{\Gamma}}(t) = \left[\underset{\Rightarrow}{\tilde{\boldsymbol{R}}}'_{s_1 s_1}(t)\underset{\Rightarrow}{\tilde{\boldsymbol{R}}}'_{s_2 s_2}(t)\right]^{-1/2}\underset{\Rightarrow}{\boldsymbol{R}}'_{s_1 s_2}(t),\ \text{coherence matrix} \\
& \underset{\Rightarrow}{\boldsymbol{R}}'_m(t) = \underset{\Rightarrow}{\tilde{\boldsymbol{R}}}'_{s_m s_m}(t)\left[\mathbf{I}_{2L\times 2L} - \underset{\Rightarrow}{\boldsymbol{\Gamma}}^H(t)\underset{\Rightarrow}{\boldsymbol{\Gamma}}(t)\right],\ m = 1,2 \\
& \underrightarrow{\hat{\boldsymbol{h}}}_{2L,1}(t) = \underrightarrow{\hat{\boldsymbol{h}}}_{2L,1}(t-1) \\
& \quad + \alpha_{\mathrm{f}}(1-\lambda_{\mathrm{f}})\,\underset{\Rightarrow}{\boldsymbol{W}}\,\underset{\Rightarrow}{\boldsymbol{R}}'^{-1}_1(t)\left[\underset{\Rightarrow}{\boldsymbol{D}}^*_{s_1}(t) - \underset{\Rightarrow}{\boldsymbol{R}}'_{s_1 s_2}(t)\underset{\Rightarrow}{\tilde{\boldsymbol{R}}}'^{-1}_{s_2 s_2}(t)\underset{\Rightarrow}{\boldsymbol{D}}^*_{s_2}(t)\right]\underrightarrow{\boldsymbol{e}}_{2L}(t) \\
& \underrightarrow{\hat{\boldsymbol{h}}}_{2L,2}(t) = \underrightarrow{\hat{\boldsymbol{h}}}_{2L,2}(t-1) \\
& \quad + \alpha_{\mathrm{f}}(1-\lambda_{\mathrm{f}})\,\underset{\Rightarrow}{\boldsymbol{W}}\,\underset{\Rightarrow}{\boldsymbol{R}}'^{-1}_2(t)\left[\underset{\Rightarrow}{\boldsymbol{D}}^*_{s_2}(t) - \underset{\Rightarrow}{\boldsymbol{R}}'_{s_2 s_1}(t)\underset{\Rightarrow}{\tilde{\boldsymbol{R}}}'^{-1}_{s_1 s_1}(t)\underset{\Rightarrow}{\boldsymbol{D}}^*_{s_1}(t)\right]\underrightarrow{\boldsymbol{e}}_{2L}(t)
\end{aligned}$$

$$\underrightarrow{\boldsymbol{p}}(t) = \lambda_{\mathrm{f}}\underrightarrow{\boldsymbol{p}}(t-1) + (1-\lambda_{\mathrm{f}})(\underset{\Rightarrow}{\boldsymbol{W}}^{10}_{2ML\times ML})^H\underset{\Rightarrow}{\boldsymbol{D}}^H_{\mathrm{b},s}(t)\underrightarrow{\boldsymbol{x}}_{2L_{\mathrm{b}}}(t) \qquad (5.70)$$

is an $(ML \times 1)$ vector. Assuming that the M input signals are not perfectly pairwise coherent, the normal equations have a unique solution which is the optimal Wiener solution.

Define the following variables:

$$\underset{\Rightarrow}{\boldsymbol{W}}^{10}_{2ML\times 2ML} = \mathrm{diag}\left[\ \underset{\Rightarrow}{\boldsymbol{W}}^{10}_{2L_{\mathrm{b}}\times 2L_{\mathrm{b}}}\ \cdots\ \underset{\Rightarrow}{\boldsymbol{W}}^{10}_{2L_{\mathrm{b}}\times 2L_{\mathrm{b}}}\ \right]$$

and

$$\underrightarrow{\hat{\boldsymbol{h}}}_{2ML}(t) = \underset{\Rightarrow}{\boldsymbol{W}}^{10}_{2ML\times ML}\underrightarrow{\hat{\boldsymbol{h}}}(t).$$

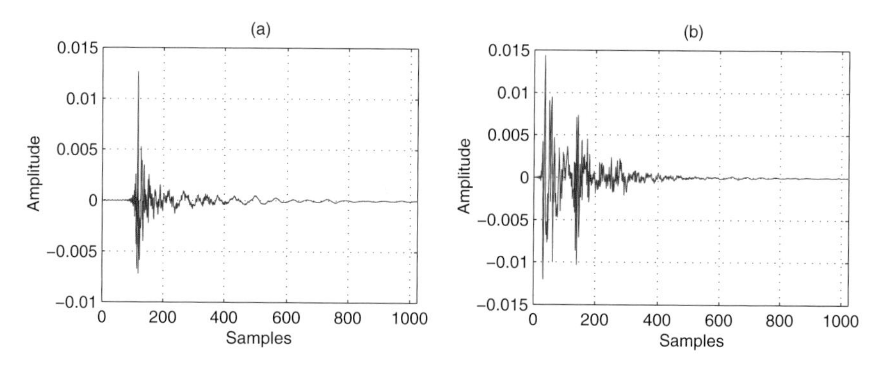

Fig. 5.1. Acoustic impulse responses used in simulations.

Using the same approach and definitions as in Sect. 5.2, we get the multichannel frequency-domain adaptive algorithm:

$$\begin{aligned}
\underset{\Rightarrow}{\boldsymbol{R}}_{\mathrm{d}}(t) &= \lambda_{\mathrm{f}} \underset{\Rightarrow}{\boldsymbol{R}}_{\mathrm{d}}(t-1) + (1-\lambda_{\mathrm{f}}) \underset{\Rightarrow}{\boldsymbol{D}}^{H}_{\mathrm{b},s}(t)\, \underset{\Rightarrow}{\boldsymbol{W}}^{01}_{2L_{\mathrm{b}}\times 2L_{\mathrm{b}}} \underset{\Rightarrow}{\boldsymbol{D}}_{\mathrm{b},s}(t), && (5.71)\\
\underset{\rightarrow}{\boldsymbol{e}}_{2L_{\mathrm{b}}}(t) &= \underset{\rightarrow}{\boldsymbol{x}}_{2L_{\mathrm{b}}}(t) - \underset{\Rightarrow}{\boldsymbol{W}}^{01}_{2L_{\mathrm{b}}\times 2L_{\mathrm{b}}} \underset{\Rightarrow}{\boldsymbol{D}}_{\mathrm{b},s}(t) \underset{\rightarrow}{\hat{\boldsymbol{h}}}_{2ML}(t-1), && (5.72)\\
\underset{\rightarrow}{\hat{\boldsymbol{h}}}_{2ML}(t) &= \underset{\rightarrow}{\hat{\boldsymbol{h}}}_{2ML}(t-1)\\
&\quad + (1-\lambda_{\mathrm{f}})\, \underset{\Rightarrow}{\boldsymbol{W}}^{10}_{2ML\times 2ML} \underset{\Rightarrow}{\boldsymbol{R}}^{-1}_{\mathrm{d}}(t) \underset{\Rightarrow}{\boldsymbol{D}}^{H}_{\mathrm{b},s}(t) \underset{\rightarrow}{\boldsymbol{e}}_{2L_{\mathrm{b}}}(t). && (5.73)
\end{aligned}$$

Now, to have a more efficient algorithm, we use the approximation given in Sect. 5.3. Finally, we have:

$$\begin{aligned}
\underset{\Rightarrow}{\boldsymbol{R}}'(t) &= \lambda_{\mathrm{f}} \underset{\Rightarrow}{\boldsymbol{R}}'(t-1) + (1-\lambda_{\mathrm{f}}) \underset{\Rightarrow}{\boldsymbol{D}}^{H}_{\mathrm{b},s}(t) \underset{\Rightarrow}{\boldsymbol{D}}_{\mathrm{b},s}(t), && (5.74)\\
\underset{\rightarrow}{\boldsymbol{e}}_{2L_{\mathrm{b}}}(t) &= \underset{\rightarrow}{\boldsymbol{x}}_{2L_{\mathrm{b}}}(t) - \underset{\Rightarrow}{\boldsymbol{W}}^{01}_{2L_{\mathrm{b}}\times 2L_{\mathrm{b}}} \underset{\Rightarrow}{\boldsymbol{D}}_{\mathrm{b},s}(t) \underset{\rightarrow}{\hat{\boldsymbol{h}}}_{2ML}(t-1), && (5.75)\\
\underset{\rightarrow}{\hat{\boldsymbol{h}}}_{2ML}(t) &= \underset{\rightarrow}{\hat{\boldsymbol{h}}}_{2ML}(t-1)\\
&\quad + \alpha_{\mathrm{f}}(1-\lambda_{\mathrm{f}})\, \underset{\Rightarrow}{\boldsymbol{W}}^{10}_{2ML\times 2ML} \underset{\Rightarrow}{\boldsymbol{R}}'^{-1}(t) \underset{\Rightarrow}{\boldsymbol{D}}^{H}_{\mathrm{b},s}(t) \underset{\rightarrow}{\boldsymbol{e}}_{2L_{\mathrm{b}}}(t). && (5.76)
\end{aligned}$$

As an example, we give in Table 5.3 the two-channel FLMS ($L_{\mathrm{b}} = L$) [33] which happens to be very useful in the context of stereophonic acoustic echo cancellation [34], [105]. The two-channel MDF is described in detail in [28], [30].

5.6 Numerical Examples

The two room acoustic impulse responses, $\mathbf{h}$, that we try to identify are the same as the ones used in the two previous chapters. For convenience, we show them again in Fig. 5.1. The impulse response of Fig. 5.1(a) is used for a SISO system while the two of them are used for a 2-channel MISO system. They are both of length $L = 1024$. The sampling rate is 8 kHz and a white Gaussian noise signal with 30 dB SNR (signal-to-noise ratio) is added to the output

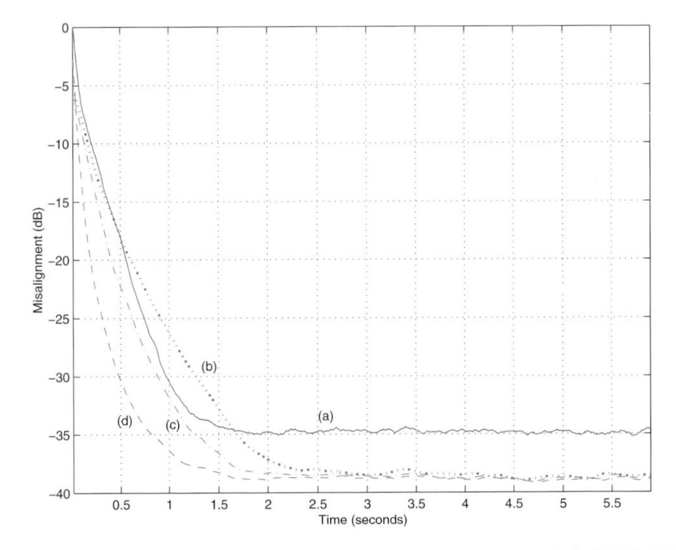

Fig. 5.2. Convergence of the normalized misalignment of the (a) SISO NLMS algorithm and the SISO FLMS algorithm for different values of the overlapping factor; (b) $o = 1$, (c) $o = 2$, and (d) $o = 4$.

$x(k)$. The normalized misalignment is utilized for the performance measure (see Chap. 3).

We first consider a SISO system whose input $s(k)$ is a zero-mean white Gaussian signal. We would like to compare the NLMS and the constrained FLMS ($L_\mathrm{b} = L$) algorithms for different values of the overlapping factors o. The parameter settings are:

- $\hat{\mathbf{h}} = \mathbf{0}$.
- $\alpha = 0.5$ (the normalized step-size parameter for the NLMS), $\delta_\mathrm{NLMS} = 10\sigma_s^2$.
- $\alpha_\mathrm{f} = 1.5$, $\delta_\mathrm{FLMS} = 10\sigma_s^2$.
- $o = 1, 2, 4$; $\lambda_\mathrm{f} = [1 - 1/(3L)]^{L/o}$.

Figure 5.2 compares the initial convergence and final MSE between the NLMS and FLMS algorithms with different overlapping factors. For $o = 4$, for example, we see that the FLMS is much better than NLMS and also requires less number of operations for its implementation.

We now consider a 2-channel MISO system and compare the MISO NLMS and MISO FLMS ($L_\mathrm{b} = L$) algorithms when the coherence between the two input signals is pretty high. We generate $s_1(k)$ and $s_2(k)$ as follows,

$$s_1(k) = w(k) + nl\frac{w(k) + |w(k)|}{2}, \tag{5.77}$$

$$s_2(k) = w(k) + nl\frac{w(k) - |w(k)|}{2}, \tag{5.78}$$

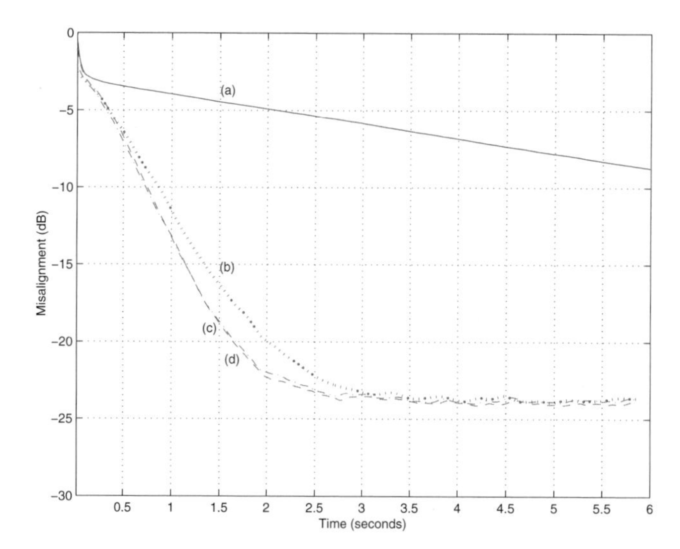

Fig. 5.3. Convergence of the normalized misalignment of the (a) MISO NLMS algorithm and the MISO FLMS algorithm for different values of the overlapping factor; (b) $o = 1$, (c) $o = 2$, and (d) $o = 4$.

where $w(k)$ is a zero-mean white Gaussian random process and nl is the amount of non-linearity (positive and negative half-wave rectifiers) added to the signals. In all our simulations, the mean of the processes $s_1(k)$ and $s_2(k)$ is removed. The parameter settings are:

- $\hat{\mathbf{h}}_1 = \hat{\mathbf{h}}_2 = \mathbf{0}$.
- $\alpha = 1$ (the normalized step-size parameter for the NLMS), $\delta_{\text{NLMS}} = 10\sigma_s^2$.
- $\alpha_{\text{f}} = 1.5$, $\delta_1 = 10\sigma_{s_1}^2$, $\delta_2 = 10\sigma_{s_2}^2$.
- $o = 1, 2, 4$; $\lambda_{\text{f}} = [1 - 1/(3L)]^{L/o}$.
- $nl = 0.5$.

Figure 5.3 compares the convergence of the normalized misalignment of the MISO NLMS and MISO FLMS algorithms with different overlapping factors. As we already know, the MISO NLMS with highly coherent signals has trouble to converge to its final MSE in a reasonable amount of time. Fortunately, as clearly shown in this figure, this is not the case for the MISO FLMS algorithm.

5.7 Summary

In many applications when an adaptive filter is required, frequency-domain algorithms (when well optimized) can be extremely good alternatives to time-domain algorithms or adaptive algorithms in subbands. First, the complexity can be made low by utilizing the computational efficiency of the FFT. The delay can be kept as small as desired since the block size is independent of the length of the adaptive filter. Finally, the convergence rate can be relatively good if some parameters of these algorithms such as the exponential window, regularization factor, and adaptation step size are properly chosen. The objective of this chapter was to present a general framework for frequency-domain adaptive filtering. We have shown that an exact algorithm can be derived from the normal equations after minimizing a block least-squares criterion in the frequency domain. We have shown the convergence conditions for this algorithm and have introduced various approximations that lead to well-known algorithms such as the FLMS, UFLMS, and MDF. Other approximations are possible that may lead to some interesting algorithms. We have also shown that APA and MDF are strongly related. Finally, we have generalized some of these ideas to the multichannel case.

6

Blind Identification of Acoustic MIMO Systems

The three previous chapters have investigated the problem of adaptively identifying an acoustic MIMO system with a reference signal. But what if no reference signal is available as in many acoustic applications? Is the problem still solvable? If so, how accurate is the estimate? These questions are going to be addressed in this chapter and we now study blind channel identification (BCI) techniques, particularly with emphasis on adaptive BCI algorithms whenever available.

6.1 Introduction

Blind channel identification is desirable in acoustic signal processing systems since *a priori* knowledge of the speech source is either inaccessible or very expensive to acquire. But historically BCI research did not originate in acoustic applications. The problem can trace back to 1970s in the attempt of designing more efficient communication systems by avoiding a training phase.

The innovative idea of BCI was first proposed by Sato in [265]. Early studies of blind channel identification and equalization focused primarily on higher (than second) order statistics (HOS) based methods [145], [295], [48] (see [223] for a tutorial on this class of approaches). Because HOS cannot be accurately computed from a small number of observations, slow convergence is the critical drawback of all existing HOS methods. In addition, a cost function based on the HOS is barely convex and an HOS algorithm can be easily misled to a local minimum by corrupting noise in the observations. Therefore, after it was recognized that the problem can be solved in the light of only second-order statistics (SOS) of system outputs [293], the focus of the BCI research has shifted to SOS methods, which are also the subject of this chapter.

There already has been a rich literature about SOS-based BCI algorithms. Celebrated work includes the cross relation (CR) algorithm [215], [152], [11], [326], the subspace algorithm [237], the linear prediction-based subspace algorithm [277], and the two-step maximum likelihood algorithm [170]. For better understanding of the characteristics of such a vast variety of algorithms, we suggest the reader to refer to [294] for a more comprehensive survey and a more systematic classification of SOS-based BCI algorithms.

The aforementioned SOS-based BCI algorithms are all batch methods. They can accurately determine the impulse responses of an identifiable multichannel system (or equivalently an oversampled single-channel system [215]) using a finite number of samples when additive noise in the system outputs is weak. However, they are in general computationally intensive and it was believed that they would be difficult to implement in an adaptive mode [294]. For practical systems, particularly those running in real time, there are three design requirements for a satisfactory BCI algorithm: quick convergence, adaptivity, and low complexity [331]. Thereafter, a great deal of efforts has been made to develop computationally efficient adaptive BCI algorithms since the end of last century. One example is the adaptive algorithm using a neural network proposed in [99]. Another attempt was the least squares smoothing (LSS) algorithm [331], which is recursive in order and can be implemented in part using a lattice filter. We believe that these adaptive BCI algorithms are algebraically complicated in development and computationally demanding in operation, which not only obstruct research efforts for further performance improvement but also make these algorithms unattractive in practical implementations.

In the first half of this chapter, we will develop a class of adaptive algorithms for blind identification of SIMO systems by using the cross relations (second order statistics) between the system outputs. This class of adaptive algorithms originated from [173] and [32], in which the adaptive eigenvalue decomposition algorithm was proposed to blindly identify a single-input two-output acoustic FIR system for time delay estimation in reverberant environments. The idea was later used by K. Berberidis et al. for channel equalization in communications [49]. Even though the generalization of the adaptive algorithm from a two-channel to a multichannel system seems straightforward, its derivation is not easy since the algebraic complexity increases dramatically with the number of channels. In [177], an error signal based on the cross relations between different channels was constructed in a systematic way for a SIMO FIR system and the corresponding cost function was concise. As a result, applying adaptive filtering techniques becomes much easier. A multichannel LMS (MCLMS) algorithm and a multichannel Newton (MCN) method were then proposed. Their convergence in the mean to the real channel impulse responses was theoretically proved and empirically justified by numerical simulations. However, none of these time-domain adaptive algorithms is perfect. The MCLMS algorithm converges slowly though steadily and the MCN method needs to invert a non-diagonal Hessian matrix which involves intensive computation. Consequently, the adaptive BCI algorithms can be improved from two perspectives: accelerating convergence of the computationally efficient MCLMS algorithm or reducing computational complexity of the fast-converging MCN algorithm. The former perspective leads to an unconstrained MCLMS algorithm with optimal step-size control [179] and in the latter perspective we developed the frequency-domain unnormalized and normalized multichannel LMS algorithms [178]. We will discussed all of these gradient-based adaptive algorithms in detail. Recently it was suggested that sparseness in acoustic channel impulse responses can be exploited in adaptive algorithms to accelerate their initial convergence and tracking by making the updates proportionate with the magnitude of the model filter coefficients [102]. We intend to explain how to apply this idea to the problem of blind SIMO identification and study the multichannel exponentiated gradient algorithm with positive and negative weights (MCEG$\pm$ algorithm) [44].

While there has been a large number of batch and adaptive algorithms for blindly identifying a SIMO system, blind MIMO identification is still a challenging problem and the current research in this area still remains at the stage of feasibility investigations. In the second half of this chapter, we will study this problem with emphasis on the discussions about whether a MIMO system can be blindly identified with second-order statistics and why there still lacks of a satisfactory solution. We will also comment on the future direction of the research on this problem.

6.2 Blind SIMO Identification

6.2.1 Identifiability and Principle

Before developing approaches to blind SIMO identification, we need to be aware of what SIMO systems are *blindly identifiable*. A multichannel FIR system can be blindly identified primarily because of the channel diversity. As an extreme counter-example, if all channels of a SIMO system are identical, the system reduces to a SISO system, becoming unidentifiable thereafter. In addition, the source signal needs to have sufficient modes to make the channels fully excited. According to [326], two inductive conditions (one on the channel diversity and the other on the input signals) are necessary and sufficient to ensure system identifiability, which are shared by all SOS-based BCI methods:

1. The polynomials formed from $\mathbf{h}_n, n = 1, 2, \cdots, N$, are co-prime, i.e., the channel transfer functions $H_n(z)$ do not share any common zeros;
2. The autocorrelation matrix $\mathbf{R}_{ss} = E\left\{\mathbf{s}(k)\mathbf{s}^T(k)\right\}$ of the input signal is of full rank (such that the SIMO system can be fully excited).

In the following if otherwise specified, these two conditions are assumed to hold so that we will be dealing with a blindly identifiable FIR SIMO system.

The easiest way to show how a SIMO system can be blindly identified probably is to use the cross-relation method. For a SIMO system as described in Sect. 2.1.2, we can take advantage of the fact that in the absence of noise

$$x_i * h_j = s * h_i * h_j = x_j * h_i, \quad i, j = 1, 2, \cdots, N, \ i \neq j, \tag{6.1}$$

and have the following relation at time k:

$$\mathbf{x}_i^T(k)\mathbf{h}_j = \mathbf{x}_j^T(k)\mathbf{h}_i, \quad i,j = 1,2,\cdots,N, \ i \neq j, \tag{6.2}$$

where

$$\mathbf{x}_n(k) = \begin{bmatrix} x_n(k) & x_n(k-1) & \cdots & x_n(k-L+1) \end{bmatrix}^T, \quad n = 1,2,\cdots,N.$$

Multiplying (6.2) by $\mathbf{x}_i(k)$ and taking expectation yields,

$$\mathbf{R}_{x_i x_i}\mathbf{h}_j = \mathbf{R}_{x_i x_j}\mathbf{h}_i, \quad i,j = 1,2,\cdots,N, \ i \neq j, \tag{6.3}$$

where $\mathbf{R}_{x_i x_j} = E\left\{\mathbf{x}_i(k)\mathbf{x}_j^T(k)\right\}$. Formula (6.3) comprises $N(N-1)$ distinct equations. By summing up the $N-1$ cross relations associated with one particular channel h_j, we get

$$\sum_{i=1, i\neq j}^{N} \mathbf{R}_{x_i x_i}\mathbf{h}_j = \sum_{i=1, i\neq j}^{N} \mathbf{R}_{x_i x_j}\mathbf{h}_i, \quad j = 1,2,\cdots,N. \tag{6.4}$$

Over all channels, we then have a total of N equations. In matrix form, this set of equations is written as:

$$\mathbf{R}_{x+}\mathbf{h} = \mathbf{0}, \tag{6.5}$$

where

$$\mathbf{R}_{x+} = \begin{bmatrix} \sum_{n\neq 1}\mathbf{R}_{x_n x_n} & -\mathbf{R}_{x_2 x_1} & \cdots & -\mathbf{R}_{x_N x_1} \\ -\mathbf{R}_{x_1 x_2} & \sum_{n\neq 2}\mathbf{R}_{x_n x_n} & \cdots & -\mathbf{R}_{x_N x_2} \\ \vdots & \vdots & \ddots & \vdots \\ -\mathbf{R}_{x_1 x_N} & -\mathbf{R}_{x_2 x_N} & \cdots & \sum_{n\neq N}\mathbf{R}_{x_n x_n} \end{bmatrix}.$$

If the SIMO system is blindly identifiable, the matrix $\mathbf{R}_{x+}$ is rank deficient by 1 (in the absence of noise) and the channel impulse responses can be uniquely determined.

When additive noise is present, the right-hand side of (6.5) is no longer zero and an error vector is produced:

$$\mathbf{e} = \mathbf{R}_{x+}\mathbf{h}. \tag{6.6}$$

This error can be used to define a cost function

$$J = \|\mathbf{e}\|^2 = \mathbf{e}^T\mathbf{e}. \tag{6.7}$$

We can then determine a vector $\hat{\mathbf{h}}$ as the solution by minimizing this cost function in the least-squares sense:

$$\hat{\mathbf{h}} = \arg\min_{\mathbf{h}} J = \arg\min_{\mathbf{h}} \mathbf{h}^T\mathbf{R}_{x+}^T\mathbf{R}_{x+}\mathbf{h}. \tag{6.8}$$

In this case, $\mathbf{R}_{x+}$ is positive definite rather than positive semidefinite and the desired estimate $\hat{\mathbf{h}}$ would be the eigenvector of $\mathbf{R}_{x+}$ corresponding to its smallest eigenvalue (here we assume that the noise is white, incoherent or uncorrelated, and weaker than the source signal).

Note that the estimated channel impulse response vector is aligned to the true one, but up to a non-zero scale. This inherent scale ambiguity is usually harmless in most, if not all, of acoustic signal processing applications. But in the development of an adaptive algorithm, attention needs to be paid to prevent from converging to a trivial all-zero estimate. This will become clearer after the reader finishes reading this chapter.

6.2.2 Constrained Time-Domain Multichannel LMS and Newton Algorithms

In this section, we intend to develop two time-domain *adaptive* algorithms for blind SIMO identification, namely multichannel LMS (MCLMS) and multichannel Newton (MCN) algorithms.

Constrained Multichannel LMS Algorithm

To begin, we use the cross relations between the ith and jth outputs given by (6.2). When noise is present and/or the estimate of channel impulse responses deviates from the true value, an *a priori* error signal is produced:

$$e_{ij}(k+1) = \mathbf{x}_i^T(k+1)\hat{\mathbf{h}}_j(k) - \mathbf{x}_j^T(k+1)\hat{\mathbf{h}}_i(k), \quad i,j = 1,2,\cdots,N, \tag{6.9}$$

where $\hat{\mathbf{h}}_i(k)$ is the model filter for the ith channel at time k. Assuming that these error signals are equally important, we now define a cost function as follows:

$$\chi(k+1) = \sum_{i=1}^{N-1} \sum_{j=i+1}^{N} e_{ij}^2(k+1), \tag{6.10}$$

where we exclude the cases of $e_{ii}(k) = 0$ $(i = 1,2,\cdots,N)$ and count the $e_{ij}(k) = -e_{ij}(k)$ pair only once.

In order to avoid a trivial estimate with all zero elements, a constraint can be imposed on the model filters $\hat{\mathbf{h}}(k)$. Two constraints can be found in the literature. One is the easily understood unit-norm constraint, i.e., $\|\hat{\mathbf{h}}(k)\| = 1$. The other is the component-normalization constraint [10], i.e., $\mathbf{c}^T\hat{\mathbf{h}}(k) = 1$, where $\mathbf{c}$ is a constant vector. As an example, if we know that one coefficient, say $h_{n,l}$ $(n = 1,2,\cdots,M,\ l = 0,1,\cdots,L-1)$, is equal to α which is not zero, then we may properly specify $\mathbf{c} = [0,...,1/\alpha,...,0]^T$ with $1/\alpha$ being the $(nL+l)$th element of $\mathbf{c}$. Even though the component-normalization constraint can be more robust to noise than the unit-norm constraint [10], the knowledge of the location of the constrained component $h_{n,l}$ and its value α may not be available in practice. So the unit-norm constraint was more widely used and will be employed in this book.

With the unit-norm constraint enforced on $\hat{\mathbf{h}}(k)$, the normalized error signal is obtained:

$$\epsilon_{ij}(k+1) = e_{ij}(k+1)/\|\hat{\mathbf{h}}(k)\|. \tag{6.11}$$

Accordingly, the cost function is formulated as:

$$J(k+1) = \sum_{i=1}^{N-1} \sum_{j=i+1}^{N} \epsilon_{ij}^2(k+1) = \frac{\chi(k+1)}{\|\hat{\mathbf{h}}(k)\|^2}. \tag{6.12}$$

The update equation of the MCLMS algorithm is then given by

$$\hat{\mathbf{h}}(k+1) = \hat{\mathbf{h}}(k) - \mu \nabla J(k+1), \tag{6.13}$$

where μ is a small positive step size and $\nabla J(k+1)$ denotes the gradient of $J(k+1)$ with respect to $\hat{\mathbf{h}}(k)$. The gradient is determined as follows:

$$\begin{aligned}
\nabla J(k+1) &= \frac{\partial J(k+1)}{\partial \hat{\mathbf{h}}(k)} = \frac{\partial}{\partial \hat{\mathbf{h}}(k)} \left[\frac{\chi(k+1)}{\|\hat{\mathbf{h}}(k)\|^2} \right] \\
&= \frac{\partial}{\partial \hat{\mathbf{h}}(k)} \left[\frac{\chi(k+1)}{\hat{\mathbf{h}}^T(k)\hat{\mathbf{h}}(k)} \right] \\
&= \frac{1}{\|\hat{\mathbf{h}}(k)\|^2} \left[\frac{\partial \chi(k+1)}{\partial \hat{\mathbf{h}}(k)} - 2J(k+1)\hat{\mathbf{h}}(k) \right],
\end{aligned} \tag{6.14}$$

where

$$\frac{\partial \chi(k+1)}{\partial \hat{\mathbf{h}}(k)} = \left[\left(\frac{\partial \chi(k+1)}{\partial \hat{\mathbf{h}}_1(k)} \right)^T \left(\frac{\partial \chi(k+1)}{\partial \hat{\mathbf{h}}_2(k)} \right)^T \cdots \left(\frac{\partial \chi(k+1)}{\partial \hat{\mathbf{h}}_N(k)} \right)^T \right]^T.$$

Let's now evaluate the partial derivative of $\chi(k+1)$ with respect to the coefficients of the nth ($n = 1, 2, \cdots, N$) channel impulse response:

$$\begin{aligned}
\frac{\partial \chi(k+1)}{\partial \hat{\mathbf{h}}_n(k)} &= \frac{\partial \left[\sum_{i=1}^{N-1} \sum_{j=i+1}^{N} e_{ij}^2(k+1) \right]}{\partial \hat{\mathbf{h}}_n(k)} \\
&= \sum_{i=1}^{n-1} 2e_{in}(k+1)\mathbf{x}_i(k+1) + \sum_{j=n+1}^{N} 2e_{nj}(k+1)[-\mathbf{x}_j(k+1)] \\
&= \sum_{i=1}^{n-1} 2e_{in}(k+1)\mathbf{x}_i(k+1) + \sum_{j=n+1}^{N} 2e_{jn}(k+1)\mathbf{x}_j(k+1) \\
&= \sum_{i=1}^{N} 2e_{in}(k+1)\mathbf{x}_i(k+1),
\end{aligned} \tag{6.15}$$

where the last step follows from the fact $e_{nn}(k+1) = 0$. We may express this equation concisely in matrix form as follows:

$$\begin{aligned}\frac{\partial \chi(k+1)}{\partial \hat{\mathbf{h}}_n(k)} &= 2\mathbf{X}(k+1)\mathbf{e}_n(k+1) \\ &= 2\mathbf{X}(k+1)[\mathbf{C}_n(k+1) - \mathbf{D}_n(k+1)]\hat{\mathbf{h}}(k),\end{aligned} \tag{6.16}$$

where we have defined, for convenience,

$$\begin{aligned}
\mathbf{X}(k+1) &= \begin{bmatrix} \mathbf{x}_1(k+1) & \mathbf{x}_2(k+1) & \cdots & \mathbf{x}_N(k+1) \end{bmatrix}_{L\times N}, \\
\mathbf{e}_n(k+1) &= \begin{bmatrix} e_{1n}(k+1) & e_{2n}(k+1) & \cdots e_{Nn}(k+1) \end{bmatrix}^T \\
&= \begin{bmatrix} \mathbf{x}_1^T(k+1)\hat{\mathbf{h}}_n(k) - \mathbf{x}_n^T(k+1)\hat{\mathbf{h}}_1(k) \\ \mathbf{x}_2^T(k+1)\hat{\mathbf{h}}_n(k) - \mathbf{x}_n^T(k+1)\hat{\mathbf{h}}_2(k) \\ \vdots \\ \mathbf{x}_N^T(k+1)\hat{\mathbf{h}}_n(k) - \mathbf{x}_n^T(k+1)\hat{\mathbf{h}}_N(k) \end{bmatrix} \\
&= [\mathbf{C}_n(k+1) - \mathbf{D}_n(k+1)]\hat{\mathbf{h}}(k), \\
\mathbf{C}_n(k+1) &= \begin{bmatrix} \mathbf{0} & \cdots & \mathbf{0} & \mathbf{x}_1^T(k+1) & \mathbf{0} & \cdots & \mathbf{0} \\ \mathbf{0} & \cdots & \mathbf{0} & \mathbf{x}_2^T(k+1) & \mathbf{0} & \cdots & \mathbf{0} \\ \vdots & \cdots & \vdots & \vdots & \vdots & \cdots & \vdots \\ \underbrace{\mathbf{0} \;\cdots\; \mathbf{0}}_{(n-1)L} & & & \mathbf{x}_N^T(k+1) & \underbrace{\mathbf{0} \;\cdots\; \mathbf{0}}_{(N-n)L} & & \end{bmatrix}_{N\times NL} \\
&= \begin{bmatrix} \mathbf{0}_{N\times(n-1)L} & \mathbf{X}^T(k+1) & \mathbf{0}_{N\times(N-n)L} \end{bmatrix}, \\
\mathbf{D}_n(k+1) &= \begin{bmatrix} \mathbf{x}_n^T(k+1) & \mathbf{0} & \cdots & \mathbf{0} \\ \mathbf{0} & \mathbf{x}_n^T(k+1) & \cdots & \mathbf{0} \\ \vdots & \vdots & \ddots & \vdots \\ \mathbf{0} & \mathbf{0} & \cdots & \mathbf{x}_n^T(k+1) \end{bmatrix}_{N\times NL}.
\end{aligned}$$

Continuing, we evaluate the two matrix products in (6.16) individually as follows:

$$\begin{aligned}
&\mathbf{X}(k+1)\mathbf{C}_n(k+1) \\
&\quad = \mathbf{X}(k+1)\begin{bmatrix} \mathbf{0}_{N\times(n-1)L} & \mathbf{X}^T(k+1) & \mathbf{0}_{N\times(N-n)L} \end{bmatrix}_{N\times NL} \\
&\quad = \begin{bmatrix} \mathbf{0}_{L\times(n-1)L} & \sum_{i=1}^{N} \tilde{\mathbf{R}}_{x_i x_i}(k+1) & \mathbf{0}_{L\times(N-n)L} \end{bmatrix}_{L\times NL},
\end{aligned} \tag{6.17}$$

$$\begin{aligned}
&\mathbf{X}(k+1)\mathbf{D}_n(k+1) \\
&\quad = \begin{bmatrix} \tilde{\mathbf{R}}_{x_1 x_n}(k+1) & \tilde{\mathbf{R}}_{x_2 x_n}(k+1) & \cdots \tilde{\mathbf{R}}_{x_N x_n}(k+1) \end{bmatrix}_{L\times NL},
\end{aligned} \tag{6.18}$$

where $\tilde{\mathbf{R}}_{x_i x_j}(k+1) = \mathbf{x}_i(k+1)\mathbf{x}_j^T(k+1)$ $(i, j = 1, 2, \cdots, N)$. Here we put a tilde in $\tilde{\mathbf{R}}_{x_i x_j}$ to distinguish this instantaneous value from its mathematical expectation $\mathbf{R}_{x_i x_j}$.

Next, substituting (6.17) and (6.18) into (6.16) yields

$$\frac{\partial \chi(k+1)}{\partial \hat{\mathbf{h}}_n(k)} = 2\Big[-\tilde{\mathbf{R}}_{x_1 x_n}(k+1) \ -\tilde{\mathbf{R}}_{x_2 x_n}(k+1) \ \cdots \ \sum_{i \neq n} \tilde{\mathbf{R}}_{x_i x_i}(k+1) \ \cdots \ -\tilde{\mathbf{R}}_{x_N x_n}(k+1) \Big] \hat{\mathbf{h}}(k). \quad (6.19)$$

Therefore, we incorporate (6.19) into (6.14) and obtain

$$\frac{\partial \chi(k+1)}{\partial \hat{h}(k)} = 2\tilde{\mathbf{R}}_{x+}(k+1)\hat{\mathbf{h}}(k), \quad (6.20)$$

$$\nabla J(k+1) = \frac{1}{\|\hat{\mathbf{h}}(k)\|^2}\Big[2\tilde{\mathbf{R}}_{x+}(k+1)\hat{\mathbf{h}}(k) - 2J(k+1)\hat{\mathbf{h}}(k)\Big], \quad (6.21)$$

where

$$\tilde{\mathbf{R}}_{x+}(k) = \begin{bmatrix} \sum_{n \neq 1} \tilde{\mathbf{R}}_{x_n x_n}(k) & -\tilde{\mathbf{R}}_{x_2 x_1}(k) & \cdots & -\tilde{\mathbf{R}}_{x_N x_1}(k) \\ -\tilde{\mathbf{R}}_{x_1 x_2}(k) & \sum_{n \neq 2} \tilde{\mathbf{R}}_{x_n x_n}(k) & \cdots & -\tilde{\mathbf{R}}_{x_N x_2}(k) \\ \vdots & \vdots & \ddots & \vdots \\ -\tilde{\mathbf{R}}_{x_1 x_N}(k) & -\tilde{\mathbf{R}}_{x_2 x_N}(k) & \cdots & \sum_{n \neq N} \tilde{\mathbf{R}}_{x_n x_n}(k) \end{bmatrix}. \quad (6.22)$$

Finally, we substitute (6.21) into (6.13) and deduce the update equation of the constrained MCLMS algorithm:

$$\hat{\mathbf{h}}(k+1) = \hat{\mathbf{h}}(k) - \frac{2\mu}{\|\hat{\mathbf{h}}(k)\|^2}\Big[\tilde{\mathbf{R}}_{x+}(k+1)\hat{\mathbf{h}}(k) - J(k+1)\hat{\mathbf{h}}(k)\Big]. \quad (6.23)$$

If the channel estimate is always normalized after each update, then we have the simplified constrained MCLMS algorithm:

$$\hat{\mathbf{h}}(k+1) = \frac{\hat{\mathbf{h}}(k) - 2\mu\Big[\tilde{\mathbf{R}}_{x+}(k+1)\hat{\mathbf{h}}(k) - \chi(k+1)\hat{\mathbf{h}}(k)\Big]}{\Big\|\hat{\mathbf{h}}(k) - 2\mu\Big[\tilde{\mathbf{R}}_{x+}(k+1)\hat{\mathbf{h}}(k) - \chi(k+1)\hat{\mathbf{h}}(k)\Big]\Big\|}. \quad (6.24)$$

The time-domain constrained MCLMS adaptive algorithm is summarized in Table 6.1.

Now we would like to briefly comment on the convergence of the constrained MCLMS algorithm. Assuming that the independence assumption [160] holds, it can be easily shown that the MCLMS algorithm converges in the mean if the step size satisfies

$$0 < \mu < \frac{1}{\lambda_{\max}}, \quad (6.25)$$

Table 6.1. The constrained multichannel LMS adaptive algorithm for the blind identification of a SIMO FIR system.

Parameters:	$\hat{\mathbf{h}} = \left[\hat{\mathbf{h}}_1^T \ \hat{\mathbf{h}}_2^T \ \cdots \ \hat{\mathbf{h}}_N^T \right]^T$, model filter
	$\mu > 0$, step size
Initialization:	$\hat{\mathbf{h}}_n(0) = [\, 1 \ 0 \ \cdots \ 0 \,]^T$, $n = 1, 2, \cdots, N$
	$\hat{\mathbf{h}}(0) = \hat{\mathbf{h}}(0)/\sqrt{N}$ (normalization)
Computation:	For $k = 0, 1, 2, \cdots$, compute
	(a) $e_{ij}(k+1) = \mathbf{x}_i^T(k+1)\hat{\mathbf{h}}_j(k) - \mathbf{x}_j^T(k+1)\hat{\mathbf{h}}_i(k)$
	$i, j = 1, 2, \cdots, N$
	(b) $\chi(k+1) = \sum_{i=1}^{N-1} \sum_{j=i+1}^{N} e_{ij}^2(k+1)$
	(c) Construct the matrix $\tilde{\mathbf{R}}_{x+}(k+1)$ following (6.22)
	(d) $\hat{\mathbf{h}}(k+1) = \dfrac{\hat{\mathbf{h}}(k) - 2\mu \left[\tilde{\mathbf{R}}_{x+}(k+1)\hat{\mathbf{h}}(k) - \chi(k+1)\hat{\mathbf{h}}(k) \right]}{\left\| \hat{\mathbf{h}}(k) - 2\mu \left[\tilde{\mathbf{R}}_{x+}(k+1)\hat{\mathbf{h}}(k) - \chi(k+1)\hat{\mathbf{h}}(k) \right] \right\|}$

where $\lambda_{\max}$ is the largest eigenvalue of the matrix $E\{\tilde{\mathbf{R}}_{x+}(k) - J(k)\mathbf{I}_{NL \times NL}\}$ and $\mathbf{I}_{NL \times NL}$ is the identity matrix of size NL by NL. After convergence, taking the expectation of (6.23) produces

$$\mathbf{R}_{x+} \frac{\hat{\mathbf{h}}(\infty)}{\|\hat{\mathbf{h}}(\infty)\|} = E\{J(\infty)\} \frac{\hat{\mathbf{h}}(\infty)}{\|\hat{\mathbf{h}}(\infty)\|}, \tag{6.26}$$

which is the desired result: $\hat{\mathbf{h}}$ converges in the mean to the eigenvector of $\mathbf{R}_{x+}$ corresponding to the smallest eigenvalue $E\{J(\infty)\}$.

Constrained Multichannel Newton Algorithm

It has been proved that the constrained MCLMS algorithm developed above can converge in the mean to the desired channel impulse responses. But one of the difficulties in the design and implementation of the adaptive multichannel LMS filters is the selection of the step size μ. For selecting the step size μ in an LMS algorithm, there is a trade-off, as pointed out in many studies, between the rate of convergence, the amount of excess mean-square error, and the ability of the algorithm to track the system as its impulse responses change. In order to achieve a good balance of the three competing design objectives, we present here the constrained multichannel Newton algorithm (see [229] for the Newton method) with a variable step size during adaptation:

$$\hat{\mathbf{h}}(k+1) = \hat{\mathbf{h}}(k) - E^{-1}\left\{\nabla^2 J(k+1)\right\} \nabla J(k+1), \tag{6.27}$$

where $\nabla^2 J(k+1)$ is the Hessian matrix of $J(k+1)$ with respect to $\hat{\mathbf{h}}(k)$. Taking derivative of (6.21) with respect to $\hat{\mathbf{h}}(k)$ and using the formula

$$\begin{aligned} \frac{\partial}{\partial \hat{\mathbf{h}}(k)}[J(k+1)\hat{\mathbf{h}}(k)] &= \hat{\mathbf{h}}(k)\left[\frac{\partial J(k+1)}{\partial \hat{\mathbf{h}}(k)}\right]^T + J(k+1)\mathbf{I}_{NL\times NL} \\ &= \hat{\mathbf{h}}(k)[\nabla J(k+1)]^T + J(k+1)\mathbf{I}_{NL\times NL}, \end{aligned} \quad (6.28)$$

we obtain

$$\begin{aligned} &\nabla^2 J(k+1) = \\ &\quad \frac{2\left\{\tilde{\mathbf{R}}_{x+}(k+1) - \left[\hat{\mathbf{h}}(k)\left(\nabla J(k+1)\right)^T + J(k+1)\mathbf{I}_{NL\times NL}\right]\right\}}{\|\hat{\mathbf{h}}(k)\|^2} - \\ &\quad \frac{4\left[\tilde{\mathbf{R}}_{x+}(k+1)\hat{\mathbf{h}}(k) - J(k+1)\hat{\mathbf{h}}(k)\right]\hat{\mathbf{h}}^T(k)}{\|\hat{\mathbf{h}}(k)\|^4}. \end{aligned} \quad (6.29)$$

With the unit-norm constraint $\|\hat{\mathbf{h}}(k)\| = 1$, (6.29) can be simplified as follows:

$$\begin{aligned} &\nabla^2 J(k+1) = \\ &\quad 2\left\{\tilde{\mathbf{R}}_{x+}(k+1) - \hat{\mathbf{h}}(k)\left[\nabla J(k+1)\right]^T - J(k+1)\mathbf{I}_{NL\times NL}\right\} - \\ &\quad 4\left[\tilde{\mathbf{R}}_{x+}(k+1)\hat{\mathbf{h}}(k) - J(k+1)\hat{\mathbf{h}}(k)\right]\hat{\mathbf{h}}^T(k). \end{aligned} \quad (6.30)$$

Taking mathematical expectation of (6.30) and invoking the independence assumption [160] produces

$$\begin{aligned} E\{\nabla^2 J(k+1)\} &= 2\mathbf{R}_{x+} - 4\hat{\mathbf{h}}(k)\hat{\mathbf{h}}^T(k)\mathbf{R}_{x+} - 4\mathbf{R}_{x+}\hat{\mathbf{h}}(k)\hat{\mathbf{h}}^T(k) \\ &\quad -2E\{J(k+1)\}\left[\mathbf{I}_{NL\times NL} - 4\hat{\mathbf{h}}(k)\hat{\mathbf{h}}^T(k)\right]. \end{aligned} \quad (6.31)$$

In practice, $\mathbf{R}_{x+}$ and $E\{J(k+1)\}$ are not known such that we have to estimate their values. Since $J(k+1)$ decreases as adaptation proceeds and is relatively small particularly after convergence, we can neglect the term $E\{J(k+1)\}$ in (6.31) for simplicity and with appropriate accuracy, as suggested by simulations. The matrix $\mathbf{R}_{x+}$ is estimated recursively in a conventional way as follows:

$$\begin{aligned} \hat{\mathbf{R}}_{x+}(1) &= \mathrm{diag}\{\sigma_{x_1}^2, \cdots, \sigma_{x_1}^2, \sigma_{x_2}^2, \cdots, \sigma_{x_2}^2, \cdots, \sigma_{x_N}^2, \cdots, \sigma_{x_N}^2\}, \\ \hat{\mathbf{R}}_{x+}(k+1) &= \lambda\hat{\mathbf{R}}_{x+}(k) + \tilde{\mathbf{R}}_{x+}(k+1), \text{ for } k \geq 1, \end{aligned} \quad (6.32)$$

where $\sigma_{x_n}^2$ $(n = 1, 2, \cdots, N)$ is the power of $x_n(k)$ and λ $(0 < \lambda < 1)$ is an exponential forgetting factor.

By using these approximations, we finally get an estimate $\mathbf{W}(k+1)$ for the mean Hessian matrix of $J(k+1)$

Table 6.2. The constrained multichannel Newton adaptive algorithm for the blind identification of a SIMO FIR system.

Parameters:	$\hat{\mathbf{h}} = \left[\hat{\mathbf{h}}_1^T \ \hat{\mathbf{h}}_2^T \ \cdots \ \hat{\mathbf{h}}_N^T \right]^T$, model filter
	$0 < \rho < 1$, step size
	$0 < \lambda < 1$, forgetting factor
Initialization:	$\hat{\mathbf{h}}_n(0) = [\, 1 \ 0 \ \cdots \ 0 \,]^T$, $n = 1, 2, \cdots, N$
	$\hat{\mathbf{h}}(0) = \hat{\mathbf{h}}(0)/\sqrt{N}$ (normalization)
Computation:	For $k = 0, 1, 2, \cdots$, compute
	(a) $e_{ij}(k+1) = \mathbf{x}_i^T(k+1)\hat{\mathbf{h}}_j(k) - \mathbf{x}_j^T(k+1)\hat{\mathbf{h}}_i(k)$, $i, j = 1, 2, \cdots, N$
	(b) $\chi(k+1) = \sum_{i=1}^{N-1} \sum_{j=i+1}^{N} e_{ij}^2(k+1)$
	(c) Construct the matrix $\tilde{\mathbf{R}}_{x+}(k+1)$ following (6.22)
	(d) $\hat{\mathbf{R}}_{x+}(k+1) = \begin{cases} \text{diag}\{\sigma_{x_1}^2, \cdots, \sigma_{x_1}^2, \cdots, \sigma_{x_N}^2, \cdots, \sigma_{x_N}^2\}, & k = 0 \\ \lambda\hat{\mathbf{R}}_{x+}(k) + \tilde{\mathbf{R}}_{x+}(k+1), & k \geq 1 \end{cases}$
	(e) $\mathbf{W}(k+1) = 2\hat{\mathbf{R}}_{x+}(k+1) - 4\hat{\mathbf{h}}(k)\hat{\mathbf{h}}(k)^T\hat{\mathbf{R}}_{x+}(k+1) - 4\hat{\mathbf{R}}_{x+}(k+1)\hat{\mathbf{h}}(k)\hat{\mathbf{h}}(k)^T$
	(f) $\hat{\mathbf{h}}(k+1) = \dfrac{\hat{\mathbf{h}}(k) - 2\rho\mathbf{W}^{-1}(k+1)\left[\tilde{\mathbf{R}}_{x+}(k+1)\hat{\mathbf{h}}(k) - \chi(k+1)\hat{\mathbf{h}}(k)\right]}{\left\|\hat{\mathbf{h}}(k) - 2\rho\mathbf{W}^{-1}(k+1)\left[\tilde{\mathbf{R}}_{x+}(k+1)\hat{\mathbf{h}}(k) - \chi(k+1)\hat{\mathbf{h}}(k)\right]\right\|}$

$$
\begin{aligned}
\mathbf{W}(k+1) &= 2\hat{\mathbf{R}}_{x+}(k+1) - 4\hat{\mathbf{h}}(k)\hat{\mathbf{h}}^T(k)\hat{\mathbf{R}}_{x+}(k+1) \\
&\quad - 4\hat{\mathbf{R}}_{x+}(k+1)\hat{\mathbf{h}}(k)\hat{\mathbf{h}}^T(k),
\end{aligned} \tag{6.33}
$$

and hence deduce the multichannel Newton algorithm:

$$
\begin{aligned}
&\hat{\mathbf{h}}(k+1) \\
&= \frac{\hat{\mathbf{h}}(k) - 2\rho\mathbf{W}^{-1}(k+1)\left[\tilde{\mathbf{R}}_{x+}(k+1)\hat{\mathbf{h}}(k) - \chi(k+1)\hat{\mathbf{h}}(k)\right]}{\left\|\hat{\mathbf{h}}(k) - 2\rho\mathbf{W}^{-1}(k+1)\left[\tilde{\mathbf{R}}_{x+}(k+1)\hat{\mathbf{h}}(k) - \chi(k+1)\hat{\mathbf{h}}(k)\right]\right\|},
\end{aligned} \tag{6.34}
$$

where ρ is a new step size, close to but less than 1. The constrained MCN algorithm for blind SIMO identification is summarized in Table 6.2.

6.2.3 Unconstrained Multichannel LMS Algorithm with Optimal Step-Size Control

As explained in the previous sections, the constrained MCLMS algorithm converges slowly while the constrained MCN algorithm is computationally intensive. To improve the performance, we can either find a way to accelerate the constrained MCLMS algorithm or reduce the complexity of the constrained MCN algorithm. We will develop algorithms along the former direction in this section and continue exploring this problem along the latter direction in the next section.

To accelerate the MCLMS algorithm, developing a scheme to determine the optimal step size during adaption is desirable and effective. We begin this endeavor with re-examining the update equation (6.13) or equivalently (6.23) of the constrained MCLMS algorithm. As the adaptive algorithm proceeds, the cost function $J(k+1)$ diminishes and its gradient with respect to $\hat{\mathbf{h}}(k)$ can be approximated as

$$\nabla J(k+1) \approx \frac{2\tilde{\mathbf{R}}_{x+}(k+1)\hat{\mathbf{h}}(k)}{\|\hat{\mathbf{h}}(k)\|^2}. \tag{6.35}$$

At this point, removing the unit-norm constraint leads to a simplified *unconstrained* MCLMS adaptive algorithm:

$$\hat{\mathbf{h}}(k+1) = \hat{\mathbf{h}}(k) - 2\mu\tilde{\mathbf{R}}_{x+}(k+1)\hat{\mathbf{h}}(k), \tag{6.36}$$

which is theoretically equivalent to the adaptive algorithm proposed in [72] although the cost functions are defined in different ways in these two adaptive blind SIMO identification algorithms.

With such a simplified adaptive algorithm, the primary concern is whether it would converge to the trivial all-zero estimate. Fortunately this will not happen as long as the initial estimate $\hat{\mathbf{h}}(0)$ is not orthogonal to the true channel impulse response vector $\mathbf{h}$, as shown in [72]. This can be easily demonstrated by pre-multiplying (6.36) with $\mathbf{h}^T$:

$$\mathbf{h}^T\hat{\mathbf{h}}(k+1) = \mathbf{h}^T\hat{\mathbf{h}}(k) - 2\mu\mathbf{h}^T\tilde{\mathbf{R}}_{x+}(k+1)\hat{\mathbf{h}}(k). \tag{6.37}$$

Using the cross relation (6.2), we know that in the absence of noise

$$\mathbf{h}^T\tilde{\mathbf{R}}_{x+}(k+1) = \mathbf{0}^T. \tag{6.38}$$

This implies that the gradient $\nabla J(k+1)$ is orthogonal to $\mathbf{h}$ at any time k. As a result, (6.37) turns out to be

$$\mathbf{h}^T\hat{\mathbf{h}}(k+1) = \mathbf{h}^T\hat{\mathbf{h}}(k). \tag{6.39}$$

This indicates that $\mathbf{h}^T\hat{\mathbf{h}}(k)$ is time-invariant for the unconstrained MCLMS algorithm. Provided that $\mathbf{h}^T\hat{\mathbf{h}}(0) \neq 0$, $\hat{\mathbf{h}}(k)$ would not converge to zero.

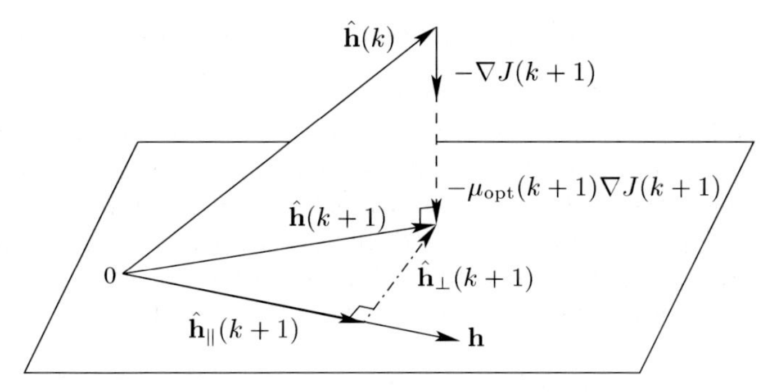

Fig. 6.1. Illustration of the optimal step size $\mu_{\mathrm{opt}}(k+1)$ for the unconstrained MCLMS blind SIMO identification algorithm in a 3-dimensional space.

Now we should feel comfortable to work on this unconstrained MCLMS algorithm and search for an optimal step size. Let's decompose the model filter $\hat{\mathbf{h}}(k)$ as follows:

$$\hat{\mathbf{h}}(k) = \hat{\mathbf{h}}_{\perp}(k) + \hat{\mathbf{h}}_{||}(k), \tag{6.40}$$

where $\hat{\mathbf{h}}_{\perp}(k)$ and $\hat{\mathbf{h}}_{||}(k)$ are perpendicular and parallel to $\mathbf{h}$, respectively. Since the gradient $\nabla J(k+1)$ is orthogonal to $\mathbf{h}$ and $\mathbf{h}$ is parallel to $\mathbf{h}_{||}(k)$, obviously $\nabla J(k+1)$ is orthogonal to $\mathbf{h}_{||}(k)$ as well. Therefore, the update equation (6.36) of the unconstrained MCLMS algorithm can be decomposed into the following two separate equations:

$$\hat{\mathbf{h}}_{\perp}(k+1) = \hat{\mathbf{h}}_{\perp}(k) - \mu\nabla J(k+1), \tag{6.41}$$

$$\hat{\mathbf{h}}_{||}(k+1) = \hat{\mathbf{h}}_{||}(k). \tag{6.42}$$

From (6.41) and (6.42), it is clear that the unconstrained MCLMS algorithm adapts the model filter only in the direction that is perpendicular to $\mathbf{h}$. The component $\hat{\mathbf{h}}_{||}(k)$ is not altered in the process of adaptation.

As far as a general system identification algorithm is concerned, the most important performance measure apparently should be the difference between the true channel impulse response and the estimate. With a BCI method, the SIMO FIR system can be blindly identified up to a scale. Therefore, the misalignment of an estimate $\hat{\mathbf{h}}(k)$ with respect to the true channel impulse response vector $\mathbf{h}$ would be:

$$d(k) = \min_{\alpha} \left\| \mathbf{h} - \alpha\hat{\mathbf{h}}(k) \right\|^2, \tag{6.43}$$

where α is an arbitrary scale. Substituting (6.40) into (6.43) and finding the minimum produces

$$d(k) = \min_{\alpha} \left[\|\hat{\mathbf{h}}(k)\|^2\alpha^2 - 2\|\hat{\mathbf{h}}_{||}(k)\|\|\mathbf{h}\|\alpha + \|\mathbf{h}\|^2 \right]$$

$$= \frac{\|\mathbf{h}\|^2}{1 + \left(\|\hat{\mathbf{h}}_{\parallel}(k)\| / \|\hat{\mathbf{h}}_{\perp}(k)\|\right)^2}. \quad (6.44)$$

Clearly the ratio of $\|\hat{\mathbf{h}}_{\parallel}(k)\|$ over $\|\hat{\mathbf{h}}_{\perp}(k)\|$ reflects how close the estimate is from the desired solution. With this feature in mind, the optimal step size $\mu_{\text{opt}}(k+1)$ for the unconstrained MCLMS algorithm at time $k+1$ would be the one that makes $\hat{\mathbf{h}}_{\perp}(k+1)$ have a minimum norm, i.e.,

$$\begin{aligned} \mu_{\text{opt}}(n+1) &= \arg\min_{\mu} \|\hat{\mathbf{h}}_{\perp}(k+1)\| \\ &= \arg\min_{\mu} \|\hat{\mathbf{h}}_{\perp}(k) - \mu \nabla J(k+1)\|. \end{aligned} \quad (6.45)$$

In order to minimize the norm of $\hat{\mathbf{h}}(n+1) = \hat{\mathbf{h}}(k) - \mu(k+1)\nabla J(k+1)$, as illustrated in Fig. 6.1, $\mu(k+1)$ should be chosen such that $\hat{\mathbf{h}}(k+1)$ is orthogonal to $\nabla J(k+1)$. Therefore, we project $\hat{\mathbf{h}}(k)$ onto $\nabla J(k+1)$ and obtain the optimal step size:

$$\mu_{\text{opt}}(k+1) = \frac{\hat{\mathbf{h}}^T(k) \nabla J(k+1)}{\|\nabla J(k+1)\|^2}. \quad (6.46)$$

Finally, this new adaptive algorithm with the optimal step size is referred to as the variable step-size unconstrained MCLMS (VSS-UMCLMS) for blind SIMO identification and is summarized in Table 6.3.

6.2.4 Frequency-Domain Unnormalized and Normalized Multichannel LMS Algorithms

In this section, we are going to study the problem of adaptive blind SIMO identification in the frequency domain. Time-frequency analysis utilizing the fast Fourier transform (FFT) is an important mathematical tool in signal processing. Since its first introduction by Dentino *et al.* [94], adaptive filtering in the frequency-domain has attracted a great deal of research interest and recently has become an essential constituent of adaptive filtering theory. By taking advantage of the computational efficiency of the FFT, a convolution of two signals can be quickly calculated. Moreover, a discrete Fourier transform processes a time sequence like a filter bank, which orthogonalizes the data, and therefore the coefficients of a frequency-domain adaptive filter can converge independently or even uniformly if the update is normalized properly. For single-channel cases, the derivation and implementation of a frequency-domain adaptive filter is relatively simple. However, when multiple channels are considered, the algorithmic complexity increases significantly with the number of channels.

Deriving a frequency-domain adaptive algorithm can involve a large number of variables in the form of both vector and matrix, in the domain of both

Table 6.3. The variable step-size unconstrained multichannel LMS adaptive algorithm for the blind identification of a SIMO FIR system.

Parameters:	$\hat{\mathbf{h}} = \begin{bmatrix} \hat{\mathbf{h}}_1^T & \hat{\mathbf{h}}_2^T & \cdots & \hat{\mathbf{h}}_N^T \end{bmatrix}^T$, model filter
Initialization:	$\hat{\mathbf{h}}_n(0) = [\,1 \;\; 1 \;\; \cdots \;\; 1\,]^T,\ n = 1, 2, \cdots, N$
Computation:	For $k = 0, 1, 2, \cdots$, compute
	(a) $e_{ij}(k+1) = \mathbf{x}_i^T(k+1)\hat{\mathbf{h}}_j(k) - \mathbf{x}_j^T(k+1)\hat{\mathbf{h}}_i(k)$ $i, j = 1, 2, \cdots, N$
	(b) $\chi(k+1) = \sum_{i=1}^{N-1} \sum_{j=i+1}^{N} e_{ij}^2(k+1)$
	(c) Construct the matrix $\tilde{\mathbf{R}}_{x+}(k+1)$ following (6.22)
	(d) $\nabla J(k+1) \approx \dfrac{2\tilde{\mathbf{R}}_{x+}(k+1)\hat{\mathbf{h}}(k)}{\lVert\hat{\mathbf{h}}(k)\rVert^2}$
	(e) $\mu_{\text{opt}}(k+1) = \dfrac{\hat{\mathbf{h}}^T(k)\nabla J(k+1)}{\lVert\nabla J(k+1)\rVert^2}$
	(f) $\hat{\mathbf{h}}(k+1) = \hat{\mathbf{h}}(k) - \mu_{\text{opt}}(k+1)\nabla J(k+1)$

time and frequency. Therefore, before formulating the problem and developing adaptive algorithms in the frequency domain, we think that it would be beneficial to the reader to clarify the notation used in the following, which is quite consistent with the other parts of this book. The notation is conventional in the time domain, but is specifically defined in the frequency domain. Uppercase and lowercase bold letters denote time-domain matrices and vectors, respectively. In the frequency domain, matrices and vectors are represented respectively by uppercase and lowercase bold, *italic* letters, and are further emphasized by an arrow underneath. But for a vector the arrow is single-barred while for a matrix the arrow is double-barred. The difference in their appearance is illustrated by the following example:

$\mathbf{x}$ a vector in the time domain (bold, lowercase),

$\mathbf{X}$ a matrix in the time domain (bold, uppercase),

$\underrightarrow{\boldsymbol{x}}$ a vector in the frequency domain (bold italic, lowercase, with a single-headed single-barred arrow underneath),

$\underRightarrow{\boldsymbol{X}}$ a matrix in the frequency domain (bold italic, uppercase, with a single-headed double-barred arrow underneath).

Frequency-Domain Multichannel LMS Algorithm

We begin with the definition of a signal $y_{ij}(k+1)$ by convolving the ith channel output $x_i(k+1)$ with the jth model filter $\hat{h}_j(k)$ $(i, j = 1, 2, \cdots, N)$:

$$y_{ij}(k+1) \stackrel{\triangle}{=} x_i(k+1) * \hat{h}_j(k). \tag{6.47}$$

As can be clearly seen in Sect. 6.2.2, $y_{ij}(k)$ is essential to calculate the error signal $e_{ij}(k+1)$ in (6.9), but involves intensive computation in the time domain. Here, we intend to perform digital filtering efficiently in the frequency domain using the overlap-save technique [243].

In the use of the overlap-save technique, the signal is processed on a frame basis. We are going to indicate the index of a frame with t. Let the vector $\tilde{\mathbf{y}}_{ij}(t+1)$ of length $2L$ denote the result of the circular convolution of $\mathbf{x}_i(t+1)$ and $\hat{\mathbf{h}}_j(t)$:

$$\tilde{\mathbf{y}}_{ij}(t+1) = \mathbf{C}_{x_i}(t+1)\hat{\mathbf{h}}_j^{10}(t), \tag{6.48}$$

where

$$\begin{aligned}
\tilde{\mathbf{y}}_{ij}(t+1) &= \begin{bmatrix} \tilde{y}_{ij}(tL) & \tilde{y}_{ij}(tL+1) & \cdots & \tilde{y}_{ij}(tL+2L-1) \end{bmatrix}^T, \\
\mathbf{C}_{x_i}(t+1) &= \begin{bmatrix} x_i(tL) & x_i(tL+2L-1) & \cdots & x_i(tL+1) \\ x_i(tL+1) & x_i(tL) & \cdots & x_i(tL+2) \\ \vdots & \vdots & \ddots & \vdots \\ x_i(tL+2L-1) & x_i(tL+2L-2) & \cdots & x_i(tL) \end{bmatrix}, \\
\hat{\mathbf{h}}_j^{10}(t) &= \begin{bmatrix} \hat{\mathbf{h}}_j^T(t) & \mathbf{0}_{L\times 1}^T \end{bmatrix}^T \\
&= \begin{bmatrix} \hat{h}_{j,0}(t) & \cdots \hat{h}_{j,L-1}(t) & 0 & \cdots & 0 \end{bmatrix}^T.
\end{aligned}$$

Note that $\mathbf{C}_{x_i}(t+1)$ is a circulant matrix. It can be seen that the last L points in the circular convolution output, i.e., $\tilde{y}_{ij}(tL+L), \cdots, \tilde{y}_{ij}(tL+2L-1)$, are identical to the results of a linear convolution:

$$\begin{aligned}
\mathbf{y}_{ij}(t+1) &= \mathbf{W}_{L\times 2L}^{01}\tilde{\mathbf{y}}_{ij}(t+1) \\
&= \mathbf{W}_{L\times 2L}^{01}\mathbf{C}_{x_i}(t+1)\hat{\mathbf{h}}_j^{10}(t) \\
&= \mathbf{W}_{L\times 2L}^{01}\mathbf{C}_{x_i}(t+1)\mathbf{W}_{2L\times L}^{10}\hat{\mathbf{h}}_j(t),
\end{aligned} \tag{6.49}$$

where

$$\begin{aligned}
\mathbf{y}_{ij}(t+1) &= \begin{bmatrix} y_{ij}(tL) & y_{ij}(tL+1) & \cdots & y_{ij}(tL+L-1) \end{bmatrix}^T, \\
\mathbf{W}_{L\times 2L}^{01} &= \begin{bmatrix} \mathbf{0}_{L\times L} & \mathbf{I}_{L\times L} \end{bmatrix}, \\
\mathbf{W}_{2L\times L}^{10} &= \begin{bmatrix} \mathbf{I}_{L\times L} & \mathbf{0}_{L\times L} \end{bmatrix}^T, \\
\hat{\mathbf{h}}_j(t) &= \begin{bmatrix} \hat{h}_{j,0}(t) & \hat{h}_{j,1}(t) & \cdots & \hat{h}_{j,L-1}(t) \end{bmatrix}^T.
\end{aligned}$$

Therefore we overlap the input data blocks by L points. For each block of length L, we have $2L$ data inputs, discard the first L circular convolution results, and retain only the last L results as outputs.

With $\mathbf{y}_{ij}(t+1)$ defined, a block of the *a priori* error signal based on the cross-relation between the ith and jth channel outputs is determined as:

$$\begin{aligned}\mathbf{e}_{ij}(t+1) &= \mathbf{y}_{ij}(t+1) - \mathbf{y}_{ji}(t+1) \\ &= \mathbf{W}^{01}_{L\times 2L}\Big[\mathbf{C}_{x_i}(t+1)\mathbf{W}^{10}_{2L\times L}\hat{\mathbf{h}}_j(t) - \\ &\qquad \mathbf{C}_{x_j}(t+1)\mathbf{W}^{10}_{2L\times L}\hat{\mathbf{h}}_i(t)\Big]. \end{aligned} \tag{6.50}$$

Next, we use the FFT technique to efficiently perform a circular convolution. Let $\mathbf{F}_{L\times L}$ be the Fourier matrix of size $L \times L$:

$$\mathbf{F}_{L\times L} = \begin{bmatrix} 1 & 1 & 1 & \cdots & 1 \\ 1 & e^{-j\frac{2\pi}{L}} & e^{-j\frac{4\pi}{L}} & \cdots & e^{-j\frac{2\pi(L-1)}{L}} \\ 1 & e^{-j\frac{4\pi}{L}} & e^{-j\frac{8\pi}{L}} & \cdots & e^{-j\frac{4\pi(L-1)}{L}} \\ \vdots & \vdots & \vdots & \ddots & \vdots \\ 1 & e^{-j\frac{2\pi(L-1)}{L}} & e^{-j\frac{4\pi(L-1)}{L}} & \cdots & e^{-j\frac{2\pi(L-1)^2}{L}} \end{bmatrix}, \tag{6.51}$$

where j is not the dummy variable we used above to indicate a system output, but stands for the square root of -1, i.e., $j = \sqrt{-1}$. The matrix $\mathbf{F}_{L\times L}$ and its inverse are related as follows:

$$\mathbf{F}^H_{L\times L} = L\mathbf{F}^{-1}_{L\times L}, \tag{6.52}$$

where $(\cdot)^H$ denotes the conjugate transpose or Hermitian transpose of a vector of a matrix. By using $\mathbf{F}_{2L\times 2L}$ and $\mathbf{F}^{-1}_{2L\times 2L}$, we can decompose the circulant matrix $\mathbf{C}_{x_i}(t+1)$:

$$\mathbf{C}_{x_i}(t+1) = \mathbf{F}^{-1}_{2L\times 2L}\underset{\Rightarrow}{\boldsymbol{D}}_{x_i}(t+1)\mathbf{F}_{2L\times 2L}, \tag{6.53}$$

where $\underset{\Rightarrow}{\boldsymbol{D}}_{x_i}(t+1)$ is a diagonal matrix whose diagonal elements are given by the DFT of the first column of $\mathbf{C}_{x_i}(t+1)$, i.e., the overlapped ith channel output in the $(t+1)$th block:

$$\mathbf{x}_i(t+1)_{2L\times 1} = \Big[\, x_i(tL) \;\; x_i(tL+1) \;\; \cdots \;\; x_i(tL+2L-1) \,\Big]^T. \tag{6.54}$$

Now we multiply (6.50) by $\mathbf{F}_{L\times L}$ and use (6.53) to determine the block error sequence in the frequency domain:

$$\begin{aligned}\underset{\rightarrow}{\boldsymbol{e}}_{ij}(t+1) &= \mathbf{F}_{L\times L}\mathbf{e}_{ij}(t+1) \\ &= \mathbf{F}_{L\times L}\mathbf{W}^{01}_{L\times 2L}\Big[\mathbf{C}_{x_i}(t+1)\mathbf{W}^{10}_{2L\times L}\hat{\mathbf{h}}_j(t) - \\ &\qquad \mathbf{C}_{x_j}(t+1)\mathbf{W}^{10}_{2L\times L}\hat{\mathbf{h}}_i(t)\Big] \\ &= \underset{\Rightarrow}{\boldsymbol{W}}^{01}_{L\times 2L}\Big[\underset{\Rightarrow}{\boldsymbol{D}}_{x_i}(t+1)\,\underset{\Rightarrow}{\boldsymbol{W}}^{10}_{2L\times L}\underset{\rightarrow}{\hat{\boldsymbol{h}}}_j(t) - \\ &\qquad \underset{\Rightarrow}{\boldsymbol{D}}_{x_j}(t+1)\,\underset{\Rightarrow}{\boldsymbol{W}}^{10}_{2L\times L}\underset{\rightarrow}{\hat{\boldsymbol{h}}}_i(t)\Big], \end{aligned} \tag{6.55}$$

where

$$\begin{aligned}\underset{\Rightarrow}{\boldsymbol{W}}^{01}_{L\times 2L} &= \mathbf{F}_{L\times L}\mathbf{W}^{01}_{L\times 2L}\mathbf{F}^{-1}_{2L\times 2L},\\ \underset{\Rightarrow}{\boldsymbol{W}}^{10}_{2L\times L} &= \mathbf{F}_{2L\times 2L}\mathbf{W}^{10}_{2L\times L}\mathbf{F}^{-1}_{L\times L},\end{aligned}$$

and $\underset{\rightarrow}{\hat{\boldsymbol{h}}}_i(t)$ consists of the L-point DFTs of the vector $\hat{\mathbf{h}}_i(t)$ at the tth block, i.e. $\underset{\rightarrow}{\hat{\boldsymbol{h}}}_i(t) = \mathbf{F}_{L\times L}\hat{\mathbf{h}}_i(t)$.

Having derived a frequency-domain block error signal (6.55), we can consequently construct a frequency-domain mean square error criterion analogous to its time-domain counterpart:

$$J_{\mathrm{f}} = E\left\{J_{\mathrm{f}}(t)\right\}, \tag{6.56}$$

where

$$J_{\mathrm{f}}(t) = \sum_{i=1}^{N-1}\sum_{j=i+1}^{N} \underset{\rightarrow}{\boldsymbol{e}}^H_{ij}(t)\underset{\rightarrow}{\boldsymbol{e}}_{ij}(t)$$

is the instantaneous square error at the tth block.

Taking the partial derivative of J_{f} with respect to $\underset{\rightarrow}{\hat{\boldsymbol{h}}}^*_n(t)$ [where $n = 1, 2, \cdots, N$ and $(\cdot)^*$ stands for complex conjugate] and pretending that $\underset{\rightarrow}{\hat{\boldsymbol{h}}}_n(t)$ is a constant [53] produces:

$$\frac{\partial J_{\mathrm{f}}}{\partial \underset{\rightarrow}{\hat{\boldsymbol{h}}}^*_n(t)} = E\left\{\frac{\partial J_{\mathrm{f}}(t+1)}{\partial \underset{\rightarrow}{\hat{\boldsymbol{h}}}^*_n(t)}\right\}. \tag{6.57}$$

In the LMS algorithm, the expectation is estimated with a one-point sample mean, i.e., the instantaneous value, which is given by:

$$\begin{aligned}\frac{\partial J_{\mathrm{f}}(t+1)}{\partial \underset{\rightarrow}{\hat{\boldsymbol{h}}}^*_n(t)} &= \frac{\partial}{\partial \underset{\rightarrow}{\hat{\boldsymbol{h}}}^*_n(t)}\left[\sum_{i=1}^{N-1}\sum_{j=i+1}^{N} \underset{\rightarrow}{\boldsymbol{e}}^H_{ij}(t+1)\underset{\rightarrow}{\boldsymbol{e}}_{ij}(t+1)\right]\\ &= \frac{\partial}{\partial \underset{\rightarrow}{\hat{\boldsymbol{h}}}^*_n(t)}\left[\sum_{i=1}^{n-1} \underset{\rightarrow}{\boldsymbol{e}}^H_{in}(t+1)\underset{\rightarrow}{\boldsymbol{e}}_{in}(t+1)\right] +\\ &\quad \frac{\partial}{\partial \underset{\rightarrow}{\hat{\boldsymbol{h}}}^*_n(t)}\left[\sum_{j=n+1}^{N} \underset{\rightarrow}{\boldsymbol{e}}^H_{nj}(t+1)\underset{\rightarrow}{\boldsymbol{e}}_{nj}(t+1)\right]\\ &= \sum_{i=1}^{n-1}\left[\underset{\Rightarrow}{\boldsymbol{W}}^{01}_{L\times 2L}\underset{\Rightarrow}{\boldsymbol{D}}_{x_i}(t+1)\,\underset{\Rightarrow}{\boldsymbol{W}}^{10}_{2L\times L}\right]^H \underset{\rightarrow}{\boldsymbol{e}}_{in}(t+1) -\\ &\quad \sum_{j=n+1}^{N}\left[\underset{\Rightarrow}{\boldsymbol{W}}^{01}_{L\times 2L}\underset{\Rightarrow}{\boldsymbol{D}}_{x_j}(t+1)\,\underset{\Rightarrow}{\boldsymbol{W}}^{10}_{2L\times L}\right]^H \underset{\rightarrow}{\boldsymbol{e}}_{nj}(t+1)\\ &= \sum_{i=1}^{N}\left[\underset{\Rightarrow}{\boldsymbol{W}}^{01}_{L\times 2L}\underset{\Rightarrow}{\boldsymbol{D}}_{x_i}(t+1)\,\underset{\Rightarrow}{\boldsymbol{W}}^{10}_{2L\times L}\right]^H \underset{\rightarrow}{\boldsymbol{e}}_{in}(t+1),\end{aligned} \tag{6.58}$$

where the last step follows from the fact $\underline{e}_{nn}(t) = \mathbf{0}$. Using this gradient, we get the frequency-domain unconstrained multichannel LMS algorithm:

$$\underline{\hat{h}}_n(t+1) = \underline{\hat{h}}_n(t) - \mu_{\mathrm{f}} \underline{\underline{W}}_{L\times 2L}^{10} \sum_{i=1}^{N} \underline{\underline{D}}_{x_i}^*(t+1) \underline{\underline{W}}_{2L\times L}^{01} \underline{e}_{in}(t+1), \quad (6.59)$$

where μ_{f} is a small positive step size and

$$\begin{aligned}
\underline{\underline{W}}_{L\times 2L}^{10} &= \mathbf{F}_{L\times L}\mathbf{W}_{L\times 2L}^{10}\mathbf{F}_{2L\times 2L}^{-1} = \frac{1}{2}\left(\underline{\underline{W}}_{2L\times L}^{10}\right)^H, \\
\underline{\underline{W}}_{2L\times L}^{01} &= \mathbf{F}_{2L\times 2L}\mathbf{W}_{2L\times L}^{01}\mathbf{F}_{L\times L}^{-1} = 2\left(\underline{\underline{W}}_{L\times 2L}^{01}\right)^H, \\
\mathbf{W}_{L\times 2L}^{10} &= \left[\begin{array}{cc} \mathbf{I}_{L\times L} & \mathbf{0}_{L\times L} \end{array}\right], \\
\mathbf{W}_{2L\times L}^{01} &= \left[\begin{array}{cc} \mathbf{0}_{L\times L} & \mathbf{I}_{L\times L} \end{array}\right]^T.
\end{aligned}$$

If the unit-norm constraint is employed, then we know that

$$\|\underline{\hat{h}}(t)\|^2 = \frac{\|\hat{\mathbf{h}}(t)\|^2}{L} = \frac{1}{L}, \quad (6.60)$$

where

$$\underline{\hat{h}}(t) \triangleq \left[\begin{array}{cccc} \underline{\hat{h}}_1^T(t) & \underline{\hat{h}}_2^T(t) & \cdots & \underline{\hat{h}}_N^T(t) \end{array}\right]^T.$$

Enforcing the constraint on (6.59), we finally deduce the frequency-domain constrained multichannel LMS (FCMCLMS) algorithm:

$$\begin{aligned}
&\underline{\hat{h}}_n(t+1) = \\
&\frac{\underline{\hat{h}}_n(t) - \mu_{\mathrm{f}} \underline{\underline{W}}_{L\times 2L}^{10} \sum_{i=1}^{N} \underline{\underline{D}}_{x_i}^*(t+1) \underline{\underline{W}}_{2L\times L}^{01} \underline{e}_{in}(t+1)}{\sqrt{L}\left\|\underline{\hat{h}}_n(t) - \mu_{\mathrm{f}} \underline{\underline{W}}_{L\times 2L}^{10} \sum_{i=1}^{N} \underline{\underline{D}}_{x_i}^*(t+1) \underline{\underline{W}}_{2L\times L}^{01} \underline{e}_{in}(t+1)\right\|}, \quad (6.61) \\
&\quad n = 1, 2, \cdots, N.
\end{aligned}$$

The derivation above is mathematically rigorous and the algorithm might look quite complicated to some of the readers since it involves a number of matrix multiplications. However, the implementation is not necessarily tedious. Actually many matrix multiplications are performed in the form of the Schur (element-by-element) product of two vectors. Here we want to briefly comment on the issue of how to skillfully and efficiently implement the FCMCLMS algorithm. We will use (6.55) as an example, in which the error signal is calculated. The product $\underline{\underline{W}}_{2L\times L}^{10}\underline{\hat{h}}_j(t)$ is obtained by appending L zeros to the end of $\hat{\mathbf{h}}_j(t)$ and taking the $2L$-point FFT:

$$\underline{\hat{h}}_j^{10}(t) \triangleq \underline{\underline{W}}_{2L\times L}^{10}\underline{\hat{h}}_j(t) = \mathrm{FFT}_{2L}\left\{\hat{\mathbf{h}}_j^{10}(t)\right\}, \quad (6.62)$$

where $\mathrm{FFT}_{2L}\{\cdot\}$ denotes $2L$-point fast Fourier transform and $\hat{\mathbf{h}}_j^{10}(t)$ is given in (6.48). Note that $\underset{\Rightarrow}{\boldsymbol{D}}_{x_i}(t+1)$ is a diagonal matrix whose diagonal elements are the DFT coefficients of $\mathbf{x}_i(t+1)_{2L\times 1}$ given in (6.54), i.e.,

$$\begin{aligned} \underset{\Rightarrow}{\boldsymbol{D}}_{x_i}(t+1) &= \mathrm{diag}\left[\underset{\rightarrow}{\boldsymbol{x}}_i(t+1)_{2L\times 1}\right] \\ &= \mathrm{diag}\left[\mathrm{FFT}_{2L}\left\{\mathbf{x}_i(t+1)_{2L\times 1}\right\}\right]. \end{aligned} \tag{6.63}$$

Therefore multiplying $\underset{\Rightarrow}{\boldsymbol{D}}_{x_i}(t+1)$ with $\underset{\Rightarrow}{\boldsymbol{W}}{}_{2L\times L}^{10}\underset{\rightarrow}{\hat{\boldsymbol{h}}}_j(t)$ would be straightforward:

$$\begin{aligned} \underset{\Rightarrow}{\boldsymbol{D}}_{x_i}(t+1)\,\underset{\Rightarrow}{\boldsymbol{W}}{}_{2L\times L}^{10}\underset{\rightarrow}{\hat{\boldsymbol{h}}}_j(t) &= \underset{\Rightarrow}{\boldsymbol{D}}_{x_i}(t+1)\underset{\rightarrow}{\hat{\boldsymbol{h}}}{}_j^{10}(t) \\ &= \underset{\rightarrow}{\boldsymbol{x}}_i(t+1)_{2L\times 1}\odot \underset{\rightarrow}{\hat{\boldsymbol{h}}}{}_j^{10}(t), \end{aligned} \tag{6.64}$$

where $\odot$ is the operator of the Schur product. Let $\mathbf{e}_{ij}(t+1)_{2L\times 1}$ be

$$\begin{aligned} &\mathbf{e}_{ij}(t+1)_{2L\times 1} \\ &= \mathrm{IFFT}_{2L}\left\{\underset{\Rightarrow}{\boldsymbol{D}}_{x_i}(t+1)\,\underset{\Rightarrow}{\boldsymbol{W}}{}_{2L\times L}^{10}\underset{\rightarrow}{\hat{\boldsymbol{h}}}_j(t) - \underset{\Rightarrow}{\boldsymbol{D}}_{x_j}(t+1)\,\underset{\Rightarrow}{\boldsymbol{W}}{}_{2L\times L}^{10}\underset{\rightarrow}{\hat{\boldsymbol{h}}}_i(t)\right\} \\ &= \mathrm{IFFT}_{2L}\left\{\underset{\rightarrow}{\boldsymbol{x}}_i(t+1)_{2L\times 1}\odot \underset{\rightarrow}{\hat{\boldsymbol{h}}}{}_j^{10}(t) - \underset{\rightarrow}{\boldsymbol{x}}_j(t+1)_{2L\times 1}\odot \underset{\rightarrow}{\hat{\boldsymbol{h}}}{}_i^{10}(t)\right\}, \end{aligned} \tag{6.65}$$

where $\mathrm{IFFT}_{2L}\{\cdot\}$ denotes $2L$-point *inverse* fast Fourier transform. Then

$$\mathbf{e}_{ij}(t+1) = \mathbf{W}_{L\times 2L}^{01}\mathbf{e}_{ij}(t+1)_{2L\times 1} \tag{6.66}$$

can be carried out simply by discarding the first L and keeping the last L elements of $\mathbf{e}_{ij}(t+1)_{2L\times 1}$ to construct $\mathbf{e}_{ij}(t+1)$. Finally $\underset{\rightarrow}{\boldsymbol{e}}_{ij}(t+1)$ is obtained by taking an L-point FFT of $\mathbf{e}_{ij}(t+1)$.

The tricks developed for (6.55) can be also applied to (6.59) and (6.61), which will be left to the reader. The FCMCLMS algorithm is summarized in Table 6.4. Compared to the time-domain frame-based multichannel LMS algorithm, the frequency-domain implementation is computationally more efficient since it uses the FFT to calculate block convolution and block correlation in the frequency domain. For each processed frame, it can be checked from Table 6.4 that (N^2+2N) FFTs of $2L$ points are sufficient to implement exact linear convolutions between the channel outputs and the adaptive filter coefficients, and to implement exact correlations between the outputs and the error signals.

Having developed the frame-based UMCLMS algorithm in the frequency domain, we need to demonstrate that it would converge in the mean to the desired solution in a stationary environment. To establish this property, we begin by defining

$$\underset{\Rightarrow}{\boldsymbol{S}}_{x_n}(t+1) \triangleq \underset{\Rightarrow}{\boldsymbol{W}}{}_{L\times 2L}^{01}\underset{\Rightarrow}{\boldsymbol{D}}_{x_n}(t+1)\,\underset{\Rightarrow}{\boldsymbol{W}}{}_{2L\times L}^{10},\quad n=1,2,\cdots,N. \tag{6.67}$$

Then the block error signal in the frequency domain given by (6.55) can be rewritten as

Table 6.4. The frequency-domain constrained multichannel LMS (FCMCLMS) adaptive algorithm for the blind identification of a SIMO FIR system.

Parameters:	$\hat{\mathbf{h}} = \left[\begin{array}{cccc} \hat{\mathbf{h}}_1^T & \hat{\mathbf{h}}_2^T & \cdots & \hat{\mathbf{h}}_N^T \end{array}\right]^T$, model filter
	$\mu_{\mathrm{f}} > 0$, step size
Initialization:	$\hat{\mathbf{h}}_n(0) = [\,1 \;\; 0 \;\; \cdots \;\; 0\,]^T$, $n = 1, 2, \cdots, N$
	$\hat{\mathbf{h}}(0) = \hat{\mathbf{h}}(0)/\sqrt{N}$ (normalization)
Computation:	For $t = 0, 1, 2, \cdots$, compute
	(a) $\underline{\hat{\boldsymbol{h}}}_n^{10}(t) = \mathrm{FFT}_{2L}\left\{\left[\begin{array}{cc} \hat{\mathbf{h}}_n^T(t) & \mathbf{0}_{1\times L} \end{array}\right]^T\right\}$, $\quad n = 1, 2, \cdots, N$
	(b) $\mathbf{x}_n(t+1)_{2L\times 1} = [\, x_n(tL) \;\; x_n(tL+1) \;\; \cdots \;\; x_n(tL+2L-1)\,]^T$
	(c) $\underline{\boldsymbol{x}}_n(t+1)_{2L\times 1} = \mathrm{FFT}_{2L}\{\mathbf{x}_n(t+1)_{2L\times 1}\}$
	(d) $\underline{\boldsymbol{e}}_{ij}(t+1)_{2L\times 1} = \underline{\boldsymbol{x}}_i(t+1)_{2L\times 1} \odot \underline{\hat{\boldsymbol{h}}}_j^{10}(t) -$ $\underline{\boldsymbol{x}}_j(t+1)_{2L\times 1} \odot \underline{\hat{\boldsymbol{h}}}_i^{10}(t)$, $\quad i, j = 1, 2, \cdots, N$
	(e) $\mathbf{e}_{ij}(t+1)_{2L\times 1} = \mathrm{IFFT}_{2L}\{\underline{\boldsymbol{e}}_{ij}(t+1)_{2L\times 1}\}$
	(f) Take the *last* L elements of $\mathbf{e}_{ij}(t+1)_{2L\times 1}$ to form $\mathbf{e}_{ij}(t+1)$
	(g) $\underline{\boldsymbol{e}}_{ij}^{01}(t+1) = \mathrm{FFT}_{2L}\left\{\left[\begin{array}{cc} \mathbf{0}_{1\times L} & \mathbf{e}_{ij}^T(t+1) \end{array}\right]^T\right\}$
	(h) $\triangle\underline{\hat{\boldsymbol{h}}}_n(t)_{2L\times 1} = \mu_{\mathrm{f}} \sum_{i=1}^{N} \underline{\boldsymbol{x}}_i^*(t+1)_{2L\times 1} \odot \underline{\boldsymbol{e}}_{in}^{01}(t+1)$
	(i) $\triangle\hat{\mathbf{h}}_n(t)_{2L\times 1} = \mathrm{IFFT}_{2L}\left\{\triangle\underline{\hat{\boldsymbol{h}}}_n(t)_{2L\times 1}\right\}$
	(j) Take the *first* L elements of $\triangle\hat{\mathbf{h}}_n(t)_{2L\times 1}$ to form $\triangle\hat{\mathbf{h}}_n(t)$
	(k) $\hat{\mathbf{h}}_n(t+1) = \dfrac{\hat{\mathbf{h}}_n(t) - \triangle\hat{\mathbf{h}}_n(t)}{\left\Vert\hat{\mathbf{h}}_n(t) - \triangle\hat{\mathbf{h}}_n(t)\right\Vert}$, $\quad n = 1, 2, \cdots, N$

$$\underline{\boldsymbol{e}}_{ij}(t+1) = \underline{\underline{\boldsymbol{S}}}_{x_i}(t+1)\underline{\hat{\boldsymbol{h}}}_j(t) - \underline{\underline{\boldsymbol{S}}}_{x_j}(t+1)\underline{\hat{\boldsymbol{h}}}_i(t), \tag{6.68}$$

and the instantaneous gradient (6.58) becomes

$$\frac{\partial J_{\mathrm{f}}(t+1)}{\partial \underline{\hat{\boldsymbol{h}}}_n^*(t)} = \sum_{i=1}^{N} \underline{\underline{\boldsymbol{S}}}_{x_i}^H(t+1)\underline{\boldsymbol{e}}_{in}(t+1) = \underline{\underline{\boldsymbol{S}}}^H(t+1)\underline{\boldsymbol{e}}_n(t+1), \tag{6.69}$$

where

$$\begin{aligned} \underline{\underline{\boldsymbol{S}}}(t+1) &= \left[\begin{array}{cccc} \underline{\underline{\boldsymbol{S}}}_{x_1}^H(t+1) & \underline{\underline{\boldsymbol{S}}}_{x_2}^H(t+1) & \cdots & \underline{\underline{\boldsymbol{S}}}_{x_N}^H(t+1) \end{array}\right]^H, \\ \underline{\boldsymbol{e}}_n(t+1) &= \left[\begin{array}{cccc} \underline{\boldsymbol{e}}_{1n}^T(t+1) & \underline{\boldsymbol{e}}_{2n}^T(t+1) & \cdots & \underline{\boldsymbol{e}}_{Nn}^T(t+1) \end{array}\right]^T. \end{aligned}$$

From (6.68), we then decompose the error signal $\underline{\boldsymbol{e}}_n(t+1)$ associated with the nth channel as follows:

$$\underline{\boldsymbol{e}}_n(t+1) = \begin{bmatrix} \underline{\boldsymbol{S}}_{x_1}(t+1)\hat{\underline{\boldsymbol{h}}}_n(t) - \underline{\boldsymbol{S}}_{x_n}(t+1)\hat{\underline{\boldsymbol{h}}}_1(t) \\ \underline{\boldsymbol{S}}_{x_2}(t+1)\hat{\underline{\boldsymbol{h}}}_n(t) - \underline{\boldsymbol{S}}_{x_n}(t+1)\hat{\underline{\boldsymbol{h}}}_2(t) \\ \vdots \\ \underline{\boldsymbol{S}}_{x_N}(t+1)\hat{\underline{\boldsymbol{h}}}_n(t) - \underline{\boldsymbol{S}}_{x_n}(t+1)\hat{\underline{\boldsymbol{h}}}_N(t) \end{bmatrix} = \left[\underline{\boldsymbol{U}}_n(t+1) - \underline{\boldsymbol{V}}_n(t+1)\right]\hat{\underline{\boldsymbol{h}}}(t), \tag{6.70}$$

where

$$\underline{\boldsymbol{U}}_n(t+1) = \begin{bmatrix} \mathbf{0}_{L\times L} & \cdots & \mathbf{0}_{L\times L} & \underline{\boldsymbol{S}}_{x_1}(t+1) & \mathbf{0}_{L\times L} & \cdots & \mathbf{0}_{L\times L} \\ \mathbf{0}_{L\times L} & \cdots & \mathbf{0}_{L\times L} & \underline{\boldsymbol{S}}_{x_2}(t+1) & \mathbf{0}_{L\times L} & \cdots & \mathbf{0}_{L\times L} \\ \vdots & \cdots & \vdots & \vdots & \vdots & \cdots & \vdots \\ \underbrace{\mathbf{0}_{L\times L} \quad \cdots \quad \mathbf{0}_{L\times L}}_{n-1 \text{ blocks}} & & & \underline{\boldsymbol{S}}_{x_N}(t+1) & \underbrace{\mathbf{0}_{L\times L} \quad \cdots \quad \mathbf{0}_{L\times L}}_{N-n \text{ blocks}} \end{bmatrix}_{NL\times NL}$$

$$= \begin{bmatrix} \mathbf{0}_{NL\times(n-1)L} & \underline{\boldsymbol{S}}(t+1) & \mathbf{0}_{NL\times(N-n)L} \end{bmatrix}_{NL\times NL},$$

$$\underline{\boldsymbol{V}}_n(t+1) = \begin{bmatrix} \underline{\boldsymbol{S}}_{x_n}(t+1) & \mathbf{0}_{L\times L} & \cdots & \mathbf{0}_{L\times L} \\ \mathbf{0}_{L\times L} & \underline{\boldsymbol{S}}_{x_n}(t+1) & \cdots & \mathbf{0}_{L\times L} \\ \vdots & \vdots & \ddots & \vdots \\ \mathbf{0}_{L\times L} & \mathbf{0}_{L\times L} & \cdots & \underline{\boldsymbol{S}}_{x_n}(t+1) \end{bmatrix}_{NL\times NL}.$$

Continuing, we can write

$$\begin{aligned} &\underline{\boldsymbol{S}}^H(t+1)\,\underline{\boldsymbol{U}}_n(t+1) \\ &\quad = \underline{\boldsymbol{S}}^H(t+1)\begin{bmatrix} \mathbf{0}_{NL\times(n-1)L} & \underline{\boldsymbol{S}}(t+1) & \mathbf{0}_{NL\times(N-n)L} \end{bmatrix} \\ &\quad = \begin{bmatrix} \mathbf{0}_{NL\times(n-1)L} & \sum_{i=1}^N \tilde{\underline{\boldsymbol{R}}}_{ii}(t+1) & \mathbf{0}_{NL\times(N-n)L} \end{bmatrix}, \end{aligned} \tag{6.71}$$

$$\begin{aligned} &\underline{\boldsymbol{S}}^H(t+1)\,\underline{\boldsymbol{V}}_n(t+1) \\ &\quad = \begin{bmatrix} \tilde{\underline{\boldsymbol{R}}}_{1n}(t+1) & \tilde{\underline{\boldsymbol{R}}}_{2n}(t+1) & \cdots & \tilde{\underline{\boldsymbol{R}}}_{Nn}(t+1) \end{bmatrix}, \end{aligned} \tag{6.72}$$

where

$$\tilde{\underline{\boldsymbol{R}}}_{ij}(t+1) \triangleq \underline{\boldsymbol{S}}_{x_i}^H(t+1)\underline{\boldsymbol{S}}_{x_j}(t+1), \quad i,j = 1,2,\cdots,N.$$

Incorporating (6.71) and (6.72) into (6.69) and (6.70) yields

$$\begin{aligned} \frac{\partial J_{\mathrm{f}}(t+1)}{\partial \hat{\underline{\boldsymbol{h}}}_n^*(t)} &= \underline{\boldsymbol{S}}^H(t+1)\left[\underline{\boldsymbol{U}}_n(t+1) - \underline{\boldsymbol{V}}_n(t+1)\right]\hat{\underline{\boldsymbol{h}}}(t) \\ &= \Big[-\tilde{\underline{\boldsymbol{R}}}_{1n}(t+1) \;\; -\tilde{\underline{\boldsymbol{R}}}_{2n}(t+1) \;\; \cdots \\ &\qquad \sum_{i\neq n} \tilde{\underline{\boldsymbol{R}}}_{ii}(t+1) \;\; \cdots \;\; -\tilde{\underline{\boldsymbol{R}}}_{Nn}(t+1) \Big]\hat{\underline{\boldsymbol{h}}}(t). \end{aligned} \tag{6.73}$$

Substituting this gradient estimate into (6.59) and concatenating the N impulse response vectors into a longer one produces a remarkably simple expression of the frequency-domain unconstrained MCLMS algorithm:

$$\underset{\rightarrow}{\hat{\boldsymbol{h}}}(t+1) = \underset{\rightarrow}{\hat{\boldsymbol{h}}}(t) - \mu_{\mathrm{f}} \underset{\Rightarrow}{\tilde{\boldsymbol{R}}}(t+1)\underset{\rightarrow}{\hat{\boldsymbol{h}}}(t), \tag{6.74}$$

where

$$\underset{\Rightarrow}{\tilde{\boldsymbol{R}}}(t+1) = \begin{bmatrix} \sum_{i \neq 1} \underset{\Rightarrow}{\tilde{\boldsymbol{R}}}_{ii}(t+1) & -\underset{\Rightarrow}{\tilde{\boldsymbol{R}}}_{21}(t+1) & \cdots & -\underset{\Rightarrow}{\tilde{\boldsymbol{R}}}_{N1}(t+1) \\ -\underset{\Rightarrow}{\tilde{\boldsymbol{R}}}_{12}(t+1) & \sum_{i \neq 2} \underset{\Rightarrow}{\tilde{\boldsymbol{R}}}_{ii}(t+1) & \cdots & -\underset{\Rightarrow}{\tilde{\boldsymbol{R}}}_{N2}(t+1) \\ \vdots & \vdots & \ddots & \vdots \\ -\underset{\Rightarrow}{\tilde{\boldsymbol{R}}}_{1N}(t+1) & -\underset{\Rightarrow}{\tilde{\boldsymbol{R}}}_{2N}(t+1) & \cdots & \sum_{i \neq N} \underset{\Rightarrow}{\tilde{\boldsymbol{R}}}_{ii}(t+1) \end{bmatrix}.$$

The update equation of the frequency-domain constrained MCLMS algorithm is then expressed as:

$$\underset{\rightarrow}{\hat{\boldsymbol{h}}}(t+1) = \frac{\underset{\rightarrow}{\hat{\boldsymbol{h}}}(t) - \mu_{\mathrm{f}} \underset{\Rightarrow}{\tilde{\boldsymbol{R}}}(t+1)\underset{\rightarrow}{\hat{\boldsymbol{h}}}(t)}{\sqrt{L}\left\| \underset{\rightarrow}{\hat{\boldsymbol{h}}}(t) - \mu_{\mathrm{f}} \underset{\Rightarrow}{\tilde{\boldsymbol{R}}}(t+1)\underset{\rightarrow}{\hat{\boldsymbol{h}}}(t) \right\|}. \tag{6.75}$$

Subtracting the true channel impulse responses $\underset{\rightarrow}{\boldsymbol{h}}$ from both sides of (6.74), we get the evolution of the misalignment in the frequency domain:

$$\delta\underset{\rightarrow}{\boldsymbol{h}}(t+1) = \delta\underset{\rightarrow}{\boldsymbol{h}}(t) - \mu_{\mathrm{f}} \underset{\Rightarrow}{\tilde{\boldsymbol{R}}}(t+1)\underset{\rightarrow}{\hat{\boldsymbol{h}}}(t), \tag{6.76}$$

where

$$\delta\underset{\rightarrow}{\boldsymbol{h}}(t) = \underset{\rightarrow}{\hat{\boldsymbol{h}}}(t) - \underset{\rightarrow}{\boldsymbol{h}}.$$

Taking the expectation of (6.76) and invoking the independence assumption [160], i.e., the channel output $\mathbf{x}_n(t+1), n = 1, 2, \cdots, N$, and the filter coefficients $\underset{\rightarrow}{\hat{\boldsymbol{h}}}(t)$ are independent, we have

$$E\left\{\delta\underset{\rightarrow}{\boldsymbol{h}}(t+1)\right\} = E\left\{\delta\underset{\rightarrow}{\boldsymbol{h}}(t)\right\} - \mu_{\mathrm{f}} \underset{\Rightarrow}{\boldsymbol{R}} E\left\{\underset{\rightarrow}{\hat{\boldsymbol{h}}}(t)\right\}, \tag{6.77}$$

where

$$\underset{\Rightarrow}{\boldsymbol{R}} = E\left\{\underset{\Rightarrow}{\tilde{\boldsymbol{R}}}(t)\right\}.$$

Recall that the gradient of the mean square error is zero for the desired solution, i.e.,

$$\frac{\partial J_{\mathrm{f}}}{\partial \underset{\rightarrow}{\boldsymbol{h}}} = \underset{\Rightarrow}{\boldsymbol{R}}\underset{\rightarrow}{\boldsymbol{h}} = \mathbf{0},$$

with which (6.77) can be written as

$$E\left\{\delta\underset{\rightarrow}{\boldsymbol{h}}(t+1)\right\} = \left(\mathbf{I}_{NL \times NL} - \mu_{\mathrm{f}} \underset{\Rightarrow}{\boldsymbol{R}}\right) E\left\{\delta\underset{\rightarrow}{\boldsymbol{h}}(t)\right\}. \tag{6.78}$$

Therefore, if the step size satisfies

$$0 < \mu_{\mathrm{f}} < \frac{2}{\lambda_{\mathrm{max}}}, \tag{6.79}$$

where λ_{max} is the largest eigenvalue of the matrix $\underset{\Rightarrow}{\boldsymbol{R}}$, then the misalignment mean $E\left\{\delta\underset{\rightarrow}{\boldsymbol{h}}(t)\right\}$ converges to zero and the estimated filter coefficients converge in the mean to the desired solutions.

Frequency-Domain Normalized Multichannel LMS Algorithm

The *unnormalized* frequency-domain MCLMS algorithm certainly is computationally efficient. But its convergence is still unsatisfactorily slow because of nonuniform convergence rates of the filter coefficients and cross-coupling between them. In this section, Newton's method will first be used and then necessary simplifications will be investigated to develop a frequency-domain *normalized* MCLMS algorithm for which the eigenvalue disparity is reduced and the convergence is accelerated.

To apply Newton's method, we need to evaluate the Hessian matrix of $J_{\mathrm{f}}(t+1)$ with respect to the filter coefficients, which can be computed by taking the row gradient of (6.69)

$$\begin{aligned} \underline{\underline{\boldsymbol{T}}}_n(t+1) &= \frac{\partial}{\partial \underline{\hat{\boldsymbol{h}}}_n^T(t)} \left[\frac{\partial J_{\mathrm{f}}(t+1)}{\partial \underline{\hat{\boldsymbol{h}}}_n^*(t)} \right] = \frac{\partial}{\partial \underline{\hat{\boldsymbol{h}}}_n^T(t)} \left[\underline{\underline{\boldsymbol{S}}}^H(t+1)\underline{\boldsymbol{e}}_n(t+1) \right] \\ &= \underline{\underline{\boldsymbol{S}}}^H(t+1)\frac{\partial \underline{\boldsymbol{e}}_n(t+1)}{\partial \underline{\hat{\boldsymbol{h}}}_n(t)} \\ &= \sum_{i=1,i\neq n}^{N} \underline{\underline{\boldsymbol{S}}}_{x_i}^H(t+1)\underline{\underline{\boldsymbol{S}}}_{x_i}(t+1). \end{aligned} \quad (6.80)$$

Then the filter coefficients will be updated as follows

$$\begin{aligned} &\underline{\hat{\boldsymbol{h}}}_n(t+1) \\ &= \underline{\hat{\boldsymbol{h}}}_n(t) - \rho_{\mathrm{f}} \underline{\underline{\boldsymbol{T}}}_n^{-1}(t+1)\underline{\underline{\boldsymbol{S}}}^H(t+1)\underline{\boldsymbol{e}}_n(t+1) \\ &= \underline{\hat{\boldsymbol{h}}}_n(t) - \rho_{\mathrm{f}} \underline{\underline{\boldsymbol{T}}}_n^{-1}(t+1)\, \underline{\underline{\boldsymbol{W}}}_{L\times 2L}^{10} \sum_{i=1}^{N} \underline{\underline{\boldsymbol{D}}}_{x_i}^*(t+1)\, \underline{\underline{\boldsymbol{W}}}_{2L\times L}^{01}\underline{\boldsymbol{e}}_{in}(t+1), \end{aligned} \quad (6.81)$$

where ρ_{f} is a new step size. Multiplying (6.81) from the left side by $\underline{\underline{\boldsymbol{W}}}_{2L\times L}^{10}$ yields

$$\begin{aligned} &\underline{\hat{\boldsymbol{h}}}_n^{10}(t+1) = \\ &\underline{\hat{\boldsymbol{h}}}_n^{10}(t) - \rho_{\mathrm{f}}\, \underline{\underline{\boldsymbol{W}}}_{2L\times L}^{10} \underline{\underline{\boldsymbol{T}}}_n^{-1}(t+1)\, \underline{\underline{\boldsymbol{W}}}_{L\times 2L}^{10} \sum_{i=1}^{N} \underline{\underline{\boldsymbol{D}}}_{x_i}^*(t+1)\underline{\boldsymbol{e}}_{in}^{01}(t+1), \end{aligned} \quad (6.82)$$

where the symbols

$$\begin{aligned} \underline{\hat{\boldsymbol{h}}}_n^{10}(t) &= \underline{\underline{\boldsymbol{W}}}_{2L\times L}^{10}\underline{\hat{\boldsymbol{h}}}_n(t) = \mathbf{F}_{2L\times 2L} \left[\begin{array}{cc} \hat{\mathbf{h}}_n^T(t) & \mathbf{0}_{1\times L} \end{array} \right]^T, \\ \underline{\boldsymbol{e}}_{in}^{01}(t+1) &= \underline{\underline{\boldsymbol{W}}}_{2L\times L}^{01}\underline{\boldsymbol{e}}_{in}(t+1) = \mathbf{F}_{2L\times 2L} \left[\begin{array}{cc} \mathbf{0}_{1\times L} & \mathbf{e}_{in}^T(t+1) \end{array} \right]^T, \end{aligned}$$

have been used in Table 6.4.

The frequency-domain Newton's algorithm is able to converge quickly. But unfortunately, the matrix $\underline{\underline{\boldsymbol{T}}}_n(t+1)$ is not diagonal and finding its inverse is computationally demanding, particularly when L is large. It is desirable to reduce the complexity by approximation, which leads to the normalized frequency-domain MCLMS algorithm.

Let us expand the product in (6.80) using the definition (6.67) to find:

$$\begin{aligned}\underline{\boldsymbol{S}}_{x_i}^H(t+1)\underline{\boldsymbol{S}}_{x_i}(t+1) &= \\ &\underline{\boldsymbol{W}}_{L\times 2L}^{10}\underline{\boldsymbol{D}}_{x_i}^*(t+1)\,\underline{\boldsymbol{W}}_{2L\times 2L}^{01}\underline{\boldsymbol{D}}_{x_i}(t+1)\,\underline{\boldsymbol{W}}_{2L\times L}^{10},\end{aligned} \tag{6.83}$$

where

$$\begin{aligned}\underline{\boldsymbol{W}}_{2L\times 2L}^{01} &\triangleq \underline{\boldsymbol{W}}_{2L\times L}^{01}\,\underline{\boldsymbol{W}}_{L\times 2L}^{01} \\ &= \mathbf{F}_{2L\times 2L}\begin{bmatrix}\mathbf{0}_{L\times L} & \mathbf{0}_{L\times L}\\ \mathbf{0}_{L\times L} & \mathbf{I}_{L\times L}\end{bmatrix}\mathbf{F}_{2L\times 2L}^{-1},\end{aligned}$$

whose elements on its main diagonal dominate as shown in [34]. When L is large, $2\,\underline{\boldsymbol{W}}_{2L\times 2L}^{01}$ can be well approximated by the identity matrix

$$2\,\underline{\boldsymbol{W}}_{2L\times 2L}^{01} \approx \mathbf{I}_{2L\times 2L}. \tag{6.84}$$

Therefore, (6.83) becomes

$$\underline{\boldsymbol{S}}_{x_i}^H(t+1)\underline{\boldsymbol{S}}_{x_i}(t+1) \approx \frac{1}{2}\,\underline{\boldsymbol{W}}_{L\times 2L}^{10}\underline{\boldsymbol{D}}_{x_i}^*(t+1)\underline{\boldsymbol{D}}_{x_i}(t+1)\,\underline{\boldsymbol{W}}_{2L\times L}^{10}, \tag{6.85}$$

and (6.80) turns into

$$\begin{aligned}\underline{\boldsymbol{T}}_n(t+1) &\approx \frac{1}{2}\,\underline{\boldsymbol{W}}_{L\times 2L}^{10}\left[\sum_{i=1,i\neq n}^{N}\underline{\boldsymbol{P}}_{x_i}(t+1)\right]\underline{\boldsymbol{W}}_{2L\times L}^{10} \\ &= \frac{1}{2}\,\underline{\boldsymbol{W}}_{L\times 2L}^{10}\underline{\boldsymbol{P}}_{\not n}(t+1)\,\underline{\boldsymbol{W}}_{2L\times L}^{10},\end{aligned} \tag{6.86}$$

where

$$\begin{aligned}\underline{\boldsymbol{P}}_{x_i}(t+1) &= \underline{\boldsymbol{D}}_{x_i}^*(t+1)\underline{\boldsymbol{D}}_{x_i}(t+1), \\ \underline{\boldsymbol{P}}_{\not n}(t+1) &= \sum_{i=1,i\neq n}^{N}\underline{\boldsymbol{P}}_{x_i}(t+1), \quad n = 1, 2, \cdots, N,\end{aligned}$$

which are diagonal matrices.

There is a useful relation between the inverses of the $\underline{\boldsymbol{T}}_n(t+1)$ and $\underline{\boldsymbol{P}}_{\not n}(t+1)$ matrices given by [34]

$$\underline{\boldsymbol{W}}_{2L\times L}^{10}\underline{\boldsymbol{T}}_n^{-1}(t+1)\,\underline{\boldsymbol{W}}_{L\times 2L}^{10} = 2\,\underline{\boldsymbol{W}}_{2L\times 2L}^{10}\underline{\boldsymbol{P}}_{\not n}^{-1}(t+1), \tag{6.87}$$

where

$$\underline{\boldsymbol{W}}_{2L\times 2L}^{10} = \underline{\boldsymbol{W}}_{2L\times L}^{10}\,\underline{\boldsymbol{W}}_{L\times 2L}^{10}.$$

This relation can be justified by post-multiplying both sides of the expression by $\underset{\Rightarrow}{P}_{\not{y}}(t+1)\,\underset{\Rightarrow}{W}^{10}_{2L\times L}$, using (6.86), and recognizing that

$$\underset{\Rightarrow}{W}^{10}_{2L\times 2L}\,\underset{\Rightarrow}{W}^{10}_{2L\times L} = \underset{\Rightarrow}{W}^{10}_{2L\times L}.$$

Substituting (6.87) into (6.82) produces

$$\hat{\underset{\rightarrow}{h}}{}^{10}_{n}(t+1) = \hat{\underset{\rightarrow}{h}}{}^{10}_{n}(t) - 2\rho_{\mathrm{f}}\,\underset{\Rightarrow}{W}^{10}_{2L\times 2L}\underset{\Rightarrow}{P}^{-1}_{\not{y}}(t+1)\sum_{i=1}^{N}\underset{\Rightarrow}{D}^{*}_{x_i}(t+1)\underset{\rightarrow}{e}^{01}_{in}(t+1). \quad (6.88)$$

If the matrix $2\,\underset{\Rightarrow}{W}^{10}_{2L\times 2L}$ is approximated by the identity matrix similar to (6.84), we finally deduce the the frequency-domain *normalized* MCLMS (FN-MCLMS) algorithm:

$$\hat{\underset{\rightarrow}{h}}{}^{10}_{n}(t+1) = \hat{\underset{\rightarrow}{h}}{}^{10}_{n}(t) - \rho_{\mathrm{f}}\underset{\Rightarrow}{P}^{-1}_{\not{y}}(t+1)\sum_{i=1}^{N}\underset{\Rightarrow}{D}^{*}_{x_i}(t+1)\underset{\rightarrow}{e}^{01}_{in}(t+1), \quad (6.89)$$
$$n = 1, 2, \cdots, N.$$

An interesting interpretation follows if we compare the frequency-domain normalized (6.89) and unnormalized (6.59) MCLMS algorithms. In the frequency-domain unnormalized MCLMS algorithm, the correction that is applied to $\underset{\rightarrow}{h}_n(t)$ is approximately proportional to the power spectrum $\underset{\Rightarrow}{P}_{\not{y}}(t+1)$ of the multiple channel outputs; this can be seen from (6.59) and the definition of $\underset{\Rightarrow}{P}_{\not{y}}(t+1)$, making the approximation $2\,\underset{\Rightarrow}{W}^{01}_{2L\times 2L} \approx \mathbf{I}_{2L\times 2L}$. When the magnitudes of the channel outputs are large, the gradient noise amplification can be evident. With the normalization of the correction by $\underset{\Rightarrow}{P}_{\not{y}}(t+1)$ as performed in the normalized FMCLMS algorithm, this noise amplification problem is diminished and the variability of the convergence rates due to the change of signal level is eliminated. In order to estimate a more stable power spectrum, a recursive scheme is employed in practice:

$$\underset{\Rightarrow}{P}_{\not{y}}(t+1) = \lambda_{\mathrm{p}}\underset{\Rightarrow}{P}_{\not{y}}(t) + (1-\lambda_{\mathrm{p}})\sum_{i=1, i\neq n}^{N}\underset{\Rightarrow}{D}^{*}_{x_i}(t+1)\underset{\Rightarrow}{D}_{x_i}(t+1), \quad (6.90)$$
$$n = 1, 2, \cdots, N,$$

where λ_{p} is a forgetting factor that may appropriately be set as

$$\lambda_{\mathrm{p}} = \left[1 - \frac{1}{3L}\right]^{L}$$

for this FNMCLMS algorithm. Although the FNMCLMS algorithm overcomes the problem of noise amplification, we are now faced with a similar problem that occurs when the channel outputs becomes too small. An alternative, therefore, is to insert a positive number δ into the normalization which leads to the following modification to the FNMCLMS algorithm:

$$\hat{\underline{h}}_n^{10}(t+1) = \hat{\underline{h}}_n^{10}(t) - \rho_{\mathrm{f}} \left[\underline{\underline{P}}_{\not{y}}(t+1) + \delta \mathbf{I}_{2L \times 2L} \right]^{-1} \sum_{i=1}^{N} \underline{\underline{D}}_{x_i}^{*}(t+1) \underline{e}_{in}^{01}(t+1), \quad (6.91)$$

$$n = 1, 2, \cdots, N.$$

From a computational point of view, the modified FNMCLMS algorithm can be easily implemented even for a real-time application since the normalization matrix $\underline{\underline{P}}_{\not{y}}(t+1) + \delta \mathbf{I}_{2L \times 2L}$ is diagonal and it is straightforward to find its inverse.

The FNMCLMS algorithm is summarized in Table 6.5.

6.2.5 Adaptive Multichannel Exponentiated Gradient Algorithm

We have developed in previous sections a number of adaptive algorithms for blind identification of SIMO FIR systems. As a matter of fact, all these algorithms are based on the classical stochastic gradient and their update rules do not discriminate against the model filter coefficients with significantly distinct magnitudes. Intuitively, however, it seems possible to develop a better adaptive blind SIMO identification algorithm by taking advantage of sparseness in acoustic impulse responses.

It has only very recently been recognized that sparseness of an impulse response can be exploited in adaptive algorithms to accelerate their initial convergence and improve their tracking performance. The proportionate normalized LMS (PNLMS) proposed by Duttweiler [102] was perhaps one of the first such algorithms, which was introduced for network echo cancellation. It was shown in [45] that the PNLMS is an approximation of the so-called exponentiated gradient algorithm with positive and negative weights (EG$\pm$ algorithm). The EG$\pm$ was proposed by Kivinen and Warmuth in the context of computational learning theory [204]. In [164], a general expression of the mean squared error (MSE) was derived for the EG$\pm$ algorithm illustrating that for sparse impulse responses, the EG$\pm$ algorithm, like the PNLMS, converges more quickly than the LMS for a given asymptotic MSE. Both the PNLMS and EG$\pm$ algorithms were developed for identifying a single channel with reference signals (see Chap. 4). In this section, we will investigate how to apply the concept of exponentiated gradient for blind identification of sparse acoustic SIMO systems and introduce the multichannel EG$\pm$ (MCEG$\pm$) algorithm [44]. Even though the constraint of the MCEG$\pm$ algorithm may be difficult or even impossible to determine in practice (this is why we put the study at the end of the discussion of this subject), we believe that it is a promising research topic and it is worth more effort on this front in the future.

Table 6.5. The frequency-domain normalized multichannel LMS (FNMCLMS) adaptive algorithm for the blind identification of a SIMO FIR system.

Parameters: $\hat{\mathbf{h}} = \left[\begin{array}{cccc} \hat{\mathbf{h}}_1^T & \hat{\mathbf{h}}_2^T & \cdots & \hat{\mathbf{h}}_N^T \end{array} \right]^T$, model filter

$\rho_{\mathrm{f}} > 0$, step size

$\lambda_{\mathrm{p}} = [1 - 1/(3L)]^L$, forgetting factor

$\delta > 0$, regularization factor

Initialization: $\hat{\mathbf{h}}_n(0) = [\, 1 \;\; 0 \;\; \cdots \;\; 0 \,]^T$, $n = 1, 2, \cdots, N$

$\hat{\mathbf{h}}(0) = \hat{\mathbf{h}}(0)/\sqrt{N}$ (normalization)

Computation: For $t = 0, 1, 2, \cdots$, compute

(a) $\underrightarrow{\hat{\boldsymbol{h}}}_n^{10}(t) = \mathrm{FFT}_{2L}\left\{ \left[\begin{array}{cc} \hat{\mathbf{h}}_n^T(t) & \mathbf{0}_{1\times L} \end{array} \right]^T \right\}, \quad n = 1, 2, \cdots, N$

(b) $\mathbf{x}_n(t+1)_{2L\times 1} = [\, x_n(tL) \;\; x_n(tL+1) \;\; \cdots \;\; x_n(tL+2L-1) \,]^T$

(c) $\underrightarrow{\boldsymbol{x}}_n(t+1)_{2L\times 1} = \mathrm{FFT}_{2L}\{\mathbf{x}_n(t+1)_{2L\times 1}\}$

(d) $\underrightarrow{\boldsymbol{e}}_{ij}(t+1)_{2L\times 1} = \underrightarrow{\boldsymbol{x}}_i(t+1)_{2L\times 1} \odot \underrightarrow{\hat{\boldsymbol{h}}}_j^{10}(t) -$

$\underrightarrow{\boldsymbol{x}}_j(t+1)_{2L\times 1} \odot \underrightarrow{\hat{\boldsymbol{h}}}_i^{10}(t), \quad i, j = 1, 2, \cdots, N$

(e) $\mathbf{e}_{ij}(t+1)_{2L\times 1} = \mathrm{IFFT}_{2L}\{\underrightarrow{\boldsymbol{e}}_{ij}(t+1)_{2L\times 1}\}$

(f) Take the *last* L elements of $\mathbf{e}_{ij}(t+1)_{2L\times 1}$ to form $\mathbf{e}_{ij}(t+1)$

(g) $\underrightarrow{\boldsymbol{e}}_{ij}^{01}(t+1) = \mathrm{FFT}_{2L}\left\{ \left[\begin{array}{cc} \mathbf{0}_{1\times L} & \mathbf{e}_{ij}^T(t+1) \end{array} \right]^T \right\}$

(h) $\underrightarrow{\tilde{\boldsymbol{p}}}_{\not{n}}(t+1) = \sum_{i=1, i\neq n}^{N} \underrightarrow{\boldsymbol{x}}_i^*(t+1)_{2L\times 1} \odot \underrightarrow{\boldsymbol{x}}_i(t+1)_{2L\times 1}$

(i) $\underrightarrow{\boldsymbol{p}}_{\not{n}}(t+1) = \begin{cases} \underrightarrow{\tilde{\boldsymbol{p}}}_{\not{n}}(t+1), & t = 0 \\ \lambda_{\mathrm{p}} \underrightarrow{\boldsymbol{p}}_{\not{n}}(t) + (1-\lambda_{\mathrm{p}}) \underrightarrow{\tilde{\boldsymbol{p}}}_{\not{n}}(t+1), & t > 0 \end{cases}$

(j) Take the element-by-element reciprocal of

$\left[\underrightarrow{\boldsymbol{p}}_{\not{n}}(t+1) + \delta \mathbf{1}_{2L\times 1} \right]$ to form $\underrightarrow{\boldsymbol{p}}_{\not{n}}^{-1}(t+1)$

(k) $\Delta \underrightarrow{\hat{\boldsymbol{h}}}_n(t)_{2L\times 1} = \rho_{\mathrm{f}} \underrightarrow{\boldsymbol{p}}_{\not{n}}^{-1}(t+1) \odot$

$\left[\sum_{i=1}^{N} \underrightarrow{\boldsymbol{x}}_i^*(t+1)_{2L\times 1} \odot \underrightarrow{\boldsymbol{e}}_{in}^{01}(t+1) \right].$

(l) $\Delta \hat{\mathbf{h}}_n(t)_{2L\times 1} = \mathrm{IFFT}_{2L}\left\{ \Delta \underrightarrow{\hat{\boldsymbol{h}}}_n(t)_{2L\times 1} \right\}$

(m) Take the *first* L elements of $\Delta \hat{\mathbf{h}}_n(t)_{2L\times 1}$ to form $\Delta \hat{\mathbf{h}}_n(t)$

(n) $\hat{\mathbf{h}}_n(t+1) = \frac{\hat{\mathbf{h}}_n(t) - \Delta \hat{\mathbf{h}}_n(t)}{\left\| \hat{\mathbf{h}}_n(t) - \Delta \hat{\mathbf{h}}_n(t) \right\|}, \quad n = 1, 2, \cdots, N$

We begin with the introduction of a new way to develop adaptive algorithms for blind SIMO identification, which is different from the gradient descent perspective described in Sect. 6.2.2. In addition to the unconstrained, *a priori* error signal (6.9), we define an *a posteriori* error:

$$\varepsilon_{ij}(k+1) = \mathbf{x}_i^T(k+1)\hat{\mathbf{h}}_j(k+1) - \mathbf{x}_j^T(k+1)\hat{\mathbf{h}}_j(k+1), \quad i,j = 1,2,\cdots,N. \tag{6.92}$$

Then an adaptive algorithm that adjusts the new model filter $\hat{\mathbf{h}}(k+1)$ from the old one $\hat{\mathbf{h}}(k)$ can be derived by minimizing the following function

$$J\left[\hat{\mathbf{h}}(k+1)\right] = d\left[\hat{\mathbf{h}}(k+1), \hat{\mathbf{h}}(k)\right] + \eta \sum_{i=1}^{N-1} \sum_{j=i+1}^{N} \varepsilon_{ij}^2(k+1), \tag{6.93}$$

where $d[\hat{\mathbf{h}}(k+1), \hat{\mathbf{h}}(k)]$ is a measure of distance from the old to the new model filter, and η is a positive constant. The magnitude of η represents the importance of correctiveness compared to the importance of conservativeness [204]. If η is very small, minimizing $J[\hat{\mathbf{h}}(k+1)]$ is close to minimizing $d[\hat{\mathbf{h}}(k+1), \hat{\mathbf{h}}(k)]$, so that the algorithm makes very small updates. On the other hand, if η is very large, the minimization of $J[\hat{\mathbf{h}}(k+1)]$ is almost equivalent to minimizing $d[\hat{\mathbf{h}}(k+1), \hat{\mathbf{h}}(k)]$ subject to the constraint $\varepsilon_{ij}(k+1) = 0, \forall i,j$.

To minimizing $J[\hat{\mathbf{h}}(k+1)]$, we need to set the partial derivative $\partial J[\hat{\mathbf{h}}(k+1)]/\partial\hat{\mathbf{h}}(k+1)$ to zero. Hence, the vector of channel impulse responses $\hat{\mathbf{h}}(k+1)$ will be found by solving the equation:

$$\frac{\partial d\left[\hat{\mathbf{h}}(k+1), \hat{\mathbf{h}}(k)\right]}{\partial \hat{\mathbf{h}}(k+1)} + \eta \frac{\partial}{\partial \hat{\mathbf{h}}(k+1)} \left[\sum_{i=1}^{N-1} \sum_{j=i+1}^{N} \varepsilon_{ij}^2(k+1)\right] = \mathbf{0}. \tag{6.94}$$

Similar to (6.20), it can be shown that

$$\frac{\partial}{\partial \hat{\mathbf{h}}(k+1)} \left[\sum_{i=1}^{N-1} \sum_{j=i+1}^{N} \varepsilon_{ij}^2(k+1)\right] = 2\tilde{\mathbf{R}}_{x+}(k+1)\hat{\mathbf{h}}(k+1). \tag{6.95}$$

Then (6.94) becomes

$$\frac{\partial d\left[\hat{\mathbf{h}}(k+1), \hat{\mathbf{h}}(k)\right]}{\partial \hat{\mathbf{h}}(k+1)} + 2\eta\tilde{\mathbf{R}}_{x+}(k+1)\hat{\mathbf{h}}(k+1) = \mathbf{0}. \tag{6.96}$$

Solving (6.96) is in general very difficult. However, if the new weight vector $\hat{\mathbf{h}}(k+1)$ is close to the old weight vector $\hat{\mathbf{h}}(k)$, replacing $\hat{\mathbf{h}}(k+1)$ in the second item on the left-hand side of (6.96) with $\hat{\mathbf{h}}(k)$ is a reasonable approximation and the equation

$$\frac{\partial d\left[\hat{\mathbf{h}}(k+1), \hat{\mathbf{h}}(k)\right]}{\partial \hat{\mathbf{h}}(k+1)} + 2\eta\tilde{\mathbf{R}}_{x+}(k+1)\hat{\mathbf{h}}(k) = \mathbf{0} \tag{6.97}$$

is much easier to solve for all distance measures d.

The unconstrained multichannel LMS algorithm is easily obtained from (6.97) by using the squared Euclidean distance

$$d_{\mathrm{E}}\left[\hat{\mathbf{h}}(k+1), \hat{\mathbf{h}}(k)\right] = \left\|\hat{\mathbf{h}}(k+1) - \hat{\mathbf{h}}(k)\right\|_2^2 . \tag{6.98}$$

If the unconstrained *a posteriori* error in (6.93) is replaced by a *constrained a posteriori* error signal, then the constrained multichannel LMS algorithm (6.23) can be deduced.

The exponentiated gradient (EG) algorithm with *positive* weights results from using for d the *relative entropy*, also known as *Kullback-Leibler divergence*,

$$d_{\mathrm{re}}\left[\hat{\mathbf{h}}(k+1), \hat{\mathbf{h}}(k)\right] = \sum_{l=0}^{NL-1} \hat{h}_l(k+1) \ln \frac{\hat{h}_l(k+1)}{\hat{h}_l(k)}, \tag{6.99}$$

with the constraint $\sum_{l=0}^{NL-1} \hat{h}_l(k+1) = u$, where $\hat{h}_l(k+1)$ is the lth ($l = 0, 1, \cdots, NL-1$) tap of the model filter $\hat{\mathbf{h}}(k+1)$ at time $k+1$ and u is a positive constant (note that this is different from the unit-norm constraint used in the constrained multichannel LMS algorithm). We slightly modify the function (6.93) as follows

$$J\left[\hat{\mathbf{h}}(k+1)\right] = d\left[\hat{\mathbf{h}}(k+1), \hat{\mathbf{h}}(k)\right] + \frac{\eta}{u} \sum_{i=1}^{N-1} \sum_{j=i+1}^{N} \varepsilon_{ij}^2(k+1). \tag{6.100}$$

As a result, (6.97) becomes

$$\frac{\partial d_{\mathrm{re}}\left[\hat{\mathbf{h}}(k+1), \hat{\mathbf{h}}(k)\right]}{\partial \hat{\mathbf{h}}(k+1)} + \frac{2\eta}{u} \tilde{\mathbf{R}}_{x+}(k+1)\hat{\mathbf{h}}(k) + \kappa \mathbf{1} = \mathbf{0}, \tag{6.101}$$

where κ is a Lagrange multiplier and $\mathbf{1} = [1\ 1\ \cdots\ 1]^T$ is a vector of ones. Substituting (6.99) into (6.101) and solving for $\hat{\mathbf{h}}(k+1)$ leads to the multichannel EG algorithm with positive weights (MCEG+):

$$\begin{aligned} \hat{h}_l(k+1) &= u \frac{\hat{h}_l(k) r_l(n+1)}{\sum_{i=0}^{NL-1} \hat{h}_i(k) r_i(k+1)}, \\ & \quad l = 0, 1, \cdots, NL-1, \end{aligned} \tag{6.102}$$

where

$$r_l(k+1) = \exp\left[-\frac{2\eta}{u} g_l(k+1)\right],$$

and $g_l(k+1)$ is the lth element of the gradient vector

$$\mathbf{g}(k+1) = \tilde{\mathbf{R}}_{x+}(k+1)\hat{\mathbf{h}}(k). \tag{6.103}$$

We can clearly see the motivation for the EG algorithms by ignoring the normalization in (6.102) and taking the logarithm. It shows that the log weights use the same update as the MCLMS algorithm. Alternatively, this can be interpreted as exponentiating the update, hence the name EG. This has the effect of assigning larger relative updates to larger weights, thereby deemphasizing the effect of smaller weights. This is qualitatively similar to the idea that leads to the PNLMS algorithm, which makes the update *proportional* to the magnitude of the weight. This type of behavior is desirable for sparse impulse responses where small weights do not contribute significantly to the *mean* solution but introduce an undesirable noise-like *variance*.

Since the MCEG+ algorithm works only for those SIMO systems with positive channel impulse responses, it is useless for blind identification of an acoustic SIMO system. For an acoustic SIMO system with both positive and negative filter coefficients, we can decompose the model filter $\hat{\mathbf{h}}(k)$ into two components $\hat{\mathbf{h}}^+(k)$ and $\hat{\mathbf{h}}^-(k)$ (both with positive coefficients) such that $\hat{\mathbf{h}}(k) = \hat{\mathbf{h}}^+(k) - \hat{\mathbf{h}}^-(k)$. Then the function (6.100) becomes:

$$\begin{aligned} & J\left[\hat{\mathbf{h}}^+(k+1), \hat{\mathbf{h}}^-(k+1)\right] \\ & = d_{\mathrm{re}}\left[\hat{\mathbf{h}}^+(k+1), \hat{\mathbf{h}}^+(k)\right] + d_{\mathrm{re}}\left[\hat{\mathbf{h}}^-(k+1), \hat{\mathbf{h}}^-(k)\right] \\ & + \frac{\eta}{u_1 + u_2} \sum_{i=1}^{N-1} \sum_{j=i+1}^{N} \varepsilon_{ij}^2(k+1), \end{aligned} \tag{6.104}$$

where u_1 and u_2 are two positive constants. Since $\hat{\mathbf{h}}(k+1) = \mathbf{0}$ is an undesired solution, it is necessary to ensure that $\hat{\mathbf{h}}^+(k+1)$ and $\hat{\mathbf{h}}^-(k+1)$ would not be equal to each other from initialization and throughout the process of adaptation. Among many methods that can be used to enforce that $\hat{\mathbf{h}}^+(k+1)$ and $\hat{\mathbf{h}}^-(k+1)$ would not be identical, we introduce the following constraint proposed in [44]:

$$\sum_{l=0}^{NL-1} \hat{h}_l^+(k+1) = u_1, \quad \sum_{l=0}^{NL-1} \hat{h}_l^-(k+1) = u_2, \tag{6.105}$$

where $u_1 \neq u_2$. Using the constraint and taking derivatives of (6.104) with respect to $\hat{\mathbf{h}}^+(k+1)$ and $\hat{\mathbf{h}}^-(k+1)$ respectively produces:

$$\ln \frac{\hat{h}_l^+(k+1)}{\hat{h}_l^+(k)} + 1 + \frac{2\eta}{u_1 + u_2} g_l(k+1) + \kappa_1 = 0, \tag{6.106}$$

$$\ln \frac{\hat{h}_l^-(k+1)}{\hat{h}_l^-(k)} + 1 - \frac{2\eta}{u_1 + u_2} g_l(k+1) + \kappa_2 = 0, \tag{6.107}$$

where κ_1 and κ_2 are two Lagrange multipliers. Solving (6.106) and (6.107) for $\hat{h}_l^+(k+1)$ and $\hat{h}_l^-(k+1)$ respectively leads to the MCEG$\pm$ algorithm:

Table 6.6. The multichannel exponentiated gradient algorithm with positive and negative weights (MCEG±) for the blind identification of a sparse SIMO FIR system.

Parameters:	$\hat{\mathbf{h}} = \left[\ \hat{\mathbf{h}}_1^T\ \hat{\mathbf{h}}_2^T\ \cdots\ \hat{\mathbf{h}}_N^T\ \right]^T = \hat{\mathbf{h}}^+ - \hat{\mathbf{h}}^-$, model filter
	$\eta > 0$, positive constant to balance correctiveness and conservativeness
	$u_1 > 0$, $u_2 > 0$, and $u_1 \neq u_2$
Initialization:	$\hat{\mathbf{h}}_n^+(0) = [\ u_1/N\ \ 0\ \cdots\ 0\ ]^T \quad n = 1, 2, \cdots, N$
	$\hat{\mathbf{h}}_n^-(0) = [\ u_2/N\ \ 0\ \cdots\ 0\ ]^T$
	$\hat{\mathbf{h}}(0) = \hat{\mathbf{h}}^+(0) - \hat{\mathbf{h}}^-(0)$
Computation:	For $k = 0, 1, 2, \cdots$, compute
	(a) Construct the matrix $\tilde{\mathbf{R}}_{x+}(k+1)$ following (6.22)
	(b) $\mathbf{g}(k+1) = \tilde{\mathbf{R}}_{x+}(k+1)\hat{\mathbf{h}}(k)$
	(c) $r_l^+(k+1) = \exp\left[-\frac{2\eta}{u_1+u_2} g_l(k+1)\right]$
	$r_l^-(k+1) = \frac{1}{r_l^+(k+1)}, \quad l = 0, 1, \cdots, NL-1$
	(d) $\hat{\mathbf{h}}^+(k+1) = \frac{u_1\left[\hat{\mathbf{h}}^+(k) \odot \mathbf{r}^+(k+1)\right]}{\left[\hat{\mathbf{h}}^+(k)\right]^T \mathbf{r}^+(k+1)}$
	$\hat{\mathbf{h}}^-(k+1) = \frac{u_2\left[\hat{\mathbf{h}}^-(k) \odot \mathbf{r}^-(k+1)\right]}{\left[\hat{\mathbf{h}}^-(k)\right]^T \mathbf{r}^-(k+1)}$
	(e) $\hat{\mathbf{h}}(k+1) = \hat{\mathbf{h}}^+(k+1) - \hat{\mathbf{h}}^-(k+1)$

$$\hat{h}_l^+(k+1) = u_1 \frac{\hat{h}_l^+(k) r_l^+(k+1)}{\sum_{i=0}^{NL-1} \hat{h}_i^+(k) r_i^+(k+1)}, \tag{6.108}$$

$$\hat{h}_l^-(k+1) = u_2 \frac{\hat{h}_l^-(k) r_l^-(k+1)}{\sum_{i=0}^{NL-1} \hat{h}_i^-(k) r_i^-(k+1)}, \tag{6.109}$$

where

$$\begin{aligned} r_l^+(k+1) &= \exp\left[-\frac{2\eta}{u_1+u_2} g_l(k+1)\right], \\ r_l^-(k+1) &= \exp\left[\frac{2\eta}{u_1+u_2} g_l(k+1)\right] \\ &= \frac{1}{r_l^+(k+1)}. \end{aligned}$$

The MCEG± algorithm is summarized in Table 6.6.

Before wrapping up the discussion of the MCEG± algorithm, we would like to briefly comment on how to properly choose u_1 and u_2 in (6.105). The constraint looks simple but, in practice, determining a proper set of u_1 and u_2 is actually quite difficult. Let $\mathbf{h}^{+0}$ be the non-negative component of $\mathbf{h}$, i.e.,

$$h_l^{+0} \triangleq \begin{cases} h_l, & h_l \geq 0 \\ 0, & h_l < 0 \end{cases}, \quad l = 0, 1, \cdots, NL-1, \tag{6.110}$$

and

$$\mathbf{h}^{-0} = \mathbf{h}^{+0} - \mathbf{h}. \tag{6.111}$$

Then a decomposition of $\mathbf{h}$ can be expressed as

$$\mathbf{h}^{+} = \mathbf{h}^{+0} + \mathbf{c}, \quad \mathbf{h}^{-} = \mathbf{h}^{-0} + \mathbf{c}, \tag{6.112}$$

where $\mathbf{c}$ is a constant vector. Since the coefficients of $\mathbf{h}^{+}$ and $\mathbf{h}^{-}$ are non-negative, $\mathbf{c}$ needs to satisfy

$$c_l \geq \max\left(-h_l^{+0}, -h_l^{-0}\right), \quad l = 0, 1, \cdots, NL-1. \tag{6.113}$$

Therefore, the following values for u_1 and u_2 are all valid:

$$u_1 = \sum_{l=0}^{NL-1} h_l^{+} = \sum_{l=0}^{NL-1} h_l^{+0} + \sum_{l=0}^{NL-1} c_l, \tag{6.114}$$

$$u_2 = \sum_{l=0}^{NL-1} h_l^{-} = \sum_{l=0}^{NL-1} h_l^{-0} + \sum_{l=0}^{NL-1} c_l. \tag{6.115}$$

6.2.6 Numerical Examples

In this section, a number of numerical experiments will be presented to help the reader better understand the performance of the developed adaptive algorithms for blind SIMO identification.

Performance Measure

Different from non-blind system identification methods, blind SIMO identification algorithms determine the channel impulse responses up to a scale. As a result, the normalized projection misalignment (NPM) is widely used as a performance measure for assessing a blind SIMO identification algorithm. The NPM for an estimate $\hat{\mathbf{h}}(\mathrm{k})$ of $\mathbf{h}$ at time k is given by:

$$\mathrm{NPM}(k) \triangleq \frac{\|\varsigma(k)\|}{\|\mathbf{h}\|}, \tag{6.116}$$

where

$$\varsigma(k) = \mathbf{h} - \frac{\mathbf{h}^T\hat{\mathbf{h}}(k)}{\hat{\mathbf{h}}^T(k)\hat{\mathbf{h}}(k)}\hat{\mathbf{h}}(k)$$

is the projection misalignment vector. By projecting $\mathbf{h}$ onto $\hat{\mathbf{h}}(k)$ and defining a projection error, we take into account only the undesirable misalignment of the channel estimate, disregarding an arbitrary gain factor inherently associated with it [234].

Example 6.1. **Blind Identification of a Well-Conditioned Two-Channel SIMO FIR System**

We consider a two-channel SIMO FIR system whose impulse responses in all channels are second order and are given by:

$$\begin{aligned} \mathbf{h}_1 &= \begin{bmatrix} 1 & -2\cos(\theta) & 1 \end{bmatrix}^T, \\ \mathbf{h}_2 &= \begin{bmatrix} 1 & -2\cos(\theta+\vartheta) & 1 \end{bmatrix}^T, \end{aligned} \tag{6.117}$$

where θ is the absolute phase value of the zeros of the first channel and ϑ specifies the angular distance between the zeros of the two channels on the unit circle. The source is an uncorrelated binary phase-shift-keying (BPSK) sequence and the additive noise is i.i.d. zero-mean Gaussian at a specified signal-to-noise ratio (SNR) defined as follows:

$$\text{SNR} \triangleq 10\log_{10}\frac{\sigma_s^2\|\mathbf{h}\|^2}{N\sigma_b^2}, \tag{6.118}$$

where σ_s^2 and σ_b^2 are the signal and background noise powers, respectively. The SNR for this example is 40 dB.

Such a second-order two-channel system was first introduced in [170] and was then widely employed in [331], [10], [177], and [178] among others. This system allows us to investigate the robustness of a blind multichannel identification algorithm to the system identifiability, conceptually described by the parameter ϑ.

In this example, it is given that $\theta = \pi/10$ and $\vartheta = \pi$. The system is deemed well-conditioned in the sense that the zeros of its two channels are separated far away to each other and the system is definitely blindly identifiable.

For such a well-conditioned SIMO system, blind identification is relatively easy and we compare only the MCLMS and VSS-UMCLMS algorithms. Figure 6.2 shows the NPM of these two adaptive algorithms. For the MCLMS, a number of different step sizes were tried and two results with $\mu = 0.01$ and $\mu = 0.005$ are presented. Both algorithms can quickly converge to the true channel impulse responses and the final NPM is around -40 dB which is very accurate. It is clear that the VSS-UMCLMS has a faster initial convergence than the MCLMS algorithm. Moreover, a proper step size for the MCLMS (large enough while has not yet caused the adaptive algorithm to diverge) cannot be identified before several trial runs. However, the VSS-UMCLMS does not have this limitation, which makes it more practical and much easier to use.

Example 6.2. **Blind Identification of an Ill-Conditioned Two-Channel SIMO FIR System**

In this example, we work on an ill-conditioned two-channel SIMO FIR system. The system and its parameters are the same as those in *Example 6.1* except for ϑ. Here $\vartheta = \pi/10$ is given and the zeros of the two channels are

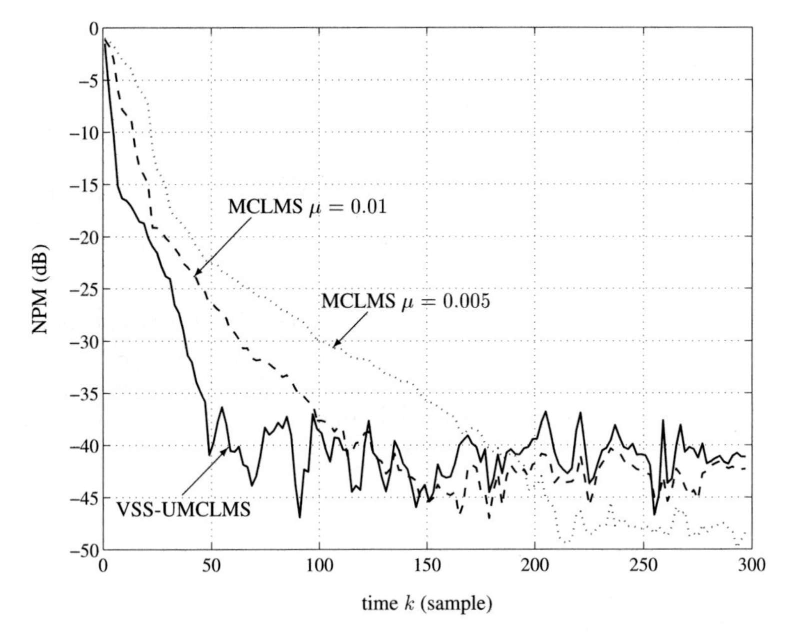

Fig. 6.2. Normalized projection misalignment of the MCLMS using step sizes of $\mu = 0.01$ (dashed line) and $\mu = 0.005$ (dotted line), and VSS-UMCLMS (solid line) algorithms for blind identification of a well-conditioned two-channel SIMO FIR system with additive white Gaussian noise at 40 dB SNR.

quite close, which is about to invalidate the identifiability assumption of no shared common zeros.

Three algorithms are used to blindly identify this system. They are the MCLMS, VSS-UMCLMS, and MCN algorithms. For the MCLMS, two step sizes were chosen: $\mu = 0.025$ and $\mu = 0.01$. For the MCN, the step size was determined as $\rho = 0.5$. Their learning curves in terms of NPM are plotted in Fig. 6.3. It is clear that the MCN converges much faster than the VSS-UMCLMS and MCLMS algorithms although its computational complexity is also significantly higher. The VSS-UMCLMS achieves a nice balance between complexity and convergence speed. The VSS-UMCLMS is always superior to the MCLMS while their complexities are still comparable.

Example 6.3. **Blind Identification of a Random Three-Channel SIMO FIR System**

The system in the previous two examples is quite simple. Let's study a more complicated three-channel SIMO system whose channel impulse responses of $L = 16$ samples long are randomly generated. The three channel impulse responses are shown in Fig. 6.4(a). Both the source signal and additive noise are white Gaussian processes and the SNR is 40 dB. Four adaptive algorithms were used: the MCLMS, VSS-UMCLMS, MCN, and FNMCLMS. The step sizes for the MCLMS, MCN, and FNMCLMS are $\mu = 0.0075$, $\rho = 0.5$,

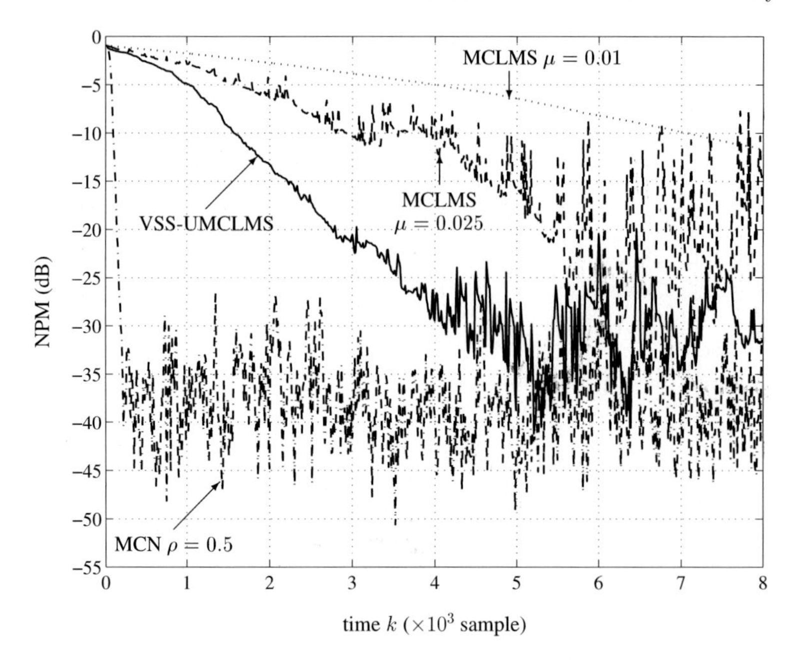

Fig. 6.3. Normalized projection misalignment of the MCLMS using step sizes of $\mu = 0.025$ (dashed line) and $\mu = 0.01$ (dotted line), VSS-UMCLMS (solid line), and MCN with a step size of $\rho = 0.5$ (dashdot line) for blind identification of an ill-conditioned two-channel SIMO FIR system with additive white Gaussian noise at 40 dB SNR.

and $\rho_f = 1.2$, respectively. Shown in Fig. 6.4(b) are the learning curves of these algorithms. Similar to what we observed in *Example 6.2*, the MCN converges much faster than the others. The VSS-UMCLMS and FNMCLMS algorithms have almost the same convergence with the VSS-UMCLMS just slightly better. Since the source signal is stationary and white, the FNMCLMS algorithm does not benefit from the normalization. Among all of these algorithms, the MCLMS has the slowest convergence.

Example 6.4. **Blind Identification of a Sparse Three-Channel SIMO FIR System**

In this example, we look at a sparse three-channel SIMO FIR system and examine the performance of the MCEG± algorithm. The channel impulse responses are randomly generated as what we did in *Example 6.3*. But several taps are magnified to make the channel impulse responses more sparse. These channel impulse responses are given in Fig. 6.5(a) and the sparseness measure (see Sect. 2.2.8 for its definition) is 0.331430. For such a system, we compare the MCEG± algorithm with the MCLMS and VSS-UMCLMS algorithms. For the MCLMS, the step size is $\mu = 0.0012$. For the MCEG±, $\eta = 0.45$, $u_1 = 306.30$, and $u_2 = 298.03$. The NPMs of the three adaptive algorithms are shown in Fig. 6.5(b). The initial convergence of the MCEG± algorithm

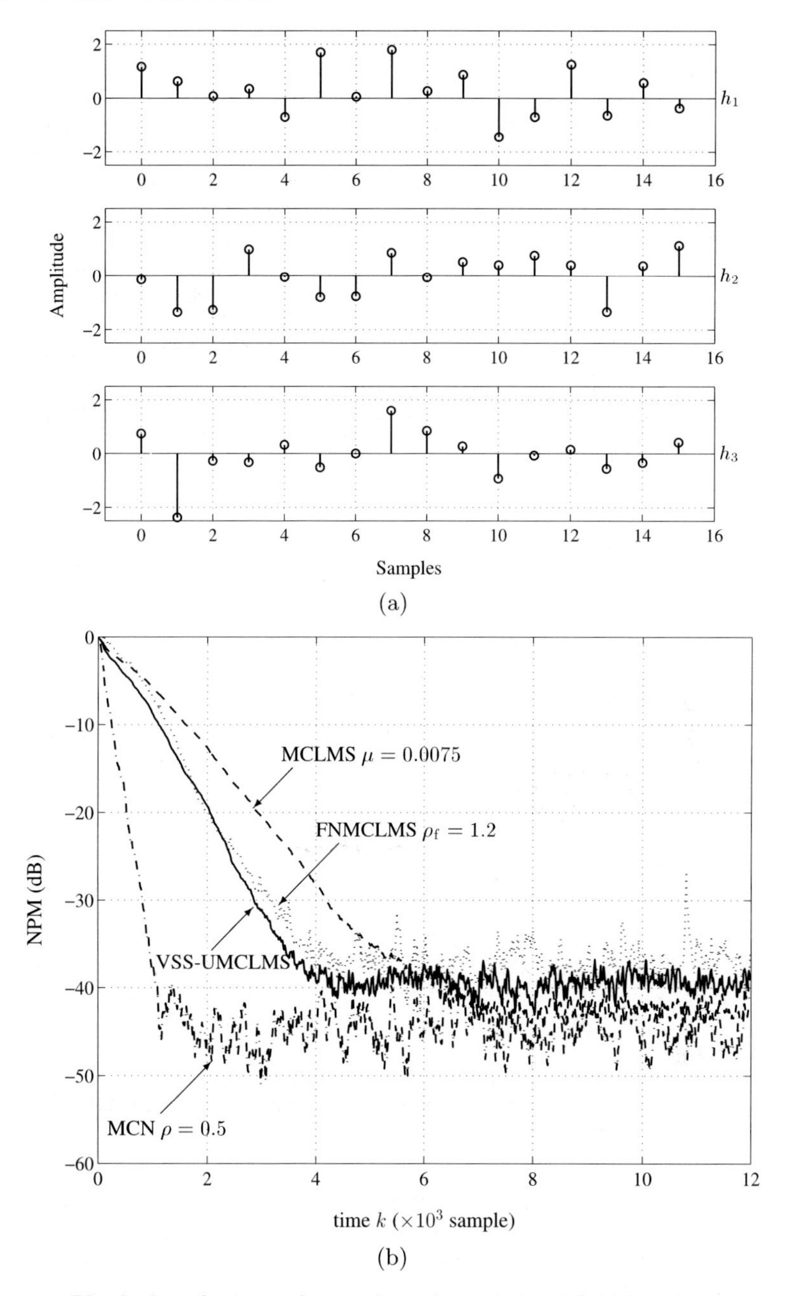

Fig. 6.4. Blind identification of a random three-channel SIMO FIR system with additive white Gaussian noise at 40 dB SNR as conducted in *Example 6.3.* (a) The channel impulse responses and (b) normalized projection misalignment of the MCLMS with a step size of $\mu = 0.0075$ (dashed line), VSS-UMCLMS (solid line), MCN with a step size of $\rho = 0.5$ (dashdot line) , and FNMCLMS with a step size of $\rho_{\mathrm{f}} = 1.2$ (dotted line).

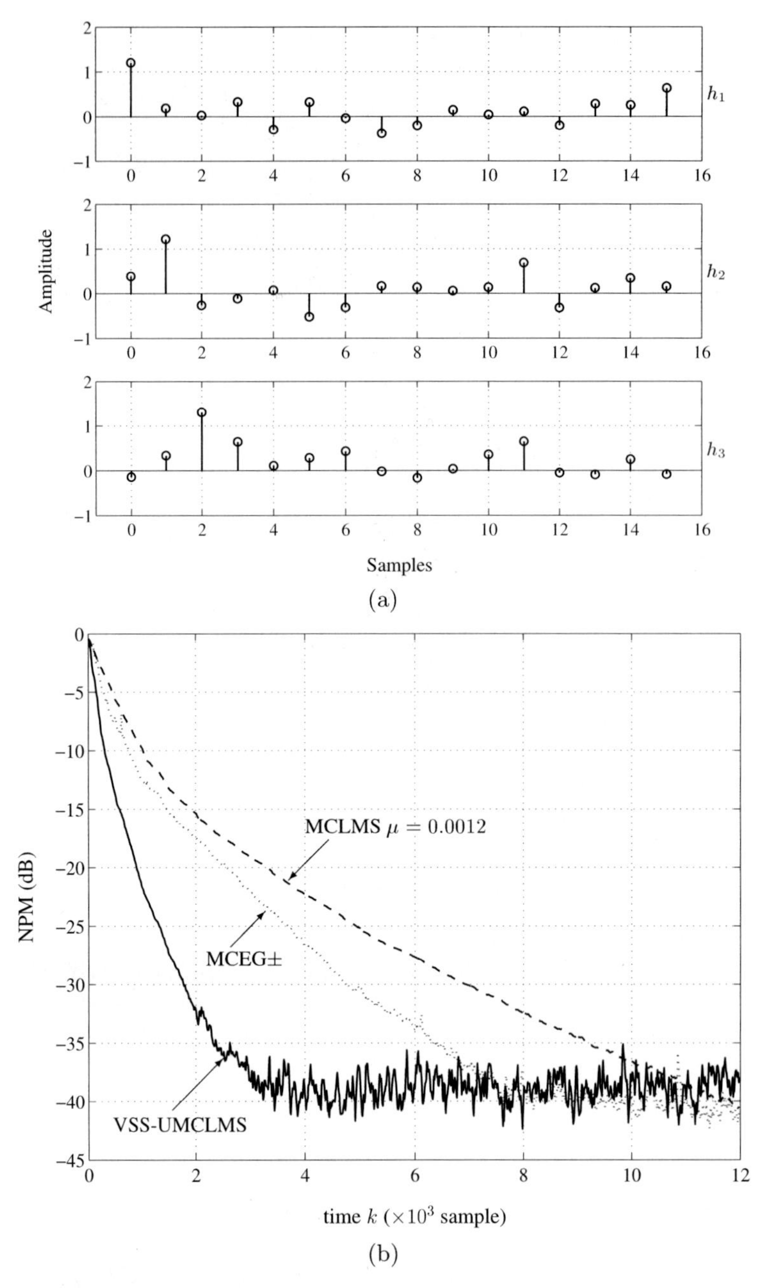

Fig. 6.5. Blind identification of a *sparse* three-channel SIMO FIR system with additive white Gaussian noise at 40 dB SNR as conducted in *Example 6.4*. (a) The channel impulse responses and (b) normalized projection misalignment of the MCLMS with a step size of $\mu = 0.0012$ (dashed line), VSS-UMCLMS (solid line), and MCEG± with $\eta = 0.45$ (dotted line).

is greater than that of the MCLMS algorithm for a given asymptotic mean square error as expected, while the VSS-UMCLMS performs much better than the other two investigated adaptive algorithms.

Example 6.5. **Blind Identification of an Acoustic Three-Channel SIMO FIR System with Speech as the Source Signal**

In this example, we investigate a real acoustic SIMO system. The channel impulse responses were measured by Härmä [156] in the varechoic chamber at Bell Labs (see Sect. 2.3.1 for a more detailed description of the chamber and measurement setup). From the 31 source (loudspeaker) positions and 22 microphones, one loudspeaker and three microphones are selected to form a 1×3 SIMO system. Their positions are illustrated in the floor plan of the varechoic chamber in Fig. 6.6(a). A male speech signal of 100 seconds long is used as the source and the first two-second sequence is plotted in Fig. 6.6(b). Since a speech signal is non-stationary and has silent periods, identifying an acoustic SIMO system driven by speech is extremely difficult. As a result, the SNR has to be high for this example: SNR = 75 dB.

The channel impulse responses were measured when 75% panels were open and the reverberation time of the chamber was about 310 ms (see Table 2.1). The measured long channel impulse responses are truncated at $L = 256$. Compared to the FNMCLMS, other adaptive algorithms are too computationally intensive. Therefore presented here is only the result of the FNMCLMS algorithm. The comparison of the true and estimated channel impulse responses is given in Fig. 6.7(a) and shown in Fig. 6.7(b) is the NPM versus time k for the FNMCLMS. Although the final NPM is only -11 dB, this result allows the FNMCLMS find an immediate application for time delay estimation as will be discussed in detail in Chap. 9.

6.3 Blind MIMO Identification

As clearly explained in the introduction of Sect. 6.1, our interest is in the SOS-based techniques for blind multichannel identification. Using SOS to blindly identify a system requires that the number of outputs should be greater than the number of inputs, implying that our discussion is with respect to only SIMO and MIMO systems. Blind identification of a SIMO system has been investigated in the previous section. In this section, we will address blind MIMO identification.

Blind identification of a MIMO FIR system with $N > M$ is not just more complicated than blind identification of a SIMO FIR system. Actually it is much more difficult or might be even unsolvable. The research on this problem remains at the stage of feasibility investigations. So far, there has been very few *practical* algorithm available. But blind MIMO identification is potentially a very important technique, particularly for acoustic signal processing. Any breakthroughs in this problem will immediately make a significant positive impact on the advancement of acoustic signal processing. Therefore it is

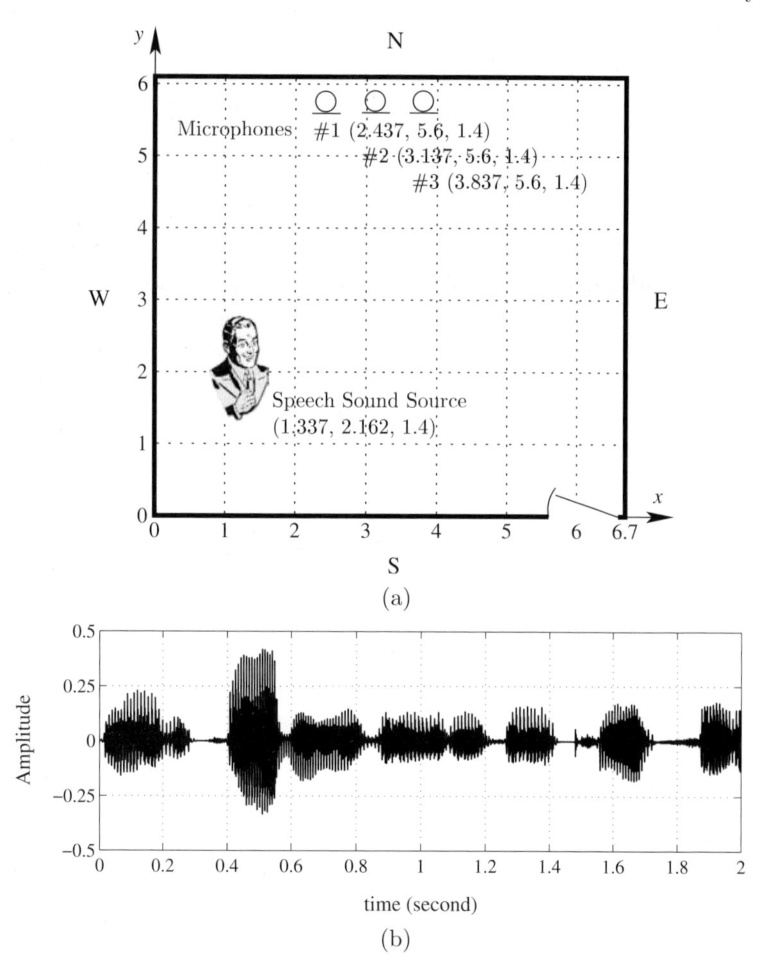

Fig. 6.6. (a) Floor plan of the varechoic chamber where the channel impulse responses used in *Example 6.5* were measured. (b) The first two-second sequence of the male speech source signal used in *Example 6.5*.

worthwhile to review the fundamentals and describe the state of the art of blind MIMO identification.

6.3.1 Problem Formulation and Background Review

We begin with a close examination of the SOS of the outputs of a MIMO system. From (2.9), we have:

$$\begin{aligned} \mathbf{R}_{xx}(\tau) &\triangleq E\left\{\mathbf{x}(k)\mathbf{x}^T(k-\tau)\right\} \\ &= \mathbf{H}\mathbf{R}_{ss}(\tau)\mathbf{H}^T + \mathbf{R}_{bb}(\tau), \end{aligned} \tag{6.119}$$

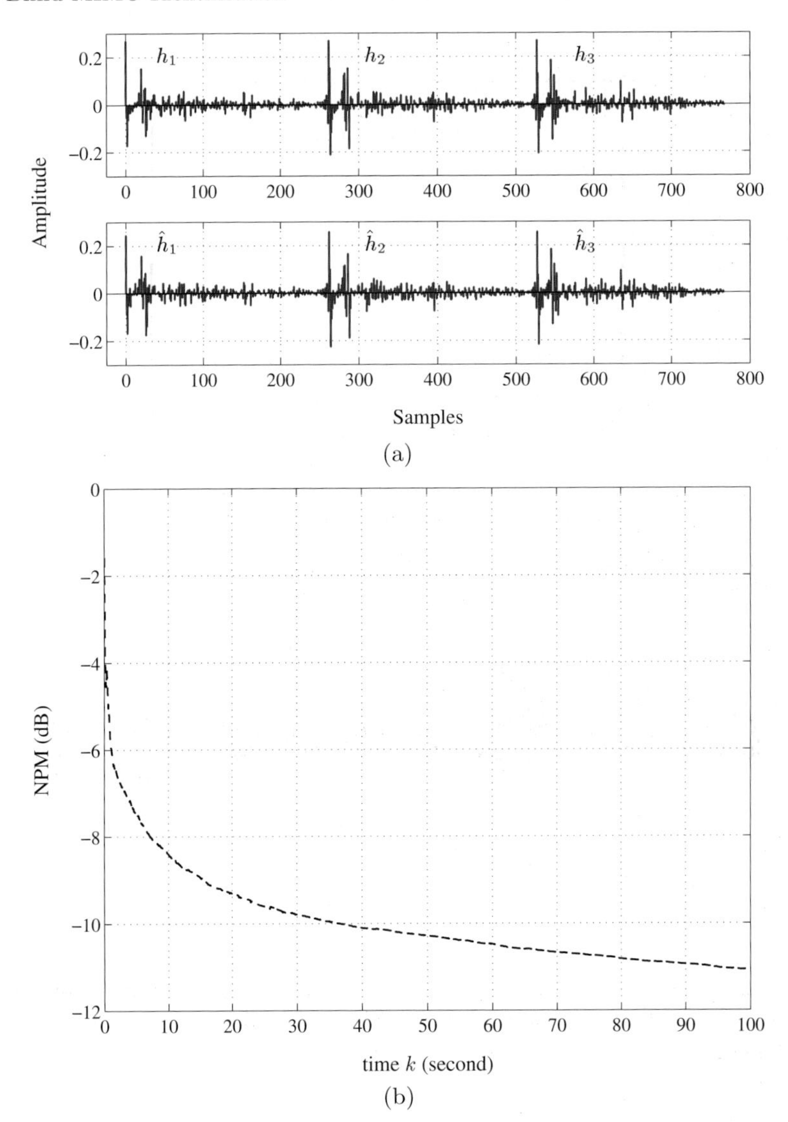

Fig. 6.7. Blind identification of an acoustic three-channel SIMO FIR system with a male speech signal as the source and additive white Gaussian noise at 75 dB SNR as conducted in *Example 6.5.* (a) The true and estimated channel impulse responses and (b) normalized projection misalignment of the FNMCLMS algorithm with $\rho_{\mathrm{f}} = 0.85$.

where $\tau \geq 0$ is a delay,

$$\mathbf{R}_{ss}(\tau) \triangleq E\left\{\mathbf{s}(k)\mathbf{s}^T(k-\tau)\right\}_{ML \times ML}, \text{ and}$$
$$\mathbf{R}_{bb}(\tau) \triangleq E\left\{\mathbf{b}(k)\mathbf{b}^T(k-\tau)\right\}_{N \times N}.$$

In many applications, it is quite common to assume that the inputs are uncorrelated with each other and with the noise as well, and the noise signals are white and spatially uncorrelated. As a result, we get:

$$\begin{aligned} \mathbf{R}_{s_i s_j}(\tau) &\triangleq E\left\{\mathbf{s}_i(k)\mathbf{s}_j^T(k-\tau)\right\} \\ &= \mathbf{0}, \text{ if } i \neq j \ (i,j = 1,2,\cdots,M), \qquad (6.120) \\ \mathbf{R}_{bb}(\tau) &= \delta(\tau)\sigma_b^2 \mathbf{I}_{N \times N}, \qquad (6.121) \end{aligned}$$

where $\delta(\tau)$ is the delta function and σ_b^2 is the noise power. Then the objective of an SOS-based blind MIMO identification algorithm is to estimate $\mathbf{H}$ from merely $\mathbf{R}_{xx}(\tau)$ $(\tau \geq 0)$.

Before we discuss whether we are able to and how we can blindly estimate $\mathbf{H}$, it is noteworthy to clarify the connotation of blind identification of a MIMO system. Similar to what we learned from blind identification of a SIMO system, a full identification of the system matrix $\mathbf{H}$ for a MIMO system is impossible. Since the exchange of a fixed scalar factor between a source signal and the corresponding columns of $\mathbf{H}$ does not affect the outputs:

$$\begin{aligned} \mathbf{x}(k) &= \mathbf{H}\mathbf{s}(k) + \mathbf{b}(k) \\ &= \begin{bmatrix} \mathbf{H}_1 & \mathbf{H}_2 & \cdots & \mathbf{H}_M \end{bmatrix} \begin{bmatrix} \mathbf{s}_1 \\ \mathbf{s}_2 \\ \vdots \\ \mathbf{s}_M \end{bmatrix} + \mathbf{b}(k) \\ &= \begin{bmatrix} \alpha_1\mathbf{H}_1 & \alpha_2\mathbf{H}_2 & \cdots & \alpha_M\mathbf{H}_M \end{bmatrix} \begin{bmatrix} \mathbf{s}_1/\alpha_1 \\ \mathbf{s}_2/\alpha_2 \\ \vdots \\ \mathbf{s}_M/\alpha_M \end{bmatrix} + \mathbf{b}(k), \qquad (6.122) \end{aligned}$$

the system matrix that associates with the mth source $\mathbf{H}_m$ can be only blindly identified up to a scale and the scales for different sources can be also different. In addition to the scale uncertainty, blind MIMO identification has a permutation issue because changing the order of sources and re-arranging the system matrix $\mathbf{H}$ accordingly does not affect the system outputs either. Therefore, a MIMO system is blindly identifiable if its channel matrix can be determined up to a scale and permutation matrix from only the outputs.

In Sect. 6.2.1 we learned that a blindly identifiable SIMO system needs to meet two requirements (one on the source and the other on the channel impulse responses). This is also true for MIMO systems while the requirements are naturally more complicated and stronger. As far as the channel impulse responses are concerned, the requirement is relatively simple and resembles that for a SIMO system:

- The channel matrices $\mathbf{H}_m$ $(m = 1, 2, \cdots, M)$ must be irreducible, i.e., the channel impulse responses in $\mathbf{H}_m$ (the row vectors) must not share any common zeros.

However, for a blindly identifiable MIMO system, a full-rank autocorrelation matrix $\mathbf{R}_{ss}(\tau)$ of the source signals is no longer enough. We will let the reader fathom this point by case studies for different types of inputs, including white or colored, stationary or non-stationary source signals. Some may not while the others will guarantee blind identifiability for a MIMO system. At the end of these studies, it will then become clear what requirements the inputs need to meet for a MIMO system to be blindly identifiable.

6.3.2 Memoryless MIMO System with White Inputs

Let's first examine the simplest MIMO system with memoryless channels[1] and with *white* inputs of the same power σ_s^2. For such a system, the second-order statistics of the system outputs $\mathbf{R}_{xx}(\tau)$ are all equal to $\mathbf{0}$ except for $\tau = 0$. From (6.119), we have

$$\mathbf{R}_{xx}(0) = \sigma_s^2 \mathbf{H}\mathbf{H}^T + \sigma_b^2 \mathbf{I}. \tag{6.123}$$

By singular value decomposition (SVD), the matrix $\mathbf{H}$ of size $N \times M$ can be written as a product

$$\mathbf{H} = \mathbf{U}\mathbf{D}\mathbf{V}^T, \tag{6.124}$$

where $\mathbf{U}$ and $\mathbf{V}$ are unitary matrices ($\mathbf{U}^T\mathbf{U} = \mathbf{U}\mathbf{U}^T = \mathbf{I}_{N\times N}$, $\mathbf{V}^T\mathbf{V} = \mathbf{V}\mathbf{V}^T = \mathbf{I}_{M\times M}$), and $\mathbf{D} = \mathrm{diag}(\lambda_1, \lambda_2, \cdots, \lambda_M)$ is an $N \times M$ diagonal matrix where $\lambda_1 \geq \lambda_2 \geq \cdots \geq \lambda_M > 0$ (here $\mathbf{H}$ has full column rank since the MIMO system is assumed to be irreducible). Using this decomposition, we obtain

$$\mathbf{H}\mathbf{H}^T = \mathbf{U}\mathbf{D}\mathbf{D}^T\mathbf{U}^T = \mathbf{U}_{1:M}\mathbf{D}_\mathrm{c}^2\mathbf{U}_{1:M}^T, \tag{6.125}$$

where $\mathbf{U}_{1:M}$ is an $N \times M$ matrix collecting the first M orthonormal columns of $\mathbf{U}$ and $\mathbf{D}_\mathrm{c} = \mathrm{diag}(\lambda_1, \lambda_2, \cdots, \lambda_M)$ is an $M \times M$ diagonal matrix by cropping $\mathbf{D}$.

The eigenvalue decomposition (EVD) of $\mathbf{R}_{xx}(0)$ is therefore expressed as

$$\mathbf{R}_{xx}(0) = \mathbf{U}\mathbf{D}_x^2\mathbf{U}^T, \tag{6.126}$$

[1] The name of memoryless channel is widely used in communications. In acoustics, a memoryless MIMO system is often referred to as a MIMO system with *instantaneous* mixtures. On the contrary, a MIMO system whose channels have memory is called a MIMO system with *convolutive* mixtures.

where $\mathbf{D}_x^2 = \text{diag}(\lambda_{x,1}^2, \lambda_{x,2}^2, \cdots, \lambda_{x,N}^2)$ is a diagonal matrix with $\lambda_{x,1}^2 \geq \lambda_{x,2}^2 \geq \cdots \geq \lambda_{x,N}^2 > 0$. From (6.123), (6.125), and (6.126), we know

$$\lambda_{x,n}^2 = \begin{cases} \sigma_s^2 \lambda_n^2 + \sigma_b^2, & n = 1, 2, \cdots, M, \\ \sigma_b^2, & n = M+1, \cdots, N. \end{cases} \tag{6.127}$$

Therefore, it is straightforward to determine the variance of the noise: the smallest eigenvalue of the received signal covariance matrix is equal to σ_b^2. Furthermore, the unitary matrix $\mathbf{U}$ in the EVD of $\mathbf{R}_{xx}(0)$ is the same as the left unitary matrix of the SVD of $\mathbf{H}$.

Define the following matrices:

$$\begin{aligned} \vec{\mathbf{R}}_{xx}(0) &\triangleq \mathbf{R}_{xx}(0) - \lambda_{x,N}^2 \mathbf{I}_{N\times N} \\ &= \mathbf{H}\mathbf{H}^T = \mathbf{U}_{1:M}\mathbf{D}_\mathrm{c}^2\mathbf{U}_{1:M}^T, \end{aligned} \tag{6.128}$$

$$\vec{\mathbf{H}} \triangleq \mathbf{U}_{1:M}\mathbf{D}_\mathrm{c}. \tag{6.129}$$

Then we have

$$\vec{\mathbf{R}}_{xx}(0) = \vec{\mathbf{H}}\vec{\mathbf{H}}^T. \tag{6.130}$$

It is easy to determine $\vec{\mathbf{H}}$ from $\mathbf{R}_{xx}(0)$. But the fact that $\vec{\mathbf{R}}_{xx}(0) = \mathbf{H}\mathbf{H}^T = \vec{\mathbf{H}}\vec{\mathbf{H}}^T$ does not imply that $\mathbf{H}$ is equal to $\vec{\mathbf{H}}$. The only thing that we can say is that:

$$\vec{\mathbf{H}} = \mathbf{H}\mathbf{V}, \tag{6.131}$$

where $\mathbf{V}$ is a unitary matrix. Equating (6.131) and (6.129), we see that this unitary matrix is, in fact, the right unitary matrix of the SVD of $\mathbf{H}$.

Clearly from the above analysis, using SOS only, we are able to determine the left unitary matrix $\mathbf{U}_{1:M}$ and the diagonal matrix $\mathbf{D}_\mathrm{c}$, but not the right unitary matrix $\mathbf{V}$, of the SVD of $\mathbf{H}$. This means that $\mathbf{H}$ can be determined up to an $M \times M$ unitary matrix, which is not acceptable for the blind identification problem. Consequently, a memoryless MIMO system with white, uncorrelated inputs of the same power is not blindly identifiable using SOS only.

6.3.3 Memoryless MIMO System with Colored Inputs

The second system that we are going to study is again a memoryless, irreducible MIMO system but with colored input signals. In this case, the input autocorrelation matrices $\mathbf{R}_{ss}(\tau)$ would not be equal to $\mathbf{0}$ for $\tau \geq 0$. Therefore the second-order statistics of the system outputs $\mathbf{R}_{xx}(\tau)$ in (6.119) can be exploited for any $\tau \geq 0$:

$$\mathbf{R}_{xx}(\tau) = \begin{cases} \mathbf{H}\mathbf{R}_{ss}(0)\mathbf{H}^T + \sigma_b^2 \mathbf{I}_{N\times N}, & \tau = 0, \\ \mathbf{H}\mathbf{R}_{ss}(\tau)\mathbf{H}^T, & \tau > 0. \end{cases} \tag{6.132}$$

As explained in Sect. 6.3.1, the exchange of a fixed scalar factor between a source signal and the corresponding columns of $\mathbf{H}$ does not affect the outputs and the system matrix $\mathbf{H}$ is blindly identified up to scaling. Therefore, *without any loss of generality*, it can be assumed that the input signals have unit variance, i.e., $\mathbf{R}_{ss}(0) = \mathbf{I}_{N\times N}$. Then (6.132) becomes

$$\mathbf{R}_{xx}(\tau) = \begin{cases} \mathbf{H}\mathbf{H}^T + \sigma_b^2 \mathbf{I}_{N\times N}, & \tau = 0, \\ \mathbf{H}\mathbf{R}_{ss}(\tau)\mathbf{H}^T, & \tau > 0. \end{cases} \tag{6.133}$$

Similar to what we derived in the previous section, $\mathbf{H}$ can be determined from $\mathbf{R}_{xx}(0)$ up to an $M \times M$ unitary matrix, which is the right unitary matrix of the SVD of $\mathbf{H}$. The remaining question is how to estimate this unitary matrix using $\mathbf{R}_{xx}(\tau)$ for $\tau > 0$.

Let $\bar{\mathbf{H}}^\dagger$ be the pseudo-inverse of $\bar{\mathbf{H}}$:

$$\bar{\mathbf{H}}^\dagger = \mathbf{D}_{\mathrm{c}}^{-1}\mathbf{U}_{1:M}^T. \tag{6.134}$$

Multiplying $\mathbf{R}_{xx}(\tau)$ by $\bar{\mathbf{H}}^\dagger$ from the left-hand side and by $\bar{\mathbf{H}}^{\dagger^T}$ from the right-and side produces

$$\begin{aligned} \bar{\mathbf{H}}^\dagger \mathbf{R}_{xx}(\tau) \bar{\mathbf{H}}^{\dagger^T} &= \bar{\mathbf{H}}^\dagger \mathbf{H}\mathbf{R}_{ss}(\tau)\mathbf{H}^T \bar{\mathbf{H}}^{\dagger^T} \\ &= \mathbf{V}^T \mathbf{R}_{ss}(\tau)\mathbf{V}. \end{aligned} \tag{6.135}$$

Let $\mathbf{R}'_{xx}(\tau)$ denote this processed covariance matrix. Then we have

$$\mathbf{V}\mathbf{R}'_{xx}(\tau)\mathbf{V}^T = \mathbf{R}_{ss}(\tau), \quad \forall \tau > 0. \tag{6.136}$$

Note that $\mathbf{V}$ is a unitary matrix and $\mathbf{R}_{ss}(\tau)$ is diagonal. As a result, $\mathbf{V}$ may be determined as a unitary diagonalizing matrix of $\mathbf{R}'_{xx}(\tau)$ for some lag $\tau > 0$ by eigenvalue decomposition. Apparently the input signals need to have distinct power spectra [171] such that $\mathbf{R}_{ss}(\tau)$ has distinct diagonal elements and consequently $\mathbf{R}'_{xx}(\tau)$ has distinct eigenvalues. Otherwise, $\mathbf{V}$ is again unable to be determined from $\mathbf{R}'_{xx}(\tau)$.

We can employ (6.136) for one specific τ [293] or for a number of τ's [18]. Both approaches will guarantee the unique determination of $\mathbf{V}$ while the latter is more robust to possible degenerate eigenvalues of $\mathbf{R}'_{xx}(\tau)$ for a pre-specified time lag τ as suggested by [18]. Since the goal of this procedure is to find a unitary matrix that diagonalizes $\mathbf{R}'_{xx}(\tau)$, the resulting diagonal matrix can be different from $\mathbf{R}_{ss}(\tau)$ by a permutation. Consequently, the unitary matrix $\mathbf{V}$ and the system matrix $\mathbf{H}$ will be determined up to a permutation matrix, which is normal and expected.

To conclude, a memoryless, irreducible MIMO system with colored input signals that have distinct power spectra can be blindly identified using only the second-order statistics of the system outputs.

6.3.4 Convolutive MIMO Systems with White Inputs

For a MIMO system with convolutive mixtures, the length of the channel impulse responses $L > 1$ and the covariance matrix of the inputs $\mathbf{R}_{ss}(\tau)$ would *not* be equal to $\mathbf{0}$ for $\tau \neq 0$ even for *white* source signals. In general, the matrix $\mathbf{R}_{s_m s_m}(\tau)$ $(m = 1, 2, \cdots, M)$ is a Toeplitz matrix:

$$\begin{aligned} &\mathbf{R}_{s_m s_m}(\tau) \\ &= E\left\{\mathbf{s}_m(k)\mathbf{s}_m^T(k-\tau)\right\} \\ &= \begin{bmatrix} r_{s_m s_m}(\tau) & r_{s_m s_m}(\tau+1) & \cdots & r_{s_m s_m}(\tau+L-1) \\ r_{s_m s_m}(\tau-1) & r_{s_m s_m}(\tau) & \cdots & r_{s_m s_m}(\tau+L-2) \\ \vdots & \vdots & \ddots & \vdots \\ r_{s_m s_m}(\tau-L+1) & r_{s_m s_m}(\tau-L+2) & \cdots & r_{s_m s_m}(\tau) \end{bmatrix}, \end{aligned} \tag{6.137}$$

where $r_{s_m s_m}(\tau) \stackrel{\triangle}{=} E\{s_m(k)s_m(k-\tau)\}$ for $\tau \geq 0$ and $r_{s_m s_m}(\tau) = r_{s_m s_m}(-\tau)$.

If the inputs are white, then we get

$$\mathbf{R}_{s_m s_m}(\tau) = \begin{cases} \sigma_{s_m}^2 \mathbf{I}_{L\times L}, & \tau = 0, \\ \sigma_{s_m}^2 \mathbf{J}_{L\times L}(\tau), & 0 < \tau \leq L-1, \\ \mathbf{0}, & \tau \geq L, \end{cases} \tag{6.138}$$

where $\mathbf{J}_{L\times L}(\tau)$ is the matrix whose elements are 1 on the τth off-diagonals and 0 elsewhere, i.e.,

$$\mathbf{J}_{L\times L}(\tau) \stackrel{\triangle}{=} \begin{bmatrix} 0 & 0 & \cdots & 0 & \cdots & 0 \\ \vdots & \vdots & \cdots & \vdots & \cdots & \vdots \\ 1 & 0 & \cdots & 0 & \cdots & 0 \\ 0 & 1 & \cdots & 0 & \cdots & 0 \\ \vdots & \vdots & \ddots & \vdots & \cdots & \vdots \\ \underbrace{0 \quad 0 \quad \cdots}_{L-\tau-1} & & & 1 & \underbrace{\cdots \quad 0}_{\tau} & \end{bmatrix}, \quad \tau = 1, 2, \cdots, L-1.$$

As clearly explained in the previous section, it can be assumed that the source signals have unit variance for blind identification of MIMO systems. Therefore, the following SOS of the system outputs can be exploited:

$$\mathbf{R}_{xx}(0) = \mathbf{H}\mathbf{H}^T + \sigma_b^2 \mathbf{I}_{N\times N}, \tag{6.139}$$

$$\begin{aligned} \mathbf{R}_{xx}(\tau) &= \sum_{m=1}^{M} \mathbf{H}_m \mathbf{J}_{L\times L}(\tau)\mathbf{H}_m^T, \\ &= \mathbf{H}\mathbf{J}_{\mathrm{d},ML\times ML}(\tau)\mathbf{H}^T, \quad \tau = 1, 2, \cdots, L-1, \end{aligned} \tag{6.140}$$

where $\mathbf{J}_{\mathrm{d},ML\times ML}(\tau)$ is a block diagonal (subscript d) matrix whose M diagonal blocks of dimension $L\times L$ are equal to $\mathbf{J}_{L\times L}(\tau)$. If such a system can be blindly identified using SOS, then $\mathbf{H}$ should be uniquely determined (up to a permutation matrix) from this set of equation.

But unfortunately we are unable to determine $\mathbf{H}$ from (6.139) and (6.140). Rather than presenting a tedious proof, we choose to use the following two examples to illustrate this point.

Example 6.6. **Blind Identifiability of an Irreducible, Convolutive, 2×2 MIMO System with $L=2$ and White Inputs Using SOS**

For such a system in the absence of noise, we have

$$\mathbf{R}_{xx}(0) = \mathbf{H}_{2\times 4}\mathbf{H}_{2\times 4}^T, \tag{6.141}$$

$$\mathbf{R}_{xx}(1) = \mathbf{H}_{2\times 4}\mathbf{J}_{\mathrm{d},4\times 4}(1)\mathbf{H}_{2\times 4}^T, \tag{6.142}$$

where

$$\mathbf{J}_{\mathrm{d},4\times 4}(1) = \begin{bmatrix} 0 & 0 & 0 & 0 \\ 1 & 0 & 0 & 0 \\ 0 & 0 & 0 & 0 \\ 0 & 0 & 1 & 0 \end{bmatrix}.$$

From (6.141), we know that $H_{2\times 4}$ can be determined up to an arbitrary, unitary matrix $\mathbf{\Omega}_{4\times 4}$:

$$\hat{\mathbf{H}}_{2\times 4} = \mathbf{H}_{2\times 4}\mathbf{\Omega}_{4\times 4}, \tag{6.143}$$

If it can be shown by using (6.142) that $\mathbf{\Omega}_{4\times 4}$ is a permutation matrix[2], then the system is blindly identifiable. Otherwise if we can find an $\mathbf{\Omega}_{4\times 4}$ that is unitary but is not a permutation matrix such that

$$\mathbf{\Omega}_{4\times 4}\mathbf{J}_{\mathrm{d},4\times 4}(1)\mathbf{\Omega}_{4\times 4}^T = \mathbf{J}_{\mathrm{d},4\times 4}(1), \tag{6.144}$$

then $\hat{\mathbf{H}}_{2\times 4}$ is also a valid solution of (6.141) and (6.142), and the system is not blindly identifiable.

Here we give an example:

$$\mathbf{\Omega}_{4\times 4} = \frac{\sqrt{2}}{2}\begin{bmatrix} 1 & 0 & 1 & 0 \\ 0 & 1 & 0 & 1 \\ 1 & 0 & -1 & 0 \\ 0 & 1 & 0 & -1 \end{bmatrix}. \tag{6.145}$$

It can be easily verified that such a $\mathbf{\Omega}_{4\times 4}^T$ is unitary and $\mathbf{\Omega}_{4\times 4}\mathbf{J}_{\mathrm{d},4\times 4}(1)\mathbf{\Omega}_{4\times 4}^T = \mathbf{J}_{\mathrm{d},4\times 4}(1)$. Therefore an irreducible, convolutive, 2×2 MIMO System with $L=2$ and white inputs using SOS is not blindly identifiable.

[2] A permutation matrix is obtained by permuting the rows (or columns) of the identity matrix. Every row and column of a permutation matrix therefore contains precisely a single 1 (0 elsewhere).

Example 6.7. **Blind Identifiability of an Irreducible, Convolutive, 2×3 MIMO System with $L = 3$ and White Inputs Using SOS**

In the previous example, the system has the same number of inputs and outputs. Here we study a more typical system with more outputs than inputs and $L > 2$. In the absence of noise, we have the following non-zero SOS of the system outputs:

$$\begin{cases} \mathbf{R}_{xx}(0) = \mathbf{H}_{3\times 6}\mathbf{H}_{3\times 6}^T, \\ \mathbf{R}_{xx}(1) = \mathbf{H}_{3\times 6}\mathbf{J}_{\mathrm{d},6\times 6}(1)\mathbf{H}_{3\times 6}^T, \\ \mathbf{R}_{xx}(2) = \mathbf{H}_{3\times 6}\mathbf{J}_{\mathrm{d},6\times 6}(2)\mathbf{H}_{3\times 6}^T, \end{cases} \tag{6.146}$$

where

$$\mathbf{J}_{\mathrm{d},6\times 6}(1) = \left[\begin{array}{ccc|ccc} 0&0&0&0&0&0\\ 1&0&0&0&0&0\\ 0&1&0&0&0&0\\ \hline 0&0&0&0&0&0\\ 0&0&0&1&0&0\\ 0&0&0&0&1&0 \end{array}\right], \quad \mathbf{J}_{\mathrm{d},6\times 6}(2) = \left[\begin{array}{ccc|ccc} 0&0&0&0&0&0\\ 0&0&0&0&0&0\\ 1&0&0&0&0&0\\ \hline 0&0&0&0&0&0\\ 0&0&0&0&0&0\\ 0&0&0&1&0&0 \end{array}\right].$$

As one example, we give

$$\mathbf{\Omega}_{6\times 6} = \frac{\sqrt{2}}{2}\begin{bmatrix} 1&0&0&1&0&0\\ 0&1&0&0&1&0\\ 0&0&1&0&0&1\\ 1&0&0&-1&0&0\\ 0&1&0&0&-1&0\\ 0&0&1&0&0&-1 \end{bmatrix}. \tag{6.147}$$

Apparently $\mathbf{\Omega}_{6\times 6}$ is not a permutation matrix. It can be checked that $\mathbf{\Omega}_{6\times 6}$ is a unitary matrix and

$$\begin{aligned} \mathbf{\Omega}_{6\times 6}\mathbf{J}_{\mathrm{d},6\times 6}(1)\mathbf{\Omega}_{6\times 6}^T &= \mathbf{J}_{\mathrm{d},6\times 6}(1), \\ \mathbf{\Omega}_{6\times 6}\mathbf{J}_{\mathrm{d},6\times 6}(2)\mathbf{\Omega}_{6\times 6}^T &= \mathbf{J}_{\mathrm{d},6\times 6}(2). \end{aligned}$$

Therefore $\mathbf{H}_{3\times 6}\mathbf{\Omega}_{6\times 6}$ is also a valid solution of (6.146). As a result, the system is blindly unidentifiable.

6.3.5 Convolutive MIMO Systems with Colored Inputs

For an acoustic MIMO systems with convolutive channels, the inputs are usually speech signals, which are always colored. Therefore it is more interesting and with more practical significance to study whether such a system can be blindly identified using only SOS and, if so, whether we can develop an effective and efficient algorithm.

Again we begin with examining the SOS of the outputs of this system. Since the inputs are colored, the covariance matrix $\mathbf{R}_{s_m s_m}(0)$ is no longer equal to the identity matrix and $\mathbf{R}_{s_m s_m}(\tau) \neq \mathbf{0}, \forall \tau > 0$, cannot be simplified in general. So we have

$$\mathbf{R}_{xx}(0) = \mathbf{H}\mathbf{R}_{ss}(0)\mathbf{H}^T + \sigma_b^2 \mathbf{I}_{N\times N}, \tag{6.148}$$
$$\mathbf{R}_{xx}(\tau) = \mathbf{H}\mathbf{R}_{ss}(\tau)\mathbf{H}^T, \quad \tau \geq 1. \tag{6.149}$$

It seems that $\mathbf{H}$ can be determined from (6.148) and (6.149), which collect an infinite number of equations. But proving it is very difficult. Unfortunately, so far there has been no success, to the best of our knowledge, reported in the literature. In addition, there has been no *practically working* algorithm that blindly identifies a convolutive MIMO system with colored inputs.

However, it is noteworthy to mention the algorithm of blind identification via decorrelating subchannels (BIDS) proposed in [172]. The BIDS algorithm requires $N > M$, i.e., more outputs than inputs. From the N system outputs, the algorithm chooses M at a time, which, together with the M inputs, forms an $M \times M$ MIMO subsystem. For each subsystem, blind source separation is conducted, producing M uncorrelated signals that are filtered version of the M source signals. Collecting the decorrelated signals from all subsystems with respect to the same source leads to a SIMO system that can be blindly identified. The BIDS algorithm is clearly based on the technique of blind source separation (BSS) using SOS. But SOS-based BSS is as challenging as blind identification and has not yet been proved theoretically. The BIDS shows a close relationship between blind MIMO identification and BSS, which will become clearer later when BSS is discussed.

6.3.6 Frequency-Domain Blind Identification of Convolutive MIMO Systems and Permutation Inconsistency

Blind identification of a convolutive MIMO system is not only difficult but also very complicated. A common way to simplify this problem is to transform the signal model into the frequency domain such that the MIMO system is decomposed into a number of memoryless MIMO systems at different frequency points. Similar to (2.10), we have the MIMO system in the frequency domain as follows

$$\underset{\rightarrow}{\boldsymbol{x}}(f) = \underset{\Rightarrow}{\boldsymbol{H}}(f)\underset{\rightarrow}{\boldsymbol{s}}(f) + \underset{\rightarrow}{\boldsymbol{b}}(f), \tag{6.150}$$

where

$$\underset{\rightarrow}{\boldsymbol{x}}(f) = \begin{bmatrix} X_1(f) & X_2(f) & \cdots & X_N(f) \end{bmatrix}^T,$$
$$\underset{\Rightarrow}{\boldsymbol{H}}(f) = \begin{bmatrix} H_{11}(f) & H_{12}(f) & \cdots & H_{1M}(f) \\ H_{21}(f) & H_{22}(f) & \cdots & H_{2M}(f) \\ \vdots & \vdots & \ddots & \vdots \\ H_{N1}(f) & H_{N2}(f) & \cdots & H_{NM}(f) \end{bmatrix},$$

$$\underline{\mathbf{s}}(f) = \begin{bmatrix} S_1(f) & S_2(f) & \cdots & S_M(f) \end{bmatrix}^T,$$
$$\underline{\mathbf{b}}(f) = \begin{bmatrix} B_1(f) & B_2(f) & \cdots & B_N(f) \end{bmatrix}^T.$$

Consequently, at each frequency point, we are confronted with a memoryless MIMO system whose inputs are white. As explained in Sect. 6.3.2, mutual independence is not sufficient for the system to be blindly identified using SOS and supplementary information needs to be used. For such source signals as speech, the quasi-stationarity can be exploited [314], [104], [248]. The SOS of the MIMO outputs are computed over a number of observation periods, making the problem solvable.

While the frequency-domain approach reduces the computational complexity, it gives rise to the so-called "permutation inconsistency" problem [185]. It is clear that a blind estimate of $\underline{\mathbf{H}}(f)$ at frequency bin f can at best be obtained up to a permutation matrix (a scale diagonal matrix as well). Since blind identification of the MIMO system at each frequency point is independently conducted, $\underline{\mathbf{H}}(f)$ is not necessarily estimated up to a consistent permutation matrix. When the channel impulse responses are long and there are a large number of frequency bins, permutation inconsistency is very harmful and is still a challenging, open research problem.

6.3.7 Convolutive MIMO Systems with White but Quasistationary Inputs

In the previous section, we learned that the quasi-stationarity of speech inputs could be exploited to help solve the problem of blind MIMO identification in the frequency domain. Strictly speaking, a blind method should assume no *a priori* knowledge of the inputs whatsoever. But in practice, any *a priori* information about the inputs is desired since it can effectively improve the accuracy, reliability, and robustness of a blind method. Although there is no problem (except complexity) to take advantage of both non-whiteness and quasi-stationarity, we choose to demonstrate the usefulness of quasi-stationarity using a convolutive MIMO system with white but quasistationary inputs in this section.

A quasistationary signal is non-stationary over a long time period of K samples, but stationary over a shorter period with fewer samples. Suppose that we divide the K samples into T periods. In the tth time period, the SOS of the MIMO outputs is expressed by

$$\begin{aligned} \mathbf{R}_{xx}(\tau, t) &\stackrel{\triangle}{=} E\left\{\mathbf{x}(k)\mathbf{x}^T(k-\tau)\right\}_t \\ &= \mathbf{H}\mathbf{R}_{ss}(\tau, t)\mathbf{H}^T + \delta(\tau)\sigma_b^2\mathbf{I}_{N\times N} \\ &= \sum_{m=1}^{M} \mathbf{H}_m\mathbf{R}_{s_m s_m}(\tau, t)\mathbf{H}_m^T + \delta(\tau)\sigma_b^2\mathbf{I}_{N\times N}, \qquad (6.151) \\ &\qquad t = 1, 2, \cdots, T, \end{aligned}$$

where t is the index of time periods within which the source signals are assumed stationary, $E\{\cdot\}_t$ denotes mathematical expectation for time index k in the tth time period,

$$\mathbf{R}_{ss}(\tau, t) \triangleq E\left\{\mathbf{s}(k)\mathbf{s}^T(k-\tau)\right\}_t,$$
$$\mathbf{R}_{s_m s_m}(\tau, t) \triangleq E\left\{\mathbf{s}_m(k)\mathbf{s}_m^T(k-\tau)\right\}_t,$$

and it has been assumed that both the system and noise are stationary over all K samples while the noise components in different channels are white and uncorrelated.

When $N \leq ML$ which is quite normal in acoustic MIMO systems, $\mathbf{R}_{xx}(0, t)$ cannot be decomposed using the subspace method, and the noise effects cannot be easily remedied. If the noise is strong, the SOS of the system outputs are seriously contaminated and it would be extremely difficult to recover the source signals in a blind manner. Therefore we confine our discussion in the following to the case in which the noise is significantly weak and can be ignored. Using the knowledge of white source signals, we get

$$\mathbf{R}_{s_m s_m}(\tau, t) = \sigma_{s_m}^2(t)\mathbf{J}_{L\times L}(\tau), \quad m = 1, 2, \cdots, M, \tag{6.152}$$

where $\sigma_{s_m}^2(t)$ is the power of the source signal s_m in the tth ($t = 1, 2, \cdots, T$) time period. Finally, with all these *a priori* knowledge incorporated, (6.151) becomes

$$\mathbf{R}_{xx}(\tau, t) = \sum_{m=1}^{M} \sigma_{s_m}^2(t)\mathbf{H}_m\mathbf{J}_{L\times L}(\tau)\mathbf{H}_m^T. \tag{6.153}$$

Let's define

$$\boldsymbol{\rho}(t) \triangleq \left[\ \sigma_{s_1}^2(t)\ \ \sigma_{s_2}^2(t)\ \ \cdots\ \ \sigma_{s_M}^2(t)\ \right]^T, \quad t = 1, 2, \cdots, T. \tag{6.154}$$

If the power of the quasistationary source signals is significantly different over time periods and T is large enough, it is plausible to assume that the set $\{\boldsymbol{\rho}(t), t = 1, 2, \cdots, T\}$ spans $\mathbb{R}^M$. Then we can find a set $\{a_m(t), t = 1, 2, \cdots, T\}$ so that

$$\sum_{t=1}^{T} a_m(t)\boldsymbol{\rho}(t) = \mathbf{e}_m, \quad m = 1, 2, \cdots, M, \tag{6.155}$$

where

$$\mathbf{e}_m \triangleq \left[\ \underbrace{0\ \cdots\ 0}_{m-1}\ \ 1\ \ \underbrace{0\ \cdots\ 0}_{M-m}\ \right]^T.$$

Equation (6.155) can also be expressed as

$$\sum_{t=1}^{T} a_m(t)\sigma_{s_i}^2(t) = \delta(m-i), \quad m, i = 1, 2, \cdots, M. \tag{6.156}$$

Multiplying (6.153) by $a_m(t)$ and summing up the T equations produces

$$\begin{aligned}\sum_{t=1}^{T} a_m(t)\mathbf{R}_{xx}(\tau,t) &= \sum_{t=1}^{T} a_m(t)\left[\sum_{i=1}^{M} \sigma_{s_i}^2(t)\mathbf{H}_i\mathbf{J}_{L\times L}(\tau)\mathbf{H}_i^T\right] \\ &= \sum_{i=1}^{M}\left[\sum_{t=1}^{T} a_m(t)\sigma_{s_i}^2(t)\right]\mathbf{H}_i\mathbf{J}_{L\times L}(\tau)\mathbf{H}_i^T \\ &= \sum_{i=1}^{M}\delta(m-i)\mathbf{H}_i\mathbf{J}_{L\times L}(\tau)\mathbf{H}_i^T \\ &= \mathbf{H}_m\mathbf{J}_{L\times L}(\tau)\mathbf{H}_m^T, \quad m=1,2,\cdots,M. \end{aligned} \tag{6.157}$$

Define

$$\mathbf{R}_{xx,m}(\tau) \triangleq \sum_{t=1}^{T} a_m(t)\mathbf{R}_{xx}(\tau,t), \tag{6.158}$$

and then (6.157) becomes

$$\begin{aligned}\mathbf{R}_{xx,m}(\tau) &= \mathbf{H}_m\mathbf{J}_{L\times L}(\tau)\mathbf{H}_m^T, \\ m &= 1,2,\cdots,M,\ \tau=0,1,\cdots,L-1, \end{aligned} \tag{6.159}$$

which implies that the blind identification of a $M \times N$ system with white, quasistationary sources can be decomposed into the blind identification of M SIMO systems. Since the MIMO system is irreducible, so are the M SIMO systems. Therefore they are blindly identifiable. In order to do so, the reader can either reformulate the problem and use the cross relation based methods (batch or adaptive) developed in the first part of this chapter, or follow the algorithm presented in the appendix to this chapter.

In the above, we have illustrated the possibility of decomposing the MIMO system into M SIMO systems by using the quasi-stationarity of such source signals as speech. Certainly a method for finding the appropriate set of $\{a_m(t)\}$ is not yet available or even possible. But this difficulty would not affect the legitimacy of our discussion on the usefulness of quasi-stationarity in blind identification of convolutive MIMO systems.

6.4 Summary

In this chapter, we developed and explored blind identification of SIMO and MIMO systems with emphasis on the methods using only second order statistics. The importance of blind acoustic system identification stems directly from the fact that in many acoustic applications source signals are not accessible. Blind identifications of SIMO and MIMO systems are significantly different in complexity and difficulty. For blind SIMO identification, the principle is well known and easy to comprehend. The interest of current research is in the development of adaptive algorithms with low complexity and fast convergence. A number of adaptive algorithms for blind SIMO identification were developed and presented in the first half of this chapter. Their performances were compared using various numerical experiments. The research of blind MIMO identification stays still on the stage of feasibility investigation. A comprehensive analysis of this challenging problem was presented in the second half of this chapter. It covers both memoryless and convolutive MIMO systems with different types of inputs, ranging from white to colored, and from stationary to quasistationary signals. This first-ever analysis can help better understand blind identifiability of a MIMO system and what *a priori* knowledge about the source signals is useful to solve the problem.

6.5 Appendix. Blind SIMO Identification: A Derivation Directly from the Covariance Matrices of the System Outputs

For an irreducible, noiseless SIMO system whose input is white, the covariance matrices of its N outputs are given as follows:

$$\mathbf{R}_{xx}(\tau) = \mathbf{H}\mathbf{J}_{L\times L}(\tau)\mathbf{H}^T, \quad \tau = 0, 1, \cdots, L-1, \tag{6.160}$$

from which we intend to determine $\mathbf{H}$ of dimension $N \times L$.

The derivations for the two cases of $N \geq L$ and $N < L$ are significantly different. We will begin with the easier case of $N \geq L$ and then continue our derivation for the case of $N < L$, which is more common in practice, meaning that the number of channels is less than the length of the channel impulse responses.

The Case of $N \geq L$

Using SVD, the channel matrix $\mathbf{H}$ can be written as a product

$$\mathbf{H} = \mathbf{U}\mathbf{D}\mathbf{V}^T, \tag{6.161}$$

where

$$\begin{aligned} \mathbf{U} &= \begin{bmatrix} \mathbf{u}_1 & \mathbf{u}_2 & \cdots & \mathbf{u}_N \end{bmatrix}_{N\times N}, \\ \mathbf{V} &= \begin{bmatrix} \mathbf{v}_1 & \mathbf{v}_2 & \cdots & \mathbf{v}_L \end{bmatrix}_{L\times L} \end{aligned}$$

are unitary matrices ($\mathbf{U}^T\mathbf{U} = \mathbf{I}_{N\times N}$, $\mathbf{V}^T\mathbf{V} = \mathbf{I}_{L\times L}$),

$$\mathbf{D} = \mathrm{diag}\{\lambda_1,\ \lambda_2,\ \cdots,\ \lambda_P\}$$

is an $N \times L$ diagonal matrix in which $\lambda_1 \geq \lambda_2 \geq \cdots \geq \lambda_P > 0$ and $P = \min(N, L)$.

Since $N \geq L$, (6.161) can be rewritten as:

$$\mathbf{H} = \mathbf{U}_{\rm c}\mathbf{D}_{\rm c}(1:L)\mathbf{V}^T, \tag{6.162}$$

where

$$\mathbf{U}_{\rm c} = \begin{bmatrix} \mathbf{u}_1 & \mathbf{u}_2 & \cdots & \mathbf{u}_L \end{bmatrix}_{N\times L}$$

is an $N \times L$ matrix with orthonormal columns and

$$\mathbf{D}_{\rm c}(1:L) = \text{diag}\{\lambda_1,\ \lambda_2,\ \cdots,\ \lambda_L\}$$

is an $L \times L$ diagonal matrix with $\lambda_1 \geq \lambda_2 \geq \cdots \geq \lambda_L > 0$. Here we have used the assumption that $\mathbf{H}$ is irreducible. As a result, $\mathbf{H}$ has a full column rank.

Now substitute the SVD of $\mathbf{H}$ (6.162) into (6.160) for $\tau = 0$ and recognize that $\mathbf{J}_{L\times L}(0) = \mathbf{I}_{L\times L}$ to obtain

$$\mathbf{R}_{xx}(0) = \mathbf{U}_{\rm c}\mathbf{D}_{\rm c}^2(1:L)\mathbf{U}_{\rm c}^T \tag{6.163}$$

Therefore, we can find $\mathbf{U}_{\rm c}$ and $\mathbf{D}_{\rm c}$ from the eigenvalue decomposition of $\mathbf{R}_{xx}(0)$. We define

$$\mathbf{Q} \triangleq \mathbf{D}_{\rm c}^{-1}(1:L)\mathbf{U}_{\rm c}^T \tag{6.164}$$

and multiply (6.160) by $\mathbf{Q}$ and $\mathbf{Q}^T$ from the left and right hand side, respectively, to obtain

$$\mathbf{Q}\mathbf{R}_{xx}(\tau)\mathbf{Q}^T = \mathbf{V}^T\mathbf{J}_{L\times L}(\tau)\mathbf{V},\quad \tau = 1, 2, \cdots, L-1. \tag{6.165}$$

Let

$$\mathbf{G}(\tau) \triangleq \mathbf{Q}\mathbf{R}_{xx}(\tau)\mathbf{Q}^T. \tag{6.166}$$

Recall that

$$\mathbf{V}^T \triangleq \begin{bmatrix} \boldsymbol{\nu}_1 & \boldsymbol{\nu}_2 & \cdots \boldsymbol{\nu}_L \end{bmatrix} \tag{6.167}$$

is a unitary matrix, where $\boldsymbol{\nu}_l$'s $(l = 1, 2, \cdots, L)$ are the column vectors of $\mathbf{V}^T$ with unit norm and are orthogonal to each other. It can be shown that

$$\begin{aligned} \mathbf{G}(\tau)\boldsymbol{\nu}_1 &= \left[\mathbf{V}^T\mathbf{J}_{L\times L}(\tau)\mathbf{V}\right]\boldsymbol{\nu}_1 \\ &= \mathbf{V}^T\mathbf{J}_{L\times L}(\tau)\cdot\begin{bmatrix} 1 & 0 & \cdots & 0 \end{bmatrix}^T \\ &= \boldsymbol{\nu}_{\tau+1},\quad \tau = 1, 2, \cdots, L-1. \end{aligned} \tag{6.168}$$

Consequently we deduce that $\mathbf{V}^T$ has the form

$$\mathbf{V}^T = \begin{bmatrix} \boldsymbol{\nu}_1 & \mathbf{G}(1)\boldsymbol{\nu}_1 & \cdots & \mathbf{G}(L-1)\boldsymbol{\nu}_1 \end{bmatrix}. \tag{6.169}$$

In order to determine $\boldsymbol{\nu}_1$, we re-examine (6.165) for $\tau = 1$

$$\mathbf{G}(1) = \mathbf{V}^T\mathbf{J}_{L\times L}(1)\mathbf{V}, \tag{6.170}$$

and we can show that

$$\begin{aligned}\mathbf{G}^T(1)\boldsymbol{\nu}_1 &= \mathbf{V}^T\mathbf{J}_{L\times L}^T(1)\mathbf{V}\cdot\boldsymbol{\nu}_1 \\ &= \mathbf{V}^T\mathbf{J}_{L\times L}^T(1)\cdot\begin{bmatrix}1 & 0 & \cdots & 0\end{bmatrix}^T \\ &= \mathbf{V}^T\cdot\mathbf{0} \\ &= \mathbf{0}.\end{aligned} \tag{6.171}$$

Clearly $\boldsymbol{\nu}_1$ is in the null space of $\mathbf{G}^T(1)$. From (6.170), we know that $\mathbf{G}(1)$ is rank deficient by 1. Therefore if $\mathbf{g}(1)$ is the eigenvector of $\mathbf{G}^T(1)$ corresponding to its unique eigenvalue 0, $\boldsymbol{\nu}_1$ can be estimated from $\mathbf{g}(1)$ up to a phase factor ψ:

$$\boldsymbol{\nu}_1 = \mathbf{g}(1)\exp(j\psi). \tag{6.172}$$

Then it follows that $\mathbf{V}^T$ and $\mathbf{H}$ will be determined up to a phase factor ψ as well. For acoustic systems whose channel impulse responses are real, ψ is either 0 or π and the MIMO system can be blindly identified up to only a scale factor.

The Case of $N < L$

For $N < L$, the general SVD of $\mathbf{H}$ given by (6.161) is simplified as follows

$$\mathbf{H} = \mathbf{U}\mathbf{D}_{\mathrm{c}}(1:N)\mathbf{V}_{\mathrm{c}}^T, \tag{6.173}$$

where

$$\mathbf{D}_{\mathrm{c}}(1:N) = \mathrm{diag}\left\{\lambda_1,\ \lambda_2,\ \cdots,\ \lambda_N\right\}$$

is an $N\times N$ matrix with $\lambda_1 \geq \lambda_2 \geq \cdots \geq \lambda_N > 0$, and

$$\mathbf{V}_{\mathrm{c}} = \begin{bmatrix}\mathbf{v}_1 & \mathbf{v}_2 & \cdots & \mathbf{v}_N\end{bmatrix}$$

is an $L\times N$ matrix with column vectors being orthonormal. But since $N < L$, the row vectors of $\mathbf{V}_{\mathrm{c}}$, i.e., the column vectors of $\mathbf{V}_{\mathrm{c}}^T$, are unnecessarily orthogonal, which makes it impossible to use the trick developed above for determining $\mathbf{V}_{\mathrm{c}}^T$. In the following, we will use a different method to show that the channel matrix $\mathbf{H}$ can be estimated from (6.160).

Let

$$\mathbf{R}_{xx} \triangleq \begin{bmatrix}\mathbf{R}_{xx}(0) & \mathbf{R}_{xx}(1) & \cdots & \mathbf{R}_{xx}(L-1) \\ \mathbf{R}_{xx}(1) & \mathbf{R}_{xx}(0) & \cdots & \mathbf{R}_{xx}(L-2) \\ \vdots & \vdots & \ddots & \vdots \\ \mathbf{R}_{xx}(L-1) & \mathbf{R}_{xx}(L-2) & \cdots & \mathbf{R}_{xx}(0)\end{bmatrix}_{NL\times NL}, \tag{6.174}$$

and construct a Sylvester matrix of dimension $NL\times(2L-1)$ from $\mathbf{H}$ as follows:

$$\mathbf{H}_{\mathrm{syl}} \triangleq \begin{bmatrix} \mathbf{h}(0) & \mathbf{h}(1) & \cdots & \mathbf{h}(L-1) & \mathbf{0} & \cdots & \mathbf{0} \\ \mathbf{0} & \mathbf{h}(0) & \cdots & \mathbf{h}(L-2) & \mathbf{h}(L-1) & \cdots & \mathbf{0} \\ \vdots & \vdots & \ddots & \ddots & \ddots & \ddots & \vdots \\ \mathbf{0} & \mathbf{0} & \cdots & \mathbf{h}(0) & \mathbf{h}(1) & \cdots & \mathbf{h}(L-1) \end{bmatrix}, \tag{6.175}$$

where

$$\mathbf{h}(l) = \begin{bmatrix} h_{1,l} & h_{2,l} & \cdots & h_{N,l} \end{bmatrix}^T, \quad l = 0, 1, \cdots, L-1.$$

Then we can rewrite (6.160) as

$$\mathbf{R}_{xx} = \mathbf{H}_{\mathrm{syl}} \mathbf{H}_{\mathrm{syl}}^T. \tag{6.176}$$

Note that $\mathbf{H}_{\mathrm{syl}}$ is a "tall" matrix since $N \geq 2$. In addition, $\mathbf{H}$ is irreducible as assumed so that $\mathbf{H}_{\mathrm{syl}}$ has full column rank. Therefore we can have an "economy" SVD of $\mathbf{H}_{\mathrm{syl}}$ as

$$\mathbf{H}_{\mathrm{syl}} = \mathbf{\Psi}_{\mathrm{c}} \mathbf{\Sigma}_{\mathrm{c}} \mathbf{\Upsilon}^T, \tag{6.177}$$

where $\mathbf{\Psi}_{\mathrm{c}}$ is an $NL \times (2L-1)$ matrix whose column vectors are orthonormal,

$$\mathbf{\Sigma}_{\mathrm{c}} = \mathrm{diag}\left\{\eta_1, \ \eta_2, \ \cdots, \ \eta_{2L-1}\right\}$$

is a $(2L-1) \times (2L-1)$ diagonal matrix with $\eta_1 \geq \eta_2 \geq \cdots \geq \eta_{2L-1} > 0$, and

$$\mathbf{\Upsilon}^T \triangleq \begin{bmatrix} \boldsymbol{v}_1 & \boldsymbol{v}_2 & \cdots & \boldsymbol{v}_{2L-1} \end{bmatrix}$$

is a $(2L-1) \times (2L-1)$ unitary matrix.

From the eigenvalue decomposition of $\mathbf{R}_{xx}$ in (6.176), we can easily obtain $\mathbf{\Psi}_{\mathrm{c}}$ and $\mathbf{\Sigma}_{\mathrm{c}}$. Let

$$\mathbf{\Phi} \triangleq \mathbf{\Psi}_{\mathrm{c}} \mathbf{\Sigma}_{\mathrm{c}} \tag{6.178}$$

and we have

$$\mathbf{H}_{\mathrm{syl}} = \mathbf{\Phi} \mathbf{\Upsilon}^T, \tag{6.179}$$

meaning that $\mathbf{H}_{\mathrm{syl}}$ has so far been estimated up to a unitary matrix $\mathbf{\Upsilon}$. Now the question is whether we can determine $\mathbf{\Upsilon}$ from (6.179) by using the specific structure of the Sylvester matrix $\mathbf{H}_{\mathrm{syl}}$. The answer is yes and we will show how to do it. Note that the Sylvester matrix $\mathbf{H}_{\mathrm{syl}}$ [see (6.175)] is determined after its first L column vectors are identified. Therefore, we only need to estimate the first L column vectors of $\mathbf{\Upsilon}^T$, i.e., $\boldsymbol{v}_1, \ \boldsymbol{v}_2, \ \cdots, \ \boldsymbol{v}_L$.

Before presenting the approach, we need to define several symbols that will be used in the derivation. We first partition $\mathbf{H}_{\mathrm{syl}}$ into L-row block matrices with each block containing N row vectors and then denote $\mathbf{H}_{\mathrm{syl}}(p,:)$ as the pth block matrix. Therefore,

$$\mathbf{H}_{\mathrm{syl}} = \begin{bmatrix} \mathbf{H}_{\mathrm{syl}}^T(1,:) & \mathbf{H}_{\mathrm{syl}}^T(2,:) & \cdots & \mathbf{H}_{\mathrm{syl}}^T(L,:) \end{bmatrix}^T, \tag{6.180}$$

and we denote $\mathbf{H}_{m,\text{syl}}(p:q,:)$ as

$$\mathbf{H}_{\text{syl}}(p:q,:) \triangleq \begin{bmatrix} \mathbf{H}_{\text{syl}}^T(p,:) & \mathbf{H}_{\text{syl}}^T(p+1,:) & \cdots & \mathbf{H}_{\text{syl}}^T(q,:) \end{bmatrix}^T, \quad 1 \leq p < q \leq L. \tag{6.181}$$

Similarly, we define $\mathbf{\Phi}(p,:)$ and $\mathbf{\Phi}(p:q,:)$ extracted from $\mathbf{\Phi}$.

Note that $\mathbf{H}_{\text{syl}}(p+1:L,:)$ is of dimension $(L-p)N \times (2L-1)$ with the first p columns being all zeros. In addition, it follows from (6.179) that

$$\mathbf{H}_{\text{syl}}(p:q,:) = \mathbf{\Phi}(p:q,:)\mathbf{\Upsilon}^T. \tag{6.182}$$

The approach presented here will recursively compute $\boldsymbol{\upsilon}_p$ $(p = 1, 2, \cdots, L)$ in a sequential way.

- For $p = 1$, we have

$$\mathbf{H}_{\text{syl}}(2:L,:) = \begin{bmatrix} \mathbf{0} & \mathbf{h}(0) & \cdots & \mathbf{h}(L-1) & \mathbf{0} & \cdots & \mathbf{0} \\ \mathbf{0} & \mathbf{0} & \mathbf{h}(0) & \cdots & \mathbf{h}(L-1) & \cdots & \mathbf{0} \\ \vdots & \vdots & \ddots & \ddots & \ddots & \ddots & \vdots \\ \mathbf{0} & \cdots & \cdots & \mathbf{0} & \mathbf{h}(0) & \cdots & \mathbf{h}(L-1) \end{bmatrix}, \tag{6.183}$$

  which is of dimension $(L-1)N \times (2L-1)$. It is a tall matrix for $N \geq 3$ but a short matrix only for $N = 2$. When $N \geq 3$ and $\mathbf{H}_{\text{syl}}(2:L,:)$ is a tall matrix, the column rank of $\mathbf{H}_{\text{syl}}(2:L,:)$ is deficient by 1, which is induced by its first zero column vector. When $N = 2$, $\mathbf{H}_{\text{syl}}(2:L,:)$ is of dimension $(2L-2) \times (2L-1)$ and its column rank is still deficient by 1.
  Although $\mathbf{\Upsilon}$ is a unitary matrix with full rank, the column rank of $\mathbf{\Phi}(2:L,:)$ will anyway be deficient by 1 and the null space of $\mathbf{\Phi}(2:L,:)$ is one dimensional. Examining the first column vector of both sides of (6.182) for $(p = 2, q = L)$ yields

$$\mathbf{\Phi}(2:L,:)\boldsymbol{\upsilon}_1 = \mathbf{0}. \tag{6.184}$$

  Since the null space of $\mathbf{\Phi}(2:L,:)$ is one-dimensional, we can uniquely determine $\boldsymbol{\upsilon}_1$ with a unit norm and hence $\mathbf{h}(0)$.
- We will then recursively compute $\boldsymbol{\upsilon}_p$ for $2 \leq p \leq L-1$. At the pth step, we suppose to already know $\boldsymbol{\upsilon}_1$ up to $\boldsymbol{\upsilon}_{p-1}$ and $\mathbf{h}(0)$ up to $\mathbf{h}(p-2)$ from the previous steps. Here, we intend to find $\boldsymbol{\upsilon}_p$ and $\mathbf{h}(p-1)$.
  Observing that the first p columns of $\mathbf{H}_{\text{syl}}(p+1:L,:)$ are all zeros, from (6.182) we know

$$\mathbf{\Phi}(p+1:L,:)\boldsymbol{\upsilon}_p = \mathbf{0}, \tag{6.185}$$

  indicating that $\boldsymbol{\upsilon}_p$ is in the null space of $\mathbf{\Phi}(p+1:L,:)$. Let

$$\mathcal{N}\left[\mathbf{\Phi}(p+1:L,:)\right] = \text{span}\left\{\mathbf{z}_1,\ \mathbf{z}_2,\ \cdots,\ \mathbf{z}_{\Omega(p)}\right\} \tag{6.186}$$

  be the null space of $\mathbf{\Phi}(p+1:L,:)$, where $\left\{\mathbf{z}_1,\ \mathbf{z}_2,\ \cdots,\ \mathbf{z}_{\Omega(p)}\right\}$ is an orthonormal basis and $\Omega(p)$ is the dimension of $\mathcal{N}\left[\mathbf{\Phi}(p+1:L,:)\right]$.

Since the first p columns of $\mathbf{H}_{\text{syl}}(p+1:L,:)$ are all zeros, $\Omega(p) \geq p$. Moreover, we know that $\mathcal{N}[\mathbf{\Phi}(2:L,:)]$ is one-dimensional and $\mathbf{\Phi}(p+1:L,:)$ is obtained from $\mathbf{\Phi}(2:L,:)$ by removing the top $(p-1)N$ rows. Therefore, the dimension of $\mathcal{N}[\mathbf{\Phi}(p+1:L,:)]$ is at most $(p-1)N$ larger than that of $\mathcal{N}[\mathbf{\Phi}(2:L,:)]$ and we have

$$p \leq \Omega(p) \leq (p-1)N+1. \tag{6.187}$$

We now express $\boldsymbol{v}_p \in \mathcal{N}[\mathbf{\Phi}(p+1:L,:)]$ as a combination of $\{\mathbf{z}_1, \mathbf{z}_2, \cdots, \mathbf{z}_{\Omega(p)}\}$

$$\boldsymbol{v}_p = \mathbf{Z}_p \mathbf{c}_p, \tag{6.188}$$

where

$$\begin{aligned} \mathbf{Z}_p &= \begin{bmatrix} \mathbf{z}_1 & \mathbf{z}_2 & \cdots & \mathbf{z}_{\Omega(p)} \end{bmatrix}, \\ \mathbf{c}_p &= \begin{bmatrix} c_1 & c_2 & \cdots & c_{\Omega(p)} \end{bmatrix}^T. \end{aligned}$$

Note that $\boldsymbol{v}_p$ is a unit-norm vector and so is $\mathbf{c}_p$.
In order to find $\boldsymbol{v}_p$ or equivalently $\mathbf{c}_p$, we have the following relations to take advantage of:

- The vector $\boldsymbol{v}_p$ is orthogonal to $\boldsymbol{v}_1, \boldsymbol{v}_2, \cdots, \boldsymbol{v}_{p-1}$ so that

$$\begin{bmatrix} \boldsymbol{v}_1^T \\ \boldsymbol{v}_2^T \\ \vdots \\ \boldsymbol{v}_{p-1}^T \end{bmatrix} \boldsymbol{v}_p = \begin{bmatrix} \boldsymbol{v}_1^T \\ \boldsymbol{v}_2^T \\ \vdots \\ \boldsymbol{v}_{p-1}^T \end{bmatrix} \mathbf{Z}_p \mathbf{c}_p = \mathbf{0}. \tag{6.189}$$

- Considering the pth column of $\mathbf{H}_{\text{syl}}$ [see (6.175)] and using (6.179), we have

$$\begin{bmatrix} \mathbf{h}(p-2) \\ \vdots \\ \mathbf{h}(0) \end{bmatrix} = \mathbf{\Phi}(2:p,:)\boldsymbol{v}_p = \mathbf{\Phi}(2:p,:)\mathbf{Z}_p \mathbf{c}_p. \tag{6.190}$$

Putting (6.189) and (6.190) together gives

$$\mathbf{A}_p \boldsymbol{v}_p = \mathbf{A}_p \mathbf{Z}_p \mathbf{c}_p = \boldsymbol{\xi}_p, \tag{6.191}$$

where

$$\mathbf{A}_p = \begin{bmatrix} \boldsymbol{v}_1^T \\ \vdots \\ \boldsymbol{v}_{p-1}^T \\ \mathbf{\Phi}(2:p,:) \end{bmatrix}$$

is a $(p-1)(N+1) \times (2L-1)$ matrix and

$$\boldsymbol{\xi}_p = \begin{bmatrix} \mathbf{0} \\ \mathbf{h}(p-2) \\ \vdots \\ \mathbf{h}(0) \end{bmatrix}$$

is a $(p-1)(N+1) \times 1$ vector.
Since

$$\Omega(p) \leq (p-1)N + 1 \leq (p-1)(N+1), \tag{6.192}$$

there are more equations than unknowns in (6.191) and we can uniquely determine $\mathbf{c}_p$ and $\boldsymbol{v}_p$ as well as $\mathbf{h}(p-1)$ for $2 \leq p \leq L-1$.

- For $p = L$, we can find an equation similar to (6.191) but with no $\mathbf{Z}_p$ as

$$\mathbf{A}_L \boldsymbol{v}_L = \boldsymbol{\xi}_L, \tag{6.193}$$

which will also let us uniquely determine $\boldsymbol{v}_L$. This then completes our approach to estimating $\mathbf{H}$ from (6.176) for $N < L$.

7

Separation and Suppression of Co-Channel and Temporal Interference

7.1 Introduction

Processing information-bearing signals is generally intended to extract useful data about the sources, while the observed signals are usually embedded in noise. An essential feature of MIMO signal processing is that the observed signals are corrupted by both co-channel interference (CCI) from multiple co-existing sources and temporal interference (TI) introduced by multipath propagation, in addition to additive noise. This makes MIMO signal processing much more challenging than its single-channel counterpart. In this chapter, we will analyze the problem of CCI and TI and develop algorithms to separate and suppress CCI and TI in MIMO systems.

Temporal interference is historically better known as inter-symbol interference (ISI), which roots from communications. But in acoustic and speech signal processing, signals do not constitute of symbols and reverberation is a more proper terminology. Since the theme of this book is acoustic MIMO signal processing, temporal interference and reverberation will be interchangeably used. While the algorithms developed in this chapter are targeted at acoustic applications, many of them are also applicable to communication problems. As a result, we believe that it is better to use a more comprehensive concept (i.e., temporal interference) than a more specific definition (either inter-symbol interference or reverberation).

Traditionally TI is suppressed by filtering the distorted signal with a temporal equalizer. When multiple sensors are used, it is possible to perform spatial filtering in the direction of a desired source, which can suppress not only CCI but also the TI that arrives from other directions. The concept of spatial filtering can be combined with temporal equalization, which leads to *spatio-temporal equalization*. In order to help the reader better understand the main idea of spatio-temporal equalization, we will present in Sect. 7.2 an analysis of the relationship between CCI and TI in MIMO systems. A method will be developed for separating CCI and TI by converting a MIMO system into a number of SIMO systems.

While combating temporal interference is by no means a new problem in speech processing and digital communications, with the use of more sensors and in the context of MIMO systems, new challenges and more importantly new opportunities, ideas, and initiatives for solving the problem appear. In Sect. 7.3, we will review traditional and emerging algorithms for suppressing TI in both single-channel and multichannel systems.

7.2 Separating Co-Channel and Temporal Interference

A MIMO system has a complex spatial temporal dependency structure. Interference in the outputs of a MIMO system involves both spatial and temporal processes. As a result, suppressing the interference is generally a difficult and computationally intensive problem. However, under most circumstances (which are quite common in practice), the spatial and temporal components of the interference, i.e., the CCI and TI, respectively, can be separated and their suppression is then carried out in a sequential manner. This makes the computational complexity of the whole procedure much lower. Therefore, we believe that separability of the CCI and TI is a desirable property of a MIMO system.

In the following, we begin our discussion with the example of a simple 2×3 MIMO system and then extend to the more general case for $M \times N$ systems. The condition for separability of the CCI and TI will be presented when it naturally arises. For the ease of presentation, we use the z-transform of the MIMO signal model given by (2.10).

7.2.1 Example: Conversion of a 2 × 3 MIMO System to Two SIMO Systems

For a 2×3 MIMO system, the CCI can be cancelled by using two output signals at a time. For instance, we can remove the interference in $X_1(z)$ and $X_2(z)$ caused by $S_2(z)$ (from the perspective of source 1) as follows:

$$\begin{aligned} X_1(z)H_{22}(z) - X_2(z)H_{12}(z) = \\ \left[H_{11}(z)H_{22}(z) - H_{21}(z)H_{12}(z)\right] S_1(z) + \\ \left[H_{22}(z)B_1(z) - H_{12}(z)B_2(z)\right]. \end{aligned} \tag{7.1}$$

Similarly, the interference caused by $S_1(z)$ (from the perspective of source 2) in these two outputs can also be cancelled. Therefore, by selecting different pairs from the three outputs, we could obtain six CCI-free signals and then could construct two separate single-input three-output systems with respect to two distinct inputs, respectively. This procedure is visualized in Fig. 7.1 and will be described in a more systematic way as follows.

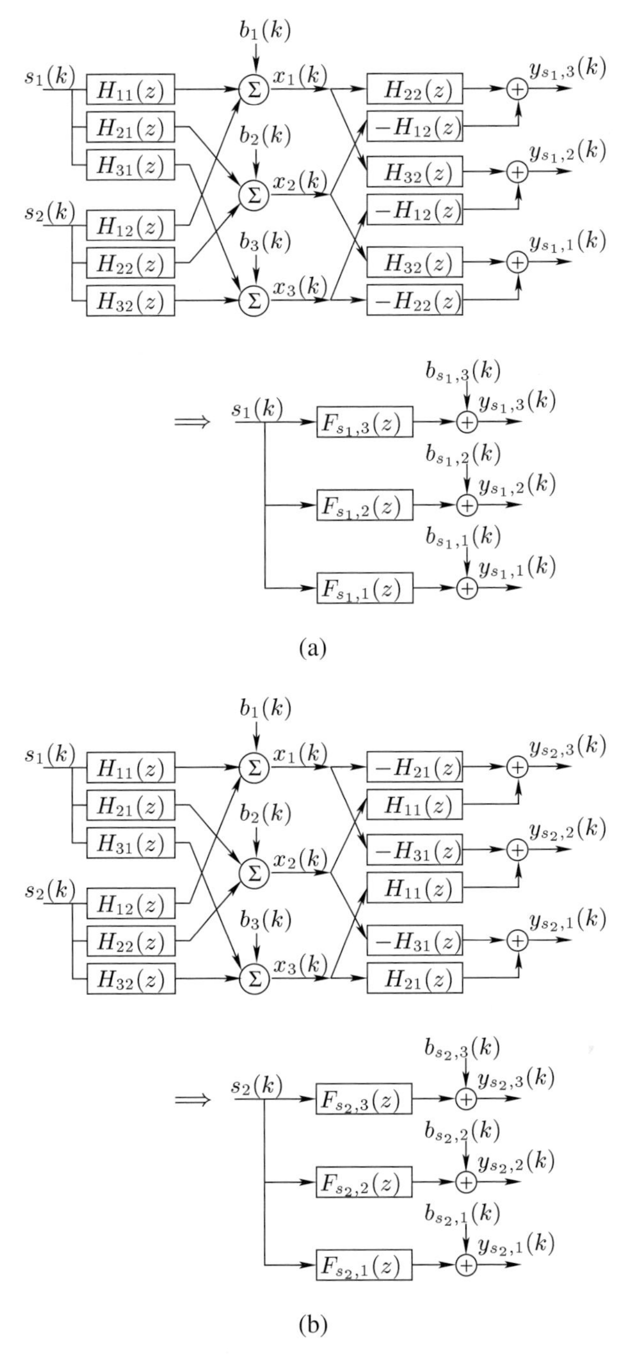

Fig. 7.1. Illustration of the conversion from a 2×3 MIMO system to two CCI-free SIMO systems with respect to (a) $s_1(k)$ and (b) $s_2(k)$.

Let's consider the following equation:

$$\begin{aligned} Y_{s_1,p}(z) &= H_{s_1,p1}(z)X_1(z) + H_{s_1,p2}(z)X_2(z) + H_{s_1,p3}(z)X_3(z) \\ &= \sum_{q=1}^{3} H_{s_1,pq}(z)X_q(z), \quad p = 1,2,3, \end{aligned} \tag{7.2}$$

where $H_{s_1,pp}(z) = 0, \ \forall p$. This means that (7.2) considers only two system outputs for each p. The objective is to find the polynomials $H_{s_1,pq}(z)$, $p, q = 1, 2, 3$, $p \neq q$, in such a way that:

$$Y_{s_1,p}(z) = F_{s_1,p}(z)S_1(z) + B_{s_1,p}(z), \quad p = 1,2,3, \tag{7.3}$$

which represents a SIMO system where s_1 is the source signal, $y_{s_1,p}$ ($p = 1, 2, 3$) are the outputs, $f_{s_1,p}$ are the corresponding channel impulse responses, and $b_{s_1,p}$ is the noise at the pth output. Substituting (2.10) in (7.2) for $X_q(z)$, we deduce that:

$$\begin{aligned} Y_{s_1,1}(z) &= \left[H_{s_1,12}(z)H_{21}(z) + H_{s_1,13}(z)H_{31}(z)\right] S_1(z) \\ &\quad + \left[H_{s_1,12}(z)H_{22}(z) + H_{s_1,13}(z)H_{32}(z)\right] S_2(z) \\ &\quad + H_{s_1,12}(z)B_2(z) + H_{s_1,13}(z)B_3(z), \end{aligned} \tag{7.4}$$

$$\begin{aligned} Y_{s_1,2}(z) &= \left[H_{s_1,21}(z)H_{11}(z) + H_{s_1,23}(z)H_{31}(z)\right] S_1(z) \\ &\quad + \left[H_{s_1,21}(z)H_{12}(z) + H_{s_1,23}(z)H_{32}(z)\right] S_2(z) \\ &\quad + H_{s_1,21}(z)B_1(z) + H_{s_1,23}(z)B_3(z), \end{aligned} \tag{7.5}$$

$$\begin{aligned} Y_{s_1,3}(z) &= \left[H_{s_1,31}(z)H_{11}(z) + H_{s_1,32}(z)H_{21}(z)\right] S_1(z) \\ &\quad + \left[H_{s_1,31}(z)H_{12}(z) + H_{s_1,32}(z)H_{22}(z)\right] S_2(z) \\ &\quad + H_{s_1,31}(z)B_1(z) + H_{s_1,32}(z)B_2(z). \end{aligned} \tag{7.6}$$

As shown in Fig. 7.1, one possibility is to choose:

$$\begin{aligned} H_{s_1,12}(z) &= H_{32}(z), & H_{s_1,13}(z) &= -H_{22}(z), \\ H_{s_1,21}(z) &= H_{32}(z), & H_{s_1,23}(z) &= -H_{12}(z), \\ H_{s_1,31}(z) &= H_{22}(z), & H_{s_1,32}(z) &= -H_{12}(z). \end{aligned} \tag{7.7}$$

In this case, we find that:

$$\begin{aligned} F_{s_1,1}(z) &= H_{32}(z)H_{21}(z) - H_{22}(z)H_{31}(z), \\ F_{s_1,2}(z) &= H_{32}(z)H_{11}(z) - H_{12}(z)H_{31}(z), \\ F_{s_1,3}(z) &= H_{22}(z)H_{11}(z) - H_{12}(z)H_{21}(z), \end{aligned} \tag{7.8}$$

and

$$\begin{aligned} B_{s_1,1}(z) &= H_{32}(z)B_2(z) - H_{22}(z)B_3(z), \\ B_{s_1,2}(z) &= H_{32}(z)B_1(z) - H_{12}(z)B_3(z), \\ B_{s_1,3}(z) &= H_{22}(z)B_1(z) - H_{12}(z)B_2(z). \end{aligned} \tag{7.9}$$

Since $\deg[H_{nm}(z)] = L - 1$, where $\deg[\cdot]$ is the degree of a polynomial, therefore $\deg[F_{s_1,p}(z)] \leq 2L - 2$. We can see from (7.8) that the polynomials $F_{s_1,1}(z)$, $F_{s_1,2}(z)$, and $F_{s_1,3}(z)$ share common zeros if $H_{12}(z)$, $H_{22}(z)$, and $H_{32}(z)$, or if $H_{11}(z)$, $H_{21}(z)$, and $H_{31}(z)$, share common zeros.

Now suppose that

$$C_2(z) = \gcd[H_{12}(z), H_{22}(z), H_{32}(z)], \tag{7.10}$$

where $\gcd[\cdot]$ denotes the greatest common divisor of the polynomials involved. We have:

$$H_{n2}(z) = C_2(z)H'_{n2}(z), \quad n = 1, 2, 3. \tag{7.11}$$

It is clear that the signal s_2 in (7.2) can be canceled by using the polynomials $H'_{n2}(z)$ [instead of $H_{n2}(z)$ as given in (7.7)], so that the SIMO system represented by (7.3) will change to:

$$Y'_{s_1,p}(z) = F'_{s_1,p}(z)S_1(z) + B'_{s_1,p}(z), \quad p = 1, 2, 3, \tag{7.12}$$

where

$$\begin{aligned} F'_{s_1,p}(z)C_2(z) &= F_{s_1,p}(z), \\ B'_{s_1,p}(z)C_2(z) &= B_{s_1,p}(z). \end{aligned}$$

It is worth noticing that $\deg\left[F'_{s_1,p}(z)\right] \leq \deg[F_{s_1,p}(z)]$ and that the polynomials $F'_{s_1,1}(z)$, $F'_{s_1,2}(z)$, and $F'_{s_1,3}(z)$ share common zeros if and only if $H_{11}(z)$, $H_{21}(z)$, and $H_{31}(z)$ share common zeros.

The second SIMO system corresponding to the source s_2 can be derived in a similar way. We can find the output signals:

$$Y_{s_2,p}(z) = F_{s_2,p}(z)S_2(z) + B_{s_2,p}(z), \quad p = 1, 2, 3, \tag{7.13}$$

by enforcing $F_{s_2,p}(z) = F_{s_1,p}(z)$ $(p = 1, 2, 3)$, which leads to:

$$\begin{aligned} B_{s_2,1}(z) &= -H_{31}(z)B_2(z) + H_{21}(z)B_3(z), \\ B_{s_2,2}(z) &= -H_{31}(z)B_1(z) + H_{11}(z)B_3(z), \\ B_{s_2,3}(z) &= -H_{21}(z)B_1(z) + H_{11}(z)B_2(z). \end{aligned}$$

This means that the two SIMO systems [for s_1 and s_2, represented by equations (7.3) and (7.13)] have identical channels but the noise at the outputs is different.

Now let's see what we can do if $H_{n1}(z)$ $(n = 1, 2, 3)$ share common zeros. Suppose that $C_1(z)$ is the greatest common divisor of $H_{11}(z)$, $H_{21}(z)$, and $H_{31}(z)$. Then we have:

$$H_{n1}(z) = C_1(z)H'_{n1}(z), \quad n = 1, 2, 3, \tag{7.14}$$

and the SIMO system of (7.13) becomes:

$$Y'_{s_2,p}(z) = F'_{s_2,p}(z)S_2(z) + B'_{s_2,p}(z), \quad p = 1, 2, 3, \tag{7.15}$$

where

$$\begin{aligned} F'_{s_2,p}(z)C_1(z) &= F_{s_2,p}(z), \\ B'_{s_2,p}(z)C_1(z) &= B_{s_2,p}(z). \end{aligned}$$

We see that

$$\begin{aligned} &\gcd\left[F'_{s_2,1}(z), F'_{s_2,2}(z), F'_{s_2,3}(z)\right] \\ &= \gcd\left[H_{12}(z), H_{22}(z), H_{32}(z)\right] \\ &= C_2(z), \end{aligned} \tag{7.16}$$

and in general $F'_{s_1,p}(z) \neq F'_{s_2,p}(z)$.

7.2.2 Generalization to $M \times N$ MIMO Systems with $M > 2$ and $M < N$

The approach to separating CCI and TI explained in the previous subsection on a simple example will be generalized here to $M \times N$ MIMO systems with $M > 2$ and $M < N$. We begin with denoting $C_m(z)$ as the greatest common divisor of $H_{1m}(z), H_{2m}(z), \cdots, H_{Nm}(z)$ $(m = 1, 2, \cdots, M)$, i.e.,

$$C_m(z) = \gcd\left[H_{1m}(z), H_{2m}(z), \cdots, H_{Nm}(z)\right], \quad m = 1, 2, \cdots, M. \tag{7.17}$$

Then, $H_{nm}(z) = C_m(z)H'_{nm}(z)$ and the channel matrix $\mathbf{H}(z)$ can be rewritten as

$$\mathbf{H}(z) = \mathbf{H}'(z)\mathbf{C}(z), \tag{7.18}$$

where $\mathbf{H}'(z)$ is an $N \times M$ matrix containing the elements $H'_{nm}(z)$ and $\mathbf{C}(z)$ is an $M \times M$ diagonal matrix with $C_m(z)$ as its nonzero components.

Let us choose M from N microphone outputs and we have

$$P = C_N^M = \frac{\Pi_{i=N-M+1}^{N} i}{\Pi_{i=1}^{M} i} \tag{7.19}$$

different ways of doing so. For the pth $(p = 1, 2, \cdots, P)$ combination, we denote the index of the M selected output signals as p_m, $m = 1, 2, \cdots, M$, which together with the M inputs form an $M \times M$ MIMO sub-system.

Consider the following equations:

$$\vec{\mathbf{Y}}_p(z) = \mathbf{H}_{s,p}(z)\vec{\mathbf{X}}_p(z), \quad p = 1, 2, \cdots, P, \tag{7.20}$$

where

$$\vec{\mathbf{Y}}_p(z) = \left[\begin{array}{cccc} Y_{s_1,p}(z) & Y_{s_2,p}(z) & \cdots & Y_{s_M,p}(z) \end{array} \right]^T,$$

$$\mathbf{H}_{s,p}(z) = \begin{bmatrix} H_{s_1,p1}(z) & H_{s_1,p2}(z) & \cdots & H_{s_1,pM}(z) \\ H_{s_2,p1}(z) & H_{s_2,p2}(z) & \cdots & H_{s_2,pM}(z) \\ \vdots & \vdots & \vdots & \vdots \\ H_{s_M,p1}(z) & H_{s_M,p2}(z) & \cdots & H_{s_M,pM}(z) \end{bmatrix},$$

$$\vec{\mathbf{X}}_p(z) = \left[\begin{array}{cccc} X_{p_1}(z) & X_{p_2}(z) & \cdots & X_{p_M}(z) \end{array} \right]^T.$$

Let $\mathbf{H}_p(z)$ be the $M \times M$ matrix obtained from the system's channel matrix $\mathbf{H}(z)$ by keeping its rows corresponding to the M selected output signals. Then similar to (2.10), we have

$$\vec{\mathbf{X}}_p(z) = \mathbf{H}_p(z)\vec{\mathbf{S}}(z) + \vec{\mathbf{B}}_p(z), \tag{7.21}$$

where

$$\vec{\mathbf{B}}_p(z) = \left[\begin{array}{cccc} B_{p_1}(z) & B_{p_2}(z) & \cdots & B_{p_M}(z) \end{array} \right]^T.$$

Substituting (7.21) into (7.20) yields

$$\vec{\mathbf{Y}}_p(z) = \mathbf{H}_{s,p}(z)\mathbf{H}_p(z)\vec{\mathbf{S}}(z) + \mathbf{H}_{s,p}(z)\vec{\mathbf{B}}_p(z). \tag{7.22}$$

In order to remove the CCI, the objective here is to find the matrix $\mathbf{H}_{s,p}(z)$ whose components are linear combinations of $H_{nm}(z)$ such that the product $\mathbf{H}_{s,p}(z)\mathbf{H}_p(z)$ would be a diagonal matrix. Consequently, we have

$$\begin{aligned} Y_{s_m,p}(z) &= F_{s_m,p}(z)S_m(z) + B_{s_m,p}(z), \\ & m = 1, 2, \cdots, M, \quad p = 1, 2, \cdots, P. \end{aligned} \tag{7.23}$$

If $\mathbf{C}_p(z)$ [obtained from $\mathbf{C}(z)$ in a similar way as $\mathbf{H}_p(z)$ is constructed] is not equal to the identity matrix, then $\mathbf{H}_p(z) = \mathbf{H}'_p(z)\mathbf{C}_p(z)$, where $\mathbf{H}'_p(z)$ has full column normal rank[1] (i.e. nrank $\left[\mathbf{H}'_p(z)\right] = M$, see [298] for a definition of normal rank), as we assume for separability of CCI and TI in a MIMO system. Thereafter, the CCI-free signals are determined as

$$\vec{\mathbf{Y}}'_p(z) = \mathbf{H}'_{s,p}(z)\mathbf{H}'_p(z)\mathbf{C}_p(z)\vec{\mathbf{S}}(z) + \mathbf{H}'_{s,p}(z)\vec{\mathbf{B}}_p(z), \tag{7.24}$$

and

$$Y'_{s_m,p}(z) = F'_{s_m,p}(z)S_m(z) + B'_{s_m,p}(z). \tag{7.25}$$

Obviously a good choice for $\mathbf{H}'_{s,p}(z)$ to make the product $\mathbf{H}'_{s,p}(z)\mathbf{H}'_p(z)$ a diagonal matrix is the adjoint of matrix $\mathbf{H}'_p(z)$, i.e., the (i, j)-th element

[1] For a square matrix $M \times M$, the normal rank is full if and only if the determinant, which is a polynomial in z, is not identically zero for all z. In this case, the rank is less than M only at a finite number of points in the z plane.

of $\mathbf{H}'_{s,p}(z)$ is the (j,i)-th cofactor of $\mathbf{H}'_p(z)$. Consequently, the polynomial $F'_{s_m,p}(z)$ would be the determinant of $\mathbf{H}'_p(z)$. Since $\mathbf{H}'_p(z)$ has full column normal rank, its determinant is not equal to zero and the polynomial $F'_{s_m,p}(z)$ is not trivial.

Since

$$F'_{s_m,p}(z) = \sum_{q=1}^{M} H'_{s_m,pq}(z) H_{p_q m}(z) \tag{7.26}$$

and $H'_{s_m,pq}(z)$ $(q = 1, 2, \cdots, M)$ are co-prime, the polynomials $F'_{s_m,p}(z)$ $(p = 1, 2, \cdots, P)$ share common zeros if and only if the polynomials $H_{nm}(z)$ $(n = 1, 2, \cdots, N)$ share common zeros. Therefore, if the channels with respect to any one input are co-prime for an (M, N) MIMO system, we can convert it into M CCI-free SIMO systems whose P channels are also co-prime, i.e., their channel matrices are irreducible.

Also, it can easily be checked that $\deg\left[F'_{s_m,p}(z)\right] \leq M(L-1)$. As a result, the length of the FIR filter $f'_{s_m,p}$ would be

$$L_f \leq M(L-1) + 1. \tag{7.27}$$

Before we finish this section, we would like to comment in a little bit more detail on the condition for separability of CCI and TI in a MIMO system. For an $M \times M$ MIMO system or an $M \times M$ subsystem of a larger $M \times N$ $(M < N)$ MIMO system, it is now clear that the *reduced* channel matrix $\mathbf{H}'_p(z)$ needs to have full column normal rank such that the CCI and TI are separable. But what happens and why is the CCI unable to be separated from the TI if $\mathbf{H}'_p(z)$ does not have full column normal rank? Let's first examine a 2×2 system and its reduced channel matrix is given by

$$\mathbf{H}'_p(z) = \begin{bmatrix} H'_{p,11}(z) & H'_{p,12}(z) \\ H'_{p,21}(z) & H'_{p,22}(z) \end{bmatrix}. \tag{7.28}$$

If $\mathbf{H}'_p(z)$ does not have full column normal rank, then there exist two non-zero polynomials $A_1(z)$ and $A_2(z)$ such that

$$\begin{bmatrix} H'_{p,11}(z) \\ H'_{p,21}(z) \end{bmatrix} A_1(z) = \begin{bmatrix} H'_{p,12}(z) \\ H'_{p,22}(z) \end{bmatrix} A_2(z), \tag{7.29}$$

or equivalently

$$\mathbf{H}'_p(z) \begin{bmatrix} A_1(z) & -A_2(z) \end{bmatrix}^T = \mathbf{0}. \tag{7.30}$$

As a result, in the absence of noise, we know that

$$X_{p,1}(z) = -\frac{A_2(z)}{A_1(z)} X_{p,2}(z), \tag{7.31}$$

which implies that the MISO systems corresponding to the two outputs are identical up to a constant filter. Therefore the 2×2 MIMO is reduced to a

2×1 MISO system where the number of inputs is greater than the number of outputs and the CCI cannot be separated from the TI.

For an $M \times M$ MIMO system with $M > 2$, if $\mathbf{H}'_p(z)$ does not have full column normal rank, then there are only $\left\{\text{nrank}\left[\mathbf{H}'_p(z)\right]\right\}$ independent MISO systems and the other $\left\{M - \text{nrank}\left[\mathbf{H}'_p(z)\right]\right\}$ MISO systems can be reduced. This indicates that the MIMO system has essentially more inputs than outputs and the CCI cannot be separated from the TI.

Extracting $C_m(z)$ $(m = 1, 2, \cdots, M)$ from the mth column of $\mathbf{H}(z)$ (if necessary) is intended to reduce the SIMO system with respect to each input. The purpose of examining the column normal rank of $\mathbf{H}'_p(z)$ is to check the dependency of the MISO systems associated with the outputs. For the $M \times N$ MIMO systems $(M < N)$, the column normal rank of $\mathbf{H}'(z)$ actually indicates how many MISO subsystems are independent. As long as nrank $[\mathbf{H}'(z)] \geq M$, there exists at least one $M \times M$ subsystem whose M MISO systems are all independent and whose CCI and TI are separable. Therefore the condition for separability of CCI and TI in an $M \times N$ MIMO system is nothing more than to require that there are more effective outputs than inputs. This condition is quite commonly met in practice, particularly in acoustic systems.

7.3 Suppressing Temporal Interference

Temporal interference caused by multipath propagation or reverberation makes source signals distorted at the channel outputs. In order to suppress the TI and recover the original source signals, an equalizer can be applied to the output signals. Channel equalization is a widely-used signal processing technique which uses a filter at the outputs to compensate for the distortions in the channels. In this section, we will review the algorithms for suppressing TI. The output signals are assumed to have no co-channel interference and consequently we consider only SISO and SIMO systems.

While there have been a number of ways to classify current channel equalization methods, an insightful classification is based on whether the knowledge of the inputs needs to be shared with the receivers in a training phase for estimating the channel impulse responses. Using this criterion, we have the family of *supervised* equalizers and the family of *unsupervised* also called *blind* equalizers. Apparently emerging blind equalization techniques are more desirable and easier to use in practice since the receivers do not have to know the source signals and a training phase is unnecessary. But if the reader has a good knowledge of blind channel identification discussed in Chap. 6 and traditional supervised equalization methods, understanding the development of a blind equalization algorithm is not a problem. In the following, we suppose that the channel impulse responses are known or have been estimated with enough accuracy, and we only explain how an equalizer suppresses TI.

There are three widely used channel equalization approaches to suppressing TI, as illustrated in Fig. 7.2. They are the direct inverse also called zero-

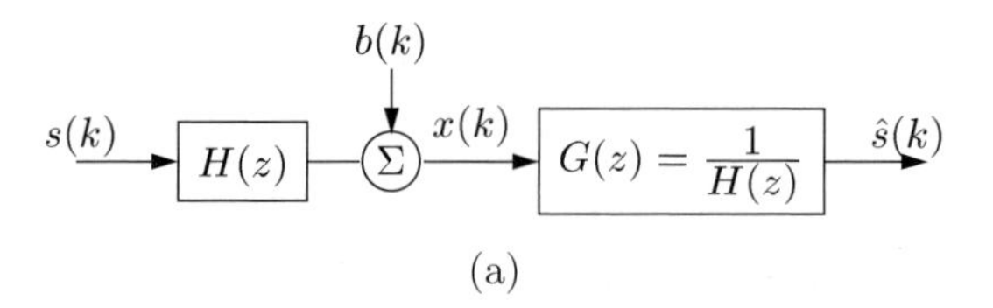

(a)

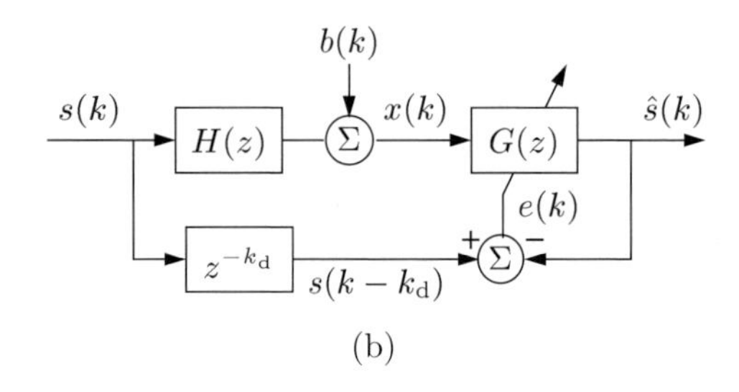

(b)

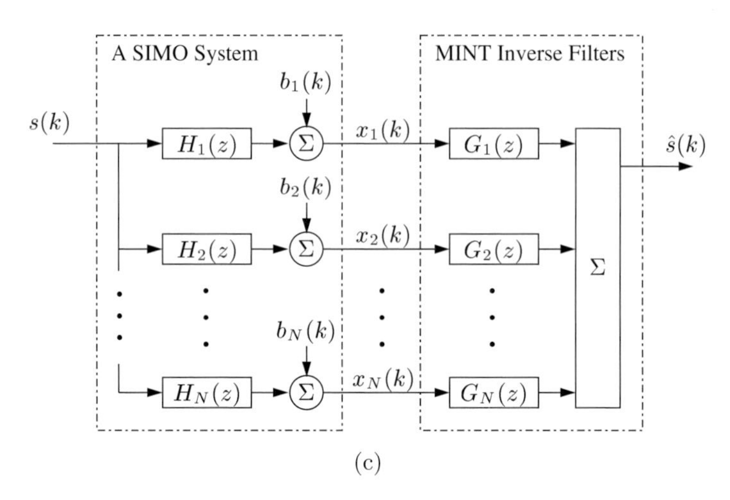

(c)

Fig. 7.2. Illustration of three widely-used channel equalization approaches to suppressing temporal interference: (a) direct inverse (or zero-forcing), (b) minimum mean square error, and (c) the MINT method.

forcing equalizer, the minimum mean square error (MMSE) equalizer, and the MINT (multichannel inverse theorem) equalizer. The first two approaches work for SISO systems and the third for SIMO systems.

7.3.1 Direct Inverse (Zero-Forcing) Equalizer

Among all available channel equalization methods, the most straightforward is the direct inverse method. As shown in Fig. 7.2(a), the equalizer filter is determined by inverting the channel transfer function $H(z)$ which is the z-

transform of the channel impulse response:

$$G(z) = \frac{1}{H(z)}. \tag{7.32}$$

In practice, the inverse filter g needs to be *stable* and *causal*. This requires that the channel impulse response of the SISO system is a minimum-phase sequence, as explained in Sect. 2.2.6. If a SISO system does not have a minimum-phase channel impulse response like most, if not all, of acoustic SISO systems, then even though a stable inverse filter can still be found by using an all-pass filter, it will be IIR which is noncausal and has a long delay. These drawbacks make direct-inverse equalizers impracticable, particularly for acoustic applications.

7.3.2 MMSE Equalizer

The second approach is the MMSE method, which estimates the equalization filter by minimizing the mean square error between the recovered signal and its reference, as shown in Fig. 7.2(b). The error signal is given by

$$\begin{aligned} e(k) &\stackrel{\triangle}{=} s(k - k_\mathrm{d}) - \hat{s}(k) \\ &= s(k - k_\mathrm{d}) - g * x(k), \end{aligned} \tag{7.33}$$

where k_d is the decision delay for the equalizer. Then the equalization filter g is determined as the one that minimizes the MMSE cost function $E\left\{e^2(k)\right\}$ as follows:

$$\hat{g}_\mathrm{MMSE} = \arg\min_g E\left\{e^2(k)\right\}, \tag{7.34}$$

which is a typical problem in estimation theory. The solution can be found with well-known adaptive or recursive algorithms.

For minimum-phase SISO systems, it can be shown that the MMSE equalizer is the same as the direct-inverse or zero-forcing equalizer. But for non-minimum-phase SISO systems, the MMSE method essentially equalizes the channel by inverting only those components whose zeros are inside the unit circle [226]. In addition, to employ the MMSE equalizer, a reference signal needs to be accessible. However, although the MMSE method has these limitations, it is quite useful in practice and has been successfully applied for many communication and acoustic systems.

7.3.3 MINT Equalizers

For a SIMO system, let's consider the polynomials $G_n(z)$ $(n = 1, 2, \cdots, N)$ and the following equation:

$$\begin{aligned}\hat{S}(z) &= \sum_{n=1}^{N} G_n(z)X_n(z) \\ &= \left[\sum_{n=1}^{N} H_n(z)G_n(z)\right] S(z) + \sum_{n=1}^{N} G_n(z)B_n(z).\end{aligned} \tag{7.35}$$

The polynomials $G_n(z)$ should be found in such a way that $\hat{S}(z) = S(z)$ in the absence of noise by using the Bezout theorem which is mathematically expressed as follows:

$$\begin{aligned}&\gcd\left[H_1(z), H_2(z), \cdots, H_N(z)\right] = 1 \\ \Leftrightarrow\ &\exists\ G_1(z), G_2(z), \cdots, G_N(z)\ :\ \sum_{n=1}^{N} H_n(z)G_n(z) = 1.\end{aligned} \tag{7.36}$$

In other words, as long as the channel impulse responses h_n are co-prime (even though they may not be minimum phase), i.e., the SIMO system is irreducible, there exists a group of g filters to completely remove the TI and perfectly recover the source signals. The idea of using the Bezout theorem for equalizing a SIMO system was first proposed in [226] in the context of room acoustics, where the principle is more widely referred to as the MINT theory.

If the channels of the SIMO system share common zeros, i.e.,

$$C(z) = \gcd\left[H_1(z), H_2(z), \cdots, H_N(z)\right] \neq 1, \tag{7.37}$$

then we have

$$H_n(z) = C(z)H_n'(z), \quad n = 1, 2, \cdots, N, \tag{7.38}$$

and the polynomials $G_n(z)$ can be found such that

$$\sum_{n=1}^{N} H_n'(z)G_p(z) = 1. \tag{7.39}$$

In this case, (7.35) becomes:

$$\hat{S}(z) = C(z)S(z) + \sum_{n=1}^{N} G_n(z)B_n(z). \tag{7.40}$$

We see that by using the Bezout theorem, the *reducible* SIMO system can be equalized up to the polynomial $C(z)$. So when there are common zeros, the MINT equalizer can only partially suppress the TI. For more complete cancellation of the TI, we have to combat the effect of $C(z)$ using either the direct inverse or MMSE methods, which depends on whether $C(z)$ is a minimum phase filter.

To find the MINT equalization filters, we write the Bezout equation (7.36) in the time domain as:

$$\mathbf{H}^{\mathrm{c}}\mathbf{g} = \sum_{n=1}^{N} \mathbf{H}_n^{\mathrm{c}}\mathbf{g}_n = \mathbf{e}_1, \tag{7.41}$$

where

$$\begin{aligned}
\mathbf{H}^{\mathrm{c}} &= \begin{bmatrix} \mathbf{H}_1^{\mathrm{c}} & \mathbf{H}_2^{\mathrm{c}} & \cdots & \mathbf{H}_N^{\mathrm{c}} \end{bmatrix}, \\
\mathbf{g} &= \begin{bmatrix} \mathbf{g}_1^T & \mathbf{g}_2^T & \cdots & \mathbf{g}_N^T \end{bmatrix}^T, \\
\mathbf{g}_n &= \begin{bmatrix} g_{n,0} & g_{n,1} & \cdots & g_{n,L_g-1} \end{bmatrix}^T,
\end{aligned}$$

L_g is the length of the FIR filter g_n,

$$\mathbf{H}_n^{\mathrm{c}} = \begin{bmatrix}
h_{n,0} & 0 & \cdots & 0 \\
h_{n,1} & h_{n,0} & \cdots & 0 \\
\vdots & \ddots & \ddots & \vdots \\
h_{n,L-1} & \cdots & \cdots & \vdots \\
0 & h_{n,L-1} & \ddots & \vdots \\
\vdots & \vdots & \ddots & \vdots \\
0 & \cdots & 0 & h_{n,L-1}
\end{bmatrix}$$

is an $(L + L_g - 1) \times L_g$ convolution matrix (superscript c), and

$$\mathbf{e}_1 = \begin{bmatrix} 1 & 0 & \cdots & 0 \end{bmatrix}^T$$

is an $(L + L_g - 1) \times 1$ vector.

In order to have a unique solution for (7.41), L_g must be chosen in such a way that $\mathbf{H}^{\mathrm{c}}$ is a square matrix. In this case, we have:

$$L_g = \frac{L-1}{N-1}. \tag{7.42}$$

However, this may not be practical since $(L-1)/(N-1)$ is not necessarily always an integer. Therefore, a larger L_g is usually chosen and solve (7.41) for $\mathbf{g}$ in the least squares sense as follows:

$$\hat{\mathbf{g}}_{\mathrm{MINT}} = \mathbf{H}^{\mathrm{c}\dagger}\mathbf{e}_1, \tag{7.43}$$

where

$$\mathbf{H}^{\mathrm{c}\dagger} = \left(\mathbf{H}^{\mathrm{c}T}\mathbf{H}^{\mathrm{c}}\right)^{-1}\mathbf{H}^{\mathrm{c}T}$$

is the pseudo-inverse of the matrix $\mathbf{H}^{\mathrm{c}}$. If a decision delay k_{d} is taken into account, then the equalization filters turn out to be

$$\hat{\mathbf{g}}_{\text{MINT}} = \mathbf{H}^{\text{c}\dagger}\mathbf{e}_{k_\text{d}}, \tag{7.44}$$

where

$$\mathbf{e}_{k_\text{d}} = \Big[\underbrace{0 \ \cdots \ 0}_{k_\text{d}} \ 1 \ \underbrace{0 \ \cdots \ 0}_{L+L_g-k_\text{d}-2}\Big]^T.$$

7.4 Summary

In MIMO systems, temporal interference (TI) and co-channel interference (CCI) are the two major sources of signal impairment and distortion. In order to recover the source signals, spatio-temporal equalization needs to be performed to suppress both TI and CCI. Through the discussion of separability of CCI and TI in the first half of this chapter, insight was offered about the relationship of CCI and TI in MIMO systems. A practical algorithm was presented to separate CCI and TI by converting an $M \times N$ ($M < N$) MIMO system into M CCI-free SIMO systems. In the second half of this chapter, we briefly reviewed traditional and emerging algorithms for combating TI in SISO and SIMO systems. They included the direct-inverse (or zero-forcing), the minimum mean square error (MMSE), and the MINT (multichannel inverse theorem) equalizers. The theory that was laid out in this chapter on separation and suppression of TI and CCI will be used in the development of algorithms for source separation and speech dereverberation in later chapters.

Part II

Applications

8

Acoustic Echo Cancellation and Audio Bridging

This chapter gives an overview on acoustic echo cancellation. Echo cancellation is a mature topic, so the emphasis here will be on recent advances such as multichannel acoustic echo cancellation and audio bridging.

8.1 Introduction

In general, conversations take place in the presence of echoes. The acoustics of rooms almost always consists of reflections from walls, floors, and ceilings. Sound reflections with short round trip delays, on the order of a few milliseconds, are perceived as spectral coloration or reverberation and are generally preferred over anechoic environments. When a sound reflection's round-trip delay exceeds a few tens of milliseconds and is unattenuated or only slightly attenuated, it is perceived as a distinct echo, and when the round-trip delay approaches a quarter of a second most people find it difficult to carry on a normal conversation.

Hands-free communications are becoming a must in teleconferencing systems (between two rooms or more), especially when there are several persons in the same room. However, in such systems, the acoustic coupling between the loudspeaker and the microphone at each end, associated with the overall delay, would produce echoes that would make conversations very difficult. Furthermore, the acoustic system could become very instable. Obviously, the echo problem is greatly amplified when the number of loudspeakers and microphones is increased.

The best solution we know so far that eliminates echoes and allows full-duplex communication is the acoustic echo canceler (AEC). This device is composed of three components: an adaptive filter, a double-talk detector, and a nonlinear processor. The role of the adaptive filter is to estimate the acoustic echo path of the system and to create a replica of the echo which is then subtracted from the microphone signal. Adaptive filters are generously covered in Part I of this book. Double-talk occurs when people at both ends

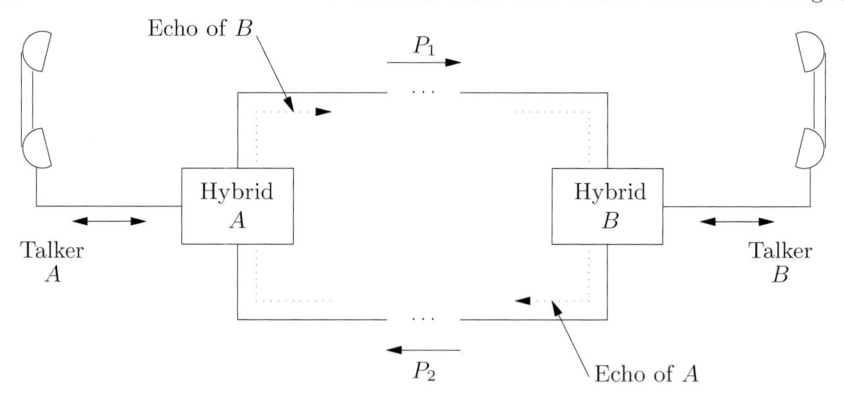

Fig. 8.1. Illustration of a long-distance connection showing local 2-wire loops connected through hybrids to a 4-wire long-line network. (This figure is borrowed from [38].)

are talking at the same time. The role of a double-talk detector is to sense this double-talk situation and to stop the updating of the filter coefficients in order to avoid divergence of the adaptive algorithm. This important function is discussed in Sect. 8.6. Finally, a nonlinear processing scheme is added after echo cancellation in order to provide further attenuation of the residual echo that the echo canceler can not remove. This last component is not further discussed here since its design is more art than science.

In the next section, we will briefly discuss the network echo problem. The progress in this area has helped considerably in finding solutions to the acoustic echo problem.

8.2 Network Echo Problem

Most of the control techniques for acoustic echoes were first developed to address the network echo problem. That is why we discuss in this section the network or line echo problem, which originates from a device called hybrid. Figure 8.1 shows a simplified long distance connection and the placement of the hybrids.

Every conventional analog telephone in a given geographical area is connected to a central office by a two-wire line, called the customer loop, which serves for communication in either direction. A local call is set up by simply connecting the two customer loops at the central office. When the distance between the two telephones exceeds about 35 miles, amplification becomes necessary. Therefore, a separate path is needed for each direction of transmission. The hybrid (or hybrid transformer) connects the four-wire part of the circuit to the two-wire portion at each end. With reference to Fig. 8.1, the purpose of the hybrids is to allow signals from Talker A to go along the path P_1 to B, and to go from Talker B along the path P_2 to A. However, Hybrid B

must prevent signals in path P_1 from returning along the path P_2 back to A. Similarly, the signal in path P_2 is to be prevented from returning along path P_1 back to B.

A hybrid is a bridge network that can achieve the aforementioned objectives, provided the impedance of the customer loop can be exactly balanced by an impedance located at the hybrid. Unfortunately, this is not possible in practice because there are far fewer four-wire circuits than there are two-wire circuits. Therefore, a hybrid may be connected to any of the customer loops served by the central office. By their very nature, customer loops have a wide variety of characteristics — various lengths, type of wire, type of telephone, number of extension phones, etc. It appears, therefore, that the echo at the hybrid cannot be completely eliminated, especially with long delays [282], [38].

In the 1920's Clark and Mathes [82] introduced the echo suppressor. Nominally, this is a voice activated switch at the 4-wire termination at each end of the long distance connection. Suppose the talker at end A (also known as the "far end") has been talking for a while. Based on the level of signals in the paths P_1 and P_2, a decision is made as to whether the signal in P_2 is an interruption by B (also known as the "near end") trying to break into the conversation or an echo of A's speech. If the decision is the latter, the circuit P_2 is opened (or a large loss is switched in). A similar switch at the other end prevents B's echo from returning to B. During so-called "double-talk" periods, when speech from speakers A and B is present simultaneously at the echo suppressor, echo suppression is inhibited so that A hears the speech from B superimposed on self echo from A.

In general, echo suppressors do not perform very well, and can cause the following degradations as observed by the far-end user [138]:

- Echo caused by slow initial far-end speech detection or false near-end speech detection.
- Clipping of the near-end speech due to slow initial interruption switching.
- Chopping of the near-end speech due to early release of the interruption state.

The most serious of these problems is the clipping and chopping of speech in the long round-trip delay type connections such as satellite links. Clearly, the echo suppressor does not allow full-duplex communication. In other words, talkers at both ends can not really speak at the same time. It's more like a half-duplex communication.

In the 1960's a new echo control device, the echo canceler [202], [280], [281], [13] was introduced. It better addresses the problems associated with the echo suppressor. A digital network echo canceler is shown in Fig. 8.2. We show only one canceler located at the end of B; a similar canceler should be used at the other end (for A). The basic idea is to generate a synthetic echo $\hat{x}(k)$ from A's speech, $s(k)$, and subtract it from the return signal, $x(k)$, to produce the outgoing signal $e(k)$ on the path P_2.

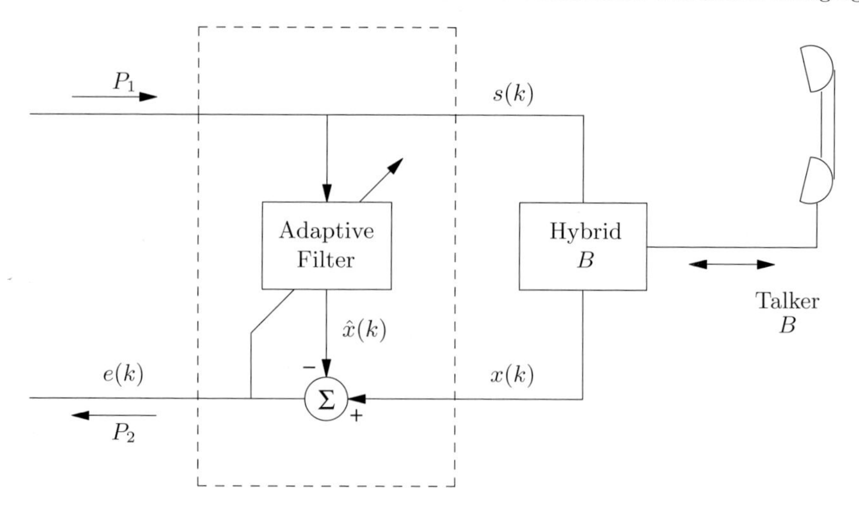

Fig. 8.2. Echo canceler.

The echo path is, in general, not stationary that is why an adaptive filter is required to estimate it. Therefore, adaptive filters (see Chaps. 3, 4, and 5) play a major role in echo cancellation. The adaptive filter will do a fine job as long as the near-end signal is silent. When the power of the near-end signal is no longer negligible, it will badly affect the adaptive algorithm. For this reason, practical echo cancelers need to inhibit adaptation of the filter taps when significant near-end signal is present. This is accomplished by a function known as a near-end speech detector (see Sect. 8.6).

The impulse response of the echo path as seen from the 4-wire termination is typically characterized by a pure delay followed by the impulse response of the hybrid and two-wire circuits whose length is about 8 ms. Adaptive filters of 32 ms in length are usually sufficient to cope with most circuits of this nature.

8.3 Single-Channel Acoustic Echo Cancellation

In the previous section, we discussed the problem of echoes in the network which is due to the impedance mismatches at hybrids. Another source of echoes is due to acoustic coupling between loudspeakers and microphones in hands-free telephony and in teleconferencing systems. The loudspeaker-enclosure-microphone system can be seen as a SISO system whose input and output are the loudspeaker and microphone, respectively. Thus, the role of a single-channel acoustic echo canceler (AEC) is to estimate the impulse response (acoustic echo path) of this system and use this estimate to make a replica of the echo that is subtracted from the microphone signal, before sending out the resulting signal (see Fig. 8.3).

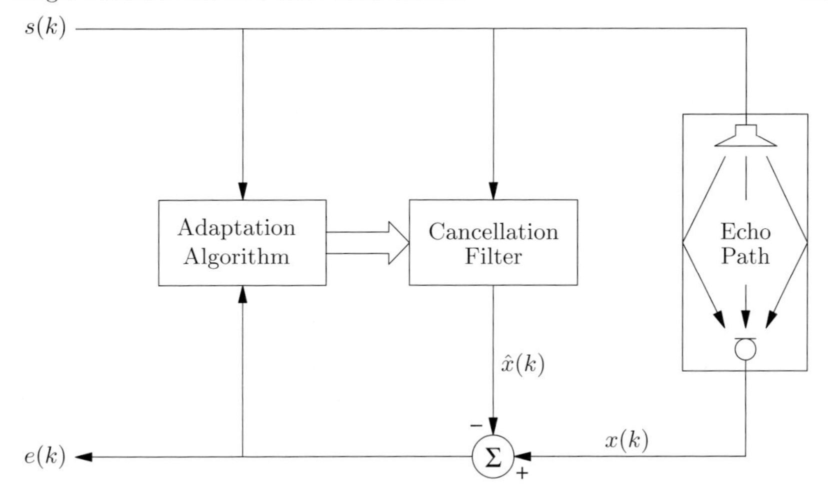

Fig. 8.3. Acoustic echo canceler.

The ultimate goal of acoustic echo control is to provide true full-duplex hands-free communication so that normal speech patterns, including polite interjections, can be made without disruption of speech. The apparent solution is echo cancellation. Unfortunately, the acoustic environment is more hostile than the network environment to echo cancelers. The main problems associated with the acoustic environment are:

- The echo path is extremely long, on the order of 125 ms.
- The echo path may rapidly change at any time during the connection (due to an opening room, a moving person, or even after a slight change of the ambient temperature).

The excessive length of the acoustic echo path in time, especially when compared to the network echo path, is mainly due to the slow speed of sound through air. Multiple reflections off walls, furniture, and other objects in the room also serve to lengthen the response. The energy of a reverberating signal in a room is a function of the room's size and the materials inside it, different materials having different reflection coefficients. For a typical office, the reverberation time is about 200 to 300 ms. Therefore, to reduce the acoustic echo of the typical office by 30 dB, a 100 to 150 ms length echo canceler is required. At an 8 KHz sample rate, this implies an adaptive filter on the order of 1000 taps is needed.

To complicate matters more, the impulse response of the room is not static over time. It varies with ambient temperature, pressure, and humidity. In addition, movement of objects, such as human bodies, doors, and the location of microphones and speakers can all dramatically and rapidly modify the acoustic impulse response [138].

Therefore, the choice of the adaptive filter in this application is crucial. Research is still very active in this area with the hope to find adaptive algorithms that can track very quickly changes in acoustic impulse responses. Some interesting ideas in that direction are presented in Chap. 4.

8.4 Multichannel Acoustic Echo Cancellation

8.4.1 Multi versus Mono

Why do we need multichannel sound for teleconferencing systems? Let's take the following example. When we are in a room with several people talking, working, laughing, or just communicating with each other, thanks to our binaural auditory system, we can concentrate on one particular talker (if several persons are talking at the same time), localize or identify a person who is talking, and somehow we are able to process a noisy or a reverberant speech signal in order to make it intelligible. On the other hand, with only one ear or, equivalently, if we record what happens in the room with one microphone and listen to this monophonic signal, it will likely make all of the above mentioned tasks much more difficult. So, multichannel sound teleconferencing systems provide a realistic presence that single-channel systems cannot offer. The ultimate goal of multichannel presentation in a tele-meeting, is to make it as if it was a local meeting where all the attendees will have the impression that they are in the same room.

In such hands-free systems, multichannel acoustic echo cancelers (MCAECs) are absolutely necessary for full-duplex communication [283]. We assume that the teleconferencing system is organized between two rooms: the "transmission" and "receiving" rooms. The transmission room is sometimes referred to as the far-end and the receiving room as the near-end. So each room needs an MCAEC for each microphone. The MCAECs model an acoustic MIMO system whose M inputs and N outputs are loudspeakers and microphones, respectively. In our teleconferencing system example, the M MCAECs consist of MN adaptive filters aiming at identifying MN echo paths from M loudspeakers to N microphones. In the following, we detail the concept on one microphone in the receiving room, knowing that the same approach applies to the other microphones. Thus, multichannel acoustic echo cancellation consists of direct identification of a MISO, unknown linear system, consisting of the parallel combination of M acoustic paths $(h_1, h_2, \cdots, h_M)$ extending through the receiving room from the loudspeakers to the microphone. The multichannel AEC tries to model this unknown system by M adaptive filters $(\hat{h}_1, \hat{h}_2, \cdots, \hat{h}_M)$. Figure 8.4 illustrates the concept of stereophonic (two-channel) echo cancellation between a transmission room (where the source is shown) and a receiving room for one microphone; similar analysis will apply to the other microphone signals.

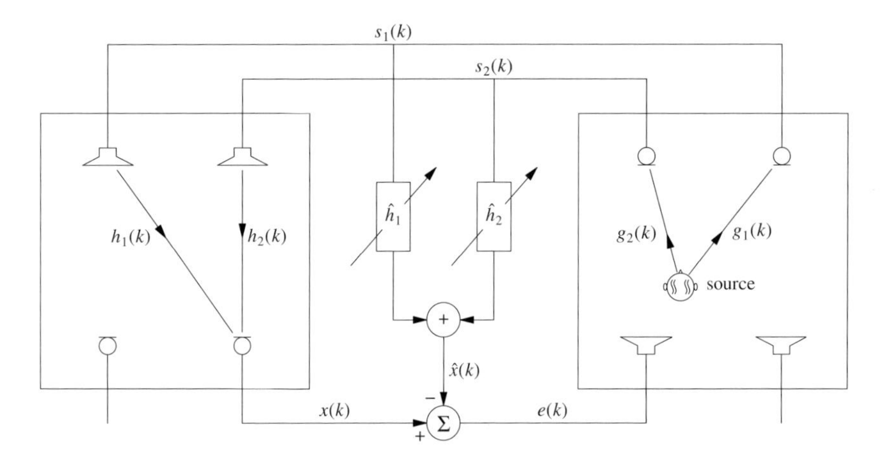

Fig. 8.4. Schematic diagram of stereophonic acoustic echo cancellation.

Although conceptually very similar, multichannel acoustic echo cancellation (MCAEC) is fundamentally different from traditional mono echo cancellation in one respect: a straightforward generalization of the mono echo canceler would not only have to track changing echo paths in the receiving room, *but also in the transmission room*! For example, the canceler would have to reconverge if one talker stops talking and another starts talking at a different location in the transmission room. There is no adaptive algorithm that can track such a change sufficiently fast and this scheme therefore results in poor echo suppression. Thus, a generalization of the mono AEC in the multichannel case does not result in satisfactory performance.

The theory explaining the problem of MCAEC was described in [283] and [24]. The fundamental problem is that the multiple channels may carry linearly related signals which in turn may make the normal equations to be solved by the adaptive algorithm singular. This implies that there is no unique solution to the equations but an infinite number of solutions, and it can be shown that all but the true one depend on the impulse responses of the transmission room. As a result, intensive studies of how to properly handle this have been conducted. It is shown in [24] that the only solution to the nonuniqueness problem is to reduce the coherence between the different signals, and an efficient low complexity method for this purpose is also given.

Lately, attention has been focused on the investigation of other methods that decrease the cross-correlation between the channels in order to get well behaved estimates of the echo paths [126], [144], [273], [5], [197]. The main problem is how to reduce the coherence sufficiently without affecting the stereo perception and the sound quality.

The performance of the MCAEC is more severely affected by the choice of algorithm than the monophonic counterpart [35], [128]. This is easily recognized since the performance of most adaptive algorithms depends on the condition number of the input signal covariance matrix. In the multichannel case, as shown in Chap. 3, the condition number depends on the coherence. So when the input signals are highly coherent, adaptive algorithms, such as the NLMS, that do not take this coherence into account, converge very slowly to the true solution. A framework for multichannel adaptive filtering can be found in [23]. A very successful frequency-domain algorithm for MCAEC is presented in Chap. 5.

8.4.2 Multichannel Identification and the Nonuniqueness Problem

In this subsection, we show that the normal equations for the multichannel identification problem do not have a unique solution as in the single-channel case. Indeed, since the M input signals are obtained by filtering from a common source, a problem of nonuniqueness is expected [283]. In the following discussion, we suppose that the length (L) of the impulse responses (in the transmission and receiving rooms) is equal to the length of the modeling filters.

Assume that the system (transmission room) is linear and time invariant; therefore, we have the following $[M(M-1)/2]$ relations [23], [24]:

$$\mathbf{s}_m^T(k)\mathbf{g}_i = \mathbf{s}_i^T(k)\mathbf{g}_m, \quad m, i = 1, 2, \cdots, M; \; m \neq i, \tag{8.1}$$

where

$$\mathbf{s}_m(k) = \begin{bmatrix} s_m(k) & s_m(k-1) & \cdots & s_m(k-L+1) \end{bmatrix}^T, \quad m = 1, 2, \cdots, M,$$

are vectors of signal samples at the microphone outputs in the transmission room and the impulse response vectors between the source and the microphones are defined as

$$\mathbf{g}_m = \begin{bmatrix} g_{m,0} & g_{m,1} & \cdots & g_{m,L-1} \end{bmatrix}^T, \quad m = 1, 2, \cdots, M.$$

Now, let us define the recursive least-squares error criterion with respect to the modeling filters:

$$J_{\mathrm{LS}}(k) = \sum_{i=0}^{k} \lambda^{k-i} e^2(i), \tag{8.2}$$

where λ $(0 < \lambda < 1)$ is a forgetting factor and

$$e(k) = x(k) - \sum_{m=1}^{M} \hat{\mathbf{h}}_m^T \mathbf{s}_m(k) \tag{8.3}$$

is the error signal at time k between the microphone output $x(k)$ in the receiving room and its estimate, where

$$\hat{\mathbf{h}}_m = \left[\begin{array}{cccc} \hat{h}_{m,0} & \hat{h}_{m,1} & \cdots & \hat{h}_{m,L-1} \end{array} \right]^T, \; m = 1, 2, \cdots, M,$$

are the M modeling filters.

The minimization of (8.2) leads to the normal equations:

$$\mathbf{R}_{ss}(k)\hat{\mathbf{h}}(k) = \mathbf{p}_{sx}(k), \tag{8.4}$$

where

$$\begin{aligned} \mathbf{R}_{ss}(k) &= \sum_{i=0}^{k} \lambda^{k-i}\mathbf{s}(i)\mathbf{s}^T(i) \\ &= \begin{bmatrix} \mathbf{R}_{s_1s_1}(k) & \mathbf{R}_{s_1s_2}(k) & \cdots & \mathbf{R}_{s_1s_M}(k) \\ \mathbf{R}_{s_2s_1}(k) & \mathbf{R}_{s_2s_2}(k) & \cdots & \mathbf{R}_{s_2s_M}(k) \\ \vdots & \vdots & \ddots & \vdots \\ \mathbf{R}_{s_Ms_1}(k) & \mathbf{R}_{s_Ms_2}(k) & \cdots & \mathbf{R}_{s_Ms_M}(k) \end{bmatrix} \end{aligned} \tag{8.5}$$

is an estimate of the input signal covariance matrix,

$$\begin{aligned} \mathbf{p}_{sx}(k) &= \sum_{i=0}^{k} \lambda^{k-i}\mathbf{s}(i)x(i) \\ &= \left[\begin{array}{cccc} \mathbf{p}_{s_1x}^T(k) & \mathbf{p}_{s_2x}^T(k) & \cdots & \mathbf{p}_{s_Mx}^T(k) \end{array} \right]^T \end{aligned} \tag{8.6}$$

is an estimate of the cross-correlation vector between the input and output signals (in the receiving room), and

$$\hat{\mathbf{h}}(k) = \left[\begin{array}{cccc} \hat{\mathbf{h}}_1^T(k) & \hat{\mathbf{h}}_2^T(k) & \cdots & \hat{\mathbf{h}}_M^T(k) \end{array} \right]^T,$$

$$\mathbf{s}(k) = \left[\begin{array}{cccc} \mathbf{s}_1^T(k) & \mathbf{s}_2^T(k) & \cdots & \mathbf{s}_M^T(k) \end{array} \right]^T.$$

Consider the following vector:

$$\mathbf{u} = \left[\begin{array}{cccc} \sum_{m=2}^{M} \varsigma_m \mathbf{g}_m^T & -\varsigma_2 \mathbf{g}_1^T & \cdots & -\varsigma_M \mathbf{g}_1^T \end{array} \right]^T,$$

where ς_m are arbitrary factors. We can verify, using (8.1), that $\mathbf{R}_{ss}(k)\mathbf{u} = \mathbf{0}_{ML\times 1}$, so $\mathbf{R}_{ss}(k)$ is not invertible. Vector $\mathbf{u}$ represents the nullspace of matrix $\mathbf{R}_{ss}(k)$. The dimension of this nullspace depends on the number of channels and is equal to $(M-2)L+1$ (for $M \geq 2$), see Chap. 3 for more details. So the problem becomes worse as M increases. Thus, there is no unique solution to the problem and an adaptive algorithm will drive to any one of many possible solutions, which can be very different from the "true" desired solution $\hat{\mathbf{h}} = \mathbf{h}$.

These nonunique "solutions" are dependent on the impulse responses in the transmission room:

$$\hat{\mathbf{h}}_1 = \mathbf{h}_1 + \beta \sum_{m=2}^{M} \varsigma_m \mathbf{g}_m, \tag{8.7}$$

$$\hat{\mathbf{h}}_m = \mathbf{h}_m - \beta \varsigma_m \mathbf{g}_1, \ m = 2, \cdots, M, \tag{8.8}$$

where β is an arbitrary factor. This, of course, is intolerable because $\mathbf{g}_m$ can change instantaneously, for example, as one person stops talking and another starts [283], [24].

8.4.3 Impulse Response Tail Effect

We first define an important measure that is very useful for MCAEC.

Definition: The quantity

$$\frac{\|\mathbf{h} - \hat{\mathbf{h}}\|_2}{\|\mathbf{h}\|_2}, \tag{8.9}$$

where $\|\cdot\|_2$ denotes the two-norm vector, is called the *normalized misalignment* and measures the mismatch between the impulse responses of the receiving room and the modeling filters. In the multichannel case, it is possible to have good echo cancellation even when the misalignment is large. However, in such a case, the cancellation will degrade if the $\mathbf{g}_m$ change. A main objective of MCAEC research is to avoid this problem.

Actually, for the practical case when the length of the adaptive filters is smaller than the length of the impulse responses in the transmission room, there is a unique solution to the normal equations, although the covariance matrix is very ill-conditioned.

On the other hand, we can easily show by using the classical normal equations that if the length of the adaptive filters is smaller than the length of the impulse responses in the receiving room, we introduce an important bias in the coefficients of these filters because of the strong cross-correlation between the input signals and the large condition number of the covariance matrix [24]. So in practice, we may have poor misalignment even if there is a unique solution to the normal equations.

The only way to decrease the misalignment is to partially decorrelate two-by-two the M input (loudspeaker) signals. As shown in Chap. 3, the correlation between two channels can be linked to ill-conditioning of the covariance matrix by means of the coherence magnitude. Ill-conditioning can therefore be monitored by the coherence function which serves as a measure of achieved decorrelation. In the next subsection, we summarize a number of approaches that have been developed recently for reducing the cross-correlation. In practice, it is always possible to choose not to distort the input signals and still have good echo cancellation. However, the MCAEC may not be very stable.

8.4.4 Some Different Solutions for Decorrelation

If we have M different channels, we need to decorrelate them partially and mutually. In the following, we show how to partially decorrelate two channels. The same process should be applied for all the channels. It is well-known that the coherence magnitude between two processes is equal to 1 if and only if they are linearly related. In order to weaken this relation, some non-linear or time-varying transformation of the stereo channels has to be made. Such a transformation reduces the coherence and hence the condition number of the covariance matrix, thereby improving the misalignment. However, the transformation has to be performed cautiously so that it is inaudible and has no effect on stereo perception.

A simple nonlinear method that gives good performance uses a half-wave rectifier [24], so that the nonlinearly transformed signal becomes

$$s'_m(k) = s_m(k) + nl\frac{s_m(k) + |s_m(k)|}{2}, \tag{8.10}$$

where nl is a parameter used to control the amount of nonlinearity. For this method, there can only be a linear relation between the nonlinearly transformed channels if $\forall k$, $s_1(k) \geq 0$ and $s_2(k) \geq 0$ or if we have $cs_1(k - \tau_1) = s_2(k - \tau_2)$ with $c > 0$. In practice however, these cases never occur because we always have zero-mean signals and $\mathbf{g}_1$, $\mathbf{g}_2$ are never related by just a simple delay.

An improved version of this technique is to use positive and negative half-wave rectifiers on each channel respectively,

$$s'_1(k) = s_1(k) + nl\frac{s_1(k) + |s_1(k)|}{2}, \tag{8.11}$$

$$s'_2(k) = s_2(k) + nl\frac{s_2(k) - |s_2(k)|}{2}. \tag{8.12}$$

This principle removes the linear relation even in the special signal cases given above.

Experiments show that stereo perception is not affected by the above methods even with nl as large as 0.5. Also, the distortion introduced for speech is hardly audible because of the nature of the speech signal and psychoacoustic masking effects [230]. This is explained by the following three reasons. First, the distorted signal s'_m depends only on the instantaneous value of the original signal s_m so that during periods of silence, no distortion is added. Second, the periodicity remains unchanged. Third, for voiced sounds, the harmonic structure of the signal induces "self-masking" of the harmonic distortion components. This kind of distortion is also acceptable for some music signals but may be objectionable for pure tones.

Other types of nonlinearities for decorrelating speech signals have also been investigated and compared [235]. The results indicate that, of the several nonlinearities considered, ideal half-wave rectification and smoothed half-wave rectification appear to be the best choices for speech. For music, the non-linearity parameter of the ideal rectifier must be readjusted. The smoothed rectifier does not require this readjustment but is a little more complicated to implement.

A subjectively meaningful measure to compare s_m and s'_m is not easy to find. A mathematical measure of distance, to be useful in speech processing, has to have a high correlation between its numerical value and the subjective distance judgment, as evaluated on real speech signals [257]. Since many psychoacoustic studies of perceived sound differences can be interpreted in terms of differences of spectral features, measurement of spectral distortion can be argued to be reasonable both mathematically and subjectively.

A very useful distortion measure is the Itakura-Saito (IS) measure [257], given as

$$d_{\mathrm{IS},i} = \frac{\mathbf{a}_i'^T \mathbf{R}_{ii} \mathbf{a}_i'}{\mathbf{a}_i^T \mathbf{R}_{ii} \mathbf{a}_i} - 1, \tag{8.13}$$

where $\mathbf{R}_{ii}$ is the Toeplitz autocorrelation matrix of the LPC model $\mathbf{a}_i$ of a speech signal frame $\mathbf{s}_i$ and $\mathbf{a}_i'$ is the LPC model of the corresponding distorted speech signal frame $\mathbf{s}_i'$. Many experiments in speech recognition show that if the IS measure is less than about 0.1, the two spectra that we compare are perceptually nearly identical. Simulations show that with a nonlinearity (half-wave) $nl = 0.5$, the IS metric is still small (about 0.03).

We could also use the ensemble interval histogram (EIH) distance (which is based on the EIH model) [142]. The interest in using this distance lies in its capability to mimic human judgement of quality. Indeed, according to [142] EIH is a very good predictor of mean opinion score (MOS), but only if the two speech observations under comparison are similar enough, which is the case here. Therefore, this measure should be a good predictor of speech signal degradation when nonlinear distortion is used. Simulations show that for a half-wave nonlinearity with $nl = 0.5$, the EIH distance is 1.8×10^{-3}, which is as good as a 32 kb/s ADPCM coder.

As an example, Fig. 8.5 shows the coherence magnitude between the right and left channels from the transmission room for a speech signal with $nl = 0$ [Fig. 8.5(a)] and $nl = 0.3$ [Fig. 8.5(b)]. We can see that a small amount of non-linearity such as the half-wave rectifier reduces the coherence. As a result, the fast two-channel RLS algorithm converges much better to the true solution with a reduced coherence (see Fig. 8.6).

In [273], a similar approach with non-linearities is proposed. The idea is expanded so that four adaptive filters operate on different non-linearly processed signals to estimate the echo paths. These non-linearities are chosen such that the input signals of two of the adaptive filters are independent, which thus represent a "perfect" decorrelation. Tap estimates are then copied to a fixed two-channel filter which performs the echo cancellation with the unprocessed signals. The advantage of this method is that the NLMS algorithm could be used instead of more sophisticated algorithms.

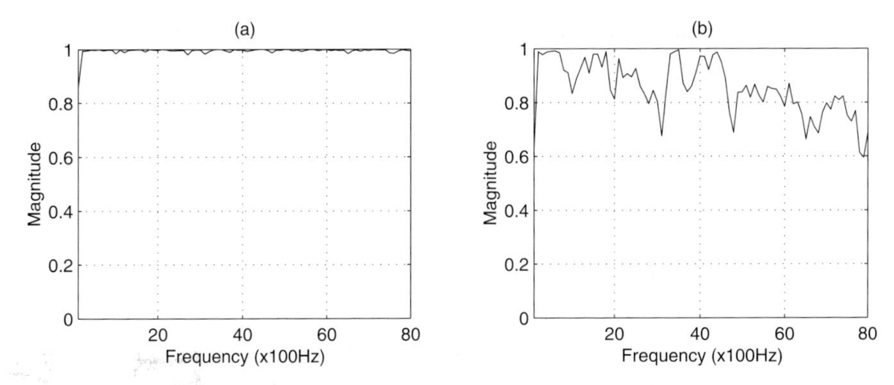

Fig. 8.5. Coherence magnitude between the right and left channels from the transmission room for a speech signal with (a) $nl = 0$ and (b) $nl = 0.3$.

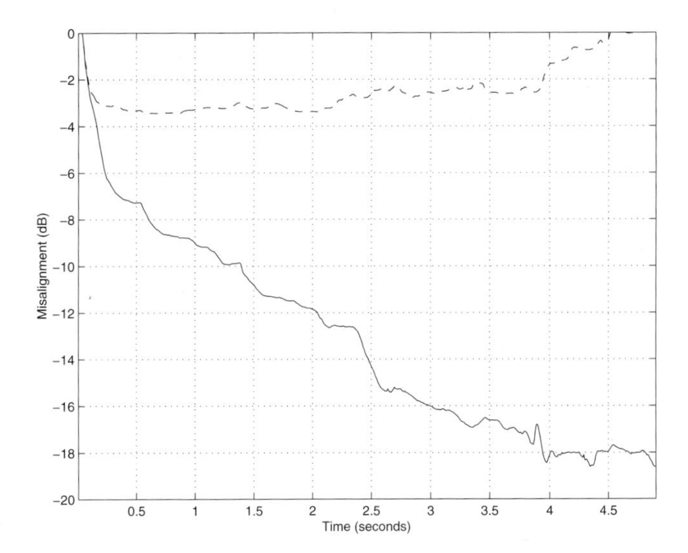

Fig. 8.6. Convergence of the misalignment using the fast two-channel RLS algorithm with $nl = 0$ (- -) and $nl = 0.3$ (—).

Another approach that makes it possible to use the NLMS algorithm is to decorrelate the channels by means of complementary comb filtering [283], [27]. The technique is based on removing the energy in a certain frequency band of the speech signal in one channel. This means the coherence would become zero in this band and thereby results in fast alignment of the estimate even when using the NLMS algorithm. Energy is removed complementarily between the channels so that the stereo perception is not severely affected for frequencies above 1 kHz. However, this method must be combined with some other decorrelation technique for lower frequencies [26].

Two methods based on introducing time-varying filters in the transmission path were presented in [5], [197]. To show that this gives the right effect we can look at the condition for perfect cancellation (under the assumption that we do not have any zeros in the input signal spectrum) in the frequency domain which can be written as,

$$\mathcal{E}_1(\omega)G_1(\omega)T_1(\omega,k) + \mathcal{E}_2(\omega)G_2(\omega)T_2(\omega,k) = 0, \; \forall\omega \in [-\pi,\pi], \quad (8.14)$$

where $\mathcal{E}_m(\omega)$, $m = 1,2$, are the Fourier transforms of the filter coefficient misalignment vectors $\mathbf{h}_m - \hat{\mathbf{h}}_m$. The responses $G_m(\omega)$, $m = 1,2$, are the Fourier transformed transmission room echo paths and $T_m(\omega,k), m = 1,2$, are the frequency representation of the introduced time-varying filters. We see again for constant $T_m(\omega)$, $m = 1,2$, the solution of (8.14) is not unique. However, for a time-varying $T_m(\omega,k), m = 1,2$, (8.14) can only be satisfied if $\mathcal{E}_m(\omega) = 0$, $\forall\omega$, $m = 1,2$, which is the idea for this method. In [5], left and right signals are filtered through two independent time-varying first-order all-pass filters. Stochastic time-variation is introduced by making the pole position of the filter a random walk process. The actual position is limited by the constraints of stability and inaudibility of the introduced distortion. While significant reduction in correlation can be achieved for higher frequencies with the imposed constraints, the lower frequencies are still fairly unaffected by the time-variation. In [197], a periodically varying filter is applied to one channel so that the signal is either delayed by one sample or passed through without delay. A transition zone between the delayed and non-delayed state is also employed in order to reduce audible discontinuities. This method may also affect the stereo perception.

Although the idea of adding independent perceptually shaped noise to the channels was mentioned in [283], [24], thorough investigations of the actual benefit of the technique was not presented. Results regarding variants of this idea can be found in [126], [144]. A pre-processing unit estimating the masking threshold and adding an appropriate amount of noise was proposed in [144]. It was also noted that adding a masked noise to each channel may affect the spatialization of the sound even if the noise is inaudible at each channel separately. This effect can be controlled through correction of the masking threshold when appropriate. In [126], the improvement of misalignment was studied in the SAEC when a perceptual audio coder was added in the transmission path. Reduced correlation between the channels was shown by means of coherence analysis, and improved convergence rate of the adaptive algorithm was observed. A low-complexity method for achieving additional decorrelation by modifying the decoder was also proposed. The encoder cannot quantize every single frequency band optimally due to rate constraints. This has the effect that there is a margin on the masking threshold which can be exploited. In the presented method, the masking threshold is estimated from the modified discrete cosine transform (MDCT) coefficients delivered by the encoder, and an appropriate inaudible amount of decorrelating noise is added to the signals.

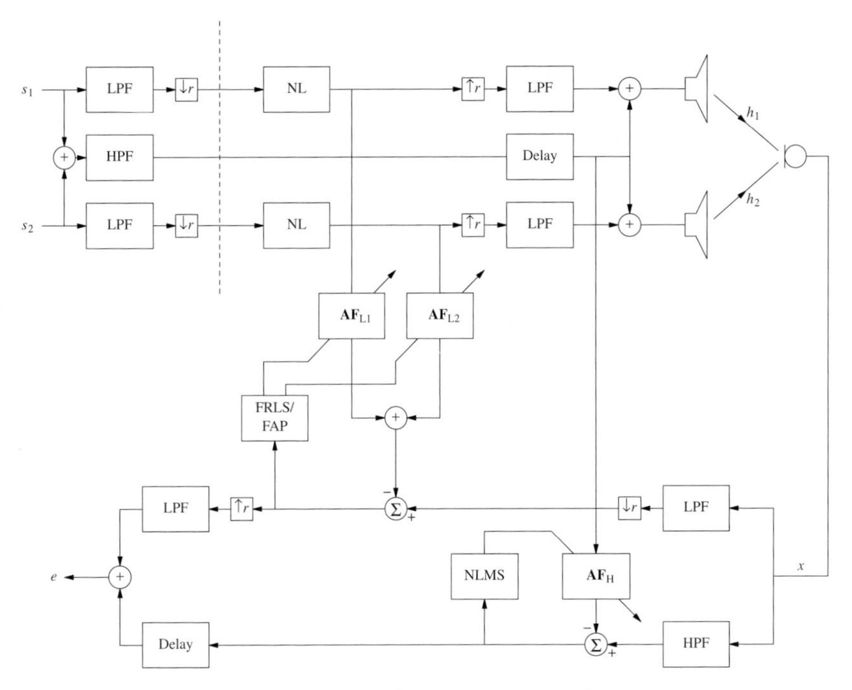

Fig. 8.7. Hybrid mono/stereo acoustic echo canceler.

8.5 Hybrid Mono/Stereo Acoustic Echo Canceler

This solution was first proposed in [26] and is a good compromise between the complexity of a fullband stereo AEC and spatial realism. This idea can obviously be easily extended to more than two channels. As a matter of fact, for a large number of channels, we definitely recommend to devise an AEC based on the approach explained below, since it is more feasible from a practical point of view.

The understanding of the stereo effect from a psychoacoustical point of view is not easy, but many experiments show that the dominant stereophonic cues are located below 1 kHz [323]. In many applications, this information could be exploited for efficient transmission of the microphone signals to the receiving room and also for devising an efficient AEC.

Based on the above psychoacoustical principle, Fig. 8.7 shows a way to transmit the two microphone signals to the receiving room and a set of AECs matched to these signals. First, the two signals s_1 and s_2 (left and right) are lowpass filtered in order to keep just the frequencies below 1 kHz for realizing the stereo effect. Then, the sum of the left and right channels is highpass filtered to keep the frequencies above 1 kHz as a monophonic signal. Finally, the left (resp. right) loudspeaker signal in the receiving room is the sum of the low frequencies of the left (resp. right) channel and high frequencies of both channels. We now note that two AECs are necessary: one mono AEC for high frequencies and one stereo AEC for low frequencies. Note that we have to put a nonlinear transformation (NL) in each low-frequency channel to help the adaptive algorithm to converge toward the "true" solution.

The above structure is much simpler than a fullband system, despite the fact that we have two different AECs. Indeed, for the stereo AEC, since the maximum frequency is $f_c = 1000$ Hz, we can subsample the signals by a factor $r = f_s/(2f_c)$, where f_s is the sampling rate of the system. As a result, the arithmetic complexity is divided by r^2 in comparison with a fullband implementation (the number of taps and the number of computations per second are both reduced by r). In this case, we can afford to use a rapidly converging adaptive algorithm like the fast two-channel RLS. On the other hand, the simple NLMS algorithm can be used to update the filter coefficients in the high frequency band. Of course, we can also use more efficient algorithms such as frequency-domain adaptive filters (see Chap. 5).

Thus, with this structure, the complexity of the system is decreased and the convergence rate of the adaptive algorithms is increased, while preserving most of the stereo effect. Based on this idea, different structures can be devised for a hybrid mono/multichannel AEC. It is up to the reader's imagination and the needs of the application.

8.6 Double-Talk Detection

Double-talk detectors (DTDs) are vital to the operation and performance of acoustic echo cancelers (AECs). In this section, we consider double-talk detection for single-channel AECs. Generalization to the multichannel case of most DTDs is often possible and some ideas are presented in [40].

8.6.1 Basics

Ideally, AECs remove undesired echoes that result from coupling between the loudspeaker and the microphone used in full-duplex hands-free telecommunication systems. The far-end speech signal $s(k)$ goes through the echo path represented by a filter $h(k)$, then it is picked up by the microphone together with the near-end talker signal $v(k)$ and ambient noise $b(k)$. The microphone signal is denoted as $x(k) = \mathbf{h}^T\mathbf{s}(k) + v(k) + b(k)$. Most often the echo path is modeled by an adaptive FIR filter, $\hat{h}(k)$, that generates a replica of the echo. This echo estimate is then subtracted from the return channel and thereby cancellation is achieved. This may look like a simple straightforward system identification task for the adaptive filter. However, in most conversation there are so called double-talk situations that make the identification much more problematic than what it might appear at a first glance. Double-talk occurs when the speech of the two talkers arrive simultaneously at the echo canceler, i.e. $s(k) \neq 0$ and $v(k) \neq 0$ (the situation with near-end talk only, $s(k) = 0$

and $v(k) \neq 0$, can be regarded as an "easy-to-detect" double-talk case). In the double-talk situation, the near-end speech acts as a large level uncorrelated noise to the adaptive algorithm. The disturbing near-end speech may cause the adaptive filter to diverge. Hence, annoying audible echo will pass through to the far-end. The usual way to alleviate this problem is to slow down or completely halt the filter adaptation when presence of near-end speech is detected. This is the very important role of the so called DTD. The basic double-talk detection scheme is based on computing a detection statistic, ζ, and comparing it with a preset threshold, T.

Double-talk detectors basically operate in the same manner. Thus, the general procedure for handling double-talk is described by the following:

1. A detection statistic ζ is formed using available signals, e.g. s, x, e, etc, and the estimated filter coefficients $\hat{h}$.
2. The detection statistic ζ is compared to a preset threshold T, and double-talk is declared if $\zeta < T$.
3. Once double-talk is declared, the detection is held for a minimum period of time T_{hold}. While the detection is held, the filter adaptation is disabled.
4. If $\zeta \geq T$ consecutively over a time T_{hold}, the filter resumes adaptation, while the comparison of ζ to T continues until $\zeta < T$ again.

The hold time T_{hold} in Step 3 and Step 4 is necessary to suppress detection dropouts due to the noisy behavior of the detection statistic. Although there are some possible variations, most of the DTD algorithms keep this basic form and only differ in how to form the detection statistic.

An "optimum" decision variable ζ for double-talk detection will behave as follows:

(i) if $v(k) = 0$ (double-talk is not present), $\zeta \geq T$.
(ii) if $v(k) \neq 0$ (double-talk is present), $\zeta < T$.
(iii) ζ is insensitive to echo path variations.

The threshold T must be a constant, independent of the data. Moreover, it is desirable that the decisions are made without introducing any delay (or minimize the introduced delay) in the updating of the adaptive filter. The delayed decisions will otherwise affect the AEC algorithm negatively.

A large number of DTD schemes have been proposed since the introduction of echo cancelers [281]. The Geigel algorithm [101] has proven successful in network echo cancelers; however, it does not always provide reliable performance when used in the acoustic situation. This is because it assumes a minimum echo path attenuation which may not be valid in the acoustic case. Other methods based on cross-correlation and coherence [329], [315], [125], [31], [37] have been studied which appear to be more appropriate for the acoustic application. Spectral comparing methods [253] and two-microphone solutions have also been proposed [207]. A DTD based on multi statistic testing in combination with modeling of the echo path by two filters is proposed in [242]. Next, we summarize some DTD proposals.

8.6.2 Double-Talk Detection Algorithms

In this subsection, we explain different DTD algorithms that can be useful for AEC. We start with the Geigel algorithm since it was the very first DTD proposal.

Geigel Algorithm

A very simple algorithm due to A. A. Geigel [101] is to declare the presence of near-end speech whenever

$$\zeta^{(\mathrm{g})} = \frac{\max\{\,|s(k)|, \cdots, |s(k-L_{\mathrm{g}}+1)|\,\}}{|x(k)|} < T, \tag{8.15}$$

where L_{g} and T are suitably chosen constants. This detection scheme is based on a waveform level comparison between the microphone signal $x(k)$ and the far-end speech $s(k)$ assuming the near-end speech $v(k)$ in the microphone signal will be typically stronger than the echo $\mathbf{h}^T\mathbf{s}$. The maximum or l_∞ norm of the L_{g} most recent samples of $s(k)$ is taken for the comparison because of the undetermined delay in the echo path. The threshold T is to compensate for the energy level of the echo path response h, and is often set to 2 for network echo cancelers because the hybrid loss is typically about 6 dB or more. For an AEC, however, it is not clear how to set a universal threshold to work reliably in all the various situations because the loss through the acoustic echo path can vary greatly depending on many factors. For L_{g}, one choice is to set it the same as the adaptive filter length L since we can assume that the echo path is covered by this length.

Cross-Correlation Method

In [329], the cross-correlation coefficient vector between $\mathbf{s}$ and e was proposed as a means for double-talk detection. A similar idea using the cross-correlation coefficient vector between $\mathbf{s}$ and x has proven more robust and reliable [315], [78]. We will therefore focus on the cross-correlation coefficient vector between $\mathbf{s}$ and x which is defined as

$$\begin{aligned}\mathbf{c}_{sx}^{(1)} &= \frac{E\{\mathbf{s}(k)x(k)\}}{\sqrt{E\{s^2(k)\}E\{x^2(k)\}}} \\ &= \frac{\mathbf{p}_{sx}}{\sigma_s\sigma_x} \\ &= \left[\begin{array}{cccc} c_{sx,0}^{(1)} & c_{sx,1}^{(1)} & \cdots & c_{sx,L-1}^{(1)} \end{array}\right]^T,\end{aligned} \tag{8.16}$$

where $c_{sx,l}^{(1)}$ is the cross-correlation coefficient between $s(k-l)$ and $x(k)$.

The idea here is to compare

$$\begin{aligned}\zeta^{(1)} &= \left\|\mathbf{c}_{sx}^{(1)}\right\|_{\infty} \\ &= \max_{l}\left|c_{sx,l}^{(1)}\right|, \; l = 0, 1, \cdots, L-1,\end{aligned} \tag{8.17}$$

to a threshold level T. The decision rule will be very simple: if $\zeta^{(1)} \geq T$, then double-talk is not present; if $\zeta^{(1)} < T$, then double-talk is present.

Although the l_∞ norm used in (8.17) is perhaps the most natural, other scalar metrics, e.g., l_1, l_2, could alternatively be used to assess the cross-correlation coefficient vectors. However, there is a fundamental problem here which is not linked to the type of metric used. The problem is that these cross-correlation coefficient vectors are not well normalized. Indeed, we can only say in general that $\zeta^{(1)} \leq 1$. If $v(k) = 0$, that does not imply that $\zeta^{(1)} = 1$ or any other known value. We do not know the value of $\zeta^{(1)}$ in general. The amount of correlation will depend a great deal on the statistics of the signals and of the echo path. As a result, the best value of T will vary a lot from one experiment to another. So there is no natural threshold level associated with the variable $\zeta^{(1)}$ when $v(k) = 0$.

The next subsection presents a decision variable that exhibits better properties than the cross-correlation algorithm. This decision variable is formed by properly normalizing the cross-correlation vector between $\mathbf{s}$ and x.

Normalized Cross-Correlation Method

There is a simple way to normalize the cross-correlation vector between a vector $\mathbf{s}$ and a scalar x in order to have a natural threshold level for ζ when $v(k) = 0$.

Suppose that $v(k) = 0$. In this case:

$$\sigma_x^2 = \mathbf{h}^T \mathbf{R}_{ss} \mathbf{h}, \tag{8.18}$$

where $\mathbf{R}_{ss} = E\{\mathbf{s}(k)\mathbf{s}^T(k)\}$. Since $x(k) = \mathbf{h}^T\mathbf{s}(k)$, we have

$$\mathbf{p}_{sx} = \mathbf{R}_{ss}\mathbf{h}, \tag{8.19}$$

and (8.18) can be re-written as

$$\sigma_x^2 = \mathbf{p}_{sx}^T \mathbf{R}_{ss}^{-1} \mathbf{p}_{sx}. \tag{8.20}$$

In general for $v(k) \neq 0$ we have,

$$\sigma_x^2 = \mathbf{p}_{sx}^T \mathbf{R}_{ss}^{-1} \mathbf{p}_{sx} + \sigma_v^2. \tag{8.21}$$

If we divide (8.20) by (8.21) and compute its square root, we obtain the decision variable [37], [31],

$$\begin{aligned}\zeta^{(2)} &= \sqrt{\mathbf{p}_{sx}^T(\sigma_x^2\mathbf{R}_{ss})^{-1}\mathbf{p}_{sx}} \\ &= \|\mathbf{c}_{sx}^{(2)}\|_2,\end{aligned} \tag{8.22}$$

where

$$\mathbf{c}_{sx}^{(2)} = (\sigma_x^2 \mathbf{R}_{ss})^{-1/2} \mathbf{p}_{sx} \tag{8.23}$$

is what we will call the normalized cross-correlation vector between $\mathbf{s}$ and x.

Substituting (8.19) and (8.21) into (8.22), we show that the decision variable is:

$$\zeta^{(2)} = \frac{\sqrt{\mathbf{h}^T \mathbf{R}_{ss} \mathbf{h}}}{\sqrt{\mathbf{h}^T \mathbf{R}_{ss} \mathbf{h} + \sigma_v^2}}. \tag{8.24}$$

We easily deduce from (8.24) that for $v(k) = 0$, $\zeta^{(2)} = 1$ and for $v(k) \neq 0$, $\zeta^{(2)} < 1$. Note also that $\zeta^{(2)}$ is not sensitive to changes of the echo path when $v = 0$.

For the particular case when x is white Gaussian noise, the autocorrelation matrix is diagonal: $\mathbf{R}_{ss} = \sigma_s^2 \mathbf{I}$. Then (8.23) becomes:

$$\begin{aligned} \mathbf{c}_{sx}^{(2)} &= \frac{\mathbf{p}_{sx}}{\sigma_s \sigma_x} \\ &= \mathbf{c}_{sx}^{(1)}. \end{aligned} \tag{8.25}$$

Hence, in general what we are doing in (8.22) is equivalent to prewhitening the signal $\mathbf{s}$, which is one of many known "generalized cross-correlation" techniques [205]. Thus, when $\mathbf{s}$ is white, no prewhitening is necessary and $\mathbf{c}_{sx}^{(2)} = \mathbf{c}_{sx}^{(1)}$. This suggests a more practical implementation, whereby matrix operations are replaced by an adaptive prewhitening filter [330].

Several efficient versions exist in the literature, in time and frequency domains, to compute the decision variable $\zeta^{(2)}$; see for example [38], [130]. A more robust version of $\zeta^{(2)}$ to noise can be found in [288].

Coherence Method

Instead of using the cross-correlation vector, a detection statistic can be formed by using the squared magnitude coherence. A DTD based on coherence was first proposed in [125]. The idea is to estimate the coherence between $s(k)$ and $x(k)$. The coherence is close to one when there is no double-talk and would be much less than one in a double-talk situation. The squared coherence is defined as,

$$|\gamma_{sx}(f)|^2 = \frac{|\underset{\Rightarrow}{R}_{sx}(f)|^2}{\underset{\Rightarrow}{R}_{ss}(f)\underset{\Rightarrow}{R}_{xx}(f)}, \tag{8.26}$$

where $\underset{\Rightarrow}{R}_{\cdot\cdot}(f)$ is the DFT based cross-power spectrum and f is the DFT frequency index. As decision parameter, an average over a few frequencies is used as detection statistics,

$$\zeta^{(3)} = \frac{1}{I} \sum_{i=0}^{I-1} |\gamma_{sx}(f_i)|^2, \tag{8.27}$$

where I is the number of intervals used. Typical choices of these parameters are $I = 3$ and f_0, f_1, f_2 are the intervals chosen such that their center correspond to approximately 300, 1200, and 1800 Hz respectively. This gives in practice a significantly better performance than averaging over the whole frequency range since there is a poorer speech-to-noise ratio in the upper frequencies (the average speech spectrum declines with about 6 dB/octave above 2 kHz).

Two-Path Model

An interesting approach to double-talk handling was proposed in [242]. This method was introduced for network echo cancellation. However, it has proven far more useful for the AEC application. In this method, two filters model the echo path, one background filter which is adaptive as in a conventional AEC solution and one foreground filter which is not adaptive. The foreground filter cancels the echo. Whenever the background filter performs better than the foreground, its coefficients are copied to the foreground. Coefficients are copied only when a set of conditions are met, which should be compared to the single statistic decision declaring "no double-talk" in a traditional DTD presented in the previous sections.

DTD Combinations with Robust Statistics

All practical double-talk detectors have a probability of miss, i.e. $P_{\mathrm{m}} \neq 0$. Requiring the probability of miss to be smaller will undoubtedly increase the probability of false alarms hence slowing down the convergence rate. As a consequence, no matter what DTD is used, undetected near-end speech will perturb the adaptive algorithm from time to time. The impact of this perturbation is governed by the echo to near-end speech ratio [131].

In practice, what has been done in the past is, first the DTD is designed to be "as good as" one can afford and then, the adaptive algorithm is slowed down so that it copes with the detection errors made by the DTD. This is natural to do since if the adaptive algorithm is very fast, it can react faster to situation changes (e.g. double-talk) than the DTD and thus can diverge. However, this approach severely penalizes the convergence rate of the AEC when the situation is good, i.e. far-end but no near-end talk is present.

In the light of these facts, it may be fruitful to look at adaptive algorithms that can handle at least a small amount of double-talk without diverging. This approach has been studied and proven very successful in the network echo canceler case [129], where the combination of outlier resistant adaptive algorithms and a Geigel DTD were studied. For the acoustic case, one could use any appropriate DTD and combine it with a robust adaptive algorithm. In fact, it is shown in [38] that the combination of the DTD based on the normalized cross-correlation vector and robust adaptive filters (both in time and frequency domains) gives excellent results.

8.6.3 Performance Evaluation of DTDs

The role of the threshold T is essential to the performance of the double-talk detector. To select the value of T and to compare different DTDs objectively one could view the DTD as a classical binary detection problem. By doing so, it is possible to rely on established detection theory. This approach to characterize DTDs was proposed in [125], [78].

The general characteristics of a binary detection scheme are:

- *Probability of False Alarm* (P_f): Probability of declaring detection when a target, in our case double-talk, is not present.
- *Probability of Detection* (P_d): Probability of successful detection when a target is present.
- *Probability of Miss* ($P_\mathrm{m} = 1 - P_\mathrm{d}$): Probability of detection failure when a target is present.

A well designed DTD maximizes P_d while minimizing P_f even in a low SNR. In general, higher P_d is achieved at the cost of higher P_f. There should be a tradeoff in performance depending on the penalty or cost function of a false alarm.

One common approach to characterize different detection methods is to represent the detection characteristic P_d (or P_m) as a function of false alarm probability P_f under a given constraint on the SNR. This is known as a receiver operating characteristic (ROC). The P_f constraint can be interpreted as the maximum tolerable false alarm rate.

Evaluation of a DTD is carried out by estimating the performance parameters, P_d (P_m) and P_f. A principle for this technique can be found in [78]. Though in the end, one should accompany these performance measures with a joint evaluation of the DTD and the AEC. This is due to the fact that the response time of the DTD can seriously affect the performance of the AEC and this is in general not shown in the ROC curve.

8.7 Audio Bridging

8.7.1 Principle

In an application like multi-participant desktop conferencing, an audio bridge is needed somewhere in the network to connect the conferees. Nowadays, all bridges handle only single-channel signals. Each conferee sends its audio single-channel signal to the bridge where they are all added. The bridge sends back a single signal to each conferee (all the added signals from the remote conferees). This simple architecture has many inconveniences and limits. Indeed, with single-channel sound, simultaneous talkers are overlaid and it is difficult to concentrate on a single voice. Moreover, the number of conferees has to be limited to a small number in order to have an acceptable quality. On

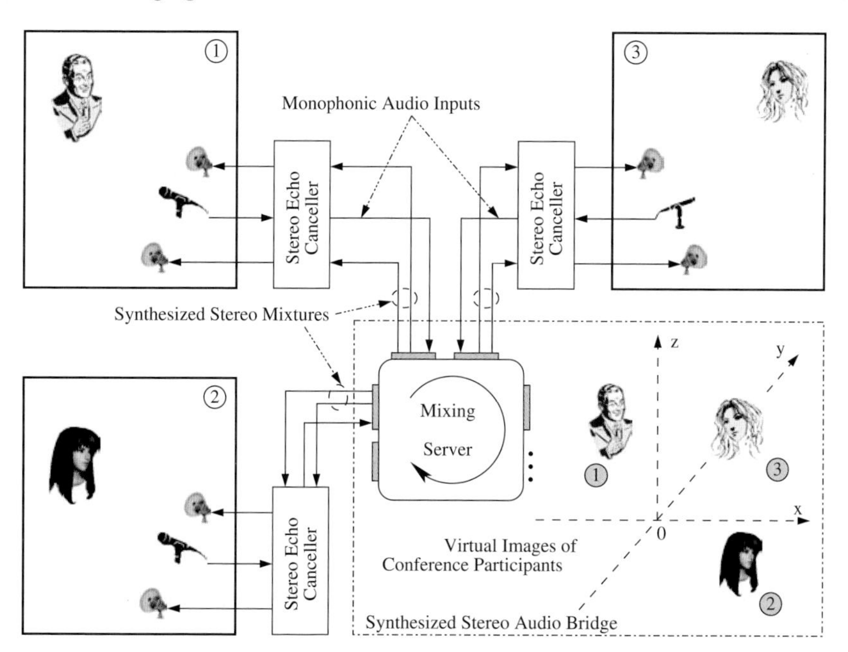

Fig. 8.8. The synthesized audio bridge combined with acoustic echo cancellation for multi-party conferencing.

the other hand, with multichannel presentation, we are able to concentrate on one source exclusively even when several sources are simultaneously active. This is a property of the human binaural auditory system commonly called the "cocktail party effect." As a result, multichannel presentation enhances intelligibility. Moreover, audio localization, enabled by multichannel audio helps us identify which particular talker is active. This is very difficult if not impossible in a monophonic presentation. The ultimate objective of multichannel sound is to give the feeling to each conferee that the remote talkers are in his/her room. This feature will make meetings far more enjoyable and efficient. To realize this, we need to build a multichannel audio bridge (MCAB). In this scenario, each conferee will send at least one microphone signal to the bridge. One function of this bridge is to synthesize stereo sound from all the remote microphone signals and send at least two signals to each conferee to give some spatialization. Another function of the MCAB is to do some noise reduction on the audio signals. Indeed, the level of noise increases rapidly as the number of talkers increases. Obviously, we can extend the functions of this bridge to video and synchronize audio and video signals.

It is important that each of the conferees have a local multichannel acoustic echo canceler. A monophonic echo canceler will simply not cancel the echo and the entire bridge will not work.

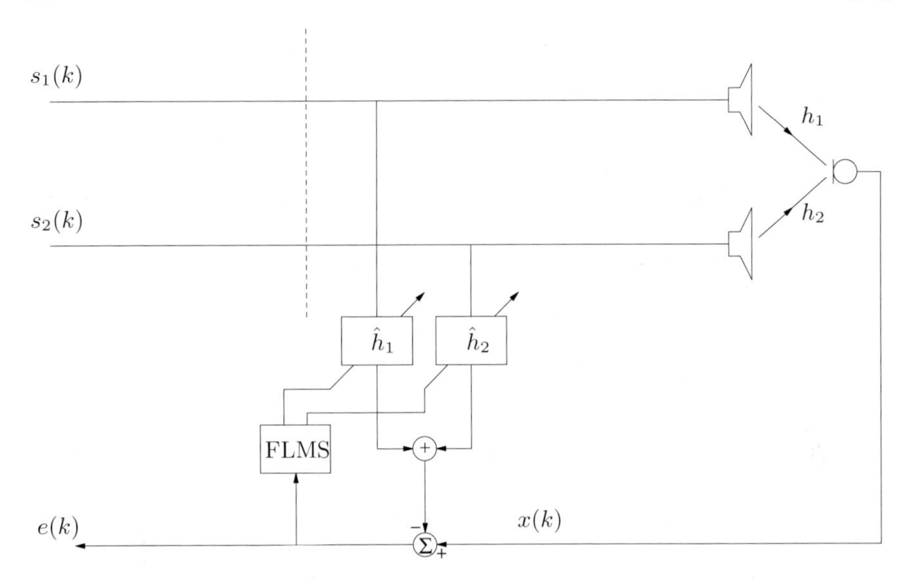

Fig. 8.9. Schematic diagram of stereophonic echo cancellation for desktop conferencing.

The general scenario is as follows (see Fig. 8.8). Several persons in different locations would like to communicate with each other, and each one has a workstation. Each participant would like to see on the screen pictures of the other participants arranged in a reasonable fashion and to hear them in perceptual space in a way that facilitates identification and understanding. For example, the voice of a participant whose picture is located on the left of the screen should appear to come from the left.

We suppose that we are located in a hands-free environment, where the composite acoustic signal is presented over loudspeakers. This study will be limited to two channels, hence to a stereo audio bridge (SAB), so we assume that each workstation is equipped with two loudspeakers (one on each side of the screen) and one microphone (somewhere on top of the screen, for example). As we will see later, a very convenient method using two loudspeakers can accomodate up to four participants. This arrangement can be generalized to create more images by using head-related transfer functions (HRTFs). The idea of using a SAB was first proposed in [25], [29].

Figure 8.9 shows the configuration for a microphone at the local site with a stereo AEC, where h_1 and h_2 represent the two echo paths between the two loudspeakers and the microphone. The two reference signals s_1 and s_2 from the audio bridge are obtained by synthesizing stereo sound from the outputs of all the remote single microphones. The nonuniqueness arises because, for each remote site, the signals are derived by filtering from a common source.

8.7.2 Interchannel Differences for Synthesizing Stereo Sound

In the following scenario, we assume that two loudspeakers are positioned symmetrically on each side of the screen and that the conferee is in front of the screen, close to and approximately centered between the loudspeakers. The location of auditory images in perceptual space is controlled by interchannel intensity and time differences and is mediated by the binaural auditory system.

In any discussion of the relationship between interchannel differences and perceptual effects, it is important to maintain a clear distinction between *interchannel* and *interaural* differences. If sounds are presented to the two ears by means of headphones, the interaural intensity and time differences ΔI_{a} and $\Delta \tau_{\mathrm{a}}$ can be controlled directly. If signals are presented over a pair of loudspeakers, each ear receives both the left- and right-channel signals. The left-channel signal arrives earlier and is more intense at the left ear than at the right, and vice versa, so that interchannel intensity and time differences ΔI_{c} and $\Delta \tau_{\mathrm{c}}$ influence ΔI_{a} and $\Delta \tau_{\mathrm{a}}$, but in general interaural intensity and time differences cannot be controlled directly. In addition to perceptual effects produced by interaural time and intensity differences, localization of sounds presented over a pair of loudspeakers is also influenced by the *precedence effect* [135]: when identical or nearly identical sounds come to a listener from two loudspeakers, the sound appears to originate at the loudspeaker from which the sound arrives first.

To arrange the acoustic images, we can manipulate interchannel intensity and time differences either separately or together. If two identical signals are presented to the two loudspeakers, so that there are no interaural differences, the image will be well fused and localized in the median plane. As the interchannel intensity ratio varies from unity, the image will move toward the loudspeaker receiving the more intense signal. If, instead, the interchannel time difference is varied, the image will in general move toward the loudspeaker receiving the leading signal [100], [324].

Pure Interchannel Intensity Difference

It is well known that the effect of introducing an interchannel intensity ratio ΔI_{c} into signals that are otherwise identical is to move the image away from the median plane toward the loudspeaker receiving the more intense signal. Recent experiments conducted at Bell Labs for a desktop configuration, as well as previous experiments with conventional loudspeaker placement in a room [51], indicate that a 20-dB interchannel intensity ratio produces almost complete lateralization.

If there are two remote conferees, experiments with headphones conducted at Bell Labs suggest that interchannel intensity difference may be the best choice for desktop conferencing in terms of auditory localization and signal separation. The suggested strategy is to present the acoustic signal from one remote participant to one loudspeaker and the acoustic signal from the other remote participant to the other loudspeaker. With three remote participants, the suggested strategy would be the same for the first two participants with the acoustic signal from the third remote participant presented equally to both loudspeakers. Thus, communication with good localization and signal separation among four conferees (one local plus three remote) appears to be feasible. The number of participants could be increased by using finer gradations of ΔI_{c}, but separating the different remote talkers would be more difficult.

Pure Interchannel Time Difference

The nature of the signal plays a more important role in localization and signal separation for interchannel time difference $\Delta\tau_{\mathrm{c}}$ than for interchannel intensity difference ΔI_{c}. For pure tones, the binaural system is insensitive to interaural time differences $\Delta\tau_{\mathrm{a}}$ for frequencies substantially above 1.5 kHz [100], and for lower frequencies, localization of the image is periodic in the interaural time difference $\Delta\tau_{\mathrm{a}}$, with a period equal to the period of the tone. For complex signals with some envelope structure, localization is influenced by both low- and high-frequency interaural time differences. Since, as discussed above, the interaural time difference $\Delta\tau_{\mathrm{a}}$ is indirectly influenced by interchannel time difference $\Delta\tau_{\mathrm{c}}$, it follows that the nature of the signal plays an important role in localization and signal separation for interchannel time difference $\Delta\tau_{\mathrm{c}}$.

If there are two remote talkers, a suggested strategy for localization with interchannel time difference is to present the acoustic signal from one remote participant with an interchannel time difference $\Delta\tau_{\mathrm{c}} = 1\,\mathrm{ms}$ and the acoustic signal from the other remote participant with an interchannel time difference $\Delta\tau_{\mathrm{c}} = -1$ ms. It has been known for a long time [100] that, with headphone presentation, lateralization increases only slightly as interaural time difference increases above 1 ms. Recent experiments with desktop loudspeakers, as well as previous experiments with conventional loudspeaker placement in a room [51], show much the same effect. We do not know how interchannel time difference specifically affects the cocktail party effect.

Combined Interchannel Intensity and Time Differences

As discussed in the previous two subsections, the localization of a sound image can be influenced by both the interchannel intensity difference ΔI_{c} and the interchannel time difference $\Delta\tau_{\mathrm{c}}$. To a certain extent, and within limits, these two types of interchannel differences are tradable in the sense that the same localization can be achieved with various combinations of the two variables. For example, one can achieve roughly the same image position with an amplitude shift, a time shift, or an appropriate combination of the two (time-intensity trading). Furthermore, under some conditions, intensity difference and time difference can be used to reinforce each other to provide a larger shift than is achievable by either one alone.

8.7.3 Choice of Interchannel Differences for Stereo AEC

In principle, for localization with three remote talkers, the best choice of interchannel difference is ΔI_c. But if we want to synthesize a remote talker on the right (resp. left), speech energy will be present only on the right (resp. left) loudspeaker, so we will be able to identify only one impulse response (from this loudspeaker to the microphone) and not the other one. From an acoustic echo cancellation point of view this situation is highly undesirable. For example, if the remote talker on the right stops talking and the remote talker on the left begins, the adaptive algorithm will have to reconverge to the corresponding acoustic path because, in the meantime, it will have "forgotten" the other acoustic path. Therefore, the adaptive algorithm will have to track the different talkers continually, reconverging for each one, so the system will become uncontrollable — especially in a nonstationary environment (changes of the acoustic paths) and in double-talk situations. As a result, we will have degraded echo cancellation much of the time.

The solution to this problem is that, for each remote talker, we must have some energy on both loudspeakers to be able to maintain identification of the two impulse responses between loudspeakers and the microphone. Thus, the optimal choice of interchannel difference from an acoustic echo cancellation point of view is pure $\Delta\tau_c$ since energy is equally presented to both loudspeakers for all remote talkers. However, in practice, this choice may not be enough for good localization. Therefore, combined $[\Delta I_c,\ \Delta\tau_c]$ seems to be the best compromise between good localization and echo cancellation.

If there are two remote talkers, a strategy for good localization and echo cancellation would be to present the acoustic signal from one remote participant to both loudspeakers with $\Delta I_c = 6$ dB, $\Delta\tau_c = 1$ ms and the acoustic signal from the other remote participant to both loudspeakers with $\Delta I_c = -6$ dB, $\Delta\tau_c = -1$ ms. With three remote participants, the suggested strategy would be the same for the first two participants with the addition of the third remote participant's microphone signal presented to both loudspeakers with $\Delta I_c = 0$ dB, $\Delta\tau_c = 0$ ms.

Thus, for any remote participant's microphone signal t_p, $p = 1, 2, \cdots, P$, the contribution to the local synthesized stereo signals is written

$$s_m(k) = g_{p,m}(k) * t_p(k),\ m = 1, 2, \tag{8.28}$$

where $*$ denotes convolution and $g_{p,m}$ are the impulse responses for realizing the desired $[\Delta I_c,\ \Delta\tau_c]$ (or HRTFs in the general case). For example, with the above suggested 6 dB, 1 ms values, a talker on the left, say, would be synthesized with

$$\begin{aligned} \mathbf{g}_{p,1} &= \begin{bmatrix} 1 & 0 & \cdots & 0 & 0 \end{bmatrix}^T, \\ \mathbf{g}_{p,2} &= \begin{bmatrix} 0 & 0 & \cdots & 0 & 0.5 \end{bmatrix}^T, \end{aligned}$$

where the number of samples of delay in $g_{p,2}$ corresponds to 1 ms.

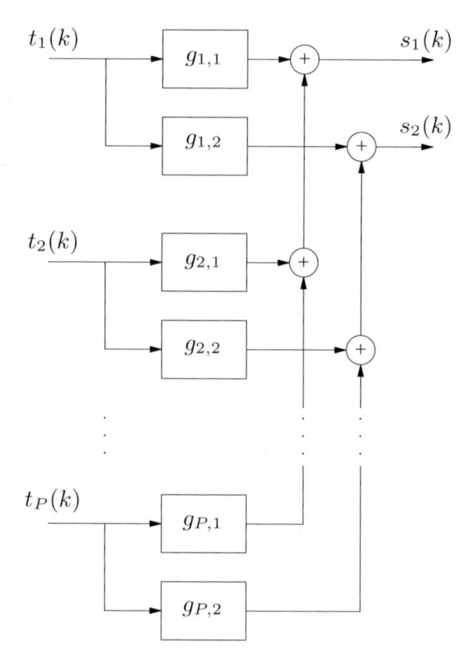

Fig. 8.10. Synthesizing local stereo signals from P remote signals.

Figure 8.10 shows how the signals from P remote conferees are combined to produce the local synthesized signals s_1, s_2. Each $g_{p,1}$, $g_{p,2}$ pair is selected as exemplified above to locate the acoustic image in some desired position. As explained previously, the synthesis function is located at the audio bridge.

A composite diagram of the synthesis, nonlinear transformation, and stereo AEC appears in Fig. 8.11. This shows the complete local signal processing suite for one conferee. This same setup is reproduced for each conferee, making obvious permutations on the remote signals and choice of synthesis filters. Further details and simulation results can be found in [25].

8.8 Summary

Researchers have been working on the problem of echo cancellation since the 1960s and the research is still going on. There are different reasons for this. First, we are still not very satisfied with current AECs. Second, progress in signal processing methods such as adaptive filers or detection benefit echo cancellation applications. Finally, there are new interesting problems such as multichannel acoustic echo cancellation, which leads to more exciting applications that take advantage of our binaural auditory system.

In this chapter, we have given an overview on acoustic echo cancellation. We discussed different aspects of this application, including the difficult prob-

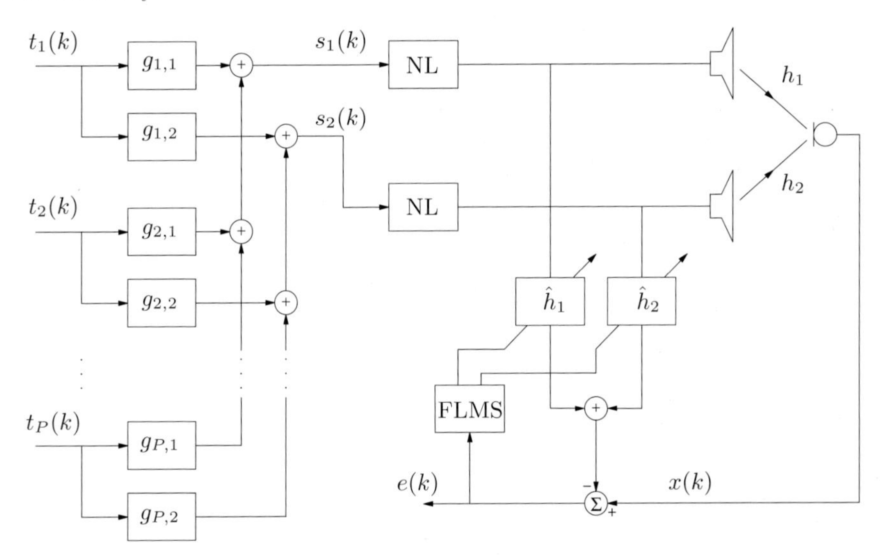

Fig. 8.11. Overall diagram of synthesis, nonlinear transformation, and stereo acoustic echo cancellation.

lem of double-talk detection. We have also shown the importance of multichannel sound and multichannel AECs.

In the last section, we have discussed the stereo audio bridge and shown the potential of such a device in future telecommunications. Today, with our phone, it's not very convenient to have a conferencing call with different people at different places. We believe that the SAB will play a major role in the way we telecommunicate.

9

Time Delay Estimation and Acoustic Source Localization

With the material presented in Chap. 2 on MIMO system modeling and that in Chap. 6 on blind channel identification, we are now ready to investigate the problem of time delay estimation (TDE) and its application for acoustic source localization. We begin this chapter by outlining the TDE problem, and then address the estimation of time-difference-of-arrival (TDOA) information by measuring the cross correlation between receivers' outputs and identifying the channel impulse responses from the radiating source to the multiple receivers. The chapter also discusses how to employ the TDOA measurements to locate radiating sources in the acoustic wavefield that emit signals to receivers.

9.1 Time Delay Estimation

In an acoustic wavefield with multiple receivers at distinct spatial locations, a wavefront, emanating from a radiating sound source, arrives at different receiving sensors at different time instances. The corresponding time differences of arrival between pairs of receivers are important parameters in acoustic MIMO signal processing, which characterize the direction and location from which the signal impinges on the receivers. The estimation of such parameters, often referred to as the time-delay-estimation (TDE) problem, has been of fundamental importance.

In an acoustic environment, speech sound often arrives at each individual sensor through multiple paths. To make the analysis easy to understand, however, it is convenient to decouple the problem into two cases, i.e, single-path and multipath problems. The former is an ideal case, where each sensor's output can be modeled as a delayed and attenuated copy of the source signal corrupted by additive noise. Assuming that there is only one (unknown) source in the sound field, we can write the output of receiver n ($n = 1, 2, \cdots, N$) as

$$x_n(k) = \alpha_n s(k - D_n) + b_n(k), \tag{9.1}$$

where α_n, which satisfies $0 \leq \alpha_n \leq 1$, is an attenuation factor due to propagation effects, D_n corresponds to the propagation time from the unknown source to receiver n, and $s(k)$, which is often speech from either a talker or a loudspeaker, is broadband in nature. In the context of TDE, it is assumed that the noise component $b_n(k)$ in (9.3) is a Gaussian random process, which is uncorrelated with both the source signal and the noise signals at other sensors.

With this signal model, the time difference of arrival (TDOA) between the ith and jth sensors can be written as

$$\tau_{ij} = D_j - D_i, \tag{9.2}$$

where $i, j = 1, 2, \cdots, N$, and $i \neq j$. The goal of TDE, then, is to obtain an estimate $\hat{\tau}_{ij}$ of τ_{ij} based on the observation signals $x_n(k)$.

In the multipath situation, the output of each receiver can be modeled as a convolution between the source signal and the corresponding channel impulse response from the source to the receiver, as given in (2.4) and (2.5). Let us rewrite (2.4) here:

$$x_n(k) = \mathbf{h}_n^T \mathbf{s}(k) + b_n(k), \quad n = 1, 2, \cdots, N. \tag{9.3}$$

Again it is assumed that $\mathbf{s}(k)$ is reasonably broadband and each $b_n(k)$ is a Gaussian random process. In addition, we assume that in $\mathbf{h}_n$ the entry corresponding to the direct path from the source to the nth receiver has the maximal magnitude out of all the elements so that once we know the impulse responses $\mathbf{h}_n$, $n = 1, 2, \cdots, N$, the TDOA information can be determined by identifying the direct paths. This assumption is reasonable from the wave propagation point of view since reflected signals take a longer path to reach the receivers and suffer energy absorption from reflection boundaries. However, multipath signals may combine at sensors both destructively and constructively in a non-coherent and unpredictable manner. As a result, it may happen occasionally that the combination of several multipath signals has a higher energy than the direct-path signal. In such a case, the correct time delay estimate can be determined by either taking into account previous estimates or incorporating some *a priori* knowledge about the source.

Comparing (9.3) with (9.1), we can see that the single-path problem is a special case of the multipath problem. As a matter of fact, if set

$$\begin{aligned}
\mathbf{h}_n &= \begin{bmatrix} h_{n,0} & h_{n,1} & \cdots & h_{n,l} & \cdots & h_{n,L-1} \end{bmatrix}^T \\
&= \begin{bmatrix} 0 & 0 & \cdots & 0 & h_{n,D_n} & 0 & \cdots & 0 \end{bmatrix}^T \\
&= \begin{bmatrix} 0 & 0 & \cdots & 0 & \alpha_n & 0 & \cdots & 0 \end{bmatrix}^T, \quad n = 1, 2, \cdots, N,
\end{aligned} \tag{9.4}$$

then (9.3) is the same as (9.1). In other words, in the single-path situation, the channel impulse response $\mathbf{h}_n$ has only one non-zero element at $l = D_n$, where $h_{n,D_n} = \alpha_n$ is governed by the propagation attenuation.

Numerous algorithms have been developed to fulfil the estimation of TDOA in the ideal propagation situation. The most widely used technique thus far is, perhaps, the generalized cross-correlation (GCC) algorithm [205], which originated from the research of passive sonar signal processing. In this approach, the delay estimate $\hat{\tau}_{ij}$ is obtained as the lag time that maximizes the generalized cross-correlation function between the filtered versions of signals $x_i(k)$ and $x_j(k)$. The GCC method has been intensively investigated and can achieve reasonably accurate estimation of time delay in the presence of weak to moderate levels of noise but absence of the multipath effect.

However, in acoustical environments in the context of speech communication, as we mentioned earlier, each sensor receives not only the direct-path signal, but also multiple attenuated and delayed replicas of the source signal due to reflections of the wavefront from boundaries and objects in the enclosure environments. This multipath propagation effect introduces echoes and spectral distortion into the observation signals, termed as reverberation, which may severely deteriorate the TDE performance of the GCC technique. In addition, the TDE problem is often complicated by conditions of poor signal-to-noise ratio, nonstationarity of speech signals, and a changing TDOA owing to the motion of the speech sources.

In the first half of this chapter, we address the problem of time delay estimation in acoustic environments with microphone receivers. Starting with the simple cross-correlation method, we will study the problem with emphasis on how to deal with strong noise and reverberation. Specifically, we discuss three ways in improving the robustness of TDE with respect to noise and reverberation. The first is to incorporate some *a priori* knowledge about the distortion sources into the cross-correlation method to ameliorate its performance. The second is to use multiple (more than two) sensors and take advantage of the redundancy to enhance the delay estimate between the two desired sensors. The third is to exploit the advanced system identification techniques to improve TDE in reverberation conditions.

9.2 Cross-Correlation Method

Suppose that we have two sensors, with their outputs being denoted as $x_1(k)$ and $x_2(k)$ respectively. The cross-correlation function (CCF) between the two signals is defined as:

$$r_{x_1x_2}(p) = E\left[x_1(k)x_2(k+p)\right]. \tag{9.5}$$

Considering the ideal propagation model given in (9.1) and substituting for $x_n(k)$ $(n = 1, 2)$ in terms of $s(k)$ and $b_n(k)$, we can readily deduce that:

$$\begin{aligned} r_{x_1x_2}(p) = \alpha_1\alpha_2 r_{ss}(p + D_1 - D_2) + \alpha_1 r_{sb_2}(p + D_1) + \\ \alpha_2 r_{sb_1}(p - D_2) + r_{b_1b_2}(p). \end{aligned} \tag{9.6}$$

With the assumption that $b_n(k)$ is Gaussian random noise, which is uncorrelated with both the signal and the noise observed at the other sensor, it can be easily checked that $r_{x_1x_2}(p)$ reaches its maximum at $p = D_2 - D_1$. Therefore, given the CCF, we can obtain an estimate of the TDOA between $x_1(k)$ and $x_2(k)$ as

$$\hat{\tau}_{12} = \arg\max_{p} r_{x_1x_2}(p), \tag{9.7}$$

where $p \in [-\tau_{\max}, \tau_{\max}]$, and $\tau_{\max}$ is the maximum possible delay.

In digital implementation of (9.7), some approximations are required because the CCF is not known and must be estimated. A normal practice is to replace the CCF defined in (9.5) by its time-averaged estimate. Suppose at time instant t we have a set of observation samples of x_n, $\{x_n(t), x_n(t+1), \cdots, x_n(t+k-1), \cdots, x_n(t+K-1)\}$, $n = 1, 2$, the corresponding CCF can be estimated either as

$$\hat{r}_{x_1x_2}(p) = \begin{cases} \dfrac{1}{K} \displaystyle\sum_{k=0}^{K-p-1} x_1(t+k)x_2(t+k+p), & p \geq 0 \\ \hat{r}_{x_2x_1}(-p), & p < 0 \end{cases}, \tag{9.8}$$

or by

$$\hat{r}_{x_1x_2}(p) = \begin{cases} \dfrac{1}{K-p} \displaystyle\sum_{k=0}^{K-p-1} x_1(t+k)x_2(t+k+p), & p \geq 0 \\ \hat{r}_{x_2x_1}(-p), & p < 0 \end{cases}, \tag{9.9}$$

where K is the block size. The difference between (9.8) and (9.9) is that the former leads to a biased estimator, while the latter is an unbiased one. However, since it has a lower estimation variance and is asymptotically unbiased, the former has been widely adopted in many applications.

Another way to estimate CCF is through the discrete Fourier transform (DFT) and the inverse discrete Fourier transform (IDFT), i.e.,

$$\hat{r}_{x_1x_2}(p) = \frac{1}{K} \sum_{k'=0}^{K-1} X_1(\omega_{k'}) X_2^*(\omega_{k'}) e^{j\omega_{k'}p}, \tag{9.10}$$

where

$$\omega_{k'} = \frac{2\pi k'}{K}, \quad k' = 0, 1, \cdots, K-1, \tag{9.11}$$

is the angular frequency,

$$X_n(\omega_{k'}) = \sum_{k=0}^{K-1} x_n(t+k) e^{-j\omega_{k'}k} \tag{9.12}$$

is the short-term DFT of $x_n(k)$ at time t. Both (9.8) and (9.10) generate the same CCF estimate. However, the latter can be implemented more efficiently

Table 9.1. The cross-correlation method for time delay estimation.

Parameter:	$\hat{\tau}_{12}$
Estimation:	For $t = 0, 1, \cdots$
	(a) Obtain a frame of observation signals at time instant t: $\{x_n(t), x_n(t+1), \cdots, x_n(t+K-1)\}$, $n = 1, 2$
	(b) Estimate the spectrum of x_n: $X_n(\omega_{k'}) = \text{FFT}\{x_n(t+k)\} = \sum_{k=0}^{K-1} x_n(t+k)e^{-j\omega_{k'}k}$
	(c) Estimate cross-correlation function: $\hat{r}_{x_1x_2}(p) = \text{IFFT}\{X_1(\omega_{k'})X_2^*(\omega_{k'})\} = \frac{1}{K}\sum_{k'=0}^{K-1} X_1(\omega_{k'})X_2^*(\omega_{k'})e^{j\omega_{k'}p}$
	(d) Obtain the time delay: $\hat{\tau}_{12} = \arg\max_p \hat{r}_{x_1x_2}(p)$

using the fast Fourier transform (FFT) and the inverse fast Fourier transform (IFFT), and therefore it has been widely adopted in systems.

The CC method for TDE is summarized in Table 9.1.

9.3 Magnitude-Difference Method

It can be seen that the cross-correlation function reaches its maximum when two realizations of the same source signal are synchronized. Therefore, the CC method indeed obtains the TDOA estimate by measuring the synchrony between the two signals. Another way to measure the synchrony is by identifying the minimum of the magnitude-difference function (MDF), which is defined as

$$\Psi_{x_1x_2}(p) = E\{|x_1(k) - x_2(k+p)|\}. \tag{9.13}$$

With the signal model given in (9.1), we can derive that [189], [190]:

$$\Psi_{x_1x_2}(p) = \sqrt{\frac{2}{\pi}\left\{E[x_1^2(k)] + E[x_2^2(k)] - 2r_{x_1x_2}(p)\right\}}. \tag{9.14}$$

It can be checked that $\Psi_{x_1x_2}(p)$ reaches its minimum at $p = D_2 - D_1$. Therefore, given the MDF, we can obtain the TDOA between $x_1(k)$ and $x_2(k)$ as

$$\hat{\tau}_{12} = \arg\min_p \Psi_{x_1x_2}(p). \tag{9.15}$$

Similar to the CC method, we have to estimate $\Psi_{x_1x_2}(p)$ in the implementation of (9.15). Normally, MDF is approximated by the average-magnitude-difference function (AMDF), which at time instance t is computed as

Table 9.2. The magnitude-difference method for time delay estimation.

Parameter: $\hat{\tau}_{12}$

Estimation: For $t = 0, 1, \cdots$

(a) Obtain a frame of observation signals at time instant t:
$\{x_n(t), x_n(t+1), \cdots, x_n(t+K-1)\}$, $n = 1, 2$

(b) Estimate the AMDF between x_1 and x_2:

If $p \geq 0$:

$$\hat{\Psi}_{x_1x_2}(p) = \frac{1}{K-p} \sum_{k=0}^{K-1-p} |x_1(t+k) - x_2(t+k+p)|$$

If $p < 0$:

$$\hat{\Psi}_{x_1x_2}(p) = \hat{\Psi}_{x_2x_1}(-p)$$

(c) Obtain the time delay:

$$\hat{\tau}_{12} = \arg\min_p \hat{\Psi}_{x_1x_2}(p)$$

$$\hat{\Psi}_{x_1x_2}(p) = \begin{cases} \frac{1}{K} \sum_{k=0}^{K-1-p} |x_1(t+k) - x_2(t+k+p)|, & p \geq 0 \\ \hat{\Psi}_{x_2x_1}(-p), & p < 0 \end{cases}. \quad (9.16)$$

The algorithm to estimate time delay based on AMDF is summarized in Table 9.2.

9.4 Maximum Likelihood Method

Maximum likelihood (ML) is one of the most popularly used statistical estimators due to its asymptotic optimal property, i.e., the estimation variance can achieve the Cramèr-Rao lower bound (CRLB) when the number of observation samples approaches infinity. For the signal model given in (9.1), if we assume that both $s(k)$ and $b_n(k)$ $(n = 1, 2)$ are mutually independent, zero-mean, stationary Gaussian random processes, D_n are constant over time, and the observation interval approximates infinity, i.e, $K \to \infty$, we can then derive the ML estimator either in the time domain or in the frequency domain [205], [67], [287]. In the frequency domain, the DFT of $x_n(k)$ is formed as:

$$\begin{aligned} X_n(\omega_{k'}) &= \sum_{k=0}^{K-1} x_n(k) e^{-j\omega_{k'} k} \\ &= \sum_{k=0}^{K-1} \left[\alpha_n s_n(k - D_n) + b_n(k)\right] e^{-j\omega_{k'} k} \\ &= \alpha_n S(\omega_{k'}) e^{-j\omega_{k'} D_n} + B_n(\omega_{k'}), \quad n = 1, 2, \end{aligned} \quad (9.17)$$

which is a Gaussian random variable. It follows that:

$$E\left[X_n(\omega_i)X_n^*(\omega_j)\right] = \begin{cases} \alpha_n^2 P_s(\omega_i) + P_{b_n}(\omega_i), & i = j \\ 0, & i \neq j \end{cases},$$

$$E\left[X_1(\omega_i)X_2^*(\omega_j)\right] = \begin{cases} \alpha_1\alpha_2 P_s(\omega_i)e^{-j\omega_i(D_1-D_2)}, & i = j \\ 0, & i \neq j \end{cases}, \tag{9.18}$$

where $n = 1, 2$, $P_s(\omega_i) = E[S(\omega_i)S^*(\omega_i)]$ and $P_{b_n}(\omega_i) = E[B_n(\omega_i)B_n^*(\omega_i)]$ are the power spectral densities of $s(k)$ and $b_n(k)$ respectively.

Now let us define three vectors:

$$\mathbf{X}_1 \triangleq \begin{bmatrix} X_1(\omega_0) & X_1(\omega_1) & \cdots & X_1(\omega_{K-1}) \end{bmatrix}^T,$$

$$\mathbf{X}_2 \triangleq \begin{bmatrix} X_2(\omega_0) & X_2(\omega_1) & \cdots & X_2(\omega_{K-1}) \end{bmatrix}^T,$$

$$\mathbf{X} \triangleq \begin{bmatrix} \mathbf{X}_1^T & \mathbf{X}_2^T \end{bmatrix}^T. \tag{9.19}$$

The covariance matrix of $\mathbf{X}$ can be written as

$$\begin{aligned} \mathbf{\Sigma} &\triangleq E\left[\mathbf{X}\mathbf{X}^H\right] \\ &= E\begin{bmatrix} \mathbf{X}_1\mathbf{X}_1^H & \mathbf{X}_1\mathbf{X}_2^H \\ \mathbf{X}_2\mathbf{X}_1^H & \mathbf{X}_2\mathbf{X}_2^H \end{bmatrix} \\ &\triangleq \begin{bmatrix} \mathbf{\Sigma}_{11} & \mathbf{\Sigma}_{12} \\ \mathbf{\Sigma}_{12}^H & \mathbf{\Sigma}_{22} \end{bmatrix}. \end{aligned}$$

From (9.18) and (9.19), we can check that $\mathbf{\Sigma}_{11}$, $\mathbf{\Sigma}_{22}$, and $\mathbf{\Sigma}_{12}$ all are diagonal matrices of size $K \times K$:

$$\mathbf{\Sigma}_{11} = E\left[\mathbf{X}_1\mathbf{X}_1^H\right] = \mathrm{diag}\Big[\, \alpha_1^2 P_s(\omega_0) + P_{b_1}(\omega_0),\ \alpha_1^2 P_s(\omega_1) + P_{b_1}(\omega_1), \\ \cdots,\ \alpha_1^2 P_s(\omega_{K-1}) + P_{b_1}(\omega_{K-1}) \Big],$$

$$\mathbf{\Sigma}_{22} = E\left[\mathbf{X}_2\mathbf{X}_2^H\right] = \mathrm{diag}\Big[\, \alpha_2^2 P_s(\omega_0) + P_{b_2}(\omega_0),\ \alpha_2^2 P_s(\omega_1) + P_{b_2}(\omega_1), \\ \cdots,\ \alpha_2^2 P_s(\omega_{K-1}) + P_{b_2}(\omega_{K-1}) \Big],$$

and

$$\mathbf{\Sigma}_{12} = E\left[\mathbf{X}_1\mathbf{X}_2^H\right] = \mathrm{diag}\Big[\, \alpha_1\alpha_2 P_s(\omega_0)e^{j\omega_0\tau_{12}},\ \alpha_1\alpha_2 P_s(\omega_1)e^{j\omega_1\tau_{12}}, \\ \cdots,\ \alpha_1\alpha_2 P_s(\omega_{K-1})e^{j\omega_{K-1}\tau_{12}} \Big],$$

where $\tau_{12} = (D_2 - D_1)$.

The log-likelihood function of $\mathbf{X}$, given the delay, attenuation factors, and the power spectral densities of both the speech and noise signals, can be written as:

$$\mathcal{L} = \ln p\left[\mathbf{X} \mid D_1, D_2, \alpha_1, \alpha_2, P_s(\omega_{k'}), P_{b_1}(\omega_{k'}), P_{b_2}(\omega_{k'})\right]. \quad (9.20)$$

Since $\mathbf{X}$ is Gaussian distributed, we have

$$\mathcal{L} = -\frac{K}{2}\ln(2\pi) - \frac{1}{2}\ln[\det(\mathbf{\Sigma})] - \frac{1}{2}\mathbf{X}^H\mathbf{\Sigma}^{-1}\mathbf{X}. \quad (9.21)$$

The determinant of the matrix $\mathbf{\Sigma}$, due to its block structure, can easily be derived as follows [43]:

$$\begin{aligned}\det(\mathbf{\Sigma}) &= \prod_{k'=0}^{K-1}\left[\alpha_1^2 P_s(\omega_{k'}) + P_{b_1}(\omega_{k'})\right] \cdot \\ &\quad \prod_{k'=0}^{K-1}\left[P_{b_2}(\omega_{k'}) + \frac{\alpha_1^2 P_s(\omega_{k'})P_{b_1}(\omega_{k'})}{\alpha_1^2 P_s(\omega_{k'}) + P_{b_1}(\omega_{k'})}\right],\end{aligned} \quad (9.22)$$

which is independent of D_1 and D_2. The inverse of $\mathbf{\Sigma}$ can be obtained as:

$$\mathbf{\Sigma}^{-1} \triangleq \mathbf{A} = \begin{bmatrix}\mathbf{A}_{11} & \mathbf{A}_{12} \\ \mathbf{A}_{12}^H & \mathbf{A}_{22}\end{bmatrix}, \quad (9.23)$$

where $\mathbf{A}_{11}$, $\mathbf{A}_{22}$ and $\mathbf{A}_{12}$ are three diagonal matrices of size $K \times K$:

$$\begin{aligned}\mathbf{A}_{11} &= \mathrm{diag}\left[\left(P_{b_1}(\omega_0) + \frac{\alpha_1^2 P_s(\omega_0)P_{b_2}(\omega_0)}{\alpha_2^2 P_s(\omega_0) + P_{b_2}(\omega_0)}\right)^{-1}, \ldots,\right. \\ &\quad \left.\left(P_{b_1}(\omega_{K-1}) + \frac{\alpha_1^2 P_s(\omega_{K-1})P_{b_2}(\omega_{K-1})}{\alpha_2^2 P_s(\omega_{K-1}) + P_{b_2}(\omega_{K-1})}\right)^{-1}\right],\end{aligned}$$

$$\begin{aligned}\mathbf{A}_{22} &= \mathrm{diag}\left[\left(P_{b_2}(\omega_0) + \frac{\alpha_2^2 P_s(\omega_0)P_{b_1}(\omega_0)}{\alpha_1^2 P_s(\omega_0) + P_{b_1}(\omega_0)}\right)^{-1}, \ldots,\right. \\ &\quad \left.\left(P_{b_2}(\omega_{K-1}) + \frac{\alpha_2^2 P_s(\omega_{K-1})P_{b_1}(\omega_{K-1})}{\alpha_1^2 P_s(\omega_{K-1}) + P_{b_1}(\omega_{K-1})}\right)^{-1}\right],\end{aligned}$$

and

$$\begin{aligned}\mathbf{A}_{12} &= \mathrm{diag}\left[-\frac{\alpha_1\alpha_2 P_s(\omega_0)e^{j\omega_0\tau_{12}}}{\alpha_1^2 P_s(\omega_0)P_{b_2}(\omega_0) + P_{b_1}(\omega_0)P_{b_2}(\omega_0)}, \ldots,\right. \\ &\quad \left.-\frac{\alpha_1\alpha_2 P_s(\omega_{K-1})e^{j\omega_{K-1}\tau_{12}}}{\alpha_1^2 P_s(\omega_{K-1})P_{b_2}(\omega_{K-1}) + P_{b_1}(\omega_{K-1})P_{b_2}(\omega_{K-1})}\right].\end{aligned}$$

Substituting the determinant and the inverse matrix of $\mathbf{\Sigma}$ into (9.21) yields

$$\mathcal{L} = \mathcal{L}_1 + \mathcal{L}_2, \tag{9.24}$$

where

$$\begin{aligned}
\mathcal{L}_1 &= -\frac{K}{2}\ln(2\pi) - \frac{1}{2}\sum_{k'=0}^{K-1}\ln[\alpha_1^2 P_s(\omega_{k'}) + P_{b_1}(\omega_{k'})] \\
&\quad -\frac{1}{2}\sum_{k'=0}^{K-1}\ln\left[P_{b_2} + \frac{\alpha_1^2 P_s(\omega_{k'})P_{b_1}(\omega_{k'})}{\alpha_1^2 P_s(\omega_{k'}) + P_{b_1}(\omega_{k'})}\right] \\
&\quad -\frac{1}{2}\sum_{k'=0}^{K-1}\left[P_{b_1}(\omega_{k'}) + \frac{\alpha_1^2 P_s(\omega_{k'})P_{b_2}(\omega_{k'})}{\alpha_2^2 P_s(\omega_{k'}) + P_{b_2}(\omega_{k'})}\right]^{-1} X_1^*(\omega_{k'})X_1(\omega_{k'}) \\
&\quad -\frac{1}{2}\sum_{k'=0}^{K-1}\left[P_{b_2}(\omega_{k'}) + \frac{\alpha_2^2 P_s(\omega_{k'})P_{b_1}(\omega_{k'})}{\alpha_1^2 P_s(\omega_{k'}) + P_{b_1}(\omega_{k'})}\right]^{-1} X_2^*(\omega_{k'})X_2(\omega_{k'}),
\end{aligned}$$

and

$$\mathcal{L}_2 = \sum_{k'=0}^{K-1}\left[\frac{\alpha_1\alpha_2 P_s(\omega_{k'})e^{j\omega_{k'}\tau_{12}}}{\alpha_1^2 P_s(\omega_{k'})P_{b_2}(\omega_{k'}) + P_{b_1}(\omega_{k'})P_{b_2}(\omega_{k'})}\right] X_1(\omega_{k'})X_2^*(\omega_{k'}).$$

Apparently, $\mathcal{L}_1$ is independent of τ_{12} and $\mathcal{L}_2$ is a function of τ_{12}. Therefore, maximizing the log likelihood function $\mathcal{L}$ with respect to τ_{12} is equal to selecting a τ_{12} that maximizes $\mathcal{L}_2$. In other words, the ML estimator for time delay is

$$\begin{aligned}
\hat{\tau}_{12} &= \arg\max_p \hat{r}_{x_1x_2}^{\mathrm{ML}}(p) \\
&= \arg\max_p \sum_{k'=0}^{K-1}\frac{\alpha_1\alpha_2 P_s(\omega_{k'})X_1(\omega_{k'})X_2^*(\omega_{k'})e^{j\omega_{k'}p}}{\alpha_1^2 P_s(\omega_{k'})P_{b_2}(\omega_{k'}) + P_{b_1}(\omega_{k'})P_{b_2}(\omega_{k'})}.
\end{aligned} \tag{9.25}$$

It can be seen from (9.25) that in order to achieve ML estimation, the attenuation factors α_1 and α_2 have to be known *a priori*. In addition, we need to know the power spectral densities of both the speech and noise signals (note that given the observation data, if the power spectral densities of the noise signals are known, we can easily estimate the power spectral density of the speech signal, and vice versa).

The ML method for TDE is summarized in Table 9.3.

9.5 Generalized Cross-Correlation Method

Comparing (9.25) with (9.10), we can see that the ML estimate is achieved by weighting the cross spectrum between sensors' outputs. As a matter of fact,

Table 9.3. The maximum likelihood method for time delay estimation.

Parameter: $\hat{\tau}_{12}$

Estimation: For $t = 0, 1, \cdots$

(a) Obtain a frame of observation signals at time instant t:
$\{x_n(t), x_n(t+1), \cdots, x_n(t+K-1)\}$, $n = 1, 2$

(b) Estimate the spectrum of $x_n(t+k)$:

$$X_n(\omega_{k'}) = \text{FFT}\{x_n(t+k)\} = \sum_{k=0}^{K-1} x_n(t+k)e^{-j\omega_{k'}k}$$

(c) Estimate the cost function:

$$\begin{aligned}\hat{r}^{\text{ML}}_{x_1x_2}(p) &= \sum_{k'=0}^{K-1} \frac{\alpha_1\alpha_2 P_s(\omega_{k'})X_1(\omega_{k'})X_2^*(\omega_{k'})e^{j\omega_{k'}p}}{\alpha_1^2 P_s(\omega_{k'})P_{b_2}(\omega_{k'}) + P_{b_1}(\omega_{k'})P_{b_2}(\omega_{k'})} \\ &= \text{IFFT}\left\{\frac{\alpha_1\alpha_2 P_s(\omega_{k'})X_1(\omega_{k'})X_2^*(\omega_{k'})}{\alpha_1^2 P_s(\omega_{k'})P_{b_2}(\omega_{k'}) + P_{b_1}(\omega_{k'})P_{b_2}(\omega_{k'})}\right\}\end{aligned}$$

(d) Obtain the time delay:
$\hat{\tau}_{12} = \arg\max_p \hat{r}^{\text{ML}}_{x_1x_2}(p)$

---

such a weighting processing has been found effective in improving TDE performance. The resulting technique is known as the generalized cross-correlation (GCC) method, which has gained its great popularity since the informative paper [205] was published by Knapp and Carter in 1976. With the signal model given in (9.1), the GCC framework estimates TDOA between $x_1(k)$ and $x_2(k)$ as:

$$\hat{\tau}_{12} = \arg\max_p \hat{r}^{\text{GCC}}_{x_1x_2}(p) \tag{9.26}$$

where

$$\begin{aligned}\hat{r}^{\text{GCC}}_{x_1x_2}(p) &= \sum_{k'=0}^{K-1} \Phi(\omega_{k'})G_{x_1x_2}(\omega_{k'})e^{j\omega_{k'}p} \\ &= \sum_{k'=0}^{K-1} \varsigma_{x_1x_2}(\omega_{k'})e^{j\omega_{k'}p}\end{aligned} \tag{9.27}$$

is the generalized cross-correlation function (GCCF),

$$G_{x_1x_2}(\omega_{k'}) = E[X_1(\omega_{k'})X_2^*(\omega_{k'})]$$

is the cross spectrum between $x_1(k)$ and $x_2(k)$, $\Phi(\omega_{k'})$ is a weighting function (sometimes called a *prefilter*), and

$$\varsigma_{x_1x_2}(\omega_{k'}) = \Phi(\omega_{k'})G_{x_1x_2}(\omega_{k'}) \tag{9.28}$$

is the weighted cross spectrum. In a practical system, the cross spectrum $G_{x_1x_2}(\omega_{k'})$ has to be estimated, which is normally achieved by replacing the expected value by its instantaneous value, i.e., $\hat{G}_{x_1x_2}(\omega_{k'}) = X_1(\omega_{k'})X_2^*(\omega_{k'})$.

Table 9.4. Commonly used weighting functions in the GCC method.

Method name	**Weighting function** $\Phi(\omega_{k'})$
Cross correlation	1
Phase transform (PHAT)	$\dfrac{1}{\lvert G_{x_1x_2}(\omega_{k'})\rvert}$
Smoothed coherence transform (SCOT)	$\dfrac{1}{\sqrt{P_{x_1}(\omega_{k'})P_{x_1}(\omega_{k'})}}$
Eckart	$\dfrac{P_s(\omega_{k'})}{P_{b_1}(\omega_{k'})P_{b_2}(\omega_{k'})}$
Maximum likelihood (ML)	$\dfrac{\alpha_1\alpha_2P_s(\omega_{k'})}{\alpha_1^2P_s(\omega_{k'})P_{b_2}(\omega_{k'})+P_{b_1}(\omega_{k'})P_{b_2}(\omega_{k'})}$

Table 9.5. The generalized cross-correlation method for time delay estimation.

Parameter: $\hat{\tau}_{12}$

Estimation: Select a weighting function $\Phi(\omega_{k'})$

For $t = 0, 1, \cdots$

(a) Obtain a frame of observation signals at time instant t:
$\{x_n(t), x_n(t+1), \cdots, x_n(t+K-1)\}$, $n = 1, 2$

(b) Estimate the spectrum of $x_n(t+k)$:

$$X_n(\omega_{k'}) = \text{FFT}\{x_n(t+k)\} = \sum_{k=0}^{K-1} x_n(t+k)e^{-j\omega_{k'}k}$$

(c) Compute $\Phi(\omega_{k'})$

(d) Estimate the generalized cross-correlation function:

$$\begin{aligned}\hat{r}_{x_1x_2}^{\text{GCC}}(p) &= \text{IFFT}\{\Phi(\omega_{k'})X_1(\omega_{k'})X_2^*(\omega_{k'})\} \\ &= \frac{1}{K}\sum_{k'=0}^{K-1}\Phi(\omega_{k'})X_1(\omega_{k'})X_2^*(\omega_{k'})e^{j\omega_{k'}p}\end{aligned}$$

(d) Obtain the time delay:

$\hat{\tau}_{12} = \arg\max_p \hat{r}_{x_1x_2}^{\text{GCC}}(p)$

There are a number of member algorithms in the GCC family depending on how the weighting function $\Phi(\omega_{k'})$ is selected [205], [66], [261], [157], [65]. Commonly used weighting functions for single-path propagation environments are summarized in Table 9.4. Combination of some of these functions was also suggested [308].

Different weighting functions possess different properties. Some can make the delay estimates more immune to additive noise, while others can improve the robustness of TDE against the multipath effect. As a result, weighting functions should be selected according to the specific application requirements and the corresponding environmental conditions.

The GCC method for time delay estimation is summarized in Table 9.5.

9.6 Adaptive Eigenvalue Decomposition Algorithm

All the algorithms outlined in the previous sections basically achieve TDOA estimate by identifying the extremum of the (generalized) cross correlation function between two channels. For an ideal acoustic system where only attenuation and delay are taken into account, an impulse would appear at the actual TDOA. In practical room acoustic environments, however, additive background noise and room reverberation make the peak no longer well-defined or even no longer dominate in the estimated cross-correlation function. Many amendments to the correlation based algorithms have been proposed. But they are still unable to deal well with reverberation.

Recently, an adaptive eigenvalue decomposition (AED) algorithm was proposed to deal with the TDE problem in room reverberant environments [32]. Unlike the cross correlation based methods, this algorithm assumes a reverberation (multipath) signal model. It first identifies the channel impulse responses from the source to the two sensors. The delay estimate is then determined by finding the direct paths from the two measured impulse responses.

For the signal model given in (9.3) with two sensors, in the absence of noise, one can easily check that:

$$x_1(k) * h_2 = s(k) * h_1 * h_2 = x_2(k) * h_1. \tag{9.29}$$

At time instant k, this relation can be rewritten in a vector-matrix form as:

$$\mathbf{x}^T(k)\mathbf{u} = \mathbf{x}_1^T(k)\mathbf{h}_2 - \mathbf{x}_2^T(k)\mathbf{h}_1 = 0, \tag{9.30}$$

where

$$\begin{aligned}
\mathbf{x}_n(k) &= \begin{bmatrix} x_n(k) & x_n(k-1) & \cdots & x_n(k-L+1) \end{bmatrix}^T, \; n = 1, 2, \\
\mathbf{x}(k) &= \begin{bmatrix} \mathbf{x}_1^T(k) & \mathbf{x}_2^T(k) \end{bmatrix}^T, \\
\mathbf{u} &= \begin{bmatrix} \mathbf{h}_2^T & -\mathbf{h}_1^T \end{bmatrix}^T.
\end{aligned}$$

Left multiplying (9.30) by $\mathbf{x}(k)$ and taking expectation yields

$$\mathbf{R}_x\mathbf{u} = \mathbf{0}, \tag{9.31}$$

where $\mathbf{R}_x = E\left[\mathbf{x}(k)\mathbf{x}^T(k)\right]$ is the covariance matrix of the two observation sensor signals. This implies that the vector $\mathbf{u}$, which consists of two impulse responses, is in the null space of $\mathbf{R}_x$. More specifically, $\mathbf{u}$ is the eigenvector of $\mathbf{R}_x$ corresponding to the eigenvalue 0. It has been shown that the two channel impulse responses (i.e., $\mathbf{h}_1$ and $\mathbf{h}_2$) can be uniquely determined (up to a scale and a common delay) from (9.31) if the following two conditions are satisfied (see Sect. 6.2):

- The polynomials formed from $\mathbf{h}_1$ and $\mathbf{h}_2$ (i.e., the z-transforms of $\mathbf{h}_1$ and $\mathbf{h}_2$) are co-prime, or they do not share any common zeros;

Table 9.6. The adaptive eigenvalue decomposition algorithm for time delay estimation.

Parameters:	$\hat{\tau}_{12}$, $\hat{\mathbf{h}}_1$, $\hat{\mathbf{h}}_2$
Estimation:	Initialize $\hat{\mathbf{h}}_1$, $\hat{\mathbf{h}}_2$
	For $k = 0, 1, \cdots$
	(a) Construct the signal vector:
	$\mathbf{x}_1(k) = \left[\begin{array}{cccc} x_1(k) & x_1(k-1) & \cdots & x_1(k-L+1) \end{array}\right]^T$
	$\mathbf{x}_2(k) = \left[\begin{array}{cccc} x_2(k) & x_2(k-1) & \cdots & x_2(k-L+1) \end{array}\right]^T$
	$\mathbf{x}(k) = \left[\begin{array}{cc} \mathbf{x}_1^T(k) & \mathbf{x}_2^T(k) \end{array}\right]^T$
	(b) Computer the error signal:
	$e(k) = \hat{\mathbf{u}}^T(k)\mathbf{x}(k)$
	(c) Update the filter coefficients:
	$\hat{\mathbf{u}}(k+1) = \frac{\hat{\mathbf{u}}(k) - \mu e(k)\mathbf{x}(k)}{\lvert\lvert\hat{\mathbf{u}}(k) - \mu e(k)\mathbf{x}(k)\rvert\rvert}$
	(d) Obtain the time delay (after convergence):
	$\hat{\tau}_{12} = \arg\max_l \lvert\hat{h}_{2,l}\rvert - \arg\max_l \lvert\hat{h}_{1,l}\rvert$

- The covariance matrix of the source signal $s(k)$, i.e., $\mathbf{R}_s = E\left[\mathbf{s}(k)\mathbf{s}^T(k)\right]$, is of full rank.

When an independent white noise signal is present at each sensor, it will regularize the covariance matrix; as a consequence, the covariance matrix $\mathbf{R}_x$ does not have a zero eigenvalue anymore. In such a case, an estimate of the impulse responses can be achieved through the following algorithm, which is an adaptive way to find the eigenvector associated with the smallest eigenvalue of $\mathbf{R}_x$:

$$\hat{\mathbf{u}}(k+1) = \frac{\hat{\mathbf{u}}(k) - \mu e(k)\mathbf{x}(k)}{\lvert\lvert\hat{\mathbf{u}}(k) - \mu e(k)\mathbf{x}(k)\rvert\rvert}, \tag{9.32}$$

with the constraint that $\lvert\lvert\hat{\mathbf{u}}(k)\rvert\rvert = 1$, where

$$e(k) = \hat{\mathbf{u}}^T(k)\mathbf{x}(k) \tag{9.33}$$

is an error signal, and μ is the adaptation step.

With the identified impulse responses $\hat{\mathbf{h}}_1$ and $\hat{\mathbf{h}}_2$, the time delay estimate is determined as the time difference between the two direct paths, i.e.,

$$\hat{\tau}_{12} = \arg\max_l \lvert\hat{h}_{2,l}\rvert - \arg\max_l \lvert\hat{h}_{1,l}\rvert. \tag{9.34}$$

The AED algorithm for time delay estimation is summarized in Table 9.6.

9.7 Multichannel Cross-Correlation Algorithm

Time delay estimation would be an easy task if the observation signals were merely delayed and attenuated copies of the source signal. In acoustical environments, however, the source signal is generally immersed in ambient noise and reverberation, making the TDE problem more complicated. In the previous section, we discussed an adaptive eigenvalue decomposition algorithm, which is formulated based on a multipath signal model and can deal well with reverberation. Another way to better cope with noise and reverberation is through the use of multiple (more than two) sensors and taking advantage of the redundant information. To illustrate the redundancy provided by multiple microphone sensors, let us consider a three-sensor linear array, which can be partitioned into three sensor pairs. Three delay measurements can then be acquired with the observation data, i.e., τ_{12} (TDOA between sensor 1 and sensor 2), τ_{23} (TDOA between sensor 2 and sensor 3), and τ_{13} (TDOA between sensor 1 and sensor 3). Apparently, these three delays are not independent. As a matter of fact, if the source is located in the far field, it is easily seen that $\tau_{13} = \tau_{12} + \tau_{23}$. Such a relation was exploited in [203] to formulate a two-stage TDE algorithm. In the preprocessing stage, three delay measurements were measured independently using the GCC method. A state equation was then formed and the Kalman filter is used in the post-processing stage to enhance the delay estimate of τ_{12}. By doing so, the estimation variance of τ_{12} can be significantly reduced. Recently, several approaches based on multiple sensor pairs were developed to deal with TDE in room acoustic environments [240], [149], [97]. Different from the Kalman filter method, these approaches fuse the estimated cost functions from multiple sensor pairs before searching for the time delay. In what follows, we discuss a multichannel cross-correlation (MCC) algorithm, which is derived from the spatial linear prediction and interpolation techniques. It can take advantage of the redundancy among multiple sensors in a more natural and coherent way than both the two-stage and fusion methods.

9.7.1 Forward Spatial Linear Prediction

Suppose that we have a microphone array consisting of N sensors positioned in a specific geometric configuration. The sensors' outputs [i.e., $x_1(k), x_2(k), \cdots, x_N(k)$] relate to the source signal through the single-path propagation model given in (9.1). We now seek to predict the signal at the first sensor from spatial samples from all the other $N-1$ sensor signals. For the ease of presentation, let us express the TDOA between the first and the nth sensors in terms of τ_{12} (the TDOA between the first and the second sensors) as:

$$\tau_{1n} = D_n - D_1 = f_n(\tau_{12}), \qquad n = 1, 2, \cdots, N, \tag{9.35}$$

where $f_1(\tau_{12}) = 0$, $f_2(\tau_{12}) = \tau_{12}$ and for $n > 2$, the function f_n depends not only on τ_{12}, but also on the array geometry. Note that in the near field, the function may also be contingent on the source position. We here consider the far-field case.

Using (9.35), we can write the signal model in (9.1) as

$$x_n(k) = \alpha_n s[k - D_1 - f_n(\tau_{12})] + b_n(k). \quad (9.36)$$

It is clear then that the signal $x_1[k - f_N(\tau_{12})]$ is aligned with the signals $x_n[k - f_N(\tau_{12}) + f_n(\tau_{12})]$, $n = 2, 3, \cdots, N$. From these observations, we define the following forward spatial prediction error signal:

$$e_1(k, q) = x_1[k - f_N(q)] - \mathbf{x}_{2:N}^T[k - f_N(q)]\mathbf{a}_q, \quad (9.37)$$

where q is a guessed relative delay for τ_{12}, and

$$\mathbf{x}_{2:N}[k - f_N(q)] = \begin{bmatrix} x_2[k - f_N(q) + f_2(q)] & x_3[k - f_N(q) + f_3(q)] \\ \cdots & x_N(k) \end{bmatrix}^T,$$

and

$$\mathbf{a}_q = \begin{bmatrix} a_{q,2} & a_{q,3} & \cdots & a_{q,N} \end{bmatrix}^T$$

is the forward spatial linear predictor. Consider the criterion:

$$J_{q,1} = E\left\{e_1^2(k, q)\right\}. \quad (9.38)$$

Minimization of (9.38) leads to the Wiener-Hopf equations:

$$\mathbf{R}_{q,2:N}\mathbf{a}_q = \mathbf{r}_{q,2:N}, \quad (9.39)$$

where

$$\begin{aligned} \mathbf{R}_{q,2:N} &= E\{\mathbf{x}_{2:N}[k - f_N(q)]\mathbf{x}_{2:N}^T[k - f_N(q)]\} \\ &= \begin{bmatrix} E\{x_2^2(k)\} & r_{x_2x_3}(q) & \cdots & r_{x_2x_N}(q) \\ r_{x_3x_2}(q) & E\{x_3^2(k)\} & \cdots & r_{x_3x_N}(q) \\ \vdots & \vdots & \ddots & \vdots \\ r_{x_Nx_2}(q) & r_{x_Nx_3}(q) & \cdots & E\{x_N^2(k)\} \end{bmatrix} \end{aligned}$$

is the spatial correlation matrix with

$$\begin{aligned} r_{x_ix_j}(q) &= E\{x_i[k - f_N(q) + f_i(q)]x_j[k - f_N(q) + f_j(q)]\} \\ &= E\{x_i[k + f_i(q)]x_j[k + f_j(q)]\} \\ &= E\{x_i[k - f_j(q)]x_j[k - f_i(q)]\}, \end{aligned} \quad (9.40)$$

and

$$\begin{aligned}
\mathbf{r}_{q,2:N} &= E\left\{\mathbf{x}_{2:N}[k - f_N(q)]x_1[k - f_N(q)]\right\} \\
&= \begin{bmatrix} E\{x_2[k - f_N(q) + f_2(q)]x_1[k - f_N(q)]\} \\ E\{x_3[k - f_N(q) + f_3(q)]x_1[k - f_N(q)]\} \\ \vdots \\ E\{x_N[k - f_N(q) + f_N(q)]x_1[k - f_N(q)]\} \end{bmatrix} \\
&= \begin{bmatrix} E\{x_2(k)x_1[k - f_2(q)]\} \\ E\{x_3(k)x_1[k - f_3(q)]\} \\ \vdots \\ E\{x_N(k)x_1[k - f_N(q)]\} \end{bmatrix}
\end{aligned}$$

is the spatial correlation vector. Note that the spatial correlation matrix is not Toeplitz in general, except in some particular cases.

For $q = \tau_{12}$ and for the noise-free case (where $b_n(k) = 0$, $n = 1, 2, \cdots, N$), it can easily be checked that with the signal model given in (9.36), the rank of matrix $\mathbf{R}_{q,2:N}$ is equal to 1. This means that the samples $x_1(k)$ can be perfectly predicted from any one of the other microphone samples. However, the noise is never absent in practice and is in general isotropic. The noise energy at different microphones is added to the main diagonal of the correlation matrix $\mathbf{R}_{q,2:N}$, which will regularize it and make it positive definite (which we suppose in the rest of this chapter). A unique solution to (9.39) is then guaranteed for any number of microphones. This solution is optimal from a Wiener theory point of view as explained in Chap. 3.

9.7.2 Backward Spatial Linear Prediction

Now we consider microphone N and we would like to align successive time samples of this microphone signal with spatial samples from the $N - 1$ other microphone signals. It is clear that $x_N(k)$ is aligned with the signals $x_n(k - f_N(\tau_{12}) + f_n(\tau_{12})]$, $n = 1, 2, \cdots, N-1$. From these observations, we define the following backward spatial prediction error signal:

$$e_N(k, q) = x_N(k) - \mathbf{x}_{1:N-1}^T[k - f_N(q)]\mathbf{b}_q, \tag{9.41}$$

where

$$\begin{aligned}
\mathbf{x}_{1:N-1}[k - f_N(q)] = &\Big[\, x_1[k - f_N(q) + f_1(q)] \;\; x_2[k - f_N(q) + f_2(q)] \\
&\cdots \; x_{N-1}[k - f_N(q) + f_{N-1}(q)] \,\Big]^T
\end{aligned}$$

and

$$\mathbf{b}_q = \begin{bmatrix} b_{q,1} & b_{q,2} & \cdots & b_{q,N-1} \end{bmatrix}^T$$

is the backward spatial linear predictor. Minimization of the criterion:

$$J_{q,N} = E\left\{e_N^2(k,q)\right\} \tag{9.42}$$

leads to the Wiener-Hopf equations:

$$\mathbf{R}_{q,1:N-1}\mathbf{b}_q = \mathbf{r}_{q,1:N-1}, \tag{9.43}$$

where

$$\begin{aligned}
\mathbf{R}_{q,1:N-1} &= E\{\mathbf{x}_{1:N-1}[k - f_N(q)]\mathbf{x}_{1:N-1}^T[k - f_N(q)]\} \\
&= \begin{bmatrix} E\{x_1^2(k)\} & r_{x_1x_2}(q) & \cdots & r_{x_1x_{N-1}}(q) \\ r_{x_2x_1}(q) & E\{x_2^2(k)\} & \cdots & r_{x_2x_{N-1}}(q) \\ \vdots & \vdots & \ddots & \vdots \\ r_{x_{N-1}x_1}(q) & r_{x_{N-1}x_2}(q) & \cdots & E\{x_{N-1}^2(k)\} \end{bmatrix}
\end{aligned}$$

and

$$\begin{aligned}
\mathbf{r}_{q,1:N-1} &= E\{\mathbf{x}_{1:N-1}[k - f_N(q)]x_N(k)\} \\
&= \begin{bmatrix} E\{x_1(k)x_N[k + f_N(q) - f_1(q)]\} \\ E\{x_2(k)x_N[k + f_N(q) - f_2(q)]\} \\ \vdots \\ E\{x_{N-1}(k)x_N[k + f_N(q) - f_{N-1}(q)]\} \end{bmatrix}.
\end{aligned}$$

9.7.3 Spatial Linear Interpolation

The ideas presented for forward/backward spatial linear prediction can easily be extended to spatial linear interpolation, where we seek to determine how any one of these signals can be interpolated from the others. For the nth sensor, the spatial interpolation error signal is defined as

$$e_n(k,q) = -\mathbf{x}_{1:N}^T[k - f_L(q)]\mathbf{c}_{q,n}, \tag{9.44}$$

where

$$\mathbf{x}_{1:N}[k - f_N(q)] = \left[x_1[k - f_N(q) + f_1(q)] \; x_2[k - f_N(q) + f_2(q)] \; \cdots \; x_N(n)\right]^T$$

and

$$\mathbf{c}_{q,n} = \begin{bmatrix} c_{q,n,1} & c_{q,n,2} & \cdots & c_{q,n,N} \end{bmatrix}^T$$

with the constraint $c_{q,n,n} = -1$. This $\mathbf{c}_{q,n}$ vector without the component $c_{q,n,n}$ is the nth interpolator. The criterion associated with (9.44) is:

$$J_{q,n} = E\left\{e_n^2(k,q)\right\}. \tag{9.45}$$

By using a Lagrange multiplier, it is easy to see that the solution to this optimization problem is:

$$\mathbf{R}_{q,1:N}\mathbf{c}_{q,n} = -\mathbf{c}_{q,n}^T\mathbf{R}_{q,1:N}\mathbf{c}_{q,n}\mathbf{u}_n, \tag{9.46}$$

where

$$\mathbf{R}_{q,1:N} = E\{\mathbf{x}_{1:N}[k - f_N(q)]\mathbf{x}_{1:N}^T[k - f_N(q)]\}$$

is the spatial correlation matrix, and

$$\mathbf{u}_n = [\, 0 \ \ldots \ 0 \ 1 \ 0 \ \ldots \ 0 \,]^T$$

is a vector of length N with its nth component equal to unity and all others zero.

9.7.4 Time Delay Estimation Using Spatial Linear Prediction

The spatial linear prediction, and more generally the spatial interpolation technique can be applied to the problem of TDE. Here, we give an example of using the forward spatial linear prediction. The idea can be easily generalized to the backward spatial linear prediction and the spatial interpolation.

Let $J_{q,1;\min}$ denote the minimum mean-squared error, for the value q, defined by

$$J_{q,1;\min} \triangleq E\left\{e_{1;\min}^2(k,q)\right\}. \tag{9.47}$$

If we replace $\mathbf{a}_q$ by $\mathbf{R}_{q,2:N}^{-1}\mathbf{r}_{q,2:N}$ in (9.37), we get:

$$e_{1;\min}(k,q) = x_1[k - f_N(q)] - \mathbf{x}_{2:N}^T[k - f_N(q)]\mathbf{R}_{q,2:N}^{-1}\mathbf{r}_{q,2:N}. \tag{9.48}$$

It follows immediately that

$$J_{q,1;\min} = E\left\{x_1^2[k - f_N(q)]\right\} - \mathbf{r}_{q,2:N}^T\mathbf{R}_{q,2:N}^{-1}\mathbf{r}_{q,2:N}. \tag{9.49}$$

The value of q that gives the minimum $J_{q,1;\min}$ corresponds to the TDOA between the first and the second sensors, i.e., τ_{12}. Mathematically, the solution to the TDE problem is given by

$$\hat{\tau}_{12} = \arg\min_q J_{q,1;\min}. \tag{9.50}$$

Particular case: Two microphones ($N = 2$). In this case, it can be checked that the solution is:

$$\begin{aligned} \hat{\tau}_{12} &= \arg\min_q \left\{ E\{x_1^2(k-q)\}\left[1 - \frac{E^2\{x_1(k-q)x_2(k)\}}{E\{x_1^2(k-q)\}E\{x_2^2(k)\}}\right]\right\} \\ &= \arg\min_q \left\{ E\{x_1^2(k-q)\}\left[1 - \frac{r_{x_1x_2}^2(q)}{E\{x_1^2(k-q)\}E\{x_2^2(k)\}}\right]\right\}. \end{aligned} \tag{9.51}$$

Since $E\{x_1^2(k-q)\}$, which is the energy of $x_1(k)$, does not affect the peak position, (9.51) can be further written as

$$\begin{aligned}\hat{\tau}_{12} &= \arg\min_q \left\{E\{x_1^2(k-q)\}\left[1-\rho^2_{x_1x_2}(q)\right]\right\} \\ &= \arg\min_q \left\{1-\rho^2_{x_1x_2}(q)\right\} \\ &= \arg\max_q \rho^2_{x_1x_2}(q),\end{aligned} \tag{9.52}$$

where

$$\rho_{x_1x_2}(q) = \frac{r_{x_1x_2}(q)}{\sqrt{E\{x_1^2(k-q)\}E\{x_2^2(k)\}}} \tag{9.53}$$

is the cross-correlation coefficient between $x_1(k-q)$ and $x_2(k)$. Apparently, this result is similar to what is obtained with the cross-correlation method. Note that in the general case with multiple microphone sensors, this approach can be viewed as an extension of the cross-correlation method from the two-channel to the multichannel cases, which can take advantage of the knowledge of the microphone array to estimate time delay between the first and the second sensors optimally in a least-mean-square sense.

9.7.5 Spatial Correlation Matrix and Its Properties

Consider the N microphone signals x_n, $n = 1, 2, \cdots, N$. The corresponding spatial correlation matrix is

$$\begin{aligned}\mathbf{R}(q) &= \mathbf{R}_{q,1:N} = E\{\mathbf{x}_{1:N}[k-f_N(q)]\mathbf{x}^T_{1:N}[k-f_N(q)]\} \\ &= \begin{bmatrix} r_{x_1x_1}(q) & r_{x_1x_2}(q) & \cdots & r_{x_1x_N}(q) \\ r_{x_2x_1}(q) & r_{x_2x_2}(q) & \cdots & r_{x_2x_N}(q) \\ \vdots & \ddots & \ddots & \vdots \\ r_{x_Nx_1}(q) & r_{x_Nx_2}(q) & \cdots & r_{x_Nx_N}(q) \end{bmatrix},\end{aligned} \tag{9.54}$$

which can be factored as:

$$\mathbf{R}(q) = \mathbf{E}\widetilde{\mathbf{R}}(q)\mathbf{E}, \tag{9.55}$$

where

$$\mathbf{E} = \begin{bmatrix} \sqrt{E[x_0^2(k)]} & 0 & \cdots & 0 \\ 0 & \sqrt{E[x_1^2(k)]} & \cdots & 0 \\ \vdots & \vdots & \ddots & \vdots \\ 0 & \cdots & 0 & \sqrt{E[x_N^2(k)]} \end{bmatrix} \tag{9.56}$$

is a diagonal matrix,

$$\widetilde{\mathbf{R}}(q) = \begin{bmatrix} 1 & \rho_{x_1x_2}(q) & \cdots & \rho_{x_1x_N}(q) \\ \rho_{x_1x_2}(q) & 1 & \cdots & \rho_{x_2x_N}(q) \\ \vdots & \vdots & \ddots & \vdots \\ \rho_{x_1x_N}(q) & \cdots & \rho_{x_{N-1}x_N}(q) & 1 \end{bmatrix} \tag{9.57}$$

is a symmetric matrix, and

$$\rho_{x_i x_j}(q) = \frac{E\{x_i[k - f_j(q)]x_j[k - f_i(q)]\}}{\sqrt{E\{x_i^2(k)\}E\{x_j^2(k)\}}}, \quad i, j = 1, 2, \cdots, N, \tag{9.58}$$

is the cross-correlation coefficient between $x_i[k - f_j(q)]$ and $x_j[k - f_i(q)]$.

The $\widetilde{\mathbf{R}}(q)$ matrix given in (9.57) has the following properties:

$$\text{(a)} \qquad 0 < \det\left[\widetilde{\mathbf{R}}(q)\right] \leq 1, \tag{9.59}$$

$$\text{(b)} \qquad \det\left[\widetilde{\mathbf{R}}(q)\right] \leq \frac{J_{q,1;\min}}{E[x_1^2(k)]} \leq 1. \tag{9.60}$$

Proofs: Since $\mathbf{R}(q)$ is symmetric and is supposed to be positive definite, it is clear that $\det[\mathbf{R}(q)] > 0$, which implies that $\det\left[\widetilde{\mathbf{R}}(q)\right] > 0$. We can easily check that:

$$\det\left[\widetilde{\mathbf{R}}(q)\right] = \det\left(\widetilde{\mathbf{R}}_{q,1:N}\right) \leq \det\left(\widetilde{\mathbf{R}}_{q,2:N}\right) \leq \cdots \leq 1. \tag{9.61}$$

As a result, $0 < \det\left[\widetilde{\mathbf{R}}(q)\right] \leq 1$.

To show Property (b), let us define

$$\underline{\mathbf{a}}_q = \begin{bmatrix} a_{q,1} & \mathbf{a}_q^T \end{bmatrix}^T. \tag{9.62}$$

Then, for $a_{q,1} = -1$, the forward prediction error signal defined in (9.37) can be rewritten as

$$e_1(k, q) = -\mathbf{x}_{1:N}^T \underline{\mathbf{a}}_q. \tag{9.63}$$

Continuing, the criterion shown in (9.38) can be expressed as

$$J_{q,1} = E\left\{e_1^2(k, q)]\right\} + \kappa(\mathbf{u}_1^T \underline{\mathbf{a}}_q + 1), \tag{9.64}$$

where κ is a Lagrange multiplier introduced to force $a_{q,1}$ to have value -1. It is then easily shown that:

$$J_{q,1;\min} = \frac{1}{\mathbf{u}_1^T \mathbf{R}^{-1}(q)\mathbf{u}_1}. \tag{9.65}$$

In this case, with (9.55), (9.65) becomes:

$$\begin{aligned} J_{q,1;\min} &= \frac{E\{x_1^2(k)\}}{\mathbf{u}_1^T \widetilde{\mathbf{R}}^{-1}(q)\mathbf{u}_1} \\ &= E\{x_1^2(k)\} \frac{\det\left[\widetilde{\mathbf{R}}(q)\right]}{\det\left(\widetilde{\mathbf{R}}_{q,2:N}\right)}. \end{aligned} \tag{9.66}$$

Using (9.61), it is clear that Property (b) is verified.

9.7.6 Multichannel Cross-Correlation Coefficient

From the previous analysis, we can see that the determinant of the spatial correlation matrix is related to the minimum mean-squared error and to the power of the signals. In the two-channel case, it is easy to see that the cross-correlation coefficient between the two signals $x_1(k)$ and $x_2(k)$ is linked to the determinant of the corresponding spatial correlation matrix:

$$\rho_{x_1 x_2}^2(q) = 1 - \det\left(\widetilde{\mathbf{R}}_{q,1:2}\right). \tag{9.67}$$

By analogy to the cross-correlation coefficient definition between two random signals, we define the multichannel cross-correlation coefficient (MCCC) among the signals x_n, $n = 1, 2, \cdots, N$, as:

$$\begin{aligned} \rho_{1:N}^2(q) &= 1 - \det\left(\widetilde{\mathbf{R}}_{q,1:N}\right) \\ &= 1 - \det\left[\widetilde{\mathbf{R}}(q)\right]. \end{aligned} \tag{9.68}$$

Basically, the coefficient $\rho_{1:N}(q)$ will measure the amount of correlation among all the channels. This coefficient possesses the following properties:

(a) $0 \leq \rho_{1:N}^2(q) \leq 1$ (the case $\rho_{1:N}^2(q) = 1$ happens when matrix $\mathbf{R}(q)$ is nonnegative definite).
(b) If two or more signals are perfectly correlated, then $\rho_{1:N}^2(q) = 1$.
(c) If all the processes are completely uncorrelated with each other, then $\rho_{1:N}^2(q) = 0$.
(d) If one of the signals is completely uncorrelated with the $N - 1$ other signals, then this coefficient will measure the correlation among those $N - 1$ remaining signals.

The proof of these properties is straightforward. We shall leave it as an exercise for the reader.

9.7.7 Time Delay Estimation Using MCCC

Obviously, the multichannel cross-correlation coefficient $\rho_{1:N}^2(q)$ can be used for time delay estimation in the following way:

$$\begin{aligned} \hat{\tau}_{12} &= \arg\max_q \left(\rho_{1:N}^2(q)\right) \\ &= \arg\max_q \left\{1 - \det\left[\widetilde{\mathbf{R}}(q)\right]\right\} \\ &= \arg\min_q \det\left[\widetilde{\mathbf{R}}(q)\right]. \end{aligned} \tag{9.69}$$

It is clear that (9.69) is equivalent to (9.50). As we mentioned in our earlier statements, this algorithm can be treated as a generalization of the cross-correlation method from the two-channel to the multichannel cases. Figure 9.1

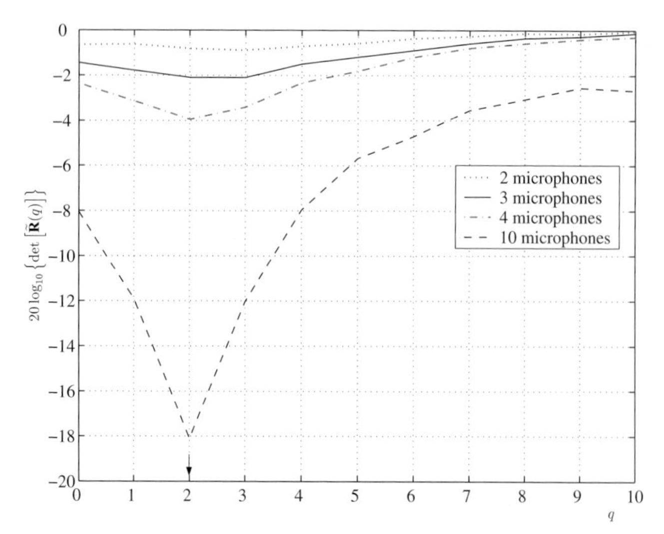

Fig. 9.1. Comparison of $\det\left[\widetilde{\mathbf{R}}(q)\right]$ for different numbers of microphones.

shows an example where we use an equi-spaced linear array consisting of 10 microphone sensors. A loudspeaker is located in the far field, playing back a speech signal recorded from a female speaker. The sensors' outputs are corrupted by white Gaussian noise with SNR $= -5$ dB. Signals are sampled at 16 kHz. The true TDOA between Sensors 1 and 2 is $\tau_{12} = 2$ (samples). When only two sensors are used, we have $\hat{\tau}_{12} = 3$ (samples), which is a wrong TDOA estimate. When more than 3 sensors are used, we see that $\hat{\tau}_{12} = 2$ (samples), which is the same as the true TDOA. It can be seen that as the number of microphones increases, the valley of the cost function is better defined, which will enable an easier search for the extremum. This example demonstrates the effectiveness of the MCC approach in taking advantage of the redundant information provided by multiple sensor to improve TDE against noise.

It is worth pointing out that a pre-filtering (or pre-whitening) process can be applied to the observation signals before computing the MCCC. In this case, this multichannel correlation algorithm can be viewed as a generalized version of the GCC method.

The multichannel correlation method for time delay estimation is summarized in Table 9.7.

9.8 Adaptive Multichannel Time Delay Estimation

In the previous section, we have shown that TDE in adverse environments can be improved by using multiple sensors and taking advantage of the redundancy. The more the number of sensors we use, the better the TDE performance will be. However, since the multichannel cross-correlation method

Table 9.7. The multichannel cross-correlation method for time delay estimation.

Parameter: $\hat{\tau}_{12}$

Estimation: For $t = 0, 1, \cdots$

(a) Obtain a frame of observation signals at time instant t:

$\{x_n(t), x_n(t+1), \cdots, x_n(t+K-1)\}$, $n = 1, 2, \cdots, N$

(b) Pre-filtering the observation signals if needed

(c) For $q = -\tau_{\max}, -\tau_{\max}+1, \cdots, \tau_{\max}$

(1) Estimate the spatial correlation matrix

$$\widetilde{\mathbf{R}}(q) = \begin{bmatrix} 1 & \rho_{x_1x_2}(q) & \cdots & \rho_{x_1x_N}(q) \\ \rho_{x_1x_2}(q) & 1 & \cdots & \rho_{x_2x_N}(q) \\ \vdots & \vdots & \ddots & \vdots \\ \rho_{x_1x_N}(q) & \cdots & \rho_{x_{N-1}x_N}(q) & 1 \end{bmatrix}$$

(2) Estimate the multichannel cross-correlation coefficient

$\rho^2_{1:N}(q) = 1 - \det\left[\widetilde{\mathbf{R}}(q)\right]$

(d) Obtain the time delay

$\hat{\tau}_{12} = \arg\max_q \left[\rho^2_{1:N}(q)\right]$

---

is derived based on the single-path propagation model, it has a fundamental weakness in its ability to deal with reverberation. In order to achieve a reasonable TDE performance in heavily reverberant environments, we may have to use a large number of receivers, which will increase both the cost and the size of the system. In this section, we discuss another multichannel algorithm based on the reverberation signal model, which can be viewed as an extension of the AED algorithm (see Sect. 9.6) from the two-channel to multichannel cases.

As shown in (9.32), the AED algorithm basically obtains the estimates of the two channel impulse responses by minimizing the following error signal:

$$e(k+1) = \mathbf{x}_1^T(k+1)\hat{\mathbf{h}}_2(k) - \mathbf{x}_2^T(k+1)\hat{\mathbf{h}}_1(k). \tag{9.70}$$

In order for the channel impulse responses to be uniquely determined, it requires that the polynomials formed from $\mathbf{h}_1$ and $\mathbf{h}_2$ do not share any common zeros. In room acoustic environments, channel impulse responses are usually very long, particularly when reverberation is strong. As a consequence, it is very likely that there are some common zeros between the two channels. One way to circumvent this common-zero problem is to use multiple sensors (channels). In the same acoustical environment, it would be less likely for all channels to share a common zero.

By analogy to the error signal defined in the AED algorithm, we can define the error signal between the ith and jth channels at time $k+1$ as:

$$e_{ij}(k+1) = \mathbf{x}_i^T(k+1)\hat{\mathbf{h}}_j(k) - \mathbf{x}_j^T(k+1)\hat{\mathbf{h}}_j(k), \quad i, j = 1, 2, \cdots, N. \tag{9.71}$$

Table 9.8. The multichannel adaptive algorithm for time delay estimation.

Parameters:	$\hat{\tau}_{ij}$, $i, j = 1, 2, \cdots, N$ and $i \neq j$
	$\hat{\mathbf{h}}_n$, $n = 1, 2, \cdots, N$
Estimation:	Initialize $\hat{\mathbf{h}}_n$, $n = 1, 2, \ldots, N$
	For $k = 0, 1, \cdots$
	(a) Estimate $\hat{\mathbf{h}}_n(k)$ using any one of the blind SIMO identification algorithms described in Chap. 6 such as the FNMCLMS algorithm
	(b) Obtain the time delays based on the estimated channel impulse responses: $\hat{\tau}_{ij} = \arg\max_l \lvert \hat{h}_{j,l} \rvert - \arg\max_l \lvert \hat{h}_{i,l} \rvert$ $i, j = 1, 2, \cdots, N$, and $i \neq j$

Assuming that these error signals are equally important, we now define a cost function as follows:

$$\chi(k+1) = \sum_{i=1}^{N-1} \sum_{j=i+1}^{N} e_{ij}^2(k+1), \tag{9.72}$$

where we exclude the cases of $e_{ii}(k) = 0$ $(i = 1, 2, \cdots, N)$ and count the $e_{ij}(k) = -e_{ij}(k)$ pair only once. With this definition of multichannel error signal, it follows immediately that many adaptive algorithms described in Chap. 6 can be exploited to estimate the channel impulse responses $\mathbf{h}_1, \mathbf{h}_2, \cdots, \mathbf{h}_N$. The TDOA estimate between any two channels can then be determined as

$$\hat{\tau}_{ij} = \arg\max_l \lvert \hat{h}_{j,l} \rvert - \arg\max_l \lvert \hat{h}_{i,l} \rvert. \tag{9.73}$$

The adaptive multichannel algorithm for time delay estimation is summarized in Table 9.8.

9.9 Acoustic Source Localization

In acoustic environments, the source location information plays an important role for applications such as automatic camera tracking for video-conferencing and beamformer steering for suppressing noise and reverberation. Estimation of source location, which is often called source-localization problem, has been of considerable interest for decades. A common method to obtain the estimate of source location is based on TDOA measurements as described previously.

Given the array geometry and TDOA measurements, the source-localization problem can be formulated mathematically as follows. The array consists of N microphones located at positions

$$\gamma_n \triangleq \begin{bmatrix} x_n & y_n & z_n \end{bmatrix}^T, \ n = 1, \cdots, N, \tag{9.74}$$

in Cartesian coordinates (see Fig. 9.2). The first microphone ($n = 1$) is regarded as the reference and is placed at the origin of the coordinate system, i.e. $\gamma_1 = [0, 0, 0]^T$. The acoustic source is located at $\gamma_\text{s} \triangleq [x_\text{s}, y_\text{s}, z_\text{s}]^T$. The distances from the origin to the nth microphone and the source are denoted by ζ_n and ζ_s, respectively, where

$$\zeta_n \triangleq \|\gamma_n\| = \sqrt{x_n^2 + y_n^2 + z_n^2}, \quad n = 1, \cdots, N, \tag{9.75}$$

$$\zeta_\text{s} \triangleq \|\gamma_\text{s}\| = \sqrt{x_\text{s}^2 + y_\text{s}^2 + z_\text{s}^2}. \tag{9.76}$$

The distance between the source and the nth microphone is denoted by

$$\eta_n \triangleq \|\gamma_n - \gamma_\text{s}\| = \sqrt{(x_n - x_\text{s})^2 + (y_n - y_\text{s})^2 + (z_n - z_\text{s})^2}\ . \tag{9.77}$$

The difference in the distances of microphones n and j from the source is given by

$$d_{nj} \triangleq \eta_n - \eta_j, \quad n, j = 1, \cdots, N. \tag{9.78}$$

This difference is usually termed the *range difference.* It is proportional to the time difference of arrival τ_{nj}. If the speed of sound is c, then

$$d_{nj} = c \cdot \tau_{nj}. \tag{9.79}$$

The speed of sound (in m/s) can be estimated from the air temperature t_air (in degrees Celsius) according to the following approximate (first-order) formula,

$$c \approx 331 + 0.610 \cdot t_\text{air}. \tag{9.80}$$

The localization problem is then to estimate γ_s given the set of γ_n and τ_{nj}. Note that there are $N(N-1)/2$ distinct TDOA estimates τ_{nj}, which exclude the case $n = j$ and count the $\tau_{nj} = -\tau_{jn}$ pair only once. However, in the absence of noise, the space spanned by these TDOA estimates is $(N-1)$-dimensional. Any $N-1$ linearly independent TDOAs determine all of the others. In a noisy environment, the TDOA redundancy can be used to improve the accuracy of the source localization algorithms, but this would increase their computational complexity. For simplicity and also without loss of generality, we choose $\tau_{n1}, n = 2, \cdots, N$ as the basis for this $\mathbb{R}^{N-1}$ space.

9.10 Measurement Model and Cramèr-Rao Lower Bound

When the source localization problem is examined using estimation theory, the measurements of the range differences are modeled by:

$$d_{n1} = g_n(\gamma_\text{s}) + \epsilon_n, \ n = 2, \cdots, N, \tag{9.81}$$

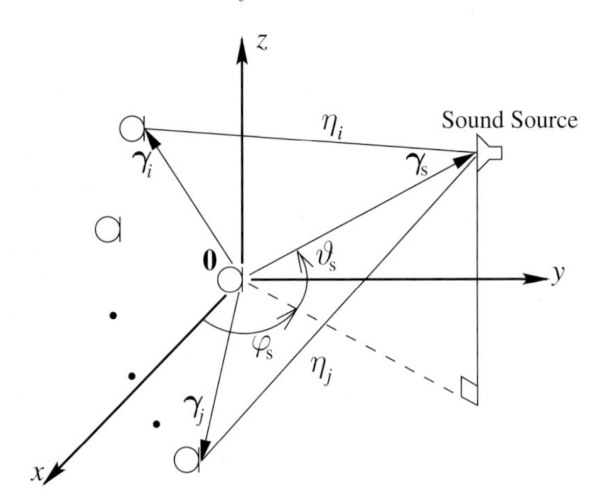

Fig. 9.2. Spatial diagram illustrating variables defined in the source-localization problem.

where

$$g_n(\boldsymbol{\gamma}_\text{s}) = \|\boldsymbol{\gamma}_n - \boldsymbol{\gamma}_\text{s}\| - \|\boldsymbol{\gamma}_\text{s}\|,$$

and the ϵ_n's are measurement errors. In a vector form, such an additive measurement error model becomes,

$$\mathbf{d} = \mathbf{g}(\boldsymbol{\gamma}_\text{s}) + \boldsymbol{\epsilon}, \tag{9.82}$$

where

$$\begin{aligned} \mathbf{d} &= \begin{bmatrix} d_{21} & d_{31} & \cdots & d_{N1} \end{bmatrix}^T, \\ \mathbf{g}(\boldsymbol{\gamma}_\text{s}) &= \begin{bmatrix} g_2(\boldsymbol{\gamma}_\text{s}) & g_3(\boldsymbol{\gamma}_\text{s}) & \cdots & g_N(\boldsymbol{\gamma}_\text{s}) \end{bmatrix}^T, \\ \boldsymbol{\epsilon} &= \begin{bmatrix} \epsilon_2 & \epsilon_3 & \cdots & \epsilon_N \end{bmatrix}^T. \end{aligned}$$

Further, we postulate that the additive measurement errors have mean zero and are independent of the range difference observation, as well as the source location $\boldsymbol{\gamma}_\text{s}$. For a continuous-time estimator, the corrupting noise, as indicated in [154], is jointly Gaussian distributed. The probability density function (PDF) of $\mathbf{d}$ conditioned on $\boldsymbol{\gamma}_\text{s}$ is subsequently given by,

$$p(\mathbf{d}|\boldsymbol{\gamma}_\text{s}) = \frac{\exp\left\{-\frac{1}{2}\left[\mathbf{d} - \mathbf{g}(\boldsymbol{\gamma}_\text{s})\right]^T \mathbf{C}_{\boldsymbol{\epsilon}}^{-1}\left[\mathbf{d} - \mathbf{g}(\boldsymbol{\gamma}_\text{s})\right]\right\}}{\sqrt{(2\pi)^N \det(\mathbf{C}_{\boldsymbol{\epsilon}})}}, \tag{9.83}$$

where $\mathbf{C}_{\boldsymbol{\epsilon}}$ is the covariance matrix of $\boldsymbol{\epsilon}$. Note that $\mathbf{C}_{\boldsymbol{\epsilon}}$ is independent of $\boldsymbol{\gamma}_\text{s}$ by assumption. Since digital equipment is used to sample the microphone waveforms and estimate the TDOAs, the error introduced by discrete-time processing also has to be taken into account. When this is done, the measurement error is no longer Gaussian and is more properly modeled as a mixture of a Gaussian noise and a noise that is uniformly distributed over $[-T_\mathrm{s}c/2, T_\mathrm{s}c/2]$, where T_s is the sampling period. As an example, for a digital source location estimator with an 8 KHz sampling rate operating at room temperature (25 degrees Celsius, i.e. $c \approx 346.25$ meters per second), the maximum error in range difference estimates due to sampling is about ± 2.164 cm, which leads to considerable errors in the location estimate, especially when the source is far from the microphone array.

Under the measurement model (9.82), we are now faced with the parameter estimation problem of extracting the source location information from the mismeasured range differences or the equivalent TDOAs. For an *unbiased* estimator, a Cramèr-Rao lower bound (CRLB) can be placed on the variance of each estimated coordinate of the source location. However, since the range difference function $\mathbf{g}(\boldsymbol{\gamma}_\mathrm{s})$ in the measurement model is nonlinear in the parameters under estimation, it is very difficult (or even impossible) to find an unbiased estimator that is mathematically simple and attains the CRLB. The CRLB is usually used as a benchmark against which the statistical efficiency of any unbiased estimators can be compared.

In general, without any assumptions made about the PDF of the measurement error $\boldsymbol{\epsilon}$, the CRLB of the ith ($i = 1, 2, 3$) parameter variance is found as the $[i, i]$ element of the inverse of the Fisher information matrix defined by [200]:

$$[\mathbf{I}(\boldsymbol{\gamma}_\mathrm{s})]_{ij} \triangleq -E\left[\frac{\partial^2 \ln p(\mathbf{d}|\boldsymbol{\gamma}_\mathrm{s})}{\partial \gamma_{\mathrm{s},i} \partial \gamma_{\mathrm{s},j}}\right], \tag{9.84}$$

where the three parameters of $\boldsymbol{\gamma}_\mathrm{s}$, i.e., $\gamma_{\mathrm{s},1}$, $\gamma_{\mathrm{s},2}$, and $\gamma_{\mathrm{s},3}$, are respectively x, y, and z coordinates of the source location.

In the case of a Gaussian measurement error, the Fisher information matrix turns into [68]

$$\mathbf{I}(\boldsymbol{\gamma}_\mathrm{s}) = \left[\frac{\partial \mathbf{g}(\boldsymbol{\gamma}_\mathrm{s})}{\partial \boldsymbol{\gamma}_\mathrm{s}}\right]^T \mathbf{C}_{\boldsymbol{\epsilon}}^{-1} \left[\frac{\partial \mathbf{g}(\boldsymbol{\gamma}_\mathrm{s})}{\partial \boldsymbol{\gamma}_\mathrm{s}}\right], \tag{9.85}$$

where $\partial \mathbf{g}(\boldsymbol{\gamma}_\mathrm{s})/\partial \boldsymbol{\gamma}_\mathrm{s}$ is an $(N-1) \times 3$ Jacobian matrix defined as,

$$\begin{aligned}
\frac{\partial \mathbf{g}(\boldsymbol{\gamma}_\mathrm{s})}{\partial \boldsymbol{\gamma}_\mathrm{s}} &= \begin{bmatrix}
\frac{\partial g_2(\boldsymbol{\gamma}_\mathrm{s})}{\partial x_\mathrm{s}} & \frac{\partial g_2(\boldsymbol{\gamma}_\mathrm{s})}{\partial y_\mathrm{s}} & \frac{\partial g_2(\boldsymbol{\gamma}_\mathrm{s})}{\partial z_\mathrm{s}} \\
\frac{\partial g_3(\boldsymbol{\gamma}_\mathrm{s})}{\partial x_\mathrm{s}} & \frac{\partial g_3(\boldsymbol{\gamma}_\mathrm{s})}{\partial y_\mathrm{s}} & \frac{\partial g_3(\boldsymbol{\gamma}_\mathrm{s})}{\partial z_\mathrm{s}} \\
\vdots & \vdots & \vdots \\
\frac{\partial g_N(\boldsymbol{\gamma}_\mathrm{s})}{\partial x_\mathrm{s}} & \frac{\partial g_N(\boldsymbol{\gamma}_\mathrm{s})}{\partial y_\mathrm{s}} & \frac{\partial g_N(\boldsymbol{\gamma}_\mathrm{s})}{\partial z_\mathrm{s}}
\end{bmatrix} \\
&= \begin{bmatrix}
(\mathbf{u}_2 - \mathbf{u}_1)^T \\
(\mathbf{u}_3 - \mathbf{u}_1)^T \\
\vdots \\
(\mathbf{u}_N - \mathbf{u}_1)^T
\end{bmatrix},
\end{aligned} \tag{9.86}$$

and

$$\mathbf{u}_n = \frac{\boldsymbol{\gamma}_\mathrm{s} - \boldsymbol{\gamma}_n}{\|\boldsymbol{\gamma}_\mathrm{s} - \boldsymbol{\gamma}_n\|} = \frac{\boldsymbol{\gamma}_\mathrm{s} - \boldsymbol{\gamma}_n}{\eta_n}, \; n = 1, 2, \cdots, N, \tag{9.87}$$

is the normalized vector of unit length pointing from the nth microphone to the sound source.

9.11 Algorithm Overview

There is a rich literature of source localization techniques that use the additive measurement error model given in the previous section. Important distinctions between these methods include likelihood-based versus least-squares and linear approximation versus direct numerical optimization, as well as iterative versus closed-form algorithms.

In early research of source localization with passive sensor arrays, the maximum likelihood (ML) principle was widely utilized [154], [313], [286], [75] because of the proven asymptotic consistency and efficiency of an ML estimator (MLE). However, the number of microphones in an array for camera pointing or beamformer steering in multimedia communication systems is always limited, which makes acoustic source localization a finite-sample rather than a large-sample problem. Moreover, ML estimators require additional assumptions about the distributions of the measurement errors. One approach is to invoke the central limit theorem and assumes a Gaussian approximation, which makes the likelihood function easy to formulate. Although a Gaussian error was justified by Hahn and Tretter [154] for continuous-time processing, it can be difficult to verify and the MLE is no longer optimal when sampling introduces additional errors in discrete-time processing. To compute the solution to the MLE, a linear approximation and iterative numerical techniques have to be used because of the nonlinearity of the hyperbolic equations. The Newton-Raphson iterative method [12], the Gauss-Newton method [121], and the least-mean-square (LMS) algorithm are among possible choices. But for these iterative approaches, selecting a good initial guesstimate to avoid a local minimum is difficult and convergence to the optimal solution cannot be guaranteed. Therefore, an ML-based estimator is usually difficult to implement in real-time source-localization systems.

For real-time applications, closed-form estimators have also gained wider attention. Of the closed-form estimators, triangulation is the most straightforward [308]. However, with triangulation it is difficult to take advantage of extra sensors and the TDOA redundancy. Nowadays most closed-form algorithms exploit a least-squares principle, which makes no additional assumption about the distribution of measurement errors. To construct a least-squares estimator, one needs to define an error function based on the measured TDOAs. Different error functions will result in different estimators with different complexity and performance. Schmidt showed that the TDOAs to three sensors whose positions are known provide a straight line of possible source locations in two dimensions and a plane in three dimensions. By intersecting the lines/planes specified by different sensor triplets, he obtained an estimator called plane intersection. Another closed-form estimator, termed spherical intersection (SX), employed a spherical LS criterion [267]. The SX algorithm is mathematically simple, but requires an *a priori* solution for the source range, which may not exist or may not be unique in the presence of measurement errors. Based on the same criterion, Smith and Abel [279] proposed the spherical interpolation (SI) method, which also solved for the source range, again in the LS sense. Although the SI method has less bias, it is not efficient and it has a large standard deviation relative to the Cramèr-Rao lower bound (CRLB). With the SI estimator, the source range is a byproduct that is assumed to be independent of the location coordinates. Chan and Ho [68] improved the SI estimation with a second LS estimator that accommodates the information redundancy from the SI estimates and updates the squares of the coordinates. We shall refer to this method as the quadratic-correction least-squares (QCLS) approach. In the QCLS estimator, the covariance matrix of measurement errors is used. But this information can be difficult to properly assume or accurately estimate, which results in a performance degradation in practice. When the SI estimate is analyzed and the quadratic correction is derived in the QCLS estimation procedure, perturbation approaches are employed and, presumptively, the magnitude of measurement errors has to be small. It has been indicated in [69] that the QCLS estimator yields an unbiased solution with a small standard deviation that is close to the CRLB at a moderate noise level. But when noise is practically strong, its bias is considerable and its variance could no longer approach the CRLB according to our Monte-Carlo simulations. Recently a linear-correction least-squares (LCLS) algorithm has been proposed in [176]. This method applies the additive measurement error model and employs the technique of Lagrange multipliers. It makes no assumption on the covariance matrix of measurement errors.

In the rest of this chapter, we will discuss various source-localization algorithms in detail and evaluate them in terms of estimation accuracy and efficiency, computational complexity, implementation flexibility, and adaptation capabilities to different and varying environments.

9.12 Maximum Likelihood Estimator

The measurement model for the source localization problem was investigated and the CRLB for any unbiased estimator was determined in Sect. 9.10. Since the measurement model is highly nonlinear, an efficient estimator that attains the CRLB may not exist or might be impossible to find even if it does exist. In practice, the maximum likelihood estimator is often used since it has the well-proven advantage of asymptotic efficiency for a large sample space.

To apply the maximum likelihood principle, the statistical characteristics of the measurements need to be known or properly assumed prior to any processing. From the central limit theorem and also for mathematical simplicity, the measurement error is usually modeled as Gaussian and the likelihood function is given by (9.83), which is considered as a function of the source position γ_s under estimation.

Since the exponential function is monotonically increasing, the MLE is equivalent to minimizing a (log-likelihood) cost function defined as,

$$\mathcal{L}(\gamma_\mathrm{s}) \triangleq [\mathbf{d} - \mathbf{g}(\gamma_\mathrm{s})]^T \mathbf{C}_\epsilon^{-1} [\mathbf{d} - \mathbf{g}(\gamma_\mathrm{s})] . \tag{9.88}$$

Direct estimation of the minimizer is generally not practical. If the noise signals at different microphones are assumed to be uncorrelated, the covariance matrix is diagonal:

$$\mathbf{C}_\epsilon = \mathrm{diag}(\sigma_2^2, \sigma_2^2, \cdots, \sigma_N^2), \tag{9.89}$$

where σ_n^2 $(n = 2, 3, \cdots, N)$ is the variance of ϵ_n, and the cost function (9.88) becomes,

$$\mathcal{L}(\gamma_\mathrm{s}) = \sum_{n=2}^{N} \frac{[d_{n1} - g_n(\gamma_\mathrm{s})]^2}{\sigma_n^2}. \tag{9.90}$$

Among other approaches, the steepest descent algorithm can be used to find $\hat{\gamma}_{\mathrm{s,MLE}}$ iteratively with

$$\hat{\gamma}_\mathrm{s}(k+1) = \hat{\gamma}_\mathrm{s}(k) - \frac{1}{2}\mu \nabla \mathcal{L}[\hat{\gamma}_\mathrm{s}(k)], \tag{9.91}$$

where μ is the step size.

The foregoing MLE can be determined and is asymptotically optimal for this problem only if its two assumptions (Gaussian and uncorrelated measurement noise) hold. However, this is not the case in practice as discussed in Sect. 9.10. Furthermore, the number of microphones in an array for camera pointing or beamformer steering is always limited, which makes the source localization a finite-sample rather than a large-sample problem. In addition, the cost function (9.90) is generally not strictly convex. In order to avoid a local minimum with the steepest descent algorithm, we need to select a good initial guesstimate of the source location, which is difficult to do in practice, and convergence of the iterative algorithm to the desired solution cannot be guaranteed.

9.13 Least-Squares Estimators

Two limitations of the MLE are that probabilistic assumptions have to be made about the measured range differences and that the iterative algorithm to find the solution is computationally intensive. An alternative method is the well-known least-squares estimator (LSE). The LSE makes no probabilistic assumptions about the data and hence can be applied to the source localization problem in which a precise statistical characterization of the data is hard to determine. Furthermore, an LSE usually produces a closed-form estimate that is desirable in real-time applications.

9.13.1 Least-Squares Error Criteria

In the LS approach, we attempt to minimize a squared error function that is zero in the absence of noise and model inaccuracies. Different error functions can be defined for closeness from the assumed (noiseless) signal based on hypothesized parameters to the observed data. When these are applied, different LSEs can be derived. For the source localization problem two LS error criteria can be constructed.

Hyperbolic LS Error Function

The first LS error function is defined as the difference between the observed range difference and that generated by a signal model depending upon the unknown parameters. Such an error function is routinely used in many LS estimators

$$\mathbf{e}_{\mathrm{h}}(\boldsymbol{\gamma}_{\mathrm{s}}) \stackrel{\triangle}{=} \mathbf{d} - \mathbf{g}(\boldsymbol{\gamma}_{\mathrm{s}}), \tag{9.92}$$

and the corresponding LS criterion is given by

$$J_{\mathrm{h}} = \mathbf{e}_{\mathrm{h}}^{T}\mathbf{e}_{\mathrm{h}} = [\mathbf{d} - \mathbf{g}(\boldsymbol{\gamma}_{\mathrm{s}})]^{T}[\mathbf{d} - \mathbf{g}(\boldsymbol{\gamma}_{\mathrm{s}})]. \tag{9.93}$$

In the source localization problem, an observed range difference d_{n1} defines a hyperboloid in 3-D space. All points lying on such a hyperboloid are potential source locations and all have the same range difference d_{n1} to the two microphones n and 1. Therefore, a sound source that is located by minimizing the hyperbolic LS error criterion (9.93) has the shortest distance to all hyperboloids associated with different microphone pairs and specified by the estimated range differences.

In (9.92), the signal model $\mathbf{g}(\boldsymbol{\gamma}_{\mathrm{s}})$ consists of a set of hyperbolic functions. Since they are nonlinear, minimizing (9.93) leads to a mathematically intractable solution as N gets large. Moreover, the hyperbolic function is very sensitive to noise, especially for far-field sources. As a result, it is rarely used in practice.

When the statistical characteristics of the corrupting noise are unknown, uncorrelated *white* Gaussian noise is one reasonable assumption. In this case, it is not surprising that the hyperbolic LSE and the MLE minimize (maximize) similar criteria.

Spherical LS Error Function

The second LS criterion is based on the errors found in the distances from a hypothesized source location to the microphones. In the absence of measurement errors, the correct source location is preferably at the intersection of a group of spheres centered at the microphones. When measurement errors are present, the best estimate of the source location would be the point that yields the shortest distance to those spheres defined by the range differences and the hypothesized source range.

Consider the distance η_n from the nth microphone to the source. From the definition of the range difference (9.78) and the fact that $\eta_1 = \zeta_s$, we have:

$$\hat{\eta}_n = \zeta_s + d_{n1}, \tag{9.94}$$

where $\hat{\eta}_n$ denotes an observation based on the measured range difference. From the inner product, we can derive the true value for η_n^2, the square of the noise-free distance generated by a spherical signal model:

$$\eta_n^2 = \|\boldsymbol{\gamma}_n - \boldsymbol{\gamma}_s\|^2 = \zeta_n^2 - 2\boldsymbol{\gamma}_n^T \boldsymbol{\gamma}_s + \zeta_s^2. \tag{9.95}$$

The spherical LS error function is then defined as the difference between the measured and hypothesized values

$$\begin{aligned} e_{\mathrm{sp},n}(\boldsymbol{\gamma}_s) &\triangleq \frac{1}{2}\left(\hat{\eta}_n^2 - \eta_n^2\right) \\ &= \boldsymbol{\gamma}_n^T \boldsymbol{\gamma}_s + d_{n1}\zeta_s - \frac{1}{2}(\zeta_n^2 - d_{n1}^2), \quad n = 2, 3, \cdots, N. \end{aligned} \tag{9.96}$$

Putting the N errors together and writing them in a vector form gives,

$$\mathbf{e}_{\mathrm{sp}}(\mathbf{r}_s) = \mathbf{A}\boldsymbol{\theta} - \mathbf{b}_r, \tag{9.97}$$

where

$$\mathbf{A} \triangleq \left[\mathbf{S} \,\middle|\, \mathbf{d}\right], \quad \mathbf{S} \triangleq \begin{bmatrix} x_2 & y_2 & z_2 \\ x_3 & y_3 & z_3 \\ & \vdots & \\ x_N & y_N & z_N \end{bmatrix},$$

$$\boldsymbol{\theta} \triangleq \begin{bmatrix} x_s \\ y_s \\ z_s \\ \zeta_s \end{bmatrix}, \quad \mathbf{b}_r \triangleq \frac{1}{2} \begin{bmatrix} \zeta_2^2 - d_{21}^2 \\ \zeta_3^2 - d_{31}^2 \\ \vdots \\ \zeta_N^2 - d_{N1}^2 \end{bmatrix},$$

and $\left[\mathbf{S} \,\middle|\, \mathbf{d}\right]$ indicates that $\mathbf{S}$ and $\mathbf{d}$ are stacked side-by-side. The corresponding LS criterion is then given by:

$$J_{\mathrm{sp}} = \mathbf{e}_{\mathrm{sp}}^T \mathbf{e}_{\mathrm{sp}} = [\mathbf{A}\boldsymbol{\theta} - \mathbf{b}_r]^T [\mathbf{A}\boldsymbol{\theta} - \mathbf{b}_r]. \tag{9.98}$$

In contrast to the hyperbolic error function (9.92), the spherical error function (9.97) is linear in $\boldsymbol{\gamma}_s$ given ζ_s and vice versa. Therefore, the computational complexity to find a solution will *not* dramatically increase as N gets large.

9.13.2 Spherical Intersection (SX) Estimator

The SX source location estimator employs the spherical error and solves the problem in two steps [267]. It first finds the least-squares solution for $\boldsymbol{\gamma}_\mathrm{s}$ in terms of ζ_s,

$$\boldsymbol{\gamma}_\mathrm{s} = \mathbf{S}^\dagger (\mathbf{b}_\mathrm{r} - \zeta_\mathrm{s}\mathbf{d}), \tag{9.99}$$

where

$$\mathbf{S}^\dagger = \left(\mathbf{S}^T\mathbf{S}\right)^{-1}\mathbf{S}^T$$

is the pseudo-inverse of matrix $\mathbf{S}$. Then, substituting (9.99) into the constraint $\zeta_\mathrm{s}^2 = \boldsymbol{\gamma}_\mathrm{s}^T\boldsymbol{\gamma}_\mathrm{s}$ yields a quadratic equation as follows

$$\zeta_\mathrm{s}^2 = \left[\mathbf{S}^\dagger (\mathbf{b}_\mathrm{r} - \zeta_\mathrm{s}\mathbf{d})\right]^T \left[\mathbf{S}^\dagger (\mathbf{b}_\mathrm{r} - \zeta_\mathrm{s}\mathbf{d})\right]. \tag{9.100}$$

After expansion, it becomes

$$\alpha_1\zeta_\mathrm{s}^2 + \alpha_2\zeta_\mathrm{s} + \alpha_3 = 0, \tag{9.101}$$

where

$$\alpha_1 = 1 - \|\mathbf{S}^\dagger\mathbf{d}\|^2,\ \alpha_2 = 2\mathbf{b}_\mathrm{r}^T{\mathbf{S}^\dagger}^T\mathbf{S}^\dagger\mathbf{d},\ \alpha_3 = -\|\mathbf{S}^\dagger\mathbf{b}_\mathrm{r}\|^2.$$

The valid (real, positive) root is taken as an estimate of the source range ζ_s and is then substituted into (9.99) to calculate the SX estimate $\hat{\boldsymbol{\gamma}}_{\mathrm{s,SX}}$ of the source location.

In the SX estimation procedure, the solution of the quadratic equation (9.101) for the source range ζ_s is required. This solution must be a positive value by all means. If a real positive root is not available, the SX solution does not *exist*. On the contrary, if both of the roots are real and greater than 0, then the SX solution is not *unique*. In both cases, the SX source location estimator fails to produce a reliable estimate, which is not desirable for a real-time implementation.

9.13.3 Spherical Interpolation (SI) Estimator

In order to overcome the drawback of the SX algorithm, a spherical interpolation estimator was proposed in [2] which attempts to relax the restriction $\zeta_\mathrm{s} = \|\boldsymbol{\gamma}_\mathrm{s}\|$ by estimating ζ_s in the least-squares sense.

To begin, we substitute the least-squares solution (9.99) into the original spherical equation $\mathbf{A}\boldsymbol{\theta} = \mathbf{b}_\mathrm{r}$ to obtain

$$\zeta_\mathrm{s}\mathbf{P}_{\mathbf{S}^\perp}\mathbf{d} = \mathbf{P}_{\mathbf{S}^\perp}\mathbf{b}_\mathrm{r}, \tag{9.102}$$

where

$$\mathbf{P}_{\mathbf{S}^\perp} \triangleq \mathbf{I}_{N\times N} - \mathbf{S}\mathbf{S}^\dagger, \tag{9.103}$$

and $\mathbf{I}_{N\times N}$ is an $N \times N$ identity matrix. Matrix $\mathbf{P}_{\mathbf{S}^\perp}$ is a projection matrix that projects a vector, when multiplied by the matrix, onto a space that is orthogonal to the column space of $\mathbf{S}$. Such a projection matrix is symmetric (i.e. $\mathbf{P}_{\mathbf{S}^\perp} = \mathbf{P}_{\mathbf{S}^\perp}^T$) and idempotent (i.e. $\mathbf{P}_{\mathbf{S}^\perp} = \mathbf{P}_{\mathbf{S}^\perp} \cdot \mathbf{P}_{\mathbf{S}^\perp}$). Then the least-squares solution to (9.102) is given by

$$\hat{\zeta}_{\mathrm{s,SI}} = \frac{\mathbf{d}^T \mathbf{P}_{\mathbf{S}^\perp} \mathbf{b}_\mathrm{r}}{\mathbf{d}^T \mathbf{P}_{\mathbf{S}^\perp} \mathbf{d}}. \tag{9.104}$$

Substituting this solution into (9.99) yields the SI estimate

$$\hat{\gamma}_{\mathrm{s,SI}} = \mathbf{S}^\dagger \left[\mathbf{I}_{N\times N} - \left(\frac{\mathbf{d}\mathbf{d}^T \mathbf{P}_{\mathbf{S}^\perp}}{\mathbf{d}^T \mathbf{P}_{\mathbf{S}^\perp} \mathbf{d}} \right) \right] \mathbf{b}_\mathrm{r}. \tag{9.105}$$

In practice, the SI estimator performs better, but is computationally a little bit more complex, than the SX estimator.

9.13.4 Linear-Correction Least-Squares Estimator

Finding the LSE based on the spherical error criterion (9.98) is a linear minimization problem, i.e.,

$$\hat{\boldsymbol{\theta}}_{\mathrm{LSE}} = \arg\min_{\boldsymbol{\theta}} \ (\mathbf{A}\boldsymbol{\theta} - \mathbf{b}_\mathrm{r})^T (\mathbf{A}\boldsymbol{\theta} - \mathbf{b}_\mathrm{r}) \tag{9.106}$$

subject to a quadratic constraint

$$\boldsymbol{\theta}^T \boldsymbol{\Xi} \boldsymbol{\theta} = 0, \tag{9.107}$$

where $\boldsymbol{\Xi} \stackrel{\triangle}{=} \mathrm{diag}(1, 1, 1, -1)$ is a diagonal and orthonormal matrix.

For such a constrained minimization problem, the technique of Lagrange multipliers will be used and the source location is determined by minimizing the Lagrangian

$$\begin{aligned} \mathcal{L}(\boldsymbol{\theta}, \kappa) &= J_{\mathrm{sp}} + \kappa \boldsymbol{\theta}^T \boldsymbol{\Xi} \boldsymbol{\theta} \\ &= (\mathbf{A}\boldsymbol{\theta} - \mathbf{b}_\mathrm{r})^T (\mathbf{A}\boldsymbol{\theta} - \mathbf{b}_\mathrm{r}) + \kappa \boldsymbol{\theta}^T \boldsymbol{\Xi} \boldsymbol{\theta}, \end{aligned}$$

where κ is a Lagrange multiplier. Expanding this expression yields

$$\mathcal{L}(\boldsymbol{\theta}, \kappa) = \boldsymbol{\theta}^T (\mathbf{A}^T \mathbf{A} + \kappa \boldsymbol{\Xi}) \boldsymbol{\theta} - 2\mathbf{b}_\mathrm{r}^T \mathbf{A} \boldsymbol{\theta} + \mathbf{b}_\mathrm{r}^T \mathbf{b}_\mathrm{r}. \tag{9.108}$$

Necessary conditions for minimizing (9.108) can be obtained by taking the gradient of $\mathcal{L}(\boldsymbol{\theta}, \kappa)$ with respect to $\boldsymbol{\theta}$ and equating the result to zero. This produces:

$$\frac{\partial \mathcal{L}(\boldsymbol{\theta}, \kappa)}{\partial \boldsymbol{\theta}} = 2 \left(\mathbf{A}^T \mathbf{A} + \kappa \boldsymbol{\Xi} \right) \boldsymbol{\theta} - 2\mathbf{A}^T \mathbf{b}_\mathrm{r} = \mathbf{0}. \tag{9.109}$$

Solving for $\boldsymbol{\theta}$ yields the constrained least squares estimate

$$\hat{\boldsymbol{\theta}} = \left(\mathbf{A}^T \mathbf{A} + \kappa \boldsymbol{\Xi} \right)^{-1} \mathbf{A}^T \mathbf{b}_\mathrm{r}, \tag{9.110}$$

where κ is yet to be determined.

In order to find κ, we can impose the quadratic constraint directly by substituting (9.110) into (9.107), which leads to

$$\mathbf{b}_{\mathrm{r}}^T \mathbf{A} \left(\mathbf{A}^T \mathbf{A} + \kappa \boldsymbol{\Xi}\right)^{-1} \boldsymbol{\Xi} \left(\mathbf{A}^T \mathbf{A} + \kappa \boldsymbol{\Xi}\right)^{-1} \mathbf{A}^T \mathbf{b}_{\mathrm{r}} = 0. \quad (9.111)$$

With eigenvalue analysis, the matrix $\mathbf{A}^T \mathbf{A} \boldsymbol{\Xi}$ can be decomposed as

$$\mathbf{A}^T \mathbf{A} \boldsymbol{\Xi} = \mathbf{U} \boldsymbol{\Lambda} \mathbf{U}^{-1}, \quad (9.112)$$

where $\boldsymbol{\Lambda} = \mathrm{diag}(\lambda_1, \cdots, \lambda_4)$ and λ_i, $i = 1, \cdots, 4$, are the eigenvalues of the matrix $\mathbf{A}^T \mathbf{A} \boldsymbol{\Xi}$. Substituting (9.112) into (9.111), we may rewrite the constraint as:

$$\mathbf{p}^T \left(\boldsymbol{\Lambda} + \kappa \mathbf{I}\right)^{-2} \mathbf{q} = 0, \quad (9.113)$$

where

$$\begin{aligned} \mathbf{p} &= \mathbf{U}^T \boldsymbol{\Xi} \mathbf{A}^T \mathbf{b}_{\mathrm{r}}, \\ \mathbf{q} &= \mathbf{U}^T \mathbf{A}^T \mathbf{b}_{\mathrm{r}}. \end{aligned}$$

Define a function of the Lagrange multiplier as follows

$$\begin{aligned} f(\kappa) &\stackrel{\triangle}{=} \mathbf{p}^T \left(\boldsymbol{\Lambda} + \kappa \mathbf{I}\right)^{-2} \mathbf{q} \\ &= \sum_{i=1}^{4} \frac{p_i q_i}{(\kappa + \lambda_i)^2}. \end{aligned} \quad (9.114)$$

This is a polynomial of degree eight and because of its complexity numerical methods need to be used for root searching. Since the root of (9.114) for κ is not unique, a two-step procedure will be followed such that the desired source location could be found.

Unconstrained Spherical Least Squares Estimator

In the first step, we assume that x_{s}, y_{s}, z_{s}, and ζ_{s} are mutually independent or equivalently disregard the quadratic constraint (9.107) in purpose. Then the LS solution minimizing (9.98) for $\boldsymbol{\theta}$ (the source location as well as its range) is given by

$$\hat{\boldsymbol{\theta}}_1 = \mathbf{A}^{\dagger} \mathbf{b}_{\mathrm{r}}, \quad (9.115)$$

where

$$\mathbf{A}^{\dagger} = \left(\mathbf{A}^T \mathbf{A}\right)^{-1} \mathbf{A}^T$$

is the pseudo-inverse of the matrix $\mathbf{A}$.

A good parameter estimator first and foremost needs to be unbiased. For such an unconstrained spherical least squares estimator, the bias and covariance matrix can be approximated by using the following perturbation analysis method.

When measurement errors are present in the range differences, $\mathbf{A}$, $\mathbf{b}_\mathrm{r}$, and the parameter estimate $\hat{\boldsymbol{\theta}}_1$ deviate from their true values and can be expressed as:

$$\mathbf{A} = \mathbf{A}^\mathrm{t} + \triangle\mathbf{A}, \ \mathbf{b}_\mathrm{r} = \mathbf{b}_\mathrm{r}^\mathrm{t} + \Delta\mathbf{b}_\mathrm{r}, \ \hat{\boldsymbol{\theta}}_1 = \boldsymbol{\theta}^\mathrm{t} + \triangle\boldsymbol{\theta}, \tag{9.116}$$

where variables with superscript t denote the true values which also satisfy

$$\boldsymbol{\theta}^\mathrm{t} = \mathbf{A}^{\mathrm{t}\dagger}\mathbf{b}_\mathrm{r}^\mathrm{t}. \tag{9.117}$$

If the magnitudes of the perturbations are small, the second-order errors are insignificant compared to their first-order counterparts and therefore can be neglected for simplicity, which then yields:

$$\triangle\mathbf{A} = \left[\mathbf{0} \,\middle|\, \boldsymbol{\epsilon}\right], \quad \triangle\mathbf{b}_\mathrm{r} \approx -\mathbf{d}^\mathrm{t} \odot \boldsymbol{\epsilon}, \tag{9.118}$$

where $\odot$ denotes the Schur (element-by-element) product. Substituting (9.116) into (9.115) gives,

$$\left(\mathbf{A}^\mathrm{t} + \triangle\mathbf{A}\right)^T \left(\mathbf{A}^\mathrm{t} + \triangle\mathbf{A}\right)\left(\boldsymbol{\theta}^\mathrm{t} + \triangle\boldsymbol{\theta}\right) = \left(\mathbf{A}^\mathrm{t} + \triangle\mathbf{A}\right)^T \left(\mathbf{b}_\mathrm{r}^\mathrm{t} + \triangle\mathbf{b}_\mathrm{r}\right). \tag{9.119}$$

Retaining only the linear perturbation terms and using (9.117) and (9.118) produces:

$$\triangle\boldsymbol{\theta} \approx -\mathbf{A}^{\mathrm{t}\dagger}\mathbf{D}\boldsymbol{\epsilon}, \tag{9.120}$$

where

$$\mathbf{D} \stackrel{\triangle}{=} \mathrm{diag}(\eta_2, \eta_3, \cdots, \eta_N)$$

is a diagonal matrix. Since the measurement error $\boldsymbol{\epsilon}$ in the range differences has zero mean, $\hat{\boldsymbol{\theta}}_1$ is an unbiased estimate of $\boldsymbol{\theta}^\mathrm{t}$ when the small error assumption holds:

$$E\{\triangle\boldsymbol{\theta}\} \approx E\left\{-\mathbf{A}^{\mathrm{t}\dagger}\mathbf{D}\boldsymbol{\epsilon}\right\} = \mathbf{0}_{4\times 1}. \tag{9.121}$$

The covariance matrix of $\triangle\boldsymbol{\theta}$ is then found as,

$$\mathbf{C}_{\triangle\boldsymbol{\theta}} = E\{\triangle\boldsymbol{\theta}\triangle\boldsymbol{\theta}^T\} = \mathbf{A}^{\mathrm{t}\dagger}\mathbf{D}\mathbf{C}_{\boldsymbol{\epsilon}}\mathbf{D}\mathbf{A}^{\mathrm{t}\dagger T}, \tag{9.122}$$

where $\mathbf{C}_{\boldsymbol{\epsilon}}$ is known or is properly assumed *a priori*. Theoretically, the covariance matrix $\mathbf{C}_{\triangle\boldsymbol{\theta}}$ cannot be calculated since it contains true values. Nevertheless, it can be approximated by using the values in $\hat{\boldsymbol{\theta}}_1$ with sufficient accuracy, as suggested by our numerical studies.

In the first unconstrained spherical LS estimate (9.115), the range information is redundant because of the independence assumption on the source location and range. If that information is simply discarded, the source location estimate is the same as the SI estimate but with less computational complexity [175]. To demonstrate this, we first write (9.115) into a block form as

$$\hat{\boldsymbol{\theta}}_1 = \begin{bmatrix} \mathbf{S}^T\mathbf{S} & \mathbf{S}^T\mathbf{d} \\ \mathbf{d}^T\mathbf{S} & \mathbf{d}^T\mathbf{d} \end{bmatrix}^{-1} \begin{bmatrix} \mathbf{S}^T \\ \mathbf{d}^T \end{bmatrix} \mathbf{b}_\mathrm{r}. \tag{9.123}$$

It can easily be shown that:

$$\begin{bmatrix} \mathbf{S}^T\mathbf{S} & \mathbf{S}^T\mathbf{d} \\ \mathbf{d}^T\mathbf{S} & \mathbf{d}^T\mathbf{d} \end{bmatrix}^{-1} = \begin{bmatrix} \mathbf{Q} & \mathbf{v} \\ \mathbf{v}^T & \aleph \end{bmatrix}, \tag{9.124}$$

where

$$\begin{aligned} \mathbf{v} &= -\left(\mathbf{S}^T\mathbf{S} - \frac{\mathbf{S}^T\mathbf{d}\mathbf{d}^T\mathbf{S}}{\mathbf{d}^T\mathbf{d}}\right)^{-1} \frac{\mathbf{S}^T\mathbf{d}}{\mathbf{d}^T\mathbf{d}}, \\ \mathbf{Q} &= \left(\mathbf{S}^T\mathbf{S}\right)^{-1}\left[\mathbf{I} - \left(\mathbf{S}^T\mathbf{d}\right)\mathbf{v}^T\right], \\ \aleph &= \frac{1 - (\mathbf{d}^T\mathbf{S})\mathbf{v}}{\mathbf{d}^T\mathbf{d}}. \end{aligned}$$

Next, we define another projection matrix $\mathbf{P}_{\mathbf{d}^\perp}$ associated with the $\mathbf{d}$-orthogonal space:

$$\mathbf{P}_{\mathbf{d}^\perp} \triangleq \mathbf{I} - \frac{\mathbf{d}\mathbf{d}^T}{\mathbf{d}^T\mathbf{d}}, \tag{9.125}$$

and find

$$\mathbf{v} = -\left(\mathbf{S}^T\mathbf{P}_{\mathbf{d}^\perp}\mathbf{S}\right)^{-1} \frac{\mathbf{S}^T\mathbf{d}}{\mathbf{d}^T\mathbf{d}}, \tag{9.126}$$

$$\mathbf{Q} = \left(\mathbf{S}^T\mathbf{P}_{\mathbf{d}^\perp}\mathbf{S}\right)^{-1}. \tag{9.127}$$

Substituting (9.124) together with (9.126) and (9.127) into (9.123) yields the unconstrained spherical LS estimate for source coordinates,

$$\hat{\gamma}_{\mathrm{s},1} = \left(\mathbf{S}^T\mathbf{P}_{\mathbf{d}^\perp}\mathbf{S}\right)^{-1}\mathbf{S}^T\mathbf{P}_{\mathbf{d}^\perp}\mathbf{b}_\mathrm{r}, \tag{9.128}$$

which is the minimizer of

$$J_1(\gamma_\mathrm{s}) = \|\mathbf{P}_{\mathbf{d}^\perp}\mathbf{b}\mathbf{b}_\mathrm{r} - \mathbf{P}_{\mathbf{d}^\perp}\mathbf{S}\gamma_\mathrm{s}\|^2, \tag{9.129}$$

or the least-squares solution to the linear equation

$$\mathbf{P}_{\mathbf{d}^\perp}\mathbf{S}\gamma_\mathrm{s} = \mathbf{P}_{\mathbf{d}^\perp}\mathbf{b}_\mathrm{r}. \tag{9.130}$$

In fact, the first unconstrained spherical LS estimator tries to approximate the projection of the observation vector $\mathbf{b}_\mathrm{r}$ with the projections of the column vectors of the microphone location matrix $\mathbf{S}$ onto the $\mathbf{d}$-orthogonal space. The source location estimate is the coefficient vector associated with the *best* approximation. Clearly from (9.130), this estimation procedure is the generalization of the plane intersection (PI) method proposed in [268].

By using the Sherman-Morrison formula [229]

$$\left(\mathbf{A}+\mathbf{x}\mathbf{y}^T\right)^{-1}=\mathbf{A}^{-1}-\frac{\mathbf{A}^{-1}\mathbf{x}\mathbf{y}^T\mathbf{A}^{-1}}{1+\mathbf{y}^T\mathbf{A}^{-1}\mathbf{x}}, \tag{9.131}$$

we can expand the item in (9.128) as

$$\left(\mathbf{S}^T\mathbf{P}_{\mathbf{d}^\perp}\mathbf{S}\right)^{-1}=\left[\mathbf{S}^T\mathbf{S}-\left(\frac{\mathbf{S}^T\mathbf{d}}{\mathbf{d}^T\mathbf{d}}\right)(\mathbf{S}^T\mathbf{d})^T\right]^{-1},$$

and finally can show that the unconstrained spherical LS estimate (9.128) is equivalent to the SI estimate (9.105), i.e. $\hat{\gamma}_{\mathrm{s},1}\equiv\hat{\gamma}_{\mathrm{s,SI}}$.

Although the unconstrained spherical LS and the SI estimators are mathematically equivalent, they are quite different in computational efficiency due to different approaches to the source localization problem. The complexities of the SI and unconstrained spherical LS estimators are in $\mathcal{O}\left(N^3\right)$ and $\mathcal{O}(N)$, respectively. In comparison, the unconstrained spherical LS estimator reduces the complexity of the SI estimator by a factor of N^2, which is significant when N is large (more microphones are used).

Linear Correction

In the previous subsection, we developed the unconstrained spherical LS estimator (USLSE) for source localization and demonstrated that it is mathematically equivalent to the SI estimator but with less computational complexity. Although the USLSE/SI estimates can be accurate as indicated in [175] among others, it is helpful to exploit the redundancy of source range for improving the statistical efficiency (i.e., to reduce the variance of source location estimates) of the overall estimation procedure. Therefore, in the second step, we intend to correct the USLS estimate $\hat{\boldsymbol{\theta}}_1$ to make a better estimate $\hat{\boldsymbol{\theta}}_2$ of $\boldsymbol{\theta}$. This new estimate should be in the neighborhood of $\hat{\boldsymbol{\theta}}_1$ and should obey the constraint (9.107). We expect that the corrected estimate would still be unbiased and would have a smaller variance.

To begin, we substitute $\hat{\boldsymbol{\theta}}_1=\boldsymbol{\theta}^{\mathrm{t}}+\triangle\boldsymbol{\theta}$ into (9.109) and expand the expression to find

$$\mathbf{A}^T\mathbf{A}\hat{\boldsymbol{\theta}}_1+\kappa\boldsymbol{\Xi}\hat{\boldsymbol{\theta}}_1-(\mathbf{A}^T\mathbf{A}+\kappa\boldsymbol{\Xi})\triangle\boldsymbol{\theta}=\mathbf{A}^T\mathbf{b}_{\mathrm{r}}. \tag{9.132}$$

Combined with (9.115), (9.132) becomes

$$(\mathbf{A}^T\mathbf{A}+\kappa\boldsymbol{\Xi})\triangle\boldsymbol{\theta}=\kappa\boldsymbol{\Xi}\hat{\boldsymbol{\theta}}_1, \tag{9.133}$$

and hence

$$\triangle\boldsymbol{\theta}=\kappa\left(\mathbf{A}^T\mathbf{A}\right)^{-1}\boldsymbol{\Xi}\boldsymbol{\theta}^{\mathrm{t}}. \tag{9.134}$$

Substituting (9.134) into $\hat{\boldsymbol{\theta}}_1=\boldsymbol{\theta}^{\mathrm{t}}+\triangle\boldsymbol{\theta}$ yields

$$\hat{\boldsymbol{\theta}}_1=\left[\mathbf{I}+\kappa\left(\mathbf{A}^T\mathbf{A}\right)^{-1}\boldsymbol{\Xi}\right]\boldsymbol{\theta}^{\mathrm{t}}. \tag{9.135}$$

Solving for $\boldsymbol{\theta}^{\mathrm{t}}$ produces the corrected estimate $\hat{\boldsymbol{\theta}}_2$ and also the final output of the linear-correction least-squares (LCLS) estimator:

$$\hat{\boldsymbol{\theta}}_2 = \left[\mathbf{I} + \kappa\left(\mathbf{A}^T\mathbf{A}\right)^{-1}\boldsymbol{\Xi}\right]^{-1}\hat{\boldsymbol{\theta}}_1. \tag{9.136}$$

Equation (9.136) suggests how the second-step processing updates the source location estimate based on the first unconstrained spherical least squares result, or equivalently the SI estimate. If the regularity condition [224]

$$\lim_{i\to\infty}\left(\kappa(\mathbf{A}^T\mathbf{A})^{-1}\boldsymbol{\Xi}\right)^i = \mathbf{0} \tag{9.137}$$

is satisfied, then the estimate $\hat{\boldsymbol{\theta}}_2$ can be expanded in a Neumann series:

$$\begin{aligned}\hat{\boldsymbol{\theta}}_2 &= \left[\mathbf{I} + \left(-\kappa\left(\mathbf{A}^T\mathbf{A}\right)^{-1}\boldsymbol{\Xi}\right) + \left(-\kappa\left(\mathbf{A}^T\mathbf{A}\right)^{-1}\boldsymbol{\Xi}\right)^2 + \cdots\right]\hat{\boldsymbol{\theta}}_1 \\ &= \hat{\boldsymbol{\theta}}_1 + \sum_{i=1}^{\infty}\left[-\kappa\left(\mathbf{A}^T\mathbf{A}\right)^{-1}\boldsymbol{\Xi}\right]^i\hat{\boldsymbol{\theta}}_1,\end{aligned} \tag{9.138}$$

where the second term is the linear correction. Equation (9.137) implies that in order to avoid divergence, the Lagrange multiplier κ should be small. In addition, κ needs to be determined carefully such that $\hat{\boldsymbol{\theta}}_2$ obeys the quadratic constraint (9.107).

Because the function $f(\kappa)$ is smooth near $\kappa = 0$ (corresponding to the neighborhood of $\hat{\boldsymbol{\theta}}_1$), as suggested by numerical experiments, the secant method [254] can be used to determine its desired root. Two reasonable initial points can be chosen as:

$$\kappa_0 = 0, \quad \kappa_1 = \beta, \tag{9.139}$$

where the small number β is dependent on the array geometry. Five iterations should be sufficient to give an accurate approximation to the root.

The idea of exploiting the relationship between a sound source's range and its location coordinates to improve the estimation efficiency of the SI estimator was first suggested by Chan and Ho in [68] with a quadratic correction. Accordingly, they constructed a quadratic data model for $\hat{\boldsymbol{\theta}}_1$.

$$\hat{\boldsymbol{\theta}}_1 \odot \hat{\boldsymbol{\theta}}_1 = \mathbf{T}(\boldsymbol{\gamma}_{\mathrm{s}} \odot \boldsymbol{\gamma}_{\mathrm{s}}) + \mathbf{n}, \tag{9.140}$$

where

$$\mathbf{T} = \begin{bmatrix} 1 & 0 & 0 \\ 0 & 1 & 0 \\ 0 & 0 & 1 \\ 1 & 1 & 1 \end{bmatrix}$$

is a constant matrix, and $\mathbf{n}$ is the corrupting noise. In contrast to the linear correction technique based on the Lagrange multiplier, the quadratic counterpart needs to know the covariance matrix $\mathbf{C}_{\boldsymbol{\epsilon}}$ of measurement errors in the range differences *a priori*. In a real-time digital source localization system, a poorly estimated $\mathbf{C}_{\boldsymbol{\epsilon}}$ will lead to performance degradation. In addition, the quadratic-correction least squares estimation procedure uses the perturbation approaches to linearly approximate $\triangle\boldsymbol{\theta}$ and $\mathbf{n}$ in (9.116) and (9.140), respectively. Therefore, the approximations of their corresponding covariance matrices $\mathbf{C}_{\triangle\boldsymbol{\theta}}$ and $\mathbf{C}_{\mathbf{n}}$ can be good only when the noise level is low. When noise is at a practically high level, the quadratic-correction least squares estimate has a large bias and a high variance. Furthermore, since the true value of the source location which is necessary for calculating $\mathbf{C}_{\triangle\boldsymbol{\theta}}$ and $\mathbf{C}_{\mathbf{n}}$ cannot be known theoretically, the estimated source location has to be utilized for approximation. It was suggested in [68] that several iterations in the second correction stage would improve estimation accuracy. However, while the bias is suppressed after iterations, the estimate is closer to the SI solution and the variance is boosted, as demonstrated in [176]. Finally, the direct solutions of the quadratic-correction least-squares estimator are the squares of the source location coordinates ($\boldsymbol{\gamma}_{\mathrm{s}} \odot \boldsymbol{\gamma}_{\mathrm{s}}$). In 3-D space, these correspond to 8 positions, which introduce decision ambiguities. Other physical criteria, such as the domain of interest, were suggested but these are hard to define in practical situations, particularly when one of the source coordinates is close to zero.

In comparison, the linear-correction method updates the source location estimate of the first unconstrained spherical LS estimator without making any assumption about the error covariance matrix and without resort to a linear approximation. Even though we need to find a small root of function (9.114) for the Lagrange multiplier κ that satisfies the regularity condition (9.137), the function $f(\kappa)$ is smooth around zero and the solution can be easily determined using the secant method. The linear-correction method achieves a relatively better balance between computational complexity and estimation accuracy.

9.14 Example System Implementation

Acoustic source localization systems are not necessarily complicated and need not use computationally powerful and consequently expensive devices for running in real time, as the implementation described briefly in this section demonstrates. The real-time acoustic source localization system with passive microphone arrays for video camera steering in teleconferencing environments was developed by the authors at Bell Laboratories. Figure 9.3 shows a signal-flow diagram of the system.

This system is based on a personal computer that is powered by an Intel Pentium® III 1 GHz general-purposed processor and that runs a Microsoft Windows® operation system. Sonorus AD/24 converter and STUDI/O® digital audio interface card are employed to simultaneously capture multiple microphone signals. The camera is a Sony EVI-D30 with pan, tilt, and zoom capabilities. These motions can be harmoniously performed at the same time by separate motors, providing good coverage of a normal conference room.

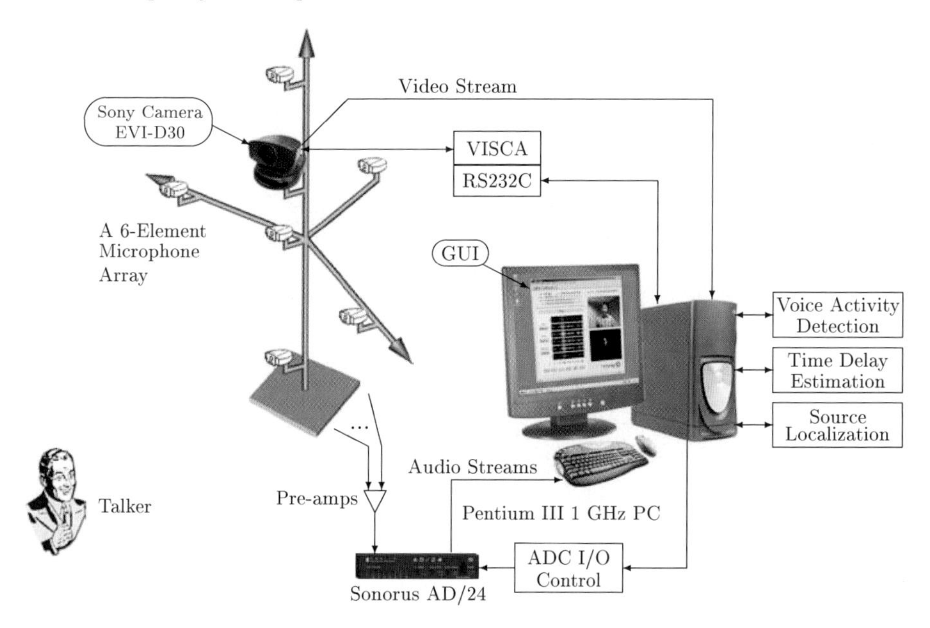

Fig. 9.3. Illustration of the real-time acoustic source localization system for video camera steering.

The host computer drives the camera via two layers of protocols, namely the RS232C serial control protocol and the Video System Control Architecture (VISCA®) protocol. The focus of the video camera is updated four times a second and the video stream is fed into the computer through a video capture card at a rate of 30 frames per second.

The microphone array uses six Lucent Speech Tracker Directional® hypercardioid microphones, as illustrated in Fig. 9.4. The frequency response of these microphones is 200-6000 Hz and beyond 4 kHz there is negligible energy. Therefore microphone signals are sampled at 8 kHz and a one-stage pre-amplifier with the fixed gain 37 dB is used prior to sampling. The reference microphone 1 is located at the center (the origin of the coordinate) and the rest microphones are in the same distance of 40 cm from the reference.

The empirical bias and standard deviation data in Figs. 9.5 and 9.6 show the results of two source localization examples using four different source localization algorithms where TDOA is estimated using the AED method (a more comprehensive numerical study can be found in [176]). In the graphs of standard deviation, the CRLBs are also plotted. For the QCLS algorithm, the true value of the source location needs to be known to calculate the covariance matrix of the first-stage SI estimate. But this knowledge is practically inaccessible and the estimated source location has to be used for approximation. It is suggested in [68] that several iterations in the second correction stage could improve the estimation accuracy. In the following, we refer to the one

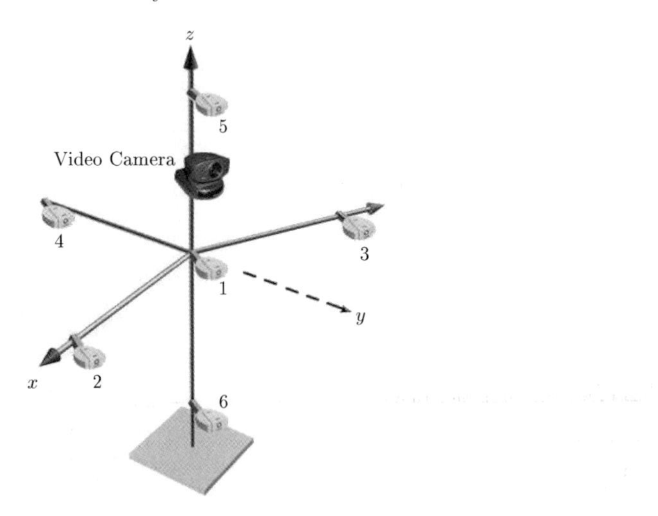

Fig. 9.4. Microphone array of the acoustic source localization system for video camera steering.

without iterations as the QCLS-i estimator and the other with iterations as the QCLS-ii estimator.

The microphone array designed for the real-time system presented above was used in these examples. As illustrated in Fig. 9.4, the six microphones are located at (distances in centimeters):

$$\begin{aligned} \gamma_1 &= (0,0,0), & \gamma_2 &= (40,0,0), & \gamma_3 &= (-40,0,0), \\ \gamma_4 &= (0,0,40), & \gamma_5 &= (0,-40,0), & \gamma_6 &= (0,0,-40). \end{aligned} \tag{9.141}$$

For such an array, the value of β in (9.139) was empirically set as $\beta = 1$. The source was positioned 300 cm away from the array with a fixed azimuth angle $\varphi_\mathrm{s} = 45°$ and varying elevation angles ϑ_s. At each location, the empirical bias and standard deviation of each estimator were obtained by averaging the results of 2000-trial Monte-Carlo runs.

In the first example, errors in time delay estimates are i.i.d. Gaussian with zero mean and 1 cm standard deviation. As seen clearly from Fig. 9.5, the QCLS-i estimator has the largest bias. Performing several iterations in the second stage can effectively reduce the estimation bias, but the solution is more like an SI estimate and the variance is boosted. In terms of standard deviation, all correction estimators perform better than the SI estimator (without correction). Among these four studied LS estimators, the QCLS-ii and the LCLS achieve the lowest standard deviation and their values approach the CRLS at most source locations.

In the second example, measurement errors are mutually dependent and their covariance matrix is given by [68]:

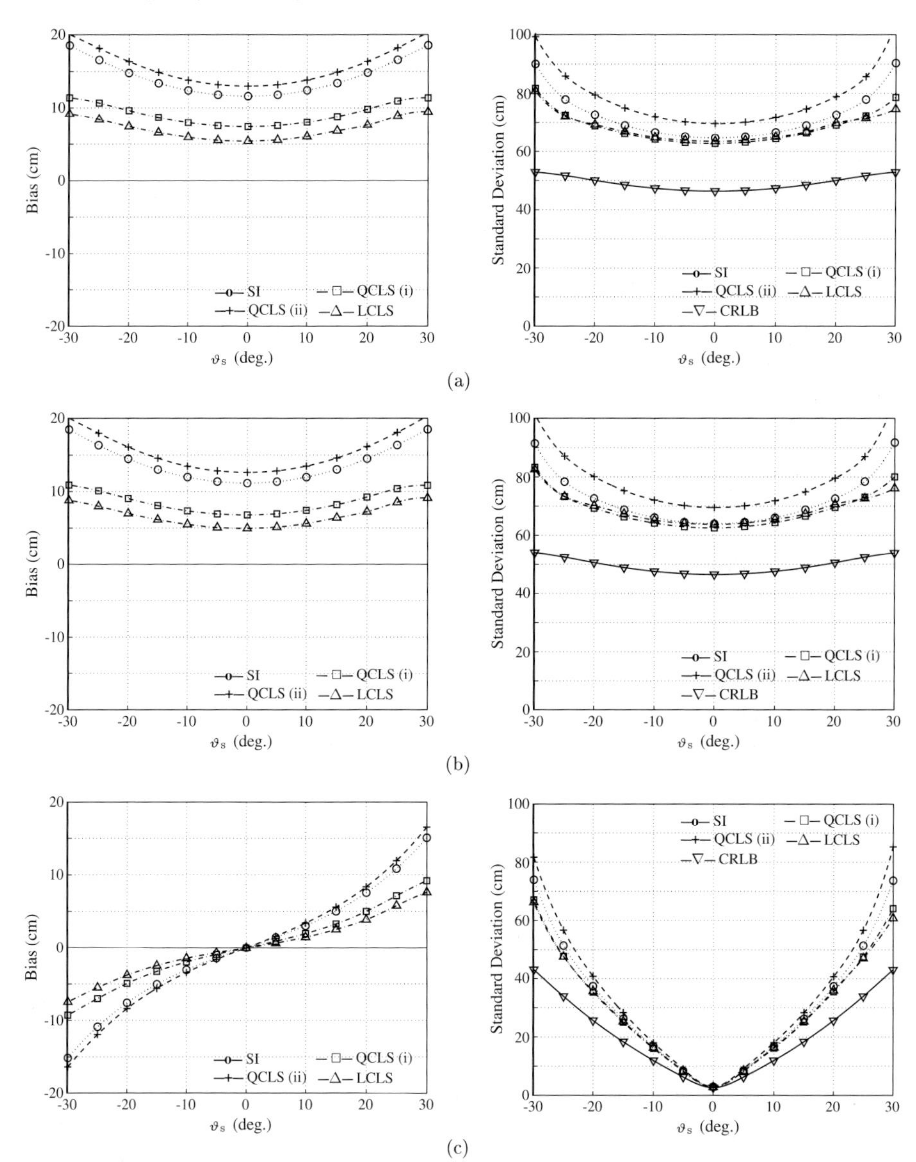

Fig. 9.5. Comparisons of empirical bias and standard deviation among the SI, QCLS-i, QCLS-ii, and LCLS estimators with zero mean i.i.d. Gaussian errors of standard deviation $\sigma_\epsilon = 1$ cm. (a) Estimators of x_s, (b) estimators of y_s, (c) estimators of z_s.

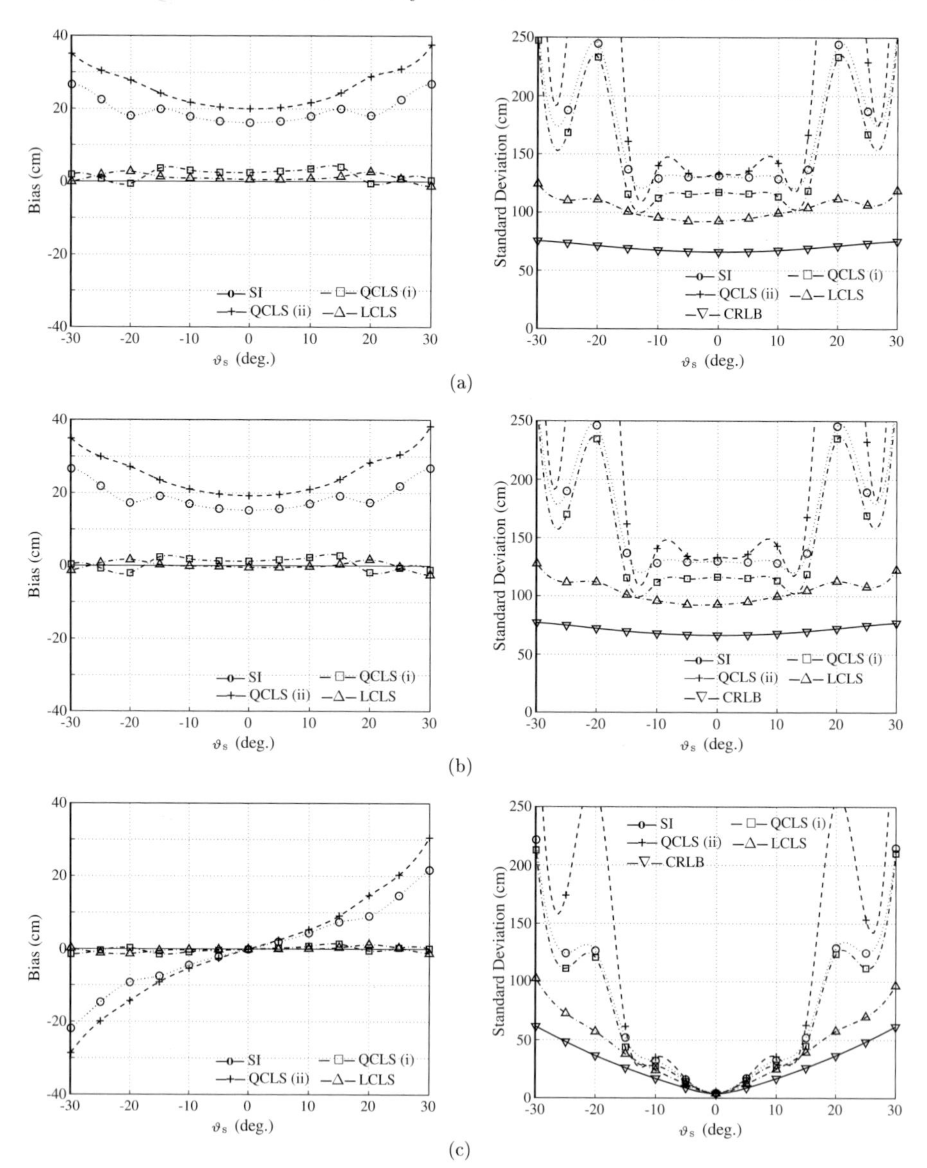

Fig. 9.6. Comparisons of empirical bias and standard deviation among the SI, QCLS-i, QCLS-ii, and LCLS estimators with zero mean *colored* Gaussian errors of standard deviation $\sigma_\epsilon = 1$ cm. (a) Estimators of x_s, (b) estimators of y_s, (c) estimators of z_s.

$$\mathbf{C}_{\boldsymbol{\epsilon}} = \frac{\sigma_\epsilon^2}{2} \begin{bmatrix} 2 & 1 & \cdots & 1 \\ 1 & 2 & \cdots & 1 \\ \vdots & \vdots & \ddots & \vdots \\ 1 & 1 & \cdots & 2 \end{bmatrix}, \tag{9.142}$$

where again $\sigma_\epsilon = 1$ cm. For a more realistic simulation, all estimators are provided with no information of the error distribution. From Fig. 9.6, we see that the performance of each estimator deteriorates because errors are no longer independent. At such a noise level, the linear approximation used by the QCLS estimators is inaccurate and the estimation procedure fails. However, the LCLS estimation procedure makes no assumption about $\mathbf{C}_{\boldsymbol{\epsilon}}$ and does not depend on a linear approximation. It produces an estimate whose bias and variance are always the smallest.

9.15 Summary

This chapter consists of two parts. The first one (from Sect. 9.1 to Sect. 9.8) was devoted to the time-delay-estimation problem in room acoustic environments. Starting from the problem formulation, we summarized the state of the art of the TDE algorithms ranging from the cross correlation based methods to the blind system identification based techniques. Since reverberation is the most difficult effect to cope with, we paid significant attention to the robustness of TDE against this effect. Fundamentally, there are three approaches to improving the robustness of TDE. When we have access to some *a priori* knowledge about the distortion sources, this *a priori* knowledge can be used to improve TDE. In case that reverberation is the most contaminant distortions, blind system identification techniques would be a good choice for TDE. If we have an array of sensors, we can improve TDE by taking advantage of the redundant information provided by multiple sensors.

The second part discussed the acoustic source localization techniques, which will play a significant role in the next-generation multimedia communication systems. The problem was postulated from a perspective of the estimation theory and the Cramèr-Rao lower bound for unbiased location estimators was derived. After an insightful review of conventional approaches ranging from maximum likelihood to least squares estimators, we presented a recently developed linear-correction least-squares algorithm that is more robust to measurement errors and that is more computationally as well as statistically efficient. As an example, we presented an acoustic source localization system for video camera steering in teleconferencing, which used a cheap Intel Pentium® III general-purposed processor.

10

Speech Enhancement and Noise Reduction

This chapter discusses how to estimate a desired speech signal from its corrupted observations and reduce the effects of unwanted additive noise. In an acoustic MIMO system, speech enhancement/noise reduction can be achieved by processing the waveform received by a single sensor, but often it is advantageous to use multiple sensors. This chapter will cover the techniques for speech enhancement/noise reduction from using a single sensor to using an array of sensors.

10.1 Introduction

Speech signals that carry the information we want can seldom be recorded in pure form since we are living in a natural environment where noise or interference is inevitable and ubiquitous. Over several decades, a significant amount of research attention has been focused on the signal processing techniques that can extract a desired speech signal and reduce the effects of unwanted noise. This is indeed one of the most important problems in acoustic MIMO signal processing.

With the MIMO signal model given in (2.9), the ultimate goal of signal estimation is to extract $s_m(k)$ ($m = 1, 2, \cdots, M$) from the observation signals $x_n(k)$ ($n = 1, 2, \cdots, N$). Since there are both convolutional and additive distortion, the estimation process would involve a wide variety of signal processing techniques such as source number estimation, noise reduction, source separation, beamforming, channel identification, dereverberation, etc. Many of these techniques have already been addressed throughout the text. This chapter will focus on elimination of additive noise. For the ease of presentation, we simplify the signal model into the following form:

$$x_n(k) = \alpha_n s(k - \tau_{n1}) + b_n(k), \quad n = 1, 2, \cdots, N, \tag{10.1}$$

where α_n is the relative attenuation factor, and τ_{n1} is the time difference of arrival (TDOA) between the nth and reference sensors. Without loss of generality, we select the 1st sensor as the reference sensor. In this case, we have $\alpha_1 = 1$, and $\tau_{11} = 0$. We assume that $s(k)$ and $b_n(k)$ $(n = 1, 2, \ldots, N)$ are mutually uncorrelated. The objective of noise reduction is then formulated as estimation of the signal component $s(k)$ from the observations $x_n(k)$ $(n = 1, 2, \cdots, N)$.

Various techniques have been developed over the past few decades. Depending on the number of sensors used in the system, these approaches can be classified into three basic categories, namely, *temporal filtering* techniques using only a single microphone, *adaptive noise cancellation* utilizing a primary sensor to pick up the noisy signal and a reference sensor to measure the noise field, and *beamforming* techniques exploiting an array of sensors.

Temporal filtering: with the use of spatial, temporal, and spectral cues, our human auditory system is remarkably accurate in extracting speech of interest from interference and noise. The same philosophy can be applied to signal processing techniques to mitigate the deleterious noise effect. In the single-channel situation, the mitigation of noise can be achieved by passing the noise corrupted signal through a temporal filter. Such a filter can be designed to significantly suppress the noise level while leaving the desired speech signal relatively unchanged, thanks to the distinction between speech and noise characteristics. Since it only uses one microphone, this technique has the potential to provide a very economic solution to the noise-reduction problem and can, in principle, be used in any existing communication device that has a microphone embedded in it. However, since the only resource that can be accessed in this situation is a mixture of the desired speech and noise, how to obtain a reliable estimate of the noise signal is a critical issue. Fortunately, a noisy signal is not occupied by speech all the time. For a large percentage of the time, it is exclusively occupied by noise alone. As a result, as long as we can detect the presence and absence of the speech, noise can then be estimated or updated during periods when the desired speech signal is absent. Apparently, this approach assumes that the noise is stationary or at least slowly varying between two consecutive updating periods so that the noise estimated during the absence of speech can represent what will be in the presence of speech. If this assumption does not hold or only partially holds, the filtering process may introduce noticeable or even severe amount of signal distortion.

Adaptive noise cancellation: this method typically uses a primary sensor to pick up noisy signal and an auxiliary or reference input derived from one or more sensors located at points in the noise field where the desired signal is weak or undetectable. It achieves noise cancellation by adaptively recreating a replica of the interfering noise using the reference input and subtracting the replica from the corrupted signal. The strength of this approach derives from the use of closed-loop adaptive filtering algorithms. It is particularly effective when the reference input can anticipate the noise field at the primary sensor so that processing delay in the filter and any propagation delay between reference input and primary sensor are easily compensated for. Furthermore, it can deal with even non-stationary noise. However, the success of this method relies on two fundamental requirements. The first is that the desired signal is collected only by the primary sensor while little or none is detected by the reference microphone. Any inclusion of the desired signal in the reference input will result in signal cancellation, and eventually leads to distortion into speech. The second requirement is that the noise measured by the reference sensor should be highly correlated with the noise that interferes with the desired signal in the primary sensor. In an extreme case, if noise components at both sensors are completely uncorrelated, no reduction of noise can be achieved.

Beamforming: this technique was developed to make use of a sensor array to explicitly enhance the desired signal as well as suppress the unwanted noise. Such an array consists of a set of sensors that are spatially distributed at known locations. These sensors collect signals from sources in their field of view. By weighting the sensor outputs, a beam can be formed and steered to a specified direction. Consequently, a signal propagating from the look direction is reinforced, while sound sources originating from directions other than the look direction are attenuated. The beamforming approach has many appealing properties. For example, in diffuse noise conditions, it can achieve a broadband SNR gain that is approximately equal to the number of sensors. In general, as long as the steering direction coincides with the incident angle of the signal of interest, beamforming does not introduce much distortion. However, the noise reduction performance of a beamformer relies on many factors such as the number of sensors, the geometry of the array, the characteristics of the noise, and so forth. In addition, the estimation of direction of arrival (DOA) is also a critical issue. While pointing along a specified direction, all signals arriving at the array along any other directions become undesirable. As a result, if the array is steered along some other directions, the signal of interest becomes an undesired signal and is suppressed. The degree of suppression depends on the angular separation between the direction of the signal and the current look direction, as well as on the array geometry, and so forth.

This chapter is devoted to signal processing techniques for speech enhancement and noise reduction. We cannot, of course, cover all the techniques and methods; rather, we focus on signal processing principles, illustrating how certain schemes can be used as a basis for speech enhancement and what tradeoff has to be made between the amount of noise reduction and the amount of distortion that will be brought into the speech of interest.

10.2 Noise-Reduction and Speech-Distortion Measures

A key question in speech enhancement and noise reduction is what can be gained and what will be sacrificed with the use of an algorithm. Depending on the specific application requirements, different criteria can be defined to assess the performance improvement. Before addressing noise reduction algorithms, we begin with a brief discussion of some popularly used performance measures in noise reduction. Some of these are objective in the sense that they are independent of listener's opinion and the results can be verified and repeated by other experimenters. The others are subjective because they rely on human listeners' judgements, i.e., the results are affected by individual listener's view and preference.

10.2.1 Noise-Reduction Factor and Noise-Reduction Gain Function

The primary issue that we must determine with noise reduction is how much noise is actually removed. To measure this, we introduce the concept of noise-reduction factor, which is defined as the ratio between the original noise intensity and the intensity of the residual noise after applying the algorithm. Consider the single-channel case, if the observation noise is $b(k)$, and the residual noise is denoted as $b_{\mathrm{nr}}(k)$, the noise-reduction factor is mathematically written as:

$$\xi_{\mathrm{nr}} \triangleq \frac{E\left[b^2(k)\right]}{E\left[b_{\mathrm{nr}}^2(k)\right]}. \tag{10.2}$$

Apparently, this factor should be lower bounded by 1. The larger the parameter ξ_{nr}, the more is the noise reduction. When multiple sensors are used, the noise observed at any one of the sensors can be treated as the original noise; but often, we define the noise-reduction factor as the ratio between the intensity of the noise at the reference sensor and that of the residual noise.

It is known that a speech waveform consists of a sequence of different events. Its characteristics, therefore, fluctuate over time and frequency. Often, the characteristics of ambient noise also vary across frequency bands. This indicates that we cannot achieve a uniform noise-reduction performance over the frequency range of interest, rather, it changes from one frequency band to another. Although the noise-reduction factor gives us an idea of the overall noise attenuation, it does not tell us the noise-reduction behavior from frequency to frequency. A more useful and insightful frequency-dependent measure of noise reduction can be defined as the ratio between the spectral densities of the original and residual noise, i.e.,

$$\begin{aligned}\Psi_{\mathrm{nr}}(\omega) &\triangleq \frac{E\left[|B(j\omega)|^2\right]}{E\left[|B_{\mathrm{nr}}(j\omega)|^2\right]} \\ &= \frac{P_b(\omega)}{P_{b_{\mathrm{nr}}}(\omega)},\end{aligned} \tag{10.3}$$

where $B(j\omega)$, $B_{\mathrm{nr}}(j\omega)$, $P_b(\omega)$, and $P_{b_{\mathrm{nr}}}(\omega)$ are the Fourier spectra and power spectral densities of $b(k)$ and $b_{\mathrm{nr}}(k)$ respectively, and ω is the angular frequency. This measure reflects the noise-reduction gain as a function of frequency. It is therefore logical to call it the noise-reduction gain function.

10.2.2 Speech-Distortion Index and Attenuation Frequency Distortion

While noise reduction is feasible with signal processing techniques, the price to pay for this is distortion of the clean speech. The degree of speech distortion is a very complicated issue, which depends on many factors such as the amount of *a priori* knowledge about the speech and noise that can be accessed, the distinction between the speech and noise characteristics, how many sensors can be used, etc. To quantify the speech distortion due to a noise-reduction algorithm, we now introduce a speech-distortion index. Suppose that the clean speech is $s(k)$, and after noise reduction, the speech component in the estimated signal is denoted as $\hat{s}_{\mathrm{nr}}(k)$, the speech-distortion index is defined as:

$$\upsilon_{\mathrm{sd}} \triangleq \frac{E\left\{\left[s(k) - \hat{s}_{\mathrm{nr}}(k)\right]^2\right\}}{E\left[s^2(k)\right]}. \tag{10.4}$$

This parameter is lower bounded by 0 and expected to be upper bounded by 1. Unlike the noise-reduction factor, a larger value of υ_{sd} means more speech distortion. This index, therefore, should be kept as small as possible.

To measure the speech distortion over frequency, we can borrow the concept of attenuation frequency distortion from the telecommunication theory. The attenuation frequency distortion is a measure that was developed to assess how a telephone channel preserves the fidelity of a speech signal. It is defined as the change in amplitude of the transmitted signal over a voice band. Here we adjust this concept and define the attenuation frequency distortion for noise reduction as

$$\begin{aligned} \Upsilon_{\mathrm{sd}}(\omega) &\triangleq \frac{E\left[|S(j\omega)|^2 - |\hat{S}_{\mathrm{nr}}(j\omega)|^2\right]}{E\left[|S(j\omega)|^2\right]} \\ &= \frac{P_s(\omega) - P_{\hat{s}_{\mathrm{nr}}}(\omega)}{P_s(\omega)}, \end{aligned} \tag{10.5}$$

where $S(j\omega)$ and $P_s(\omega)$ are the Fourier spectrum and power spectral density of the clean speech $s(k)$, and $\hat{S}_{\mathrm{nr}}(j\omega)$ and $P_{\hat{s}_{\mathrm{nr}}}(\omega)$ are, respectively, the Fourier spectrum and power spectral density of the speech component in the noise-reduced signal.

10.2.3 Signal-to-Noise Ratio

Signal-to-noise ratio (SNR) is a measure of signal intensity relative to the intensity of additive background noise. The ratio is usually measured in decibels (dB). With the signal model shown in (10.1), the SNR at the nth sensor is given by the following formula:

$$\mathrm{SNR} \triangleq \alpha_n^2 \frac{\sigma_s^2}{\sigma_{b_n}^2} = \alpha_n^2 \frac{E\left[s^2(k)\right]}{E\left[b_n^2(k)\right]}. \tag{10.6}$$

A frequency-domain counterpart of the above definition can also be formulated according to the well-known Parseval's relation:

$$\mathrm{SNR} = \alpha_n^2 \frac{\int_{-\pi}^{\pi} P_s(\omega) d\omega}{\int_{-\pi}^{\pi} P_{b_n}(\omega) d\omega}, \tag{10.7}$$

where $P_s(\omega)$ and $P_{b_n}(\omega)$ are the power spectral densities of signal $s(k)$ and noise $b_n(k)$, respectively. Apparently, SNR is an objective measure, which quantifies how noisy the signal is. The higher the SNR value, the cleaner is the signal.

In the context of noise reduction, both the terms *a priori* SNR and *a posteriori* SNR are often used. The former usually refers to the SNR of the observed noisy signal, and the latter regards the SNR of the noise-reduced signal. The difference between the two is called SNR improvement. The higher the SNR improvement, the more effective is the noise-reduction algorithm.

10.2.4 Log-Spectral Distance

Log-spectral distance is a measurement of similarity (or dissimilarity) between two signals. Consider two speech signals $s_1(k)$ and $s_2(k)$ with their power spectral densities being denoted as $P_{s_1}(\omega)$ and $P_{s_2}(\omega)$. The difference between the two spectra on a log magnitude versus frequency scale is defined by

$$\mathcal{D}_{s_1 s_2}(\omega) \triangleq \ln P_{s_1}(\omega) - \ln P_{s_2}(\omega). \tag{10.8}$$

One logical choice for a distance measure between the spectra P_{s_1} and P_{s_2} is the set of l_p norms defined as

$$\mathrm{LSD}_{s_1 s_2} \triangleq \left[\frac{1}{2\pi} \int_{-\pi}^{\pi} |\mathcal{D}_{s_1 s_2}(\omega)|^p d\omega \right]^{1/p}. \tag{10.9}$$

For $p = 1$, (10.9) defines the mean absolute log-spectral distance. For $p = 2$, (10.9) defines the root mean square (RMS) log-spectral distance, which has been widely adopted in many speech-processing systems. For the limiting case as p approaches infinity, (10.9) reduces to the peak log-spectral distances. Since perceived loudness of a speech signal is approximately logarithmic, the log-spectral distance family appears to be closely related to the subjective assessment of speech differences; as a result, it has been widely used to evaluate speech distortion in the context of speech coding, speech recognition, speech enhancement, and noise reduction.

10.2.5 Itakura Distance

In many aspects of speech processing, it is necessary to measure the closeness of the spectral envelope of one speech signal to that of another speech signal.

One way of doing this is through comparing their linear prediction (LP) coefficients. Consider two speech signals $s_1(k)$ and $s_2(k)$, we construct the two corresponding signal vectors as follows:

$$\mathbf{s}_1(k) = \begin{bmatrix} s_1(k) & s_1(k-1) & \dots & s_1(k-P+1) \end{bmatrix}^T,$$
$$\mathbf{s}_2(k) = \begin{bmatrix} s_2(k) & s_2(k-1) & \dots & s_2(k-P+1) \end{bmatrix}^T. \tag{10.10}$$

We know from the forward linear prediction theory that [160]

$$\begin{bmatrix} r_{s_is_i}(0) & \mathbf{r}_{s_is_i}^T \\ \mathbf{r}_{s_is_i} & \mathbf{R}_{s_i} \end{bmatrix} \begin{bmatrix} 1 \\ -\mathbf{a}_{s_i} \end{bmatrix} = \begin{bmatrix} E_{s_i} \\ \mathbf{0} \end{bmatrix}, \quad i = 1, 2, \tag{10.11}$$

where

$$r_{s_is_i}(p) = E\left[s_i(k)s_i(k+p)\right]$$

is the autocorrelation function of the signal $s_i(k)$ for lag p,

$$\mathbf{r}_{s_is_i} = E\left[\mathbf{s}_i(k-1)s_i(k)\right] = \begin{bmatrix} r_{s_is_i}(1) \\ r_{s_is_i}(2) \\ \vdots \\ r_{s_is_i}(P) \end{bmatrix} \tag{10.12}$$

is the correlation vector between $\mathbf{s}_i(k-1)$ and $s_i(k)$,

$$\mathbf{R}_{s_i} = E\left[\mathbf{s}_i(k-1)\mathbf{s}_i^T(k-1)\right] = \begin{bmatrix} r_{s_is_i}(0) & r_{s_is_i}(1) & \dots & r_{s_is_i}(P-1) \\ r_{s_is_i}(1) & r_{s_is_i}(0) & \dots & r_{s_is_i}(P-2) \\ \vdots & \vdots & \ddots & \vdots \\ r_{s_is_i}(P-1) & r_{s_is_i}(P-2) & \dots & r_{s_is_i}(0) \end{bmatrix} \tag{10.13}$$

is the correlation matrix of $s_i(k)$,

$$\mathbf{a}_{s_i} = \begin{bmatrix} a_{s_i,1} & a_{s_i,2} & \dots a_{s_i,P} \end{bmatrix}^T \tag{10.14}$$

is the forward linear predictor of order P, $\mathbf{0}$ is the $P \times 1$ null vector, and E_{s_i} is the corresponding prediction error energy. Now if we define

$$\breve{\mathbf{R}}_{s_i} \triangleq \begin{bmatrix} r_{s_is_i}(0) & \mathbf{r}_{s_is_i}^T \\ \mathbf{r}_{s_is_i} & \mathbf{R}_{s_i} \end{bmatrix} \tag{10.15}$$

and

$$\breve{\mathbf{a}}_{s_i} \stackrel{\triangle}{=} \begin{bmatrix} 1 & -\mathbf{a}_{s_i}^T \end{bmatrix}^T, \tag{10.16}$$

it can easily be checked that

$$\breve{\mathbf{a}}_{s_i}^T \breve{\mathbf{R}}_{s_i} \breve{\mathbf{a}}_{s_i} = E_{s_i}. \tag{10.17}$$

A key question is then, how close is the linear predictor $\mathbf{a}_{s_1}$ to the linear predictor $\mathbf{a}_{s_2}$? One way of answering this question is to assess the logarithmic ratio between E_{s_2} and E_{s_1}, resulting in the widely known Itakura distance:

$$\mathrm{ID}_{s_1 s_2} \stackrel{\triangle}{=} \ln \frac{E_{s_2}}{E_{s_1}} = \ln \frac{\breve{\mathbf{a}}_{s_2}^T \breve{\mathbf{R}}_{s_2} \breve{\mathbf{a}}_{s_2}}{\breve{\mathbf{a}}_{s_1}^T \breve{\mathbf{R}}_{s_1} \breve{\mathbf{a}}_{s_1}}. \tag{10.18}$$

Note that the Itakura distance is not symmetric, i.e.,

$$\mathrm{ID}_{s_1 s_2} \neq \mathrm{ID}_{s_2 s_1}, \tag{10.19}$$

therefore, it is not a true distance metric. However, asymmetry should not cause a problem in evaluating speech quality.

10.2.6 Itakura-Saito Distance

This distance, originally known as the *error matching measure*, was proposed by Itakura and Saito in their formulation of linear prediction as an approximate maximum likelihood estimation [188]. Again, let us consider two signals $s_1(k)$ and $s_2(k)$. With the augmented correlation matrices $\breve{\mathbf{R}}_{s_i}$ $(i = 1, 2)$ and the linear prediction vectors $\breve{\mathbf{a}}_{s_i}$ being defined, respectively, in (10.15) and (10.16), the Itakura-Saito distance between the two signals is defined as

$$\mathrm{ISD}_{s_1 s_2} \stackrel{\triangle}{=} \frac{\breve{\mathbf{a}}_{s_1}^T \breve{\mathbf{R}}_{s_1} \breve{\mathbf{a}}_{s_1}}{\breve{\mathbf{a}}_{s_2}^T \breve{\mathbf{R}}_{s_2} \breve{\mathbf{a}}_{s_2}} - \ln \frac{\breve{\mathbf{a}}_{s_1}^T \breve{\mathbf{R}}_{s_1} \breve{\mathbf{a}}_{s_1}}{\breve{\mathbf{a}}_{s_2}^T \breve{\mathbf{R}}_{s_2} \breve{\mathbf{a}}_{s_2}} - 1. \tag{10.20}$$

Same as the Itakura distance, this measure is not symmetric either; therefore, it is not a true metric. However, as we mentioned previously, the asymmetry should not cause a problem for speech quality evaluation since the original speech and the noise-cleaned speech are very different anyway.

The Itakura-Saito distance has many interesting properties. It has been shown that this measure is highly correlated with subjective quality judgements [255]. For example, a recent report on speech codec evaluation reveals that if the Itakura-Saito measure between two processed speech signals is less than 0.5, the difference in their mean opinion score would be less than 1.6 [71]. Many other reported experiments also confirmed that when the Itakura-Saito distance between two speech signals is below 0.1, they would be perceived nearly identically by human ears. As a result, the Itakura-Saito distance is often used as an objective measure of speech quality. It is probably the most widely used measure of similarity between speech signals.

10.2.7 Mean Opinion Score

In principle, speech quality should be assessed by subjective methods using listeners' judgements because it is the listener's judgement that counts after all. Various subjective speech quality tests have been developed for evaluation of speech quality. In real application, choosing which test should be according to the objective of the task. Commonly used subjective tests include diagnostic rhyme test (DRT), modified rhyme test (MRT), diagnostic acceptability measure (DAM), naturalness test, mean opinion score (MOS), etc. The most widely used subjective method is the MOS, which is extracted from the results of an absolute category rating test where the listeners are presented with processed speech signal and asked to give a rating using a 5-point scale as follows: 1 - bad; 2 - poor; 3 - fair; 4 - good; 5 - excellent. The arithmetic mean of all the individual scores represents the MOS, ranging from 1 (worst) to 5 (best).

10.3 Single-Channel Noise-Reduction Algorithms: a Brief Overview

We now consider noise reduction in the scenarios where there is only one microphone. The first algorithm can be traced back to 40 years ago with 2 patents by Schroeder where an analog implementation of the spectral magnitude subtraction method was described [270], [271]. Since then single-channel noise reduction has become an area of active research. Over the past several decades, a variety of approaches have been developed, including *Wiener filter*, *spectral restoration*, *signal subspace*, *parametric-model-based method*, and *statistical-model-based method*.

- *Wiener filter*: this technique restores the desired speech signal by passing the noisy speech through an FIR filter whose coefficients are estimated by minimizing the mean square error (MSE) between the clean speech and its estimate [320]. Wiener filter can also be delineated in the frequency domain, resulting in various derivative techniques such as spectral subtraction [52], [222], [301], [219], parametric Wiener filter [214], [112], etc.
- *Spectral restoration*: in the frequency domain, the problem of noise reduction can be described as to estimate the clean speech spectra from the spectra of the noisy speech signal. Since a complex spectrum can be factorized into a product of two components, namely magnitude and phase, the problem can further be formulated as to design two optimal estimators that make decisions separately on magnitude and phase spectra from the observed signal. However, it has been proven that an optimal estimation of the spectral magnitude and phase of the clean speech cannot be achieved at the same time [106]. Fortunately, the human perception system is relatively insensitive to the phase corruption. Therefore, noise reduction in the

frequency domain is essentially a matter of recovering the spectral magnitude or spectral envelope of the clean speech from that of the corrupted speech [214], [301], [307], [107], [304], [70], [256], [92], [93].

- *Signal subspace*: this method decomposes the vector space of the noisy speech into two orthogonal subspaces using either the Karhunen-Loève transform (KLT) or the singular value decomposition technique: one is composed of both speech and noise; and the other consists exclusively of noise component alone. This is possible because it has been proven that clean speech can be described with a low-rank model. After decomposition, the speech signal is estimated by removing the noise subspace, and cleaning the speech-plus-noise subspace [111], [91], [155], [191], [212], [259], [225], [169].
- *Parametric-model-based method*: based on the fact that a speech signal can be modelled as a stochastic autoregressive (AR) process, this method provides an iterative way to estimate the clean speech from its corrupted observation, i.e., it first estimates the AR coefficients, and then applies the Kalman filter to extract the desired speech signal [246], [143], [210], [133].
- *Statistical-model-based methods*: this method originated from the statistical speech recognition technique. Both the clean speech and noise are modelled with Guassian mixture hidden Markov models (HMM's). The model parameters of the clean speech signal are estimated from the training sequences. The enhancement of the noisy speech is achieved by reestimating the clean speech waveform using the expectation-maximization (EM) algorithm, extended Kalman filtering technique, particle filtering technique, or the Markov Chain Monte Carlo method [108], [109], [110], [83], [216], [303].

In the following sections, we discuss how to estimate clean speech from its noise corrupted observations. As we mentioned earlier, we cannot cover all the techniques. Instead, we will focus on signal processing principles, illustrating what is the best we can achieve in this context.

10.4 Time-Domain Wiener Filter

Wiener filter is obtained by minimizing the MSE between the signal of interest and its estimate. It is, therefore, an optimal filter in the minimum-mean-square-error (MMSE) sense.

10.4.1 Estimation of the Clean Speech Samples

In the single-channel case, the observation signal, according to (10.1), is modelled as

$$x(k) = s(k) + b(k). \tag{10.21}$$

Passing the noisy signal $x(k)$ through a temporal filter, we can obtain an estimate of the clean speech $s(k)$, i.e.,

$$\hat{s}(k) = \mathbf{h}^T \mathbf{x}(k), \tag{10.22}$$

where

$$\mathbf{h} = \begin{bmatrix} h_0 & h_1 & \cdots & h_{L-1} \end{bmatrix}^T \tag{10.23}$$

is an FIR filter of length L, and

$$\mathbf{x}(k) = \begin{bmatrix} x(k) & x(k-1) & \cdots & x(k-L+1) \end{bmatrix}^T \tag{10.24}$$

is a vector containing the L most recent samples of the observation signal $x(k)$. We now can define the error signal between the clean speech sample at time k and its estimate:

$$e_s(k) \stackrel{\triangle}{=} s(k) - \hat{s}(k) = s(k) - \mathbf{h}^T \mathbf{x}(k). \tag{10.25}$$

The MSE criterion can then be written as:

$$J_s(\mathbf{h}) \stackrel{\triangle}{=} E\left[e_s^2(k)\right]. \tag{10.26}$$

The optimal estimate $\hat{s}_{\mathrm{o}}(k)$ of the clean speech sample $s(k)$ tends to contain less noise than the observation sample $x(k)$, and the optimal filter that forms $\hat{s}_{\mathrm{o}}(k)$ is the Wiener filter, which is obtained as follows,

$$\mathbf{h}_{\mathrm{o}} = \arg\min_{\mathbf{h}} J_s(\mathbf{h}). \tag{10.27}$$

Consider the particular filter,

$$\mathbf{h} = \mathbf{u}_1 = \begin{bmatrix} 1 & 0 & \cdots & 0 \end{bmatrix}^T. \tag{10.28}$$

This means that the observed signal $x(k)$ will pass this filter unaltered (no noise reduction), thus the corresponding MSE is

$$\begin{aligned} J_s(\mathbf{u}_1) &= E\left\{\left[s(k) - \mathbf{u}_1^T \mathbf{x}(k)\right]^2\right\} \\ &= E\left\{\left[s(k) - x(k)\right]^2\right\} \\ &= E\left\{b^2(k)\right\} = \sigma_b^2. \end{aligned} \tag{10.29}$$

In principle, for the optimal filter $\mathbf{h}_{\mathrm{o}}$, we should have,

$$J_s(\mathbf{h}_{\mathrm{o}}) < J_s(\mathbf{u}_1) = \sigma_b^2. \tag{10.30}$$

In other words, the Wiener filter will be able to reduce the level of noise in the noisy speech signal $x(k)$.

From (10.27), we easily find the Wiener-Hopf equations:

$$\mathbf{R}_x \mathbf{h}_\mathrm{o} = \mathbf{r}_{xs}, \tag{10.31}$$

where

$$\mathbf{R}_x = E\left\{\mathbf{x}(k)\mathbf{x}^T(k)\right\} \tag{10.32}$$

is the correlation matrix of the observed signal $x(k)$ and

$$\mathbf{r}_{xs} = E\left\{\mathbf{x}(k)s(k)\right\} \tag{10.33}$$

is the cross-correlation vector between the noisy and clean speech signals. However, $s(k)$ is unobservable; as a result, an estimation of $\mathbf{r}_{xs}$ may seem difficult to obtain. But using the facts that $s(k) = x(k) - b(k)$ and $\mathbf{x}(k) = \mathbf{s}(k) + \mathbf{b}(k)$, we have

$$\begin{aligned} \mathbf{r}_{xs} &= E\left\{\mathbf{x}(k)\left[x(k) - b(k)\right]\right\} \\ &= E\left\{\mathbf{x}(k)x(k)\right\} - E\left\{\left[\mathbf{s}(k) + \mathbf{b}(k)\right] b(k)\right\}. \end{aligned} \tag{10.34}$$

Since $s(k)$ and $b(k)$ are uncorrelated, $\mathbf{r}_{xs}$ can further be expressed as,

$$\begin{aligned} \mathbf{r}_{xs} &= E\left\{\mathbf{x}(k)x(k)\right\} - E\{\mathbf{b}(k)b(k)\} \\ &= \mathbf{r}_{xx} - \mathbf{r}_{bb}. \end{aligned} \tag{10.35}$$

Now $\mathbf{r}_{xs}$ depends on the correlation vectors $\mathbf{r}_{xx}$ and $\mathbf{r}_{bb}$. The vector $\mathbf{r}_{xx}$ (which is also the first column of $\mathbf{R}_x$) can be easily estimated during speech and noise periods while $\mathbf{r}_{bb}$ can be estimated during noise-only intervals assuming that the statistics of the noise has not yet changed much with time.

Substituting (10.35) into (10.31) and using the fact that $\mathbf{u}_1 = \mathbf{R}_x^{-1}\mathbf{r}_{xx}$, we obtain the optimal filter:

$$\mathbf{h}_\mathrm{o} = \mathbf{u}_1 - \mathbf{R}_x^{-1}\mathbf{r}_{bb} = \left[\mathbf{I} - \mathbf{R}_x^{-1}\mathbf{R}_b\right]\mathbf{u}_1, \tag{10.36}$$

where

$$\mathbf{R}_b = E\left\{\mathbf{b}(k)\mathbf{b}^T(k)\right\}.$$

If define two normalized correlation matrices:

$$\tilde{\mathbf{R}}_s \triangleq \frac{\mathbf{R}_s}{\sigma_s^2}, \quad \tilde{\mathbf{R}}_b \triangleq \frac{\mathbf{R}_b}{\sigma_b^2}, \tag{10.37}$$

we can further write (10.36) as:

$$\mathbf{h}_\mathrm{o} = \left[\frac{\mathbf{I}}{\mathrm{SNR}} + \tilde{\mathbf{R}}_b^{-1}\tilde{\mathbf{R}}_s\right]^{-1} \tilde{\mathbf{R}}_b^{-1}\tilde{\mathbf{R}}_s\mathbf{u}_1,$$

where

$$\text{SNR} = \frac{\sigma_s^2}{\sigma_b^2}, \tag{10.38}$$

and $\mathbf{I}$ is the identity matrix. As SNR approaches infinity, it can easily be checked that

$$\lim_{\text{SNR}\to\infty} \mathbf{h}_\text{o} = \mathbf{u}_1. \tag{10.39}$$

At the other extreme case where SNR approaches 0, we have:

$$\lim_{\text{SNR}\to 0} \mathbf{h}_\text{o} = \mathbf{0}, \tag{10.40}$$

where $\mathbf{0}$ has the same size as $\mathbf{h}_\text{o}$ and consists of all zeros. The MMSE associated with $\mathbf{h}_\text{o}$ is,

$$\begin{aligned} J_s(\mathbf{h}_\text{o}) &= \sigma_s^2 - \mathbf{r}_{xs}^T \mathbf{h}_\text{o} \\ &= \sigma_b^2 - \mathbf{r}_{bb}^T \mathbf{R}_x^{-1} \mathbf{r}_{bb} \\ &= \mathbf{r}_{bb}^T \mathbf{h}_\text{o}. \end{aligned} \tag{10.41}$$

We see clearly from the previous expression that $J_s(\mathbf{h}_\text{o}) < J_s(\mathbf{u}_1)$; therefore, noise reduction is possible.

The normalized MMSE is

$$\tilde{J}_s(\mathbf{h}_\text{o}) \triangleq \frac{J_s(\mathbf{h}_\text{o})}{J_s(\mathbf{u}_1)} = \frac{J_s(\mathbf{h}_\text{o})}{\sigma_b^2}, \tag{10.42}$$

and $0 < \tilde{J}_s(\mathbf{h}_\text{o}) < 1$.

10.4.2 Estimation of the Noise Samples

Having discussed the estimation of clean speech samples, we now consider estimation of the noise samples by passing the observations $x(k)$ through an FIR filter, i.e.,

$$\hat{b}(k) = \mathbf{g}^T \mathbf{x}(k), \tag{10.43}$$

where

$$\mathbf{g} = \begin{bmatrix} g_0 & g_1 & \cdots & g_{L-1} \end{bmatrix}^T \tag{10.44}$$

is an FIR filter of length L. The error signal between the noise sample at time k and its estimate is written as:

$$e_b(k) \triangleq b(k) - \hat{b}(k) = b(k) - \mathbf{g}^T \mathbf{x}(k), \tag{10.45}$$

The MSE criterion associated with (10.45) is,

$$J_b(\mathbf{g}) \triangleq E\left\{e_b^2(k)\right\}. \tag{10.46}$$

The estimation of $b(k)$ in the MSE sense will tend to attenuate the clean speech.

The minimization of (10.46) leads to the Wiener-Hopf equations:

$$\mathbf{g}_{\mathrm{o}} = \mathbf{R}_x^{-1}\mathbf{r}_{xb}. \quad (10.47)$$

Since $s(k)$ and $b(k)$ are uncorrelated, we can derive:

$$\begin{aligned} \mathbf{g}_{\mathrm{o}} &= \mathbf{R}_x^{-1}\mathbf{r}_{bb} = \mathbf{R}_x^{-1}\mathbf{R}_b\mathbf{u}_1 \\ &= \left[\mathrm{SNR}\cdot\mathbf{I} + \tilde{\mathbf{R}}_s^{-1}\tilde{\mathbf{R}}_b\right]^{-1}\tilde{\mathbf{R}}_s^{-1}\tilde{\mathbf{R}}_b\mathbf{u}_1. \end{aligned} \quad (10.48)$$

It can then be checked that

$$\lim_{\mathrm{SNR}\to\infty}\mathbf{g}_{\mathrm{o}} = \mathbf{0}, \quad (10.49)$$

$$\lim_{\mathrm{SNR}\to 0}\mathbf{g}_{\mathrm{o}} = \mathbf{u}_1. \quad (10.50)$$

The MSE for the particular filter $\mathbf{g} = \mathbf{u}_1$ (no clean speech reduction) is,

$$J_b(\mathbf{u}_1) = E\left\{s^2(k)\right\} = \sigma_s^2. \quad (10.51)$$

Therefore, the MMSE and the normalized MMSE are respectively,

$$J_b(\mathbf{g}_{\mathrm{o}}) = \sigma_b^2 - \mathbf{r}_b^T\mathbf{R}_x^{-1}\mathbf{r}_b = \sigma_b^2 - \mathbf{r}_b^T\mathbf{g}_{\mathrm{o}}, \quad (10.52)$$

$$\tilde{J}_b(\mathbf{g}_{\mathrm{o}}) \triangleq \frac{J_b(\mathbf{g}_{\mathrm{o}})}{J_b(\mathbf{u}_1)} = \frac{J_b(\mathbf{g}_{\mathrm{o}})}{\sigma_s^2}. \quad (10.53)$$

It is seen that $J_b(\mathbf{g}_{\mathrm{o}}) < J_b(\mathbf{u}_1)$, therefore, the Wiener filter is able to reduce the level of clean speech in the signal $x(k)$. As a result, $0 < \tilde{J}_b(\mathbf{g}_{\mathrm{o}}) < 1$.

In the next section, we will see that while the normalized MMSE, $\tilde{J}_s(\mathbf{h}_{\mathrm{o}})$, of the clean speech estimation plays a key role in noise reduction, the normalized MMSE, $\tilde{J}_b(\mathbf{g}_{\mathrm{o}})$, of the noise estimation plays a key role in speech distortion.

10.4.3 Noise Reduction versus Speech Distortion

Obviously, there are some important relationships between the estimation of the clean speech and noise samples. From (10.36) and (10.47), we get a relation between the two optimal filters:

$$\mathbf{h}_{\mathrm{o}} = \mathbf{u}_1 - \mathbf{g}_{\mathrm{o}}. \quad (10.54)$$

In fact, minimizing $J_s(\mathbf{h})$ or $J_b(\mathbf{u}_1 - \mathbf{h})$ with respect to $\mathbf{h}$ is equivalent. In the same manner, minimizing $J_b(\mathbf{g})$ or $J_s(\mathbf{u}_1 - \mathbf{g})$ with respect to $\mathbf{g}$ is the same thing. At the optimum, we have,

$$\begin{aligned} e_{s,\mathrm{o}}(k) &= s(k) - \mathbf{h}_{\mathrm{o}}^T\mathbf{x}(k) \\ &= s(k) - \left[\mathbf{u}_1 - \mathbf{g}_{\mathrm{o}}\right]^T\left[\mathbf{s}(k) + \mathbf{b}(k)\right] \\ &= -b(k) + \mathbf{g}_{\mathrm{o}}^T\mathbf{x}(k) \\ &= -e_{b,\mathrm{o}}(k). \end{aligned} \quad (10.55)$$

From (10.41) and (10.52), we see that the two MMSEs are equal,

$$J_s(\mathbf{h}_\text{o}) = J_b(\mathbf{g}_\text{o}). \tag{10.56}$$

However, the normalized MMSEs are not, in general. Indeed, we have a relation between the two:

$$\begin{aligned} \tilde{J}_b(\mathbf{g}_\text{o}) &= \frac{J_b(\mathbf{g}_\text{o})}{\sigma_s^2} = \frac{J_s(\mathbf{h}_\text{o})}{\sigma_s^2} \\ &= \frac{\sigma_b^2}{\sigma_s^2}\frac{J_s(\mathbf{h}_\text{o})}{\sigma_b^2} = \frac{\tilde{J}_s(\mathbf{h}_\text{o})}{\text{SNR}}. \end{aligned} \tag{10.57}$$

So the only situation where the two normalized MMSEs are equal is when the SNR is equal to 1. For SNR < 1, $\tilde{J}_s(\mathbf{h}_\text{o}) < \tilde{J}_b(\mathbf{g}_\text{o})$ and for SNR > 1, $\tilde{J}_b(\mathbf{g}_\text{o}) < \tilde{J}_s(\mathbf{h}_\text{o})$. Also, $\tilde{J}_s(\mathbf{h}_\text{o}) <$ SNR and $\tilde{J}_b(\mathbf{g}_\text{o}) < 1/\text{SNR}$.

It is given without proof that:

$$J_b(\mathbf{h}_\text{o}) = J_s(\mathbf{g}_\text{o}) = \sigma_x^2 - 3J_s(\mathbf{h}_\text{o}), \tag{10.58}$$

which implies that $J_s(\mathbf{h}_\text{o}) < \sigma_x^2/3$. We already know that $J_s(\mathbf{h}_\text{o}) < \sigma_b^2$ and $J_s(\mathbf{h}_\text{o}) < \sigma_s^2$.

The optimal estimation of the clean speech, in the Wiener sense, is in fact what we call noise reduction:

$$\hat{s}_\text{o}(k) = \mathbf{h}_\text{o}^T\mathbf{x}(k), \tag{10.59}$$

or equivalently, if the noise is estimated first:

$$\hat{b}_\text{o}(k) = \mathbf{g}_\text{o}^T\mathbf{x}(k), \tag{10.60}$$

we can use this estimate to reduce the noise from the observed signal:

$$\hat{s}_\text{o}(k) = x(k) - \hat{b}_\text{o}(k). \tag{10.61}$$

The power of the estimated clean speech signal with the optimal Wiener filter is,

$$\begin{aligned} E\left[\hat{s}_\text{o}^2(k)\right] &= \mathbf{h}_\text{o}^T\mathbf{R}_x\mathbf{h}_\text{o} = \sigma_s^2 - J_s(\mathbf{h}_\text{o}) \\ &= \mathbf{h}_\text{o}^T\mathbf{R}_s\mathbf{h}_\text{o} + \mathbf{h}_\text{o}^T\mathbf{R}_b\mathbf{h}_\text{o}, \end{aligned} \tag{10.62}$$

which is the sum of two terms. The first is the power of the attenuated clean speech and the second is the power of the residual noise (always greater than zero). While noise reduction is feasible with the Wiener filter, expression (10.62) shows that the price to pay for this is also a reduction of the clean speech [by a quantity equal to $J_s(\mathbf{h}_\text{o}) + \mathbf{h}_\text{o}^T\mathbf{R}_b\mathbf{h}_\text{o}$ and this implies distortion], since $\mathbf{h}_\text{o}^T\mathbf{R}_s\mathbf{h}_\text{o} < \sigma_s^2$. In other words, the power of the attenuated clean speech signal is, obviously, always smaller than the power of the clean speech itself;

this means that parts of the clean speech are attenuated in the process and as a result, distortion is unavoidable with this approach.

We now assess the speech distortion due to the optimal filtering operation. According to its definition given in (10.4), the speech-distortion index after the Wiener filtering, can be written as

$$\upsilon_{\mathrm{sd}}(\mathbf{h}_{\mathrm{o}}) \triangleq \frac{E\left\{[s(k) - \mathbf{h}_{\mathrm{o}}^T \mathbf{s}(k)]^2\right\}}{\sigma_s^2}. \tag{10.63}$$

Using (10.54), we can easily check that

$$\begin{aligned} \upsilon_{\mathrm{sd}}(\mathbf{h}_{\mathrm{o}}) &= \frac{\mathbf{g}_{\mathrm{o}}^T \mathbf{R}_s \mathbf{g}_{\mathrm{o}}}{\sigma_s^2} \\ &= \frac{1}{\mathrm{SNR}} \left[\tilde{J}_s(\mathbf{h}_{\mathrm{o}}) - \mathbf{h}_{\mathrm{o}}^T \tilde{\mathbf{R}}_b \mathbf{h}_{\mathrm{o}}\right] \\ &< \tilde{J}_b(\mathbf{g}_{\mathrm{o}}). \end{aligned}$$

Clearly, the speech-distortion index is always between 0 and 1 for the optimal filter. When $\upsilon_{\mathrm{sd}}(\mathbf{h}_{\mathrm{o}})$ is close to 1, the speech signal is highly distorted and when $\upsilon_{\mathrm{sd}}(\mathbf{h}_{\mathrm{o}})$ is near 0, the speech signal is slightly distorted. Also,

$$\lim_{\mathrm{SNR}\to 0} \upsilon_{\mathrm{sd}}(\mathbf{h}_{\mathrm{o}}) = 1, \tag{10.64}$$

$$\lim_{\mathrm{SNR}\to \infty} \upsilon_{\mathrm{sd}}(\mathbf{h}_{\mathrm{o}}) = 0. \tag{10.65}$$

Therefore, for low SNRs, the Wiener filter may have a disastrous effect on the speech signal.

Similarly, the noise-reduction factor due to the Wiener filter, according to (10.2), is written as,

$$\xi_{\mathrm{nr}}(\mathbf{h}_{\mathrm{o}}) \triangleq \frac{\sigma_b^2}{E\left\{[\mathbf{h}_{\mathrm{o}}^T \mathbf{b}(k)]^2\right\}}. \tag{10.66}$$

With some simple mathematical manipulation, we can deduce that

$$\xi_{\mathrm{nr}}(\mathbf{h}_{\mathrm{o}}) = \frac{\sigma_b^2}{\mathbf{h}_{\mathrm{o}}^T \mathbf{R}_b \mathbf{h}_{\mathrm{o}}} = \frac{1}{\mathrm{SNR} \cdot \left[\tilde{J}_b(\mathbf{g}_{\mathrm{o}}) - \mathbf{g}_{\mathrm{o}}^T \tilde{\mathbf{R}}_s \mathbf{g}_{\mathrm{o}}\right]}.$$

It is then easy to verify that

$$\xi_{\mathrm{nr}}(\mathbf{h}_{\mathrm{o}}) > \frac{1}{\tilde{J}_s(\mathbf{h}_{\mathrm{o}})}$$

and

$$\xi_{\mathrm{nr}}(\mathbf{h}_{\mathrm{o}}) > 1.$$

The greater is $\xi_{\mathrm{nr}}(\mathbf{h}_{\mathrm{o}})$, the more noise reduction we have. Also,

$$\lim_{\mathrm{SNR}\to 0} \xi_{\mathrm{nr}}(\mathbf{h}_{\mathrm{o}}) = \infty, \tag{10.67}$$

$$\lim_{\mathrm{SNR}\to \infty} \xi_{\mathrm{nr}}(\mathbf{h}_{\mathrm{o}}) = 1. \tag{10.68}$$

Using (10.63) and (10.66), we obtain important relationships between the speech-distortion index and the noise-reduction factor:

$$\upsilon_{\mathrm{sd}}(\mathbf{h}_{\mathrm{o}}) = \frac{1}{\mathrm{SNR}} \left[\tilde{J}_s(\mathbf{h}_{\mathrm{o}}) - \frac{1}{\xi_{\mathrm{nr}}(\mathbf{h}_{\mathrm{o}})} \right], \tag{10.69}$$

$$\xi_{\mathrm{nr}}(\mathbf{h}_{\mathrm{o}}) = \frac{1}{\mathrm{SNR} \cdot \left[\tilde{J}_b(\mathbf{g}_{\mathrm{o}}) - \upsilon_{\mathrm{sd}}(\mathbf{g}_{\mathrm{o}}) \right]}. \tag{10.70}$$

Therefore, for the optimum filter, when the SNR is very high, there is little noise reduction (which is not really needed in this situation) and little speech distortion. On the other hand, when the SNR is very low, noise reduction is large, but so is speech distortion.

10.4.4 *A Priori* SNR versus *a Posteriori* SNR

A key question in noise reduction is whether the SNR is improved. For the time-domain Wiener filter, we have the following theorem.

Theorem: With the time-domain Wiener filter in the context of noise reduction, the *a posteriori* SNR (defined after the Wiener filter) is always greater than, or at least equal to the *a priori* SNR (defined before the Wiener filter).

Proof. From the SNR definition given in (10.6), The *a priori* SNR of signal $x(k)$ can be written as

$$\mathrm{SNR} = \frac{\mathbf{u}_1^T \mathbf{R}_s \mathbf{u}_1}{\mathbf{u}_1^T \mathbf{R}_b \mathbf{u}_1}. \tag{10.71}$$

After the Wiener filtering, the *a posteriori* SNR can be expressed by

$$\mathrm{SNR}_{\mathrm{o}} = \frac{\mathbf{h}_{\mathrm{o}}^T \mathbf{R}_s \mathbf{h}_{\mathrm{o}}}{\mathbf{h}_{\mathrm{o}}^T \mathbf{R}_b \mathbf{h}_{\mathrm{o}}}. \tag{10.72}$$

From their definitions, we know that all three matrices, $\mathbf{R}_s$, $\mathbf{R}_b$, and $\mathbf{R}_x$ are symmetric, and positive semi-definite. We further assume that $\mathbf{R}_b$ is positive definite so its inverse exists. In addition, based on the assumption that speech and noise signals are uncorrelated, we have $\mathbf{R}_x = \mathbf{R}_s + \mathbf{R}_b$. In case that both $\mathbf{R}_s$ and $\mathbf{R}_b$ are diagonal matrices, or $\mathbf{R}_b$ is a scaled version of $\mathbf{R}_s$ (i.e., $\mathbf{R}_s = \mathrm{SNR} \cdot \mathbf{R}_b$), it can be easily seen that $\mathrm{SNR}_{\mathrm{o}} = \mathrm{SNR}$. Here, we consider more complicated situations where at least one of the $\mathbf{R}_s$ and $\mathbf{R}_b$ matrices is not diagonal. In this case, according to [123], there exists a linear transform that can simultaneously diagonalize $\mathbf{R}_s$, $\mathbf{R}_b$, and $\mathbf{R}_x$. The process is done as follows:

$$\begin{aligned}\mathbf{R}_s &= (\mathbf{B}^T)^{-1}\mathbf{\Lambda}\mathbf{B}^{-1},\\ \mathbf{R}_b &= (\mathbf{B}^T)^{-1}\mathbf{B}^{-1},\\ \mathbf{R}_x &= (\mathbf{B}^T)^{-1}[\mathbf{I}+\mathbf{\Lambda}]\mathbf{B}^{-1},\end{aligned} \tag{10.73}$$

where again $\mathbf{I}$ is the identity matrix,

$$\mathbf{\Lambda} = \text{diag}\left[\begin{array}{cccc}\lambda_1, & \lambda_2, & \ldots, & \lambda_L\end{array}\right] \tag{10.74}$$

is the eigenvalue matrix of $\mathbf{R}_b^{-1}\mathbf{R}_s$, with $\lambda_1 \geq \lambda_2 \geq \cdots \geq \lambda_L \geq 0$, $\mathbf{B}$ is the eigenvector matrix of $\mathbf{R}_b^{-1}\mathbf{R}_s$, and

$$\mathbf{R}_b^{-1}\mathbf{R}_s\mathbf{B} = \mathbf{B}\mathbf{\Lambda}. \tag{10.75}$$

Note that $\mathbf{B}$ is not necessarily orthogonal since $\mathbf{R}_b^{-1}\mathbf{R}_s$ is not necessarily symmetric. We then immediately have

$$\text{SNR} = \frac{\mathbf{u}_1^T(\mathbf{B}^{-1})^T\mathbf{\Lambda}\mathbf{B}^{-1}\mathbf{u}_1}{\mathbf{u}_1^T(\mathbf{B}^{-1})^T\mathbf{B}^{-1}\mathbf{u}_1}, \tag{10.76}$$

and

$$\begin{aligned}\text{SNR}_\text{o} &= \frac{\mathbf{u}_1^T\mathbf{R}_s^T\mathbf{R}_x^{-1}\mathbf{R}_s\mathbf{R}_x^{-1}\mathbf{R}_s\mathbf{u}_1}{\mathbf{u}_1^T\mathbf{R}_s^T\mathbf{R}_x^{-1}\mathbf{R}_b\mathbf{R}_x^{-1}\mathbf{R}_s\mathbf{u}_1}\\ &= \frac{\mathbf{u}_1^T(\mathbf{B}^{-1})^T\mathbf{\Lambda}(\mathbf{I}+\mathbf{\Lambda})^{-1}\mathbf{\Lambda}(\mathbf{I}+\mathbf{\Lambda})^{-1}\mathbf{\Lambda}\mathbf{B}^{-1}\mathbf{u}_1}{\mathbf{u}_1^T(\mathbf{B}^{-1})^T\mathbf{\Lambda}(\mathbf{I}+\mathbf{\Lambda})^{-1}(\mathbf{I}+\mathbf{\Lambda})^{-1}\mathbf{\Lambda}\mathbf{B}^{-1}\mathbf{u}_1}\\ &= \frac{\mathbf{u}_1^T(\mathbf{B}^{-1})^T\mathbf{\Sigma}_1\mathbf{B}^{-1}\mathbf{u}_1}{\mathbf{u}_1^T(\mathbf{B}^{-1})^T\mathbf{\Sigma}_2\mathbf{B}^{-1}\mathbf{u}_1},\end{aligned} \tag{10.77}$$

where

$$\begin{aligned}\mathbf{\Sigma}_1 &\triangleq \mathbf{\Lambda}(\mathbf{I}+\mathbf{\Lambda})^{-1}\mathbf{\Lambda}(\mathbf{I}+\mathbf{\Lambda})^{-1}\mathbf{\Lambda}\\ &= \text{diag}\left[\frac{\lambda_1^3}{(1+\lambda_1)^2}, \frac{\lambda_2^3}{(1+\lambda_2)^2}, \ldots, \frac{\lambda_L^3}{(1+\lambda_L)^2}\right]\end{aligned}$$

and

$$\begin{aligned}\mathbf{\Sigma}_2 &\triangleq \mathbf{\Lambda}(\mathbf{I}+\mathbf{\Lambda})^{-1}(\mathbf{I}+\mathbf{\Lambda})^{-1}\mathbf{\Lambda}\\ &= \text{diag}\left[\frac{\lambda_1^2}{(1+\lambda_1)^2}, \frac{\lambda_2^2}{(1+\lambda_2)^2}, \ldots, \frac{\lambda_L^2}{(1+\lambda_L)^2}\right]\end{aligned}$$

are two diagonal matrices. For the ease of expression, let us denote $\mathbf{B}^{-1}$ as $\mathbf{A} = \mathbf{B}^{-1} = [a_{ij}]$. Both SNR and SNR_o can then be rewritten as

$$\mathrm{SNR} = \frac{\sum_{i=1}^{L} \lambda_i a_{i1}^2}{\sum_{i=1}^{L} a_{i1}^2},$$
$$\mathrm{SNR_o} = \frac{\sum_{i=1}^{L} \frac{\lambda_i^3}{(1+\lambda_i)^2} a_{i1}^2}{\sum_{i=1}^{L} \frac{\lambda_i^2}{(1+\lambda_i)^2} a_{i1}^2}. \tag{10.78}$$

Since $\sum_{i=1}^{L} \frac{\lambda_i^3}{(1+\lambda_i)^2} a_{i1}^2$, $\sum_{i=1}^{L} \frac{\lambda_i^2}{(1+\lambda_i)^2} a_{i1}^2$, $\sum_{i=1}^{L} \lambda_i a_{i1}^2$, and $\sum_{i=1}^{L} a_{i1}^2$ all are non-negative numbers, as long as we can show that the inequality

$$\sum_{i=1}^{L} \frac{\lambda_i^3}{(1+\lambda_i)^2} a_{i1}^2 \sum_{i=1}^{L} a_{i1}^2 \geq \sum_{i=1}^{L} \frac{\lambda_i^2}{(1+\lambda_i)^2} a_{i1}^2 \sum_{i=1}^{L} \lambda_i a_{i1}^2 \tag{10.79}$$

holds, then $\mathrm{SNR_o} \geq \mathrm{SNR}$. Now we prove this inequality by way of induction.

- Basic Step: if $L = 2$,

$$\sum_{i=1}^{2} \frac{\lambda_i^3}{(1+\lambda_i)^2} a_{i1}^2 \sum_{i=1}^{2} a_{i1}^2 = \frac{\lambda_1^3}{(1+\lambda_1)^2} a_{11}^4 + \frac{\lambda_2^3}{(1+\lambda_2)^2} a_{21}^4 + \left[\frac{\lambda_1^3}{(1+\lambda_1)^2} + \frac{\lambda_2^3}{(1+\lambda_2)^2} \right] a_{11}^2 a_{21}^2. \tag{10.80}$$

  Since $\lambda_i \geq 0$, it is trivial to show that

$$\frac{\lambda_1^3}{(1+\lambda_1)^2} + \frac{\lambda_2^3}{(1+\lambda_2)^2} \geq \frac{\lambda_1^2 \lambda_2}{(1+\lambda_1)^2} + \frac{\lambda_1 \lambda_2^2}{(1+\lambda_2)^2}, \tag{10.81}$$

  where "=" holds when $\lambda_1 = \lambda_2$. Therefore

$$\begin{aligned} \sum_{i=1}^{2} \frac{\lambda_i^3}{(1+\lambda_i)^2} a_{i1}^2 \sum_{i=1}^{2} a_{i1}^2 &\geq \frac{\lambda_1^3}{(1+\lambda_1)^2} a_{11}^4 + \frac{\lambda_2^3}{(1+\lambda_2)^2} a_{21}^4 + \\ &\quad \left[\frac{\lambda_1^2 \lambda_2}{(1+\lambda_1)^2} + \frac{\lambda_1 \lambda_2^2}{(1+\lambda_2)^2} \right] a_{11}^2 a_{21}^2 \\ &= \sum_{i=1}^{2} \frac{\lambda_i^2}{(1+\lambda_i)^2} a_{i1}^2 \sum_{i=1}^{2} \lambda_i a_{i1}^2. \end{aligned}$$

  So the property is true for $L = 2$, where "=" holds when any one of a_{11} and a_{21} is equal to 0 (note that a_{11} and a_{21} cannot be zero at the same time since $\mathbf{A}$ is invertible) or when $\lambda_1 = \lambda_2$.
- Inductive Step: assume that the property is true for $L = n$, i.e.,

$$\sum_{i=1}^{n} \frac{\lambda_i^3}{(1+\lambda_i)^2} a_{i1}^2 \sum_{i=1}^{n} a_{i1}^2 \geq \sum_{i=1}^{n} \frac{\lambda_i^2}{(1+\lambda_i)^2} a_{i1}^2 \sum_{i=1}^{n} \lambda_i a_{i1}^2. \tag{10.82}$$

  We must prove that it is also true for $L = n + 1$. As a matter of fact,

$$\sum_{i=1}^{n+1} \frac{\lambda_i^3}{(1+\lambda_i)^2} a_{i1}^2 \sum_{i=1}^{n+1} a_{i1}^2$$
$$= \left[\sum_{i=1}^{n} \frac{\lambda_i^3}{(1+\lambda_i)^2} a_{i1}^2 + \frac{\lambda_{n+1}^3}{(1+\lambda_{n+1})^2} a_{n+11}^2 \right] \left[\sum_{i=1}^{n} a_{i1}^2 + a_{n+11}^2 \right]$$
$$= \left[\sum_{i=1}^{n} \frac{\lambda_i^3}{(1+\lambda_i)^2} a_{i1}^2 \right] \left[\sum_{i=1}^{n} a_{i1}^2 \right] + \frac{\lambda_{n+1}^3}{(1+\lambda_{n+1})^2} a_{n+11}^4 +$$
$$\sum_{i=1}^{n} \left[\frac{\lambda_i^3}{(1+\lambda_i)^2} + \frac{\lambda_{n+1}^3}{(1+\lambda_{n+1})^2} \right] a_{i1}^2 a_{n+11}^2. \tag{10.83}$$

Using the induction hypothesis, and also the fact that

$$\frac{\lambda_i^3}{(1+\lambda_i)^2} + \frac{\lambda_{n+1}^3}{(1+\lambda_{n+1})^2} \geq \frac{\lambda_i^2 \lambda_{n+1}}{(1+\lambda_i)^2} + \frac{\lambda_i \lambda_{n+1}^2}{(1+\lambda_{n+1})^2}, \tag{10.84}$$

we can deduce

$$\sum_{i=1}^{n+1} \frac{\lambda_i^3}{(1+\lambda_i)^2} a_{i1}^2 \sum_{i=1}^{n+1} a_{i1}^2$$
$$\geq \sum_{i=1}^{n} \frac{\lambda_i^2}{(1+\lambda_i)^2} a_{i1}^2 \sum_{i=1}^{n} \lambda_i a_{i1}^2 + \frac{\lambda_{n+1}^3}{(1+\lambda_{n+1})^2} a_{n+11}^4 +$$
$$\sum_{i=1}^{n} \left[\frac{\lambda_i^2 \lambda_{n+1}}{(1+\lambda_i)^2} + \frac{\lambda_i \lambda_{n+1}^2}{(1+\lambda_{n+1})^2} \right] a_{i1}^2 a_{n+11}^2$$
$$= \sum_{i=1}^{n+1} \frac{\lambda_i^2}{(1+\lambda_i)^2} a_{i1}^2 \sum_{i=1}^{n+1} \lambda_i a_{i1}^2, \tag{10.85}$$

where "=" holds when all the λ_i's corresponding to nonzero a_{i1} are equal, where $i = 1, 2, \ldots, n+1$. That completes the proof.

As seen, as long as there are some disparities among all the eigenvalues of the matrix $\mathbf{R}_b^{-1}\mathbf{R}_s$, the time-domain Wiener filter will be able to improve SNR, no matter if the noise is white or colored. When all the eigenvalues are identical, the noise subspace would be the same as the signal subspace. As a result, the Wiener filter fails to discriminate speech from noise, and in this case the *a priori* SNR is equal to the *a posteriori* SNR.

Note that even though it can improve the SNR, the Wiener filter does not maximize the *a posteriori* SNR. As a matter of fact, (10.72) is well known as the generalized Rayleigh quotient. So the filter that maximizes the *a posteriori* SNR is the eigenvector corresponding to the maximum eigenvalue of the matrix $\mathbf{R}_b^{-1}\mathbf{R}_s$. However, this filter typically gives rise to large speech distortion.

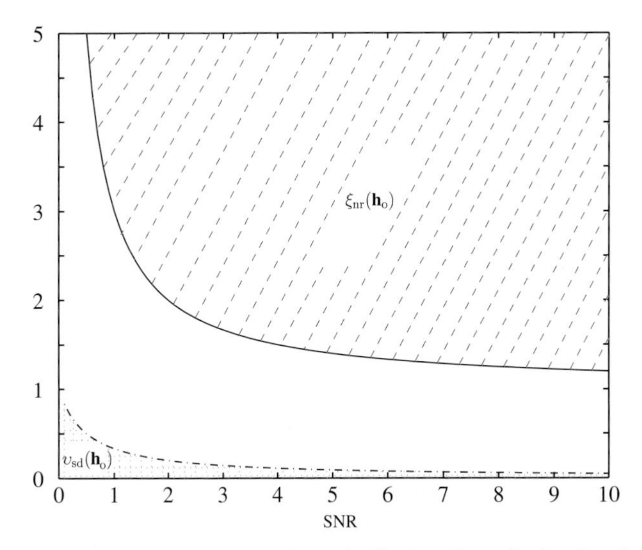

Fig. 10.1. Illustration of the areas where $\xi_{\rm nr}(\mathbf{h}_{\rm o})$ and $\upsilon_{\rm sd}(\mathbf{h}_{\rm o})$ take their values as a function of the SNR. $\xi_{\rm nr}(\mathbf{h}_{\rm o})$ can take any value above the solid line while $\upsilon_{\rm sd}(\mathbf{h}_{\rm o})$ can take any value under the dotted line.

10.4.5 Bounds for Noise Reduction and Speech Distortion

Knowing that $\mathrm{SNR}_{\rm o} \geq \mathrm{SNR}$, we now can develop the lower bound for the speech-distortion index and the upper bound for the noise-reduction factor. In fact, it can easily be checked that

$$\begin{aligned}\mathrm{SNR}_{\rm o} &= \frac{\mathbf{h}_{\rm o}^T \mathbf{R}_s \mathbf{h}_{\rm o}}{\mathbf{h}_{\rm o}^T \mathbf{R}_b \mathbf{h}_{\rm o}} = \mathrm{SNR} \cdot \frac{\mathbf{h}_{\rm o}^T \tilde{\mathbf{R}}_s \mathbf{h}_{\rm o}}{\mathbf{h}_{\rm o}^T \tilde{\mathbf{R}}_b \mathbf{h}_{\rm o}} \\ &= -1 + \mathrm{SNR} \cdot \xi_{\rm nr}(\mathbf{h}_{\rm o}) \left[1 - \tilde{J}_b(\mathbf{g}_{\rm o})\right] \\ &= -1 + \frac{1 - \tilde{J}_b(\mathbf{g}_{\rm o})}{\tilde{J}_b(\mathbf{g}_{\rm o}) - \upsilon_{\rm sd}(\mathbf{h}_{\rm o})}. \\ &\geq \mathrm{SNR}. \end{aligned} \qquad (10.86)$$

Since $\upsilon_{\rm sd}(\mathbf{h}_{\rm o}) < \tilde{J}_b(\mathbf{g}_{\rm o})$, and $0 \leq \upsilon_{\rm sd}(\mathbf{h}_{\rm o}) \leq 1$, it can be shown that

$$\xi_{\rm nr}(\mathbf{h}_{\rm o}) \geq \frac{\mathrm{SNR} + 2}{\mathrm{SNR}}. \qquad (10.87)$$

Similarly, we can derive the upper bound for $\upsilon_{\rm sd}(\mathbf{h}_{\rm o})$, i.e.,

$$\upsilon_{\rm sd}(\mathbf{h}_{\rm o}) \leq \frac{1}{2 \cdot \mathrm{SNR} + 1}. \qquad (10.88)$$

Figure 10.1 illustrates expressions (10.87) and (10.88).

We now introduce another index for noise reduction:

$$\zeta_{\mathrm{nr}}(\mathbf{h}_{\mathrm{o}}) \triangleq 1 - \tilde{J}_s(\mathbf{h}_{\mathrm{o}}) < 1. \tag{10.89}$$

The closer is $\zeta_{\mathrm{nr}}(\mathbf{h}_{\mathrm{o}})$ to 1, the more noise reduction we get. This index will be useful in the following sections.

10.4.6 Particular Case: White Gaussian Noise

In case where the additive noise is white Gaussian noise, we have:

$$\mathbf{r}_{bb} = \sigma_b^2 \mathbf{u}_1. \tag{10.90}$$

From (10.42) and (10.53), we observe that the two normalized MMSEs are

$$\tilde{J}_s(\mathbf{h}_{\mathrm{o}}) = h_{\mathrm{o},0}, \tag{10.91}$$

$$\tilde{J}_b(\mathbf{g}_{\mathrm{o}}) = \frac{1 - g_{\mathrm{o},0}}{\mathrm{SNR}} = \frac{h_{\mathrm{o},0}}{\mathrm{SNR}}, \tag{10.92}$$

where $h_{\mathrm{o},0}$ and $g_{\mathrm{o},0}$ are the first components of the vectors $\mathbf{h}_{\mathrm{o}}$ and $\mathbf{g}_{\mathrm{o}}$, respectively. Clearly, $0 < h_{\mathrm{o},0} < 1$ and $0 < g_{\mathrm{o},0} < 1$. Hence, the normalized MMSE $\tilde{J}_s(\mathbf{h}_{\mathrm{o}})$ is completely governed by the first element of the Wiener filter $\mathbf{h}_{\mathrm{o}}$.

Now, the speech-distortion index and the noise-reduction factor for the optimal filter can be simplified:

$$\begin{aligned} v_{\mathrm{sd}}(\mathbf{h}_{\mathrm{o}}) &= \frac{1}{\mathrm{SNR}} \left[h_{\mathrm{o},0} - \mathbf{h}_{\mathrm{o}}^T \mathbf{h}_{\mathrm{o}} \right] \\ &= \frac{\mathbf{g}_{\mathrm{o}}^T \mathbf{h}_{\mathrm{o}}}{\mathrm{SNR}} = \frac{1}{\mathrm{SNR}} \left[g_{\mathrm{o},0} - \mathbf{g}_{\mathrm{o}}^T \mathbf{g}_{\mathrm{o}} \right], \end{aligned} \tag{10.93}$$

$$\xi_{\mathrm{nr}}(\mathbf{h}_{\mathrm{o}}) = \frac{1}{\mathbf{h}_{\mathrm{o}}^T \mathbf{h}_{\mathrm{o}}}. \tag{10.94}$$

We also deduce from (10.93) that $h_{\mathrm{o},0} > \mathbf{h}_{\mathrm{o}}^T \mathbf{h}_{\mathrm{o}}$ and $g_{\mathrm{o},0} > \mathbf{g}_{\mathrm{o}}^T \mathbf{g}_{\mathrm{o}}$.

We know from linear prediction theory that [160],

$$\mathbf{R}_x \begin{bmatrix} 1 \\ -\mathbf{a}_x \end{bmatrix} = \begin{bmatrix} E_x \\ \mathbf{0} \end{bmatrix}, \tag{10.95}$$

where $\mathbf{a}_x$ is the forward linear predictor, E_x is the corresponding error energy, and $\mathbf{0}$ is the $(L-1) \times 1$ null vector. Replacing the previous equation in (10.36), we obtain:

$$\mathbf{h}_{\mathrm{o}} = \mathbf{u}_1 - \sigma_b^2 \mathbf{R}_x^{-1} \mathbf{u}_1 = \begin{bmatrix} h_{\mathrm{o},0} \\ \frac{\sigma_b^2}{E_x} \mathbf{a}_x \end{bmatrix}, \tag{10.96}$$

where

$$h_{\mathrm{o},0} = \tilde{J}_s(\mathbf{h}_{\mathrm{o}}) = 1 - \frac{\sigma_b^2}{E_x}. \tag{10.97}$$

Equation (10.96) shows how the Wiener filter is related to the forward predictor of the observed signal $x(n)$. This expression also gives a hint on how to choose the length of the optimal filter $\mathbf{h}_o$: it should be equal to the length of the predictor $\mathbf{a}_x$ required to have a good prediction of the observed signal $x(n)$. Equation (10.97) contains some very interesting information. Indeed, if the clean speech signal is completely predictable, this means that $E_x \approx \sigma_b^2$ and $\tilde{J}_s(\mathbf{h}_o) \approx 0$. On the other hand, if $s(n)$ is not predictable, we have $E_x \approx \sigma_x^2$ and $\tilde{J}_s(\mathbf{h}_o) \approx 1 - \sigma_b^2/\sigma_x^2$. This implies that the Wiener filter is more efficient to reduce the level of noise for predictable signals than for unpredictable ones.

10.4.7 A Suboptimal Filter

For a noise-reduction/speech-enhancement system, we always expect that it can achieve maximal noise reduction without much speech distortion. From the previous analysis, however, it follows that while noise reduction is maximized with the optimal Wiener filter, speech distortion is also maximized. A key question then is whether better ways exist to control the tradeoff between the conflicting requirements of noise reduction and speech distortion. Examining (10.63), one can see that to control the speech distortion, we need to minimize $E\left\{[s(k) - \mathbf{h}_o^T\mathbf{s}(k)]^2\right\}$. This can be achieved in three different ways. For example, a speech signal can be modelled as an AR process. If the AR coefficients are known *a priori* or can be estimated from the noisy speech, these coefficients can be exploited to minimize $E\left\{[s(k) - \mathbf{h}_o^T\mathbf{s}(k)]^2\right\}$, while simultaneously achieving a reasonable level of noise attenuation. This is often referred to as the parametric-model-based technique [246], [143]. In case where we have multiple microphones, spatial filtering techniques can be used to achieve noise reduction with less or even no speech distortion, which will be discussed latter on. In this section, we describe a suboptimal filter, which can control the compromise between noise reduction and speech distortion without requiring any *a priori* knowledge or more resources. Such a suboptimal filter is constructed from the Wiener filter as:

$$\begin{aligned} \mathbf{h}_{\text{sub}} &= \mathbf{u}_1 - \mathbf{g}_{\text{sub}} \\ &= \mathbf{u}_1 - \alpha \mathbf{g}_o, \end{aligned} \tag{10.98}$$

where α is a real number. The MSE of the clean speech estimation corresponding to $\mathbf{h}_{\text{sub}}$ is,

$$\begin{aligned} J_s(\mathbf{h}_{\text{sub}}) &= E\left\{[s(k) - \mathbf{h}_{\text{sub}}^T\mathbf{x}(k)]^2\right\} \\ &= \sigma_b^2 - \alpha(2-\alpha)\mathbf{r}_{bb}^T\mathbf{R}_x^{-1}\mathbf{r}_{bb}, \end{aligned} \tag{10.99}$$

and, obviously, $J_s(\mathbf{h}_{\text{sub}}) \geq J_s(\mathbf{h}_o)$, $\forall\alpha$; we have equality for $\alpha = 1$. In order to have noise reduction, α must be chosen in such a way that $J_s(\mathbf{h}_{\text{sub}}) < J_s(\mathbf{u}_1)$, therefore,

$$0 < \alpha < 2. \tag{10.100}$$

We can check that,

$$J_b(\mathbf{g}_{\text{sub}}) = E\left\{\left[b(k) - \alpha \mathbf{g}_{\text{o}}^T \mathbf{x}(k)\right]^2\right\} = J_s(\mathbf{h}_{\text{sub}}). \tag{10.101}$$

Let

$$\hat{s}_{\text{sub}}(k) = \mathbf{h}_{\text{sub}}^T \mathbf{x}(k) \tag{10.102}$$

denote the estimate of the clean speech at time k using $\mathbf{h}_{\text{sub}}$. The power of $\hat{s}_{\text{sub}}(k)$ is,

$$\begin{aligned} E\left\{\hat{s}_{\text{sub}}^2(k)\right\} &= \mathbf{h}_{\text{sub}}^T \mathbf{R}_x \mathbf{h}_{\text{sub}} \\ &= \left[\mathbf{u}_1 - \alpha \mathbf{R}_x^{-1} \mathbf{r}_{bb}\right]^T \left[\mathbf{r}_{xx} - \alpha \mathbf{r}_{bb}\right] \\ &= \sigma_s^2 + (1 - 2\alpha)\sigma_b^2 + \alpha^2 \mathbf{r}_{bb}^T \mathbf{R}_x^{-1} \mathbf{r}_{bb} \\ &= \mathbf{h}_{\text{sub}}^T \mathbf{R}_s \mathbf{h}_{\text{sub}} + \mathbf{h}_{\text{sub}}^T \mathbf{R}_b \mathbf{h}_{\text{sub}}. \end{aligned} \tag{10.103}$$

The speech-distortion index corresponding to the filter $\mathbf{h}_{\text{sub}}$ is,

$$\begin{aligned} \upsilon_{\text{sd}}(\mathbf{h}_{\text{sub}}) &= \frac{E\left\{\left[s(k) - \mathbf{h}_{\text{sub}}^T \mathbf{s}(k)\right]^2\right\}}{\sigma_s^2} \\ &= \alpha^2 \mathbf{g}_{\text{o}}^T \tilde{\mathbf{R}}_s \mathbf{g}_{\text{o}} = \alpha^2 \upsilon_{\text{sd}}(\mathbf{h}_{\text{o}}). \end{aligned} \tag{10.104}$$

The previous expression shows that the ratio of the speech-distortion indices corresponding to the two filters $\mathbf{g}_{\text{sub}}$ and $\mathbf{g}_{\text{o}}$ depends on α only.

In order to have less distortion with the suboptimal filter $\mathbf{h}_{\text{sub}}$ than with the Wiener filter $\mathbf{h}_{\text{o}}$, we must find α in such a way that,

$$\upsilon_{\text{sd}}(\mathbf{h}_{\text{sub}}) < \upsilon_{\text{sd}}(\mathbf{h}_{\text{o}}), \tag{10.105}$$

hence, the condition on α should be

$$-1 < \alpha < 1. \tag{10.106}$$

Finally, if α is taken such as,

$$0 < \alpha < 1, \tag{10.107}$$

the suboptimal filter $\mathbf{h}_{\text{sub}}$ can reduce the level of noise in the observed signal $x(k)$ but causes less distortion than the Wiener filter $\mathbf{h}_{\text{o}}$. For the extreme case where $\alpha = 0$, we obtain $\mathbf{h}_{\text{sub}} = \mathbf{u}_1$, so there is no noise reduction at all but no additional distortion added. At the other extreme case as $\alpha = 1$, we have $\mathbf{h}_{\text{sub}} = \mathbf{h}_{\text{o}}$; hence the noise reduction is maximized and so is the speech distortion.

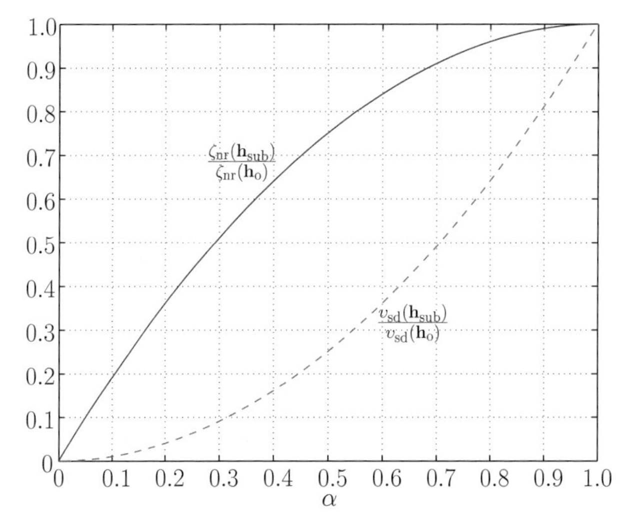

Fig. 10.2. $\upsilon_{\rm sd}(\mathbf{h}_{\rm sub})/\upsilon_{\rm sd}(\mathbf{h}_{\rm o})$ (dashed line) and $\zeta_{\rm nr}(\mathbf{h}_{\rm sub})/\zeta_{\rm nr}(\mathbf{h}_{\rm o})$ (solid line), both as a function of α.

Since

$$\begin{aligned} J_b(\mathbf{g}_{\rm sub}) &= \mathbf{g}_{\rm sub}^T \mathbf{R}_s \mathbf{g}_{\rm sub} + \mathbf{h}_{\rm sub}^T \mathbf{R}_b \mathbf{h}_{\rm sub} \\ &= \sigma_s^2 \mathbf{g}_{\rm sub}^T \tilde{\mathbf{R}}_s \mathbf{g}_{\rm sub} + \sigma_b^2 \mathbf{h}_{\rm sub}^T \tilde{\mathbf{R}}_b \mathbf{h}_{\rm sub} \\ &= J_s(\mathbf{h}_{\rm sub}), \end{aligned} \tag{10.108}$$

it follows immediately that the speech-distortion index and the noise-reduction factor due to $\mathbf{h}_{\rm sub}$ are,

$$\upsilon_{\rm sd}(\mathbf{h}_{\rm sub}) = \frac{1}{\rm SNR}\left[\tilde{J}_s(\mathbf{h}_{\rm sub}) - \frac{1}{\xi_{\rm nr}(\mathbf{h}_{\rm sub})}\right], \tag{10.109}$$

$$\xi_{\rm nr}(\mathbf{h}_{\rm sub}) = \frac{\sigma_b^2}{\mathbf{h}_{\rm sub}^T \mathbf{R}_b \mathbf{h}_{\rm sub}} = \frac{1}{{\rm SNR}\left[\tilde{J}_b(\mathbf{g}_{\rm sub}) - \upsilon_{\rm sd}(\mathbf{g}_{\rm sub})\right]}. \tag{10.110}$$

From (10.104), one can see that $\upsilon_{\rm sd}(\mathbf{h}_{\rm sub})/\upsilon_{\rm sd}(\mathbf{h}_{\rm o}) = \alpha^2$, which is a function of α only. Unlike $\upsilon_{\rm sd}(\mathbf{h}_{\rm sub})/\upsilon_{\rm sd}(\mathbf{h}_{\rm o})$, $\xi_{\rm nr}(\mathbf{h}_{\rm sub})/\xi_{\rm nr}(\mathbf{h}_{\rm o})$ depends not only on α, but also on the characteristics of both the speech and noise signals.

However, using (10.99) and (10.41), we find that,

$$\frac{\zeta_{\rm nr}(\mathbf{h}_{\rm sub})}{\zeta_{\rm nr}(\mathbf{h}_{\rm o})} = \frac{1 - \tilde{J}_s(\mathbf{h}_{\rm sub})}{1 - \tilde{J}_s(\mathbf{h}_{\rm o})} = \alpha(2-\alpha). \tag{10.111}$$

Figure 10.2 plots $\upsilon_{\rm sd}(\mathbf{h}_{\rm sub})/\upsilon_{\rm sd}(\mathbf{h}_{\rm o})$ and $\zeta_{\rm nr}(\mathbf{h}_{\rm sub})/\zeta_{\rm nr}(\mathbf{h}_{\rm o})$, both as a function of α. We can see that when $\alpha = 0.7$, the sub-optimal filter achieves 91% of the noise reduction with the Wiener filter, while the speech distortion is only 49% of that of the Wiener filter. In real applications, we may want the system to achieve maximal noise reduction, while keeping the speech distortion as low as possible. If we define a cost function to measure the compromise between the noise reduction and the speech distortion as

$$\begin{aligned} J_{\zeta v}(\alpha) &\triangleq \frac{\zeta_{\mathrm{nr}}(\mathbf{h}_{\mathrm{sub}})}{\zeta_{\mathrm{nr}}(\mathbf{h}_{\mathrm{o}})} - \frac{\upsilon_{\mathrm{sd}}(\mathbf{h}_{\mathrm{sub}})}{\upsilon_{\mathrm{sd}}(\mathbf{h}_{\mathrm{o}})} \\ &= 2\alpha - 2\alpha^2. \end{aligned} \tag{10.112}$$

It is trivial to see that the α that maximizes $J_{\zeta v}(\alpha)$ is

$$\alpha_{\mathrm{o}} = \arg\max_{\alpha} J_{\zeta v}(\alpha) = \frac{1}{2}. \tag{10.113}$$

In this case, the suboptimal filter achieves 75% of the noise reduction with the Wiener filter, while the speech-distortion is only 25% of that of the Wiener filter.

Another way to obtain an optimal α is to define a discriminative cost function between $\xi_{\mathrm{nr}}(\mathbf{h}_{\mathrm{sub}})/\xi_{\mathrm{nr}}(\mathbf{h}_{\mathrm{o}})$ and $\upsilon_{\mathrm{sd}}(\mathbf{h}_{\mathrm{sub}})/\upsilon_{\mathrm{sd}}(\mathbf{h}_{\mathrm{o}})$, i.e.,

$$\begin{aligned} J_{\xi v}(\alpha) &\triangleq \frac{\xi_{\mathrm{nr}}(\mathbf{h}_{\mathrm{sub}})}{\xi_{\mathrm{nr}}(\mathbf{h}_{\mathrm{o}})} - \beta\frac{\upsilon_{\mathrm{sd}}(\mathbf{h}_{\mathrm{sub}})}{\upsilon_{\mathrm{sd}}(\mathbf{h}_{\mathrm{o}})} \\ &= \frac{(\mathbf{u}_1 - \mathbf{g}_{\mathrm{o}})^T\mathbf{R}_b(\mathbf{u}_1 - \mathbf{g}_{\mathrm{o}})}{(\mathbf{u}_1 - \alpha\mathbf{g}_{\mathrm{o}})^T\mathbf{R}_b(\mathbf{u}_1 - \alpha\mathbf{g}_{\mathrm{o}})} - \beta\alpha^2 \\ &= \frac{\sigma_b^2 + \mathbf{g}_{\mathrm{o}}^T\mathbf{R}_b\mathbf{g}_{\mathrm{o}} - 2\mathbf{r}_{bb}^T\mathbf{g}_{\mathrm{o}}}{\sigma_b^2 + \alpha^2\mathbf{g}_{\mathrm{o}}^T\mathbf{R}_b\mathbf{g}_{\mathrm{o}} - 2\alpha\mathbf{r}_{bb}^T\mathbf{g}_{\mathrm{o}}} - \beta\alpha^2, \end{aligned} \tag{10.114}$$

where β is an application-dependent constant and determines the relative importance between the improvement in speech distortion and degradation in noise reduction (e.g., in hearing aid applications we may tune this parameter using subjective intelligibility tests).

In contrast to $J_{\zeta v}(\alpha)$, which is a function of α only, the cost function $J_{\xi v}(\alpha)$ depends not only on α, but on the characteristics of the speech and noise signals as well. Figure 10.3 plots $J_{\xi v}(\alpha)$ as a function of α in different SNR conditions, where both the signal and the noise are assumed to be Gaussian random processes and $\beta = 0.7$. It shows that for the same α, $J_{\xi v}(\alpha)$ decreases with SNR, indicating that the higher the SNR, the better the suboptimal filter is able to control the compromise between noise reduction and speech distortion.

In order for the suboptimal filter to be able to control the tradeoff between noise reduction and speech distortion, α should be chosen in such a way that $\xi_{\mathrm{nr}}(\mathbf{h}_{\mathrm{sub}})/\xi_{\mathrm{nr}}(\mathbf{h}_{\mathrm{o}}) > \upsilon_{\mathrm{sd}}(\mathbf{h}_{\mathrm{sub}})/\upsilon_{\mathrm{sd}}(\mathbf{h}_{\mathrm{o}})$. Therefore $J_{\xi v}(\alpha)$ should satisfy $J_{\xi v}(\alpha) > 0$. From Fig. 10.3, we notice that $J_{\xi v}(\alpha)$ is always positive if the SNR is above 0 dB. When the SNR drops below 0 dB, however, $J_{\xi v}(\alpha)$ may become negative, indicating that the suboptimal filter cannot work reliably in very noisy conditions (when SNR < 0 dB).

Figure 10.3 also shows the α_{o} that maximizes $J_{\xi v}(\alpha)$ in different SNR situations. It is interesting to see that the α_{o} approaches to 1 when SNR $<$

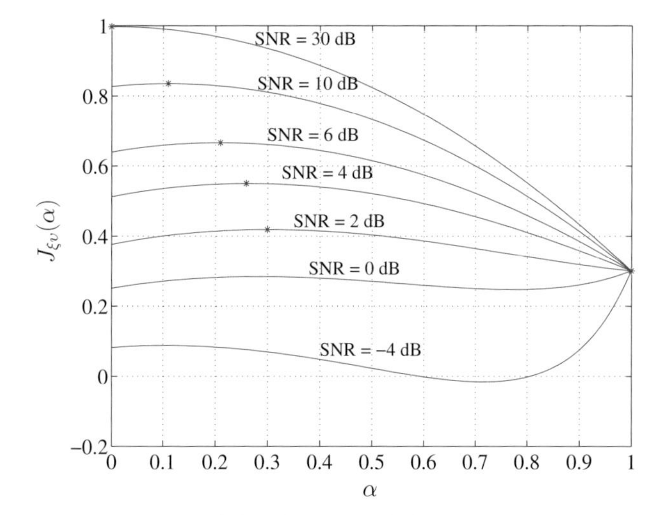

Fig. 10.3. Illustration of $J_{\xi v}(\alpha)$ in different SNR conditions, where both the signal and noise are assumed to be Gaussian random processes, and $\beta = 0.7$. The "$*$" symbol in each curve represents the maximum of $J_{\xi v}(\alpha)$ in the corresponding condition.

0 dB, which means that the suboptimal filter converges to the Wiener filter in very low SNR conditions. As we increase the SNR, the α_{o} begins to decrease. It goes to 0 when SNR is increased to 30 dB. This is understandable. When the SNR is very high, it means that the observation signal is already very clean, so filtering is not really needed. By searching the α_{o} that maximizes (10.114), the system can adaptively achieve the best tradeoff between noise reduction and speech distortion according to the characteristics of both the speech and noise signals.

10.5 Frequency-Domain Wiener Filter

We now consider Wiener filter in the frequency domain. Perhaps the simplest way to derive such a filter is to transform the time-domain Wiener filter given in (10.31) into the frequency domain using the so-called overlap-add technique. Of course, the resulting Wiener filter and its time-domain counterpart have exactly the same performance. More often, however, the frequency-domain Wiener filter is formulated from directly estimating the clean speech spectrum from the noisy speech spectrum, resulting in a subband noncausal filter.

10.5.1 Estimation of the Clean Speech Spectrum

Applying the discrete Fourier transform to both sides of (10.21), we have

$$\begin{aligned} X(j\omega_{k'}) &= \sum_{k=0}^{K-1} x(k)e^{-j\omega_{k'}k} \\ &= \sum_{k=0}^{K-1} s(k)e^{-j\omega_{k'}k} + \sum_{k=0}^{K-1} b(k)e^{-j\omega_{k'}k} \\ &= S(j\omega_{k'}) + B(j\omega_{k'}), \end{aligned} \tag{10.115}$$

where K is the DFT length and $\omega_{k'} = \frac{2\pi}{K}k'$ $(k' = 0, \cdots, K-1)$ is the angular frequency. Now suppose that we can achieve an estimate of $S(j\omega_{k'})$ through

$$\hat{S}(j\omega_{k'}) = H(j\omega_{k'})X(j\omega_{k'}). \tag{10.116}$$

The error signal between the clean speech spectrum and its estimate is written as

$$\mathcal{E}(j\omega_{k'}) = S(j\omega_{k'}) - H(j\omega_{k'})X(j\omega_{k'}), \tag{10.117}$$

where $H(j\omega_{k'})$ is a frequency-domain subband filter. We now can write the MSE criterion:

$$\begin{aligned} J_S\left[H(j\omega_{k'})\right] &= E\left[|\mathcal{E}(j\omega_{k'})|^2\right] \\ &= E\left[\mathcal{E}(j\omega_{k'})\mathcal{E}^*(j\omega_{k'})\right], \end{aligned} \tag{10.118}$$

where $(\cdot)^*$ stands for complex conjugate. The optimal estimate $\hat{S}_{\mathrm{o}}(j\omega_{k'})$ of the clean speech spectrum $S(j\omega_{k'})$ tends to contain less noise than the noisy speech spectrum $X(j\omega_{k'})$. The optimal filter that forms $\hat{S}_{\mathrm{o}}(j\omega_{k'})$ is the frequency-domain Wiener filter, which is obtained as [322]:

$$\begin{aligned} H_{\mathrm{o}}(j\omega_{k'}) &= \arg \min_{H(j\omega_{k'})} J_S\left[H(j\omega_{k'})\right] \\ &= \arg \min_{H(j\omega_{k'})} E\left\{|X(j\omega_{k'}) - H(j\omega_{k'})Y(j\omega_{k'})|^2\right\}. \end{aligned} \tag{10.119}$$

Taking the derivative of $J_S\left[H(j\omega_{k'})\right]$ with respect to $H(j\omega_{k'})$ and equating the result to zero, we can obtain the frequency-domain Wiener filter:

$$H_{\mathrm{o}}(j\omega_{k'}) = \frac{E[|S(j\omega_{k'})|^2]}{E[|X(j\omega_{k'})|^2]} = \frac{P_s(\omega_{k'})}{P_x(\omega_{k'})}, \tag{10.120}$$

where $P_s(\omega_{k'}) = \frac{1}{K}E\left[|S(j\omega_{k'})|^2\right]$ and $P_x(\omega_{k'}) = \frac{1}{K}E\left[|X(j\omega_{k'})|^2\right]$ are the power spectral densities (PSD) of $s(k)$ and $x(k)$ respectively. It can be seen from this expression that the frequency-domain Wiener filter $H_{\mathrm{o}}(j\omega_{k'})$ is real valued. Therefore, from now on we shall denote $H_{\mathrm{o}}(j\omega_{k'})$ as $H_{\mathrm{o}}(\omega_{k'})$ without introducing any confusion.

In order to compute the Wiener filter $H_{\mathrm{o}}(\omega_{k'})$, we need to know both $P_x(\omega_{k'})$ and $P_s(\omega_{k'})$. The former can be directly estimated from the noisy speech signal. However, $s(k)$ is unobservable, as a result, an estimation of

$P_s(\omega_{k'})$ may seem difficult to achieve. But since $s(k)$ and $b(k)$ are uncorrelated, we have

$$P_x(\omega_{k'}) = P_s(\omega_{k'}) + P_b(\omega_{k'}), \tag{10.121}$$

which leads to

$$P_s(\omega_{k'}) = P_x(\omega_{k'}) - P_b(\omega_{k'}). \tag{10.122}$$

This indicates that $P_s(\omega_{k'})$ can be computed from the PSDs of $P_x(\omega_{k'})$ and $P_b(\omega_{k'})$. As mentioned earlier, the former can be estimated during speech and noise periods, while the latter can be measured during noise-only intervals assuming that the statistics of the noise do not change much with time.

The optimal estimate of the clean speech, using $H_\text{o}(\omega_{k'})$, is

$$\begin{aligned} \hat{S}_\text{o}(j\omega_{k'}) &= H_\text{o}(\omega_{k'})X(j\omega_{k'}) \\ &= H_\text{o}(\omega_{k'})|X(j\omega_{k'})|e^{j\psi_X(\omega_{k'})} \\ &= \frac{P_s(\omega_{k'})}{P_x(\omega_{k'})}|X(\omega_{k'})|e^{j\psi_X(\omega_{k'})}, \end{aligned} \tag{10.123}$$

where $X(\omega_{k'})$ and $\psi_X(\omega_{k'})$ are, respectively, the magnitude and phase components of the complex speech spectrum $X(j\omega_{k'})$. Clearly, the Wiener filter estimates the clean speech spectrum by modifying the magnitude of the noisy speech spectrum, while leaving the phase component unchanged.

Comparing (10.120) with (10.31), one can readily see the difference between the frequency-domain and the time-domain Wiener filters. Briefly, the former operates on a subband basis, where each subband filter is independent on the filters obtained from other frequency bands, while the latter is a full-band technique. In addition, the time-domain Wiener filter is a causal filter, while the frequency-domain Wiener filter can be noncausal.

From the optimal estimate of the clean speech, it can easily be deduced that

$$P_{\hat{s}_\text{o}}(\omega_{k'}) = \frac{1}{K}E\left[|\hat{S}_\text{o}(j\omega_{k'})|^2\right] = \frac{P_s^2(\omega_{k'})}{P_x(\omega_{k'})}. \tag{10.124}$$

Since $P_b(\omega_{k'}) \geq 0$, we have $P_s(\omega_{k'}) \leq P_x(\omega_{k'})$. It follows immediately that

$$P_{\hat{s}_\text{o}}(\omega_{k'}) \leq P_x(\omega_{k'}), \tag{10.125}$$

which indicates that noise reduction is possible with the frequency-domain Wiener filter. In order, however, to check if it can really achieve noise reduction, it is necessary to assess the SNR improvement, which will be discussed in the next section.

10.5.2 *A Priori* SNR versus *a Posteriori* SNR

Given the PSDs of both the speech and noise signals, the *a priori* SNR can be expressed in the frequency domain as

$$\mathrm{SNR} = \frac{\sum_{k'=0}^{K-1} P_s(\omega_{k'})}{\sum_{k'=0}^{K-1} P_b(\omega_{k'})}. \tag{10.126}$$

Similarly, we can write, after the Wiener filtering, the *a posteriori* SNR according to (10.132), as

$$\mathrm{SNR_o} = \frac{\sum_{k'=0}^{K-1} \frac{P_s^2(\omega_{k'})}{P_x^2(\omega_{k'})} P_s(\omega_{k'})}{\sum_{k'=0}^{K-1} \frac{P_s^2(\omega_{k'})}{P_x^2(\omega_{k'})} P_b(\omega_{k'})}. \tag{10.127}$$

Now we give the following theorem.

Theorem: With the frequency-domain Wiener filter, the *a priori* SNR given in (10.126) and the *a posteriori* SNR defined in (10.127) satisfy

$$\mathrm{SNR_o} \geq \mathrm{SNR}. \tag{10.128}$$

Proof: Since $P_s(\omega_{k'}) \geq 0$, $P_b(\omega_{k'}) \geq 0$, and $P_x(\omega_{k'}) \geq 0$, the theorem follows immediately from the following inequality:

$$\begin{aligned} &\left\{ \sum_{k'=0}^{K-1} \frac{P_s^2(\omega_{k'})}{P_x^2(\omega_{k'})} P_s(\omega_{k'}) \right\} \left\{ \sum_{k'=0}^{K-1} P_b(\omega_{k'}) \right\} \geq \\ &\left\{ \sum_{k'=0}^{K-1} \frac{P_s^2(\omega_{k'})}{P_x^2(\omega_{k'})} P_b(\omega_{k'}) \right\} \left\{ \sum_{k'=0}^{K-1} P_s(\omega_{k'}) \right\}. \end{aligned} \tag{10.129}$$

Rewrite (10.129) into the following form:

$$\begin{aligned} &\left\{ \sum_{k'=0}^{K-1} \frac{P_s^2(\omega_{k'})}{P_x^2(\omega_{k'})} P_s(\omega_{k'}) \right\} \left\{ \sum_{k'=0}^{K-1} P_b(\omega_{k'}) \right\} - \\ &\left\{ \sum_{k'=0}^{K-1} \frac{P_s^2(\omega_{k'})}{P_x^2(\omega_{k'})} P_b(\omega_{k'}) \right\} \left\{ \sum_{k'=0}^{K-1} P_s(\omega_{k'}) \right\} \geq 0. \end{aligned} \tag{10.130}$$

The left-hand side of (10.130) can be developed as

$$\sum_{k'=0}^{K-1}\sum_{j=0}^{K-1}\frac{P_s^3(\omega_{k'})}{P_x^2(\omega_{k'})}P_b(\omega_j)-\sum_{k'=0}^{K-1}\sum_{j=0}^{K-1}\frac{P_s^2(\omega_{k'})}{P_x^2(\omega_{k'})}P_b(\omega_{k'})P_s(\omega_j)$$

$$=\sum_{k'=0}^{K-1}\sum_{j=0}^{K-1}\frac{P_s^2(\omega_{k'})}{P_x^2(\omega_{k'})}\left[P_s(\omega_{k'})P_b(\omega_j)-P_b(\omega_{k'})P_s(\omega_j)\right]$$

$$=\sum_{k'=0}^{K-1}\sum_{j>k'}^{K-1}\left[\frac{P_s^2(\omega_{k'})}{P_x^2(\omega_{k'})}-\frac{P_s^2(\omega_j)}{P_x^2(\omega_j)}\right]\left[P_s(\omega_{k'})P_b(\omega_j)-P_b(\omega_{k'})P_s(\omega_j)\right].$$

Since $s(k)$ and $b(k)$ are uncorrelated, we have $P_x(\omega_{k'}) = P_s(\omega_{k'}) + P_b(\omega_j)$. Therefore, the above expression can further be deduced as,

$$\sum_{k'=0}^{K-1}\sum_{j>k'}^{K-1}\left[\frac{P_s(\omega_{k'})}{P_x(\omega_{k'})}+\frac{P_s(\omega_j)}{P_x(\omega_j)}\right]\cdot$$

$$\left[\frac{P_s(\omega_{k'})}{P_x(\omega_{k'})}-\frac{P_s(\omega_j)}{P_x(\omega_j)}\right]\left[P_s(\omega_{k'})P_b(\omega_j)-P_b(\omega_{k'})P_s(\omega_j)\right]$$

$$=\sum_{k'=0}^{K-1}\sum_{j>k'}^{K-1}\frac{1}{P_x(\omega_{k'})P_x(\omega_j)}\left[\frac{P_s(\omega_{k'})}{P_x(\omega_{k'})}+\frac{P_s(\omega_j)}{P_x(\omega_j)}\right]\cdot$$

$$\left[P_s(\omega_{k'})P_b(\omega_j)-P_b(\omega_{k'})P_s(\omega_j)\right]^2. \quad (10.131)$$

Obviously, the right-hand side of (10.131) is greater than, or at least equal to 0. Therefore, (10.129) holds and $\text{SNR}_\text{o} \geq \text{SNR}$, where we see from (10.131) that "=" holds when $P_s(\omega_{k'}) = 0$, $k' = 0, \cdots, K-1$, or $\frac{P_s(\omega_0)}{P_b(\omega_0)} = \frac{P_s(\omega_1)}{P_b(\omega_1)} = \cdots = \frac{P_s(\omega_{K-1})}{P_b(\omega_{K-1})}$. In other words, the frequency-domain Wiener filter will be able to increase SNR unless there is no signal or all the subband SNRs are equal. It is easy to understand for the former case. In the latter situation, when all the subband SNRs are equal, it means that both speech and noise have the same PSD under the given resolution condition. In this case, the Wiener filter is not able to distinguish noise from speech, and hence not able to increase SNR. Otherwise, the subband Wiener filter can improve SNR. That completes the proof.

The power of the estimated clean speech (which is equal to $E[s_o^2(n)]$ according to the Parseval's relation) can be computed as:

$$\sigma_{\hat{S}_\text{o}}^2 = \sum_{k'=0}^{K-1}\frac{1}{K}E\left[|\hat{S}_\text{o}(j\omega_{k'})|^2\right] = \sum_{k'=0}^{K-1}\frac{P_s^2(\omega_{k'})}{P_x^2(\omega_{k'})}P_x(\omega_{k'})$$

$$= \sum_{k'=0}^{K-1}\frac{P_s^2(\omega_{k'})}{P_x^2(\omega_{k'})}P_s(\omega_{k'}) + \sum_{k'=0}^{K-1}\frac{P_s^2(\omega_{k'})}{P_x^2(\omega_{k'})}P_b(\omega_{k'}), \quad (10.132)$$

which is the sum of two terms. The first one is the power of the attenuated clean speech and the second one is the power of the residual noise, which is always greater than zero. Just like the time-domain one, the frequency-domain Wiener filter also achieves noise reduction at a price of distorting speech signal. The more the noise reduction, the more will be the speech distortion. The concrete relationship between noise reduction and speech distortion can easily be derived by following the same analysis procedure shown in Sect. 10.4, which will be left to the reader's investigation.

10.6 Noise Reduction Through Spectral Magnitude Restoration

From the previous analysis, we see that the frequency-domain Wiener filter estimates the clean speech spectrum by filtering the magnitude component of the noisy speech spectrum, while leaving the phase component unchanged. A key question then arises: is it sufficient for noise reduction if we only restore the spectral magnitudes, or can noise reduction performance be augmented by recovering the signal's phase? To answer this question, let us reformulate the noise-reduction problem in a slightly different way. Consider the decomposition of a complex spectrum into magnitude and phase components, i.e.,

$$\begin{aligned} X(j\omega_{k'}) &= X(\omega_{k'})e^{j\psi_X(\omega_{k'})}, \\ S(j\omega_{k'}) &= S(\omega_{k'})e^{j\psi_S(\omega_{k'})}. \end{aligned} \tag{10.133}$$

The problem of noise reduction then becomes to design two signal estimators that make decisions separately on the magnitude and phase spectra from the observed signal.

Now let us estimate the phase component of the clean speech spectrum from the noisy speech spectrum. Ephraim and Malah formulated in [106] an MMSE phase estimator, i.e.,

$$\hat{\psi}_S(\omega_{k'})|_{\text{MMSE}} = \arg\min_{\hat{\psi}_S(\omega_{k'})} E\left\{ \left| e^{j\psi_S(\omega_{k'})} - e^{j\hat{\psi}_S(\omega_{k'})} \right|^2 \,\middle|\, X(j\omega_{k'}) \right\}. \tag{10.134}$$

It turns out that the solution of this MMSE estimator is

$$\hat{\psi}_S(\omega_{k'})|_{\text{MMSE}} = \psi_X(\omega_{k'}). \tag{10.135}$$

That is, the noisy phase, $\psi_X(\omega_{k'})$ is an optimal estimate of the signal's phase in the MMSE sense.

Evidence from perception research shows that the human perception system is relatively insensitive to phase corruption. At least, within the context of noise reduction, it has been shown that speech distortion resulted from phase corruption is generally imperceptible when subband SNR at any $\omega_{k'}$ is greater than 6 dB [301]. Therefore, from both the signal processing and perception points of view, we see that the use of noisy phase as the signal's phase is good enough for speech enhancement in most applications. As a result, the single-channel noise reduction problem is generally formulated as to acquire an estimate of the magnitude of the clean speech spectrum, $\hat{S}(\omega_{k'})$, based on the noisy spectrum $X(j\omega_{k'})$.

10.7 Spectral Subtraction

Just like the frequency-domain Wiener filter, spectral subtraction is another widely used approach to noise reduction through recovering the spectral magnitude of the clean speech.

10.7.1 Estimation of the Spectral Magnitude of the Clean Speech

Since

$$X(j\omega_{k'}) = S(j\omega_{k'}) + B(j\omega_{k'}), \tag{10.136}$$

it follows immediately that

$$\begin{aligned} |X(j\omega_{k'})|^2 &= |S(j\omega_{k'})|^2 + |B(j\omega_{k'})|^2 + 2|S(j\omega_{k'})||B(j\omega_{k'})| \cos\theta_{k'} \\ &= S^2(\omega_{k'}) + B^2(\omega_{k'}) + 2S(\omega_{k'})B(\omega_{k'}) \cos\theta_{k'}, \end{aligned} \tag{10.137}$$

where $S(\omega_{k'})$ and $B(\omega_{k'})$ denote the magnitudes of $S(j\omega_{k'})$ and $B(j\omega_{k'})$ respectively, and $\theta_{k'}$ is the phase difference between the speech and noise signals. Since the two random processes $s(k)$ and $b(k)$ are assumed to be uncorrelated, we have

$$E\left[X^2(\omega_{k'})\right] = E\left[S^2(\omega_{k'})\right] + E\left[B^2(\omega_{k'})\right]. \tag{10.138}$$

If we replace the expectation by the *instantaneous* value, the above equality does no longer hold with rigor. However, we can still approximately have

$$X^2(\omega_{k'}) \approx S^2(\omega_{k'}) + B^2(\omega_{k'}). \tag{10.139}$$

In case we can achieve an estimate of $B^2(\omega_{k'})$, the magnitude-square spectrum of the signal, $S^2(\omega_{k'})$, can be recovered by subtracting such an estimate from $X^2(\omega_{k'})$, i.e.,

$$\begin{aligned} \hat{S}^2(\omega_{k'}) &= X^2(\omega_{k'}) - \hat{B}^2(\omega_{k'}) \\ &= S^2(\omega_{k'}) + [B^2(\omega_{k'}) - \hat{B}^2(\omega_{k'})]. \end{aligned} \tag{10.140}$$

If $\hat{B}^2(\omega_{k'})$ is an unbiased estimate of $B^2(\omega_{k'})$, taking expectations of both sides of (10.140) yields

$$E\left[\hat{S}^2(\omega_{k'})\right] = E\left[S^2(\omega_{k'})\right]. \tag{10.141}$$

This indicates that the estimated magnitude-square spectrum converges in the mean to the true magnitude-square spectrum of the clean speech signal.

Given $\hat{S}^2(\omega_{k'})$, the magnitude spectrum of the speech signal is computed as

$$\begin{aligned}\hat{S}(\omega_{k'}) &= \sqrt{\hat{S}^2(\omega_{k'})} \\ &= \sqrt{X^2(\omega_{k'}) - \hat{B}^2(\omega_{k'})}.\end{aligned} \quad (10.142)$$

Combined with the phase of the noisy signal, an estimate of the spectrum of the clean speech is written as

$$\hat{S}(j\omega_{k'}) = \left[X^2(\omega_{k'}) - \hat{B}^2(\omega_{k'})\right]^{1/2} e^{j\psi_X(\omega_{k'})}. \quad (10.143)$$

This forms the basis of a popularly used noise reduction method called spectral subtraction [52]. A similar algorithm can be developed in the magnitude spectral domain. If the uncorrelation assumption holds, it can be shown that

$$X(\omega_{k'}) \approx S(\omega_{k'}) + B(\omega_{k'}), \quad (10.144)$$

where $X(\omega_{k'})$, $S(\omega_{k'})$, and $B(\omega_{k'})$ are, respectively, the magnitudes of the spectra $X(j\omega_{k'})$, $S(j\omega_{k'})$, and $B(j\omega_{k'})$. Therefore, the magnitude of the speech spectrum can be directly estimated by

$$\begin{aligned}\hat{S}(\omega_{k'}) &= X(\omega_{k'}) - \hat{B}(\omega_{k'}) \\ &= S(\omega_{k'}) + [B(\omega_{k'}) - \hat{B}(\omega_{k'})].\end{aligned} \quad (10.145)$$

Again, as long as $\hat{B}(\omega_{k'})$ is an unbiased estimate of $B(\omega_{k'})$, we can check that $\hat{S}(\omega_{k'})$ converges in the mean to $S(\omega_{k'})$.

In a more general form, (10.140) and (10.145) can be expressed as

$$\hat{S}^p(\omega_{k'}) = X^p(\omega_{k'}) - \eta\hat{B}^p(\omega_{k'}), \quad (10.146)$$

where p is an exponent and η is a parameter introduced to control the amount of noise to be subtracted. For full subtraction, we set $\eta = 1$ and, for over subtraction we set $\eta > 1$. Consequently, the estimate of the clean speech spectrum can be constructed as

$$\hat{S}(j\omega_{k'}) = \left[X^p(\omega_{k'}) - \eta\hat{B}^p(\omega_{k'})\right]^{1/p} e^{j\psi_X(\omega_{k'})}. \quad (10.147)$$

This is often referred to as parametric spectral subtraction [112]. The magnitude-square spectral subtraction results from $p = 2$ and $\eta = 1$, and the magnitude spectral subtraction is derived from $p = 1$ and $\eta = 1$.

The technique of spectral subtraction is developed to recover the magnitude of a speech spectrum. The estimate should therefore be non-negative. However, (10.140) and (10.145), and more generally (10.146), may lead to negative estimates. This is one of the major drawbacks of spectral subtraction. A non-linear rectification process is often used to map a negative estimate into a non-negative value. But, this process introduces additional distortion into the recovered signal, which becomes more noticeable when the SNR is low.

10.7.2 Estimation of the Noise Spectrum

A paramount issue in spectral subtraction is how to obtain a good noise estimate; its accuracy greatly affects the noise reduction performance. There has been a tremendous effort in tackling this problem. Representative approaches include estimating noise during the absence of speech, minimum statistics method, quantile-based method, sequential estimation using single-pole recursion, etc.

Usually, a noisy speech signal is not occupied by speech all the time. For a large percentage of the time, it is exclusively occupied by noise alone. Therefore, if a voice activity detector (VAD) can be designed to distinguish speech and non-speech segments for a given noisy signal, the noise can then be estimated from regions where the speech signal is absent. Apparently, this basic noise-estimation technique relies on a VAD with high detection accuracy. When noise is strong and the SNR becomes rather low, the distinction of speech and noise segments could be difficult. Moreover, the noise is estimated intermittently and updated only during the speech silent periods. This may cause problems if the noise is non-stationary, which is the case in many applications.

To avoid explicit speech/non-speech detection, Martin proposed to estimate the noise via minimum statistics [219]. This technique is based on the assumption that during a speech pause, or within brief periods between words and even syllables, the speech energy is close to zero. As a result, a short-term power spectrum estimate of the noisy signal, even during speech activity, decays frequently to the noise power. Thus, by tracking the temporal spectral minimum without distinguishing between the presence and absence of speech, the noise power in a specific frequency band can be estimated. Although a VAD is not necessary in this approach, the noise estimate is often too small to provide sufficient noise reduction.

Instead of using minimum statistics, Hirsch *et al.* proposed a histogram based method which achieves a noise estimate from sub-band energy histograms [165]. A threshold is set over which peaks in the histogram profile are attributed to speech. The highest peak in the profile below this threshold is treated as noise energy. Stahl *et al.* extended this idea to a quantile-based noise estimation approach [285], which works on the assumption that even in active speech sections of the input signal not all frequency bands are permanently occupied with speech, and for a large percentage of the time the energy is at the noise level. Thus, this method computes short-term power spectra and sorts them. The noise estimate is obtained by taking a value near to the median of the resulting profiles. Evans *et al.* compared the histogram and the quantile-based noise estimation approaches for noisy speech recognition [113]. The conclusion was in favor of the latter.

More generally, noise can be estimated sequentially using a single-pole recursive average with an implicit speech/non-speech decision embedded. The noisy signal $x(k)$ is segmented into blocks of N_{b} samples. Each block is then transformed via a DFT into a block of N_b spectral samples. Successive blocks of spectral samples form a two-dimensional time-frequency matrix denoted by $X_t(j\omega_{k'})$, where subscript t is the frame index and denotes the time dimension. Then the sequential noise estimation is formulated as

$$\hat{B}_t^p(\omega_{k'}) = \begin{cases} \alpha_\mathrm{a}\hat{B}_{t-1}^p(\omega_{k'}) + (1-\alpha_\mathrm{a})X_t^p(\omega_{k'}), & \text{if } X_t^p(\omega_{k'}) \geq \hat{B}_{t-1}^p(\omega_{k'}) \\ \alpha_\mathrm{d}\hat{B}_{t-1}^p(\omega_{k'}) + (1-\alpha_\mathrm{d})X_t^p(\omega_{k'}), & \text{if } X_t^p(\omega_{k'}) < \hat{B}_{t-1}^p(\omega_{k'}) \end{cases}, \tag{10.148}$$

where α_a is the "attack" coefficient and α_d is the "decay" coefficient. This method is attractive for its simplicity and efficiency. Some variations of this method can be found in [98], [302], [119], [73].

10.7.3 Relationship Between Spectral Subtraction and Wiener Filtering

In the context of noise reduction, the Wiener filtering and spectral subtraction techniques are closely related.

From (10.120), we know that the optimal Wiener filter is

$$\begin{aligned} H_\mathrm{o}(\omega_{k'}) &= \frac{E\left[S^2(\omega_{k'})\right]}{E\left[X^2(\omega_{k'})\right]} \\ &= \frac{E\left[X^2(\omega_{k'})\right] - E\left[B^2(\omega_{k'})\right]}{E\left[X^2(\omega_{k'})\right]}, \end{aligned} \tag{10.149}$$

and the speech spectrum estimated through the Wiener filter is:

$$\hat{S}(j\omega_{k'}) = H_\mathrm{o}(\omega_{k'})X(j\omega_{k'}). \tag{10.150}$$

Through some simple manipulation, the magnitude-square spectral subtraction, i.e., (10.143), can be rewritten as

$$\begin{aligned} \hat{S}(j\omega_{k'}) &= \left[X^2(\omega_{k'}) - \hat{B}^2(\omega_{k'})\right]^{1/2} e^{j\psi_X(\omega_{k'})} \\ &= \left[\frac{X^2(\omega_{k'}) - \hat{B}^2(\omega_{k'})}{X^2(\omega_{k'})}\right]^{1/2} X(j\omega_{k'}). \end{aligned} \tag{10.151}$$

Denoting

$$H_\mathrm{MS}(\omega_{k'}) = \left[\frac{X^2(\omega_{k'}) - \hat{B}^2(\omega_{k'})}{X^2(\omega_{k'})}\right]^{1/2}, \tag{10.152}$$

we can express (10.151) as

$$\hat{S}(j\omega_{k'}) = H_\mathrm{MS}(\omega_{k'})X(j\omega_{k'}). \tag{10.153}$$

Apparently, (10.153) is akin to the Wiener filter shown in (10.150). This shows the close relation between the magnitude-square spectral subtraction and the optimal Wiener filter.

Similarly, the magnitude spectral subtraction can be reformulated as

$$\hat{S}(j\omega_{k'}) = H_{\mathrm{M}}(\omega_{k'})X(j\omega_{k'}), \tag{10.154}$$

where

$$H_{\mathrm{M}}(\omega_{k'}) = \frac{X(\omega_{k'}) - \hat{B}(\omega_{k'})}{X(\omega_{k'})}. \tag{10.155}$$

Both (10.151) and (10.155) can be unified into a general form referred to as the parametric Wiener filtering technique [214], i.e.,

$$\hat{S}(j\omega_{k'}) = H_{\mathrm{PW}}(\omega_{k'})X(j\omega_{k'}), \tag{10.156}$$

where

$$H_{\mathrm{PW}}(\omega_{k'}) = \left[\frac{X^p(\omega_{k'}) - \eta\hat{B}^p(\omega_{k'})}{X^p(\omega_{k'})}\right]^{1/q} \tag{10.157}$$

is called a gain filter, η, again, is a parameter introduced to control the amount of noise to be reduced. The magnitude-square spectral subtraction is derived from (10.157) by selecting $p = q = 2$; and the magnitude spectral subtraction results from $p = q = 1$.

Now let us evaluate the noise reduction performance for the parametric Wiener filtering technique. From (10.156), we can easily deduce the noise-reduction gain function:

$$\Psi_{\mathrm{NR}}(\omega_{k'}) = \frac{1}{H^2_{\mathrm{PW}}(\omega_{k'})}. \tag{10.158}$$

Under the ideal conditions that the speech and noise are uncorrelated and the noise estimate is equal to the true noise spectrum, (10.158) can be expressed as

$$\Psi_{\mathrm{NR}}(\omega_{k'}) = \left[\frac{\gamma^{\frac{p}{2}}(\omega_{k'}) + 1}{\gamma^{\frac{p}{2}}(\omega_{k'}) + 1 - \eta}\right]^{\frac{2}{q}}, \tag{10.159}$$

where $\gamma(\omega_{k'}) = S^2(\omega_{k'})/B^2(\omega_{k'})$ is the *a priori* SNR at frequency $\omega_{k'}$. Similarly, the attenuation frequency distortion due to the parametrical Wiener filter can be expressed as

$$\Upsilon_{\mathrm{sd}}(\omega_{k'}) = 1 - \left[\frac{\gamma^{\frac{p}{2}}(\omega_{k'}) + 1 - \eta}{\gamma^{\frac{p}{2}}(\omega_{k'}) + 1}\right]^{\frac{2}{q}}. \tag{10.160}$$

Figure 10.4 plots the noise-reduction performance of the parametric Wiener filter as a function of $\gamma(\omega_{k'})$. As can be seen, the three filters achieve

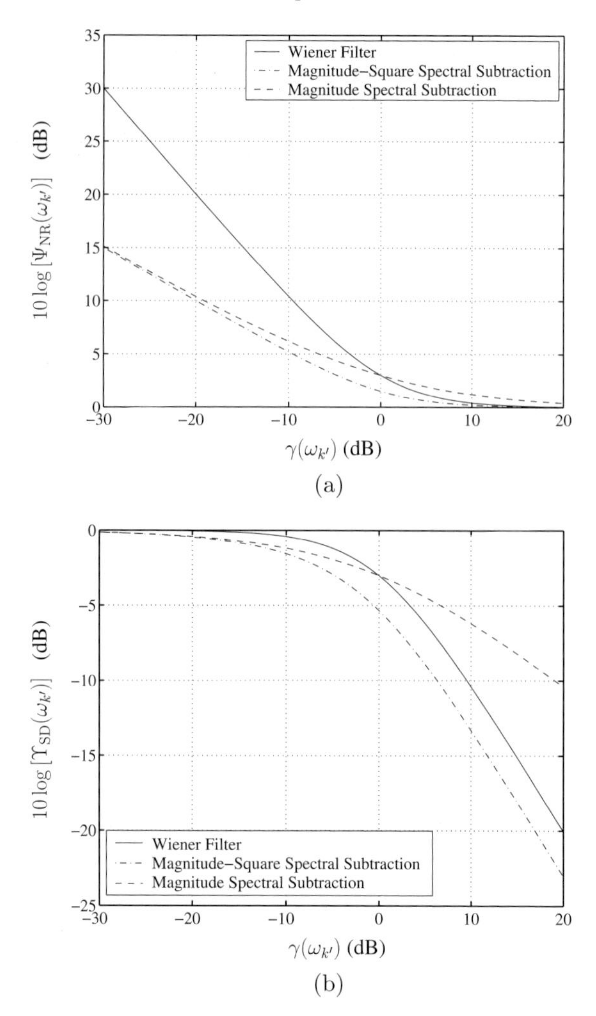

Fig. 10.4. Noise-reduction performance of the parametrical Wiener filtering techniques when $\eta = 1$: (a) noise-reduction gain function, and (b) attenuation frequency distortion.

similar noise-reduction gain for high SNRs. In low SNR conditions, the Wiener filter can lead to more noise attenuation. Notice that when SNR is above 0 dB, the three filters provide noise reduction of no more than 4 dB. If more noise reduction is needed, this could be done by setting the parameter η greater than 1. It can also be seen that not much speech distortion is introduced when SNR is high. But the amount of speech distortion increases as the *a priori* SNR decreases. Again, if we want to control speech distortion, we can do so by setting an appropriate value for η.

10.7.4 Estimation of the Wiener Gain Filter

As noise estimation is the central issue in spectral subtraction, the gain filter computation is the foremost problem for the parametric Wiener filtering method. If we define the subband speech plus noise to noise ratio (SNNR) as

$$\varrho(\omega_{k'}) = X(\omega_{k'})/B(\omega_{k'}), \tag{10.161}$$

the parametric Wiener filter given in (10.157) can be recast as

$$H_{\mathrm{PW}}(\omega_{k'}) = \left[1 - \eta \frac{1}{\varrho^{p}(\omega_{k'})}\right]^{\frac{1}{q}}. \tag{10.162}$$

The above expression shows that the parametric Wiener filter is a function of the SSNR $\varrho(\omega_{k'})$. Therefore, the estimation of the parametric Wiener filter is essentially a matter of estimating the SNNR. Note from (10.161) that $1 \leq \varrho(\omega_{k'}) < \infty$, therefore $H_{\mathrm{PW}}(\omega_{k'})$ (for $\eta = 1$) should be between 0 and 1. Any estimate greater than 1 should be mapped to 1 and any estimate less than 0 should be mapped to 0.

We have previously discussed the estimation of the noise spectrum. Substituting a noise estimate, (10.148) for instance, into (10.161), one can achieve an estimate of the SNNR $\varrho(\omega_{k'})$. However, such an estimate of SNNR fluctuates dramatically due to the large variance of the DFT spectrum [302]. Two approaches can be employed to reduce the fluctuation, namely, time-averaging and frequency-averaging the DFT spectra before computing SNNR. If we denote by $X_t(\omega_{k'})$ the short-time magnitude spectrum of the noisy signal at frame index t, the time-averaging can be implemented using a single-pole recursion,

$$\bar{X}_t(\omega_{k'}) = \beta \bar{X}_{t-1}(\omega_{k'}) + (1-\beta) X_t(\omega_{k'}), \tag{10.163}$$

where β is a parameter to control the time constant. An even better smoothing effect can be achieved by a two-sided single-pole recursion,

$$\bar{X}_t(\omega_{k'}) = \begin{cases} \beta_{\mathrm{a}} \bar{X}_{t-1}(\omega_{k'}) + (1-\beta_{\mathrm{a}}) X_t(\omega_{k'}), & \text{if } X_t(\omega_{k'}) \geq \bar{X}_{t-1}(\omega_{k'}) \\ \beta_{\mathrm{d}} \bar{X}_{t-1}(\omega_{k'}) + (1-\beta_{\mathrm{d}}) X_t(\omega_{k'}), & \text{if } X_t(\omega_{k'}) < \bar{X}_{t-1}(\omega_{k'}) \end{cases}, \tag{10.164}$$

where again β_{a} is the "attack" coefficient and β_{d} is the "decay" coefficient. Combining (10.148) and (10.164), we have an estimate of the *narrow-band* SNNR at time instance t as

$$\varrho_t^{\mathrm{N}}(\omega_{k'}) = \frac{\bar{X}_t(\omega_{k'})}{\hat{B}_t(\omega_{k'})}. \tag{10.165}$$

Further reduction of the fluctuation of SNNR can be achieved by frequency-averaging the smoothed spectrum $\bar{X}_t(\omega_{k'})$, i.e.,

$$\bar{X}_t^{\mathrm{W}}(\omega_{k'}) = \frac{\sum_{j=k-N_{\mathrm{w}}/2}^{k+N_{\mathrm{w}}/2} w_j \bar{X}_t(\omega_j)}{\sum_{j=k-N_{\mathrm{w}}/2}^{k+N_{\mathrm{w}}/2} w_j}, \tag{10.166}$$

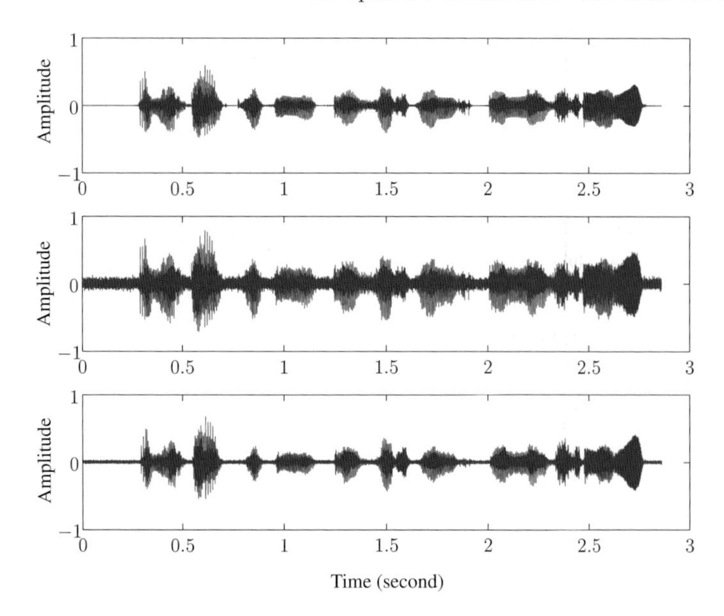

Fig. 10.5. Performance of a speech enhancement system based on the parametric Wiener filtering technique. The upper trace shows the waveform of the original clean speech signal. The middle trace plots the noisy signal. The lower trace shows the waveform after speech enhancement.

where $\{w_j\}$ defines a window, and N_{w} defines the width of the window. Based on this estimate, we can introduce a *wide-band* SNNR as

$$\varrho_t^{\mathrm{W}}(\omega_{k'}) = \frac{\bar{X}_t^{\mathrm{W}}(\omega_{k'})}{\hat{B}_t(\omega_{k'})}. \tag{10.167}$$

The final estimate of the SNNR is determined as

$$\varrho_t(\omega_{k'}) = \max[\varrho_t^{\mathrm{N}}(\omega_{k'}), \varrho_t^{\mathrm{W}}(\omega_{k'})]. \tag{10.168}$$

By doing so, the estimated SNNR approximates the true SNNR and has small fluctuation in various noise conditions.

10.7.5 Simulations

Figures 10.5 and 10.6 show the noise reduction performance for a parametric Wiener filtering whose gain is computed through (10.162) and (10.168). For this experiment, $p = q = 2$ and $\eta = 1.15$. Signal is digitized with a sampling frequency of 16 kHz. A white Gaussian noise is scaled and added to the signal so that the SNR is equal to 10 dB. Figure 10.5 shows the waveforms of the clean signal, the noise-corrupted signal, and the enhanced signal. Figure 10.6

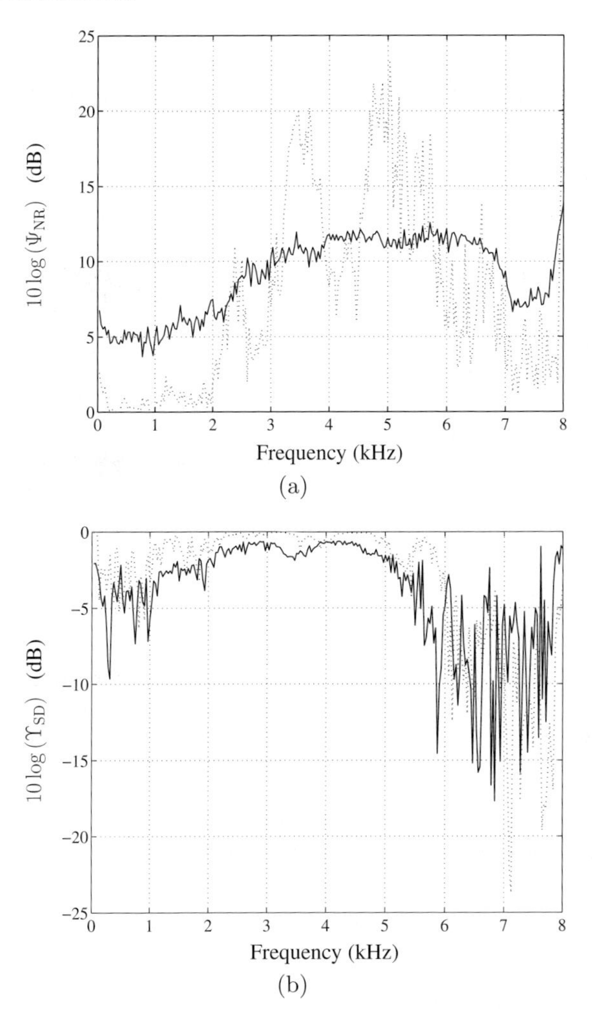

Fig. 10.6. Actual noise reduction performance of the optimal Wiener filter (dotted line) and a speech enhancement system based on the parametric Wiener filter (solid line): (a) noise-reduction gain, (b) attenuation frequency distortion.

gives the noise reduction performance. For comparison, the performance of the optimal Wiener filter is also presented. As can be seen, both the the parametric and the optimal Wiener filters lead to similar speech distortion. while the optimal filter gives more noise reduction in the frequency range between 3 to 6 kHz, the implemented parametric Wiener filter reduces more noise otherwise.

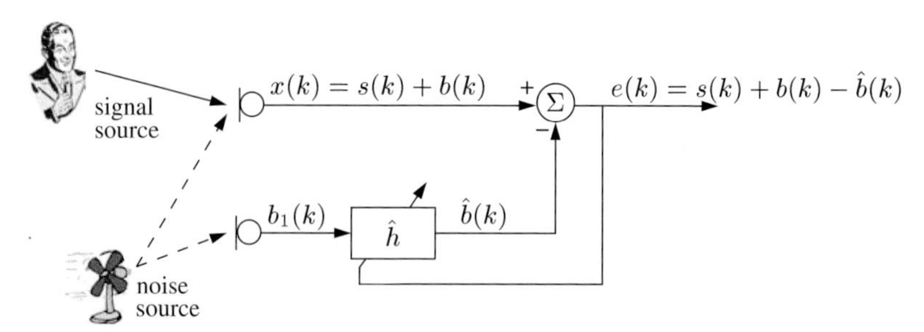

Fig. 10.7. Model for adaptive noise cancellation with primary and auxiliary inputs.

10.8 Adaptive Noise Cancellation

In the previous sections we have attempted to recover clean speech from its noise corrupted realization observed at a single microphone. With almost no exception, the algorithms introduce some speech distortion while achieving noise reduction. In order to control the amount of speech distortion, we either need some *a priori* knowledge about the speech signal or have to trade off the amount of noise reduction. We now depart from this philosophy and investigate a class of algorithms using two or more sensors where noise is eliminated by adaptively recreating a noise replica using a reference signal and then subtracting such a noise replica from the corrupted signal. The reference signal is derived from one or more sensors located at points in the noise field where the signal is weak or undetectable. This technique is often referred to as adaptive noise cancellation (ANC) [214], [320], [319], [147], [208], [3], [305].

10.8.1 Estimation of the Clean Speech

A typical ANC configuration is depicted in Fig. 10.7, where two microphone sensors are used, with the primary capturing the sum of speech signal $s(k)$ and unwanted noise $b(k)$, and the auxiliary measuring the noise signal $b_1(k)$. To cancel the noise in the primary input, we need a noise estimate $\hat{b}(k)$, usually generated by passing the auxiliary input through an FIR filter

$$\hat{\mathbf{h}} = [\hat{h}_0, \hat{h}_1, \cdots, \hat{h}_{L-1}]^T,$$

of length L, whose coefficients are estimated adaptively. We have:

$$\hat{b}(k) = \hat{\mathbf{h}}^T \mathbf{b}_1(k), \tag{10.169}$$

where

$$\mathbf{b}_1(k) = [b_1(k), b_1(k-1), \cdots, b_1(k-L+1)]^T.$$

Subtracting $\hat{b}(k)$ from the primary signal yields the error signal,

$$e(k) = x(k) - \hat{b}(k) = s(k) + [b(k) - \hat{b}(k)]. \quad (10.170)$$

It follows immediately that

$$e^2(k) = s^2(k) + [b(k) - \hat{b}(k)]^2 + 2s(k)[b(k) - \hat{b}(k)], \quad (10.171)$$

and

$$E\{e^2(k)\} = E\{s^2(k)\} + E\left\{[b(k) - \hat{b}(k)]^2\right\} + 2E\left\{s(k)[b(k) - \hat{b}(k)]\right\}. \quad (10.172)$$

The goal of the ANC is to find an FIR filter that minimizes $E\{e^2(n)\}$. Three assumptions are made in the ANC:

- Noise [$b(k)$ and $b_1(k)$] is uncorrelated with the speech signal $s(k)$.
- $b(k)$ and $b_1(k)$ are at least partially coherent.
- The auxiliary microphone is well isolated from the speech source so that it does not pick up speech.

When the noise is stationary, the optimum filter $\hat{\mathbf{h}}_o$ can be determined as the Wiener solution, i.e.,

$$\hat{\mathbf{h}}_o = \mathbf{R}_{b_1}^{-1} \mathbf{r}_{b_1 x}, \quad (10.173)$$

where $\mathbf{R}_{b_1} = E\left[\mathbf{b}_1(k)\mathbf{b}_1^T(k)\right]$ is the correlation matrix of the noise signal $b_1(k)$ and $\mathbf{r}_{b_1 x} = E\left[\mathbf{b}_1(k)x(k)\right]$ is the cross-correlation vector between $b_1(k)$ and $x(k)$.

Very often, however, the filter is estimated sequentially through adaptive algorithms to account for the time varying noise statistics. Commonly used adaptive techniques include the steepest-descent-gradient (SDG), the least-mean-square (LMS), the normalized LMS (NLMS), the recursive-least-squares (RLS) algorithms, which have been discussed in Chap. 3. For example, using the SDG algorithm, we can update the filter coefficients as:

$$\begin{aligned} \hat{\mathbf{h}}(k) &= \hat{\mathbf{h}}(k-1) - \mu \bigtriangledown E\left[e^2(k)\right] \\ &= (1 - 2\mu \mathbf{R}_{b_1})\hat{\mathbf{h}}(k-1) + 2\mu \mathbf{r}_{xb_1}, \end{aligned} \quad (10.174)$$

where $\bigtriangledown$ is the gradient with respect to $\hat{\mathbf{h}}$, and μ is the step size which controls the rate of change.

If the FIR filter is sparse, which means that a large number of coefficients in $\hat{\mathbf{h}}$ is effectively zero (or close to zero), then the proportionate NLMS (PNLMS), the IPNLMS, and the exponentiated gradient (EG) algorithms may be used to accelerate the convergence rate. We refer the reader to Chap. 4 for more details.

10.8.2 Ideal Noise Cancellation Performance

Although different adaptive algorithms may lead to different degrees of noise reduction, it can be shown that when $s(k)$, $b(k)$, and $b_1(k)$ are stationary, all algorithms are bounded to the optimal Wiener solution, i.e., (10.173). We now consider the noise cancellation performance. For the ease of analysis, we assume that there is no signal components in the reference input and the adaptive algorithm has converged. In this case, with $\hat{H}_{\mathrm{o}}(j\omega)$, $B_1(j\omega)$, and $X(j\omega)$ representing, respectively, the Fourier transform of $\hat{\mathbf{h}}_{\mathrm{o}}$, $b_1(k)$, and $x(k)$, (10.173) can be expressed as

$$\hat{H}_{\mathrm{o}}(j\omega) = \frac{E\left[B_1^*(j\omega)X(j\omega)\right]}{E\left[|B_1(j\omega)|^2\right]}. \tag{10.175}$$

The spectrum of the output signal is written as

$$\begin{aligned}\mathcal{E}(j\omega) &= X(j\omega) - \hat{H}_{\mathrm{o}}(j\omega)B_1(j\omega) \\ &= S(j\omega) + B(j\omega) - \hat{H}_{\mathrm{o}}(j\omega)B_1(j\omega),\end{aligned} \tag{10.176}$$

which is the sum of three terms. The first one is the spectrum of the desired speech signal, and the rest are associated with the residual noise that remains in the noise-reduced output signal.

It is clear from (10.176) that the speech component in the output signal is the same as that in the primary input. So there is no speech distortion. If one lets the residual noise spectrum be denoted as

$$B_{\mathrm{NR}}(j\omega) = B(j\omega) - \hat{H}_{\mathrm{o}}(j\omega)B_1(j\omega), \tag{10.177}$$

then it can easily be derived that

$$\begin{aligned}E\left[|B_{\mathrm{NR}}(j\omega)|^2\right] &= E\left[|B(j\omega)|^2\right] - \left|\frac{E\left[B_1^*(j\omega)B(j\omega)\right]}{E\left[|B_1(j\omega)|^2\right]}\right|^2 E\left[|B_1(j\omega)|^2\right], \\ &= \left[1 - |\rho_{b_1 b}(j\omega)|^2\right] E\left[|B(j\omega)|^2\right],\end{aligned} \tag{10.178}$$

where

$$\rho_{b_1 b}(j\omega) = \frac{E[B_1^*(j\omega)B(j\omega)]}{\sqrt{E[|B_1(j\omega)|^2]E[|B(j\omega)|^2]}}$$

is the coherence function between $b_1(k)$ and $b(k)$.

The ratio between $E\left[|B(j\omega)|^2\right]$ and $E\left[|B_{\mathrm{NR}}(j\omega)|^2\right]$, according to (10.3), gives the noise-reduction gain, i.e.,

$$\Psi_{\mathrm{NR}}(\omega) = \frac{E\left[|B(j\omega)|^2\right]}{E\left[|B_{\mathrm{NR}}(j\omega)|^2\right]} = \frac{1}{1 - |\rho_{b_1 b}(j\omega)|^2}. \tag{10.179}$$

It can easily be checked that $\Psi_{\mathrm{NR}}(\omega)$ is a monotonously increasing function with respect to $|\rho_{b_1 b}(\omega)|$. A plot of this function is shown in Fig. 10.8. The

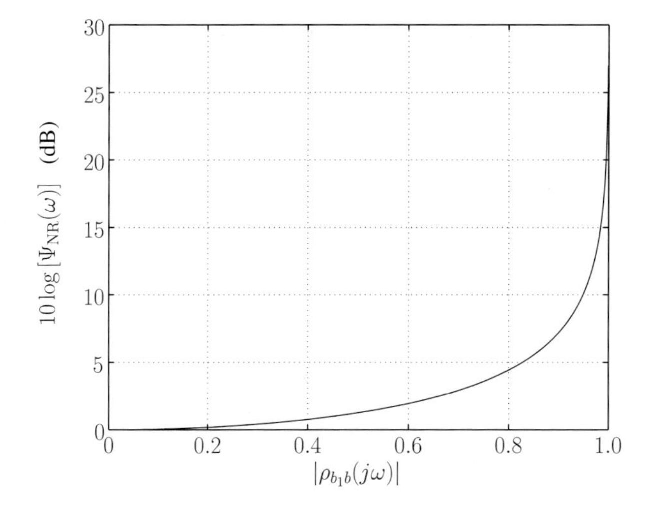

Fig. 10.8. Noise reduction performance of adaptive noise cancellation as a function of the noise coherence.

maximum noise cancellation is achieved when $|\rho_{b_1 b}(j\omega)| = 1$, namely $b_1(k)$ and $b(k)$ are coherent. If $b_1(k)$ and $b(k)$ are uncorrelated, i.e., $|\rho_{b_1 b}(j\omega)| = 0$, no noise reduction can be gained. From Fig. 10.8, it can be seen that a significant amount of correlation between noise in the primary microphone and that in the reference microphone (e.g., $|\rho_{b_1 b}(j\omega)| > 0.8$) is necessary for even a small amount (e.g., 5 dB) of noise reduction.

10.8.3 Signal Cancellation Problem

From the previous analysis, we see that in the ideal condition where the reference input consists of no speech components, the ANC technique will not introduce any speech distortion. In practice, however, the reference sensor may not be well isolated from the speech source and it may pick up some amount of speech signal. The presence of the speech signal in the reference input will cause some cancellation of the speech in the primary input. Now we examine the noise cancellation performance in this case.

Let us assume that the signal is leaked into the reference sensor through channel $\mathbf{g}$ so the spectrum of the reference input is $G(j\omega)S(j\omega) + B_1(\omega)$. In this case, the Wiener solution of the ANC system shown in Fig. 10.7 can be derived as

$$\begin{aligned} \hat{H}_o(j\omega) &= \frac{E\left\{[G(j\omega)S(j\omega) + B_1(j\omega)]^* X(j\omega)\right\}}{E\left\{|G(j\omega)S(j\omega) + B_1(j\omega)|^2\right\}} \\ &= \frac{G^*(j\omega)E\left[|S(j\omega)|^2\right] + E\left[B_1^*(j\omega)B(j\omega)\right]}{|G(j\omega)|^2 E\left[|S(j\omega)|^2\right] + E\left[|B_1(j\omega)|^2\right]}. \end{aligned} \quad (10.180)$$

The spectrum of the output signal can then be written as

$$\mathcal{E}(j\omega) = \hat{S}_{\mathrm{NR}}(j\omega) + \hat{B}_{\mathrm{NR}}(j\omega), \tag{10.181}$$

where

$$\hat{S}_{\mathrm{NR}}(j\omega) = \left[1 - \hat{H}_{\mathrm{o}}(j\omega)G(j\omega)\right] S(j\omega) \tag{10.182}$$

is the signal component, and

$$\hat{B}_{\mathrm{NR}}(j\omega) = B(j\omega) - \hat{H}_{\mathrm{o}}(j\omega)B_1(j\omega) \tag{10.183}$$

is the residual noise component. With some mathematical manipulation, we can derive

$$\begin{aligned} E\left[|\hat{S}_{\mathrm{NR}}(j\omega)|^2\right] &= \left|1 - \hat{H}_{\mathrm{o}}(j\omega)G(j\omega)\right|^2 E\left[|S(j\omega)|^2\right] \\ &= \left|\frac{E\left[|B_1(j\omega)|^2\right] - G(j\omega)E\left[B_1^*(j\omega)B(j\omega)\right]}{|G(j\omega)|^2 E\left[|S(j\omega)|^2\right] + E\left[|B_1(j\omega)|^2\right]}\right|^2 E\left[|S(j\omega)|^2\right], \end{aligned} \tag{10.184}$$

and

$$\begin{aligned} E\left[|\hat{B}_{\mathrm{NR}}(j\omega)|^2\right] &= E\left[|B(j\omega) - \hat{H}_{\mathrm{o}}(j\omega)B_1(j\omega)|^2\right] \\ &= \left|\frac{E\left[|B_1(j\omega)|^2\right] - G(j\omega)E\left[B_1^*(j\omega)B(j\omega)\right]}{|G(j\omega)|^2 E\left[|S(j\omega)|^2\right] + E\left[|B_1(j\omega)|^2\right]}\right|^2 \cdot \\ &\quad \frac{|G(j\omega)|^2 \left\{E\left[|S(j\omega)|^2\right]\right\}^2}{E\left[|B_1(j\omega)|^2\right]}. \end{aligned} \tag{10.185}$$

According to its definition presented in Sect. 10.2.3, the *a posteriori* SNR at the output can be written as follows,

$$\varrho(\omega_{k'}) = \frac{E\left[|\hat{S}_{\mathrm{NR}}(j\omega)|^2\right]}{E\left[|\hat{B}_{\mathrm{NR}}(j\omega)|^2\right]} = \frac{E\left[|B_1(j\omega)|^2\right]}{|G(j\omega)|^2 E\left[|S(j\omega)|^2\right]}. \tag{10.186}$$

Similarly, the (*a priori*) SNR at the reference sensor is given as

$$\gamma_{\mathrm{ref}}(\omega_{k'}) = \frac{|G(j\omega)|^2 E\left[|S(j\omega)|^2\right]}{E\left[|B_1(j\omega)|^2\right]}. \tag{10.187}$$

Therefore, we have

$$\varrho(\omega_{k'}) = \frac{1}{\gamma_{\mathrm{ref}}(\omega_{k'})}. \tag{10.188}$$

In other words, the SNR at the canceller's output is the reciprocal of the SNR at the reference sensor. This phenomenon is often referred to as the power-inversion problem, which is well explained in the informative paper by Widrow et al [319]. Clearly, any inclusion of the speech signal in the reference input will result in cancellation of speech at the output.

The speech distortion in this case can easily be deduced from (10.182):

$$\begin{aligned}\Upsilon_{\mathrm{sd}}(\omega_{k'}) &= \frac{E\left[|S(\omega_{k'})|^2\right] - E\left[|\hat{S}_{\mathrm{NR}}(j\omega)|^2\right]}{E\left[|S(\omega_{k'})|^2\right]} \\ &= 1 - \left|\frac{E\left[|B_1(j\omega)|^2\right] - G(j\omega)E\left[B_1^*(j\omega)B(j\omega)\right]}{|G(j\omega)|^2 E\left[|S(j\omega)|^2\right] + E\left[|B_1(j\omega)|^2\right]}\right|^2 \\ &= 1 - \frac{\left|1 - G(j\omega)\rho_{b_1 b}(\omega)\sqrt{E\left[|B(j\omega)|^2\right]/E\left[|B_1(j\omega)|^2\right]}\right|^2}{\left[\gamma_{\mathrm{ref}}(\omega_{k'}) + 1\right]^2}. \end{aligned} \tag{10.189}$$

Clearly, the amount of speech distortion is dependent on the channel that leaks the signal, the SNR at the reference sensor, and the coherence between the two noise signals. The higher the signal level at the reference sensor, the more will be the speech distortion.

The noise-reduction gain can be easily formulated from (10.185), which will be left as an exercise for the reader.

10.8.4 Simulations

Figures 10.9 and 10.10 show the noise reduction performance of the ANC with different adaptive algorithms. The original speech signal is digitized with a 16 kHz sampling frequency. The reference signal used here is white Gaussian noise. It was passed through a digital filter whose transfer function (the coefficients are random numbers normalized to the first coefficient) is

$$\begin{aligned} H(z) = 1 &+ 0.53z^{-1} - 0.57z^{-2} + 2.2z^{-3} - 1.8z^{-4} \\ &+ 6.7z^{-5} - 0.4z^{-6} + 0.3z^{-7}, \end{aligned}$$

and was added to the original signal to form the primary signal. The SNR of the primary signal is 0 dB. An adaptive filter is then applied to cancel the noise in the primary signal. Since it is not known *a priori*, the order of the adaptive filter is tentatively set to 10. Figure 10.9 plots the waveform of the original clean signal, primary microphone signal, and the signal estimated with the Wiener filter [eq. (10.173)], LMS, NLMS, and RLS algorithms. Since the noise is stationary, the Wiener filter gives the maximum reduction. The LMS algorithm takes a longer time to converge and the NLMS converges faster than the LMS. The RLS algorithm outperforms both LMS and NLMS. Figure 10.10 presents the noise reduction performances. The RLS, due to its fast convergence speed, performs close to the Wiener filter.

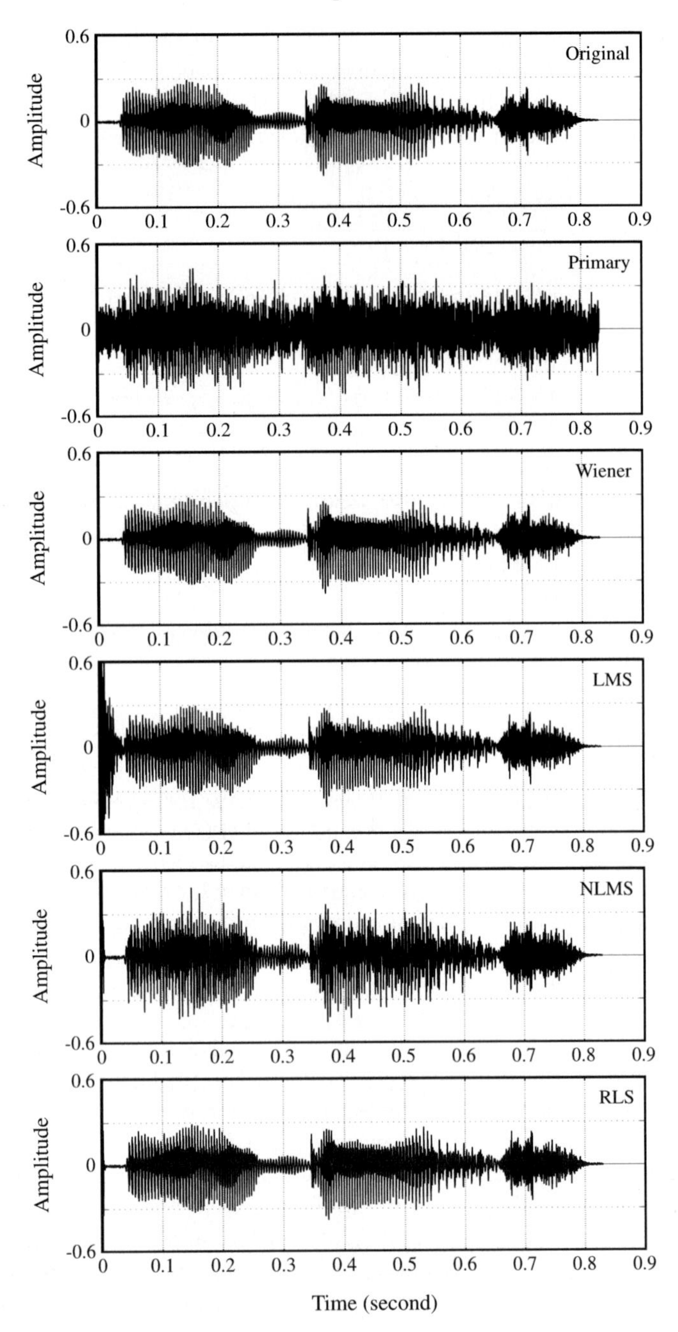

Fig. 10.9. Waveforms of the original speech signal, primary microphone signal, and the signal estimated with Wiener, LMS, NLMS, and RLS algorithms.

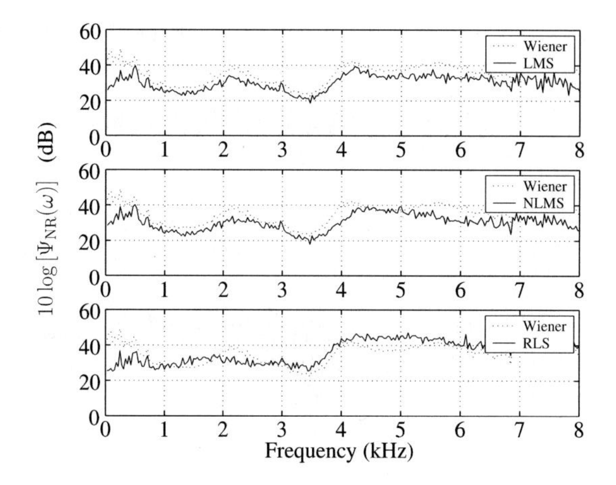

Fig. 10.10. Noise reduction performance of LMS (top), NLMS (middle), and RLS (bottom) algorithms (solid line) as compared to the ideal Wiener solution (dashed line).

10.9 Noise Reduction with a Microphone Array

We now consider to use an array of microphones. The array technique has been in use for several decades in many practical signal processing applications to detect the presence of a desired signal and reduce the effects of unwanted noise. Its potential in speech enhancement has gained special attention since the early 1960s. The underlying idea can be described as synchronizing-and-adding. Consider a single source situation where the speech waveform produced by each sensor consists of a speech signal, which is a time-delayed or time-advanced version of the signal at the reference sensor, and random noise, statistically independent from sensor to sensor. By advancing or delaying the sensor outputs to make the signal components in-phase and then adding them together, the signal at each sensor reinforces, while the noise components tend to remain at the original level because of their random, uncorrelated nature.

The synchronizing-and-adding is accomplished through a widely known technique called *beamforming*. Beamforming algorithms vary according to the location of the speech source relative to an array. If the source is located close to the array, the wavefront of the propagating wave is perceptively curved with respect to the dimensions of the array. Such a case is referred to as a near-field scenario. If the direction of propagation is approximately equal at each sensor, then the source is located in the array's far-field, and the propagating field consists of plane waves. Beamforming methodologies differ for far-field and near-field cases. Over several decades, a large number of beamforming algorithms have been developed [317], [60], [122], [124], [87], [192]. A detailed discussion on these techniques are provided in Chap. 11. This section will

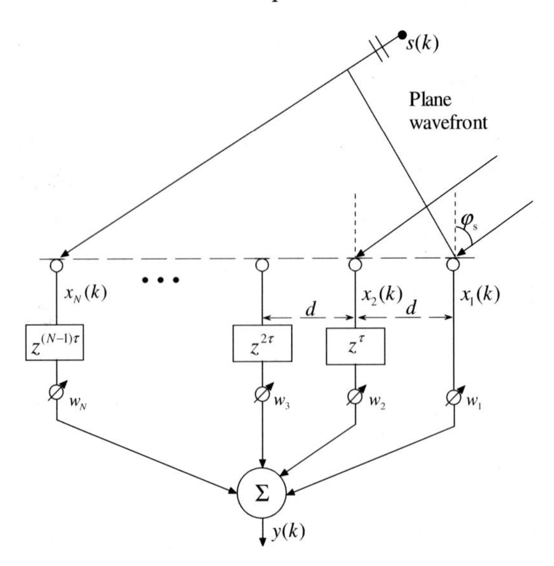

Fig. 10.11. Illustration of a delay-and-sum beamformer with a uniformly spaced linear microphone array when the sound source is in the far-field.

focus on illustrating the advantages of using multiple sensors over using a single microphone for the task of noise reduction.

10.9.1 Delay-and-Sum Algorithm

For the ease of performance analysis, we start with the simple delay-and-sum algorithm which is the basis of other beamformers. We then show that when the direction of arrival (DOA) is known (either given *a priori* or estimated using the techniques described in Chap. 9), adaptive algorithms can be employed to achieve greater noise reduction.

A typical delay-and-sum beamforming algorithm is illustrated in Fig. 10.11. A plane wave, propagating from a far-field source, arrives at a uniformly spaced linear array, which consists of N microphones. The sensor outputs can be modeled as

$$\mathbf{x}(k) = \begin{bmatrix} x_1(k) \\ x_2(k) \\ \vdots \\ x_N(k) \end{bmatrix} = \begin{bmatrix} s(k) \\ s(k-\tau) \\ \vdots \\ s(k-(N-1)\tau) \end{bmatrix} + \begin{bmatrix} b_1(k) \\ b_2(k) \\ \vdots \\ b_N(k) \end{bmatrix}, \quad (10.190)$$

where $x_n(k)$ is the signal received at the nth sensor, $\tau = d \sin\varphi_{\mathrm{s}}/c$ is the relative delay between any two neighboring microphones, c represents velocity of sound propagation in the air, φ_{s} is the incident angle of the source signal, and $b_n(k)$ is the sensor noise, which is assumed to be uncorrelated with the source signal $s(k)$ and noise signals received at the other sensors.

In the frequency domain, (10.190) can be written as

$$\begin{bmatrix} X_1(j\omega) \\ X_2(j\omega) \\ \vdots \\ X_N(j\omega) \end{bmatrix} = \begin{bmatrix} S(j\omega) \\ S(j\omega)e^{-j\omega\tau} \\ \vdots \\ S(j\omega)e^{-j\omega(N-1)\tau} \end{bmatrix} + \begin{bmatrix} B_1(j\omega) \\ B_2(j\omega) \\ \vdots \\ B_N(j\omega) \end{bmatrix}. \quad (10.191)$$

The power of the received signal at the nth microphone is given by

$$P_n(\omega) = E\left[|X_n(j\omega)|^2\right] = E\left[|S(j\omega)|^2\right] + E\left[|B_n(j\omega)|^2\right]. \quad (10.192)$$

We assume that all microphones have the same noise power:

$$E\left[|B_1(j\omega)|^2\right] = \cdots = E\left[|B_N(j\omega)|^2\right] = E\left[|B(j\omega)|^2\right], \quad (10.193)$$

so that

$$P_1(\omega) = P_2(\omega) = \cdots = P_N(\omega). \quad (10.194)$$

The narrow band signal-to-noise ratio (SNR) at each microphone can then be expressed as:

$$\gamma(\omega) = \frac{E\left[|S(j\omega)|^2\right]}{E\left[|B(j\omega)|^2\right]}. \quad (10.195)$$

The delay-and-sum beamformer consists of applying a delay τ_n and an amplitude weight w_n to the nth sensor output, then summing the N resulting signals. The delay-and-sum beamformer's output is given by

$$Y(j\omega) = \sum_{n=1}^{N} w_n X_n(j\omega) e^{j\omega\tau_n}, \quad (10.196)$$

where $\{w_n\}$ is sometimes called the array taper. Substituting (10.191) into (10.196) and further setting $w_n = 1/N$ and $\tau_n = (n-1)\tau = (n-1)d \, \sin\varphi_s/c$, we have:

$$Y(j\omega) = S(j\omega) + \frac{1}{N}\sum_{n=1}^{N} B_n(j\omega) e^{j\omega(n-1)\tau}. \quad (10.197)$$

One can see that the signal component in the beamformer's output remains the same as at the reference sensor. Therefore, there is no speech distortion. If one lets the residual noise be denoted as

$$B_{\mathrm{NR}}(j\omega) = \frac{1}{N}\sum_{n=1}^{N} B_n(j\omega) e^{j\omega(n-1)\tau}, \quad (10.198)$$

the noise-reduction gain, at frequency ω, can then be expressed as

$$\begin{aligned}\Psi_{\text{NR}}(\omega) &= \frac{E\left[|B(j\omega)|^2\right]}{E\left[|B_{\text{NR}}(j\omega)|^2\right]} \\ &= \frac{E\left[|B(j\omega)|^2\right]}{\frac{1}{N^2}E\left[\left|\sum_{n=1}^{N} B_n(j\omega)e^{j\omega(n-1)\tau}\right|^2\right]}.\end{aligned} \tag{10.199}$$

In the diffuse noise field, if the noise signals, $b_n(k)$ $(n = 1, 2, \cdots, N)$, are uncorrelated with each other, (10.199) reduces to

$$\Psi_{\text{NR}}(\omega) = \frac{E\left[|B(j\omega)|^2\right]}{\frac{1}{N}E\left[|B(j\omega)|^2\right]} = N. \tag{10.200}$$

In other words, in the diffuse-noise case, if the DOA of speech signal is know *a priori*, the delay-and-sum beamformer can achieve a frequency-independent noise-reduction gain, which is equal to the number of sensors, without distorting the speech signal.

For directional noise, *e.g.*, noise that originates from a point source, whose angle of incidence is denoted by φ_{b}, (10.199) can be expressed as

$$\begin{aligned}\Psi_{\text{NR}}(\omega) &= \frac{E\left[|B(j\omega)|^2\right]}{\frac{1}{N^2}E\left[|B(j\omega)|^2\right]E\left[\left|\sum_{n=1}^{N} e^{j\omega(n-1)d(\sin\varphi_{\text{b}}-\sin\varphi_{\text{s}})/c}\right|^2\right]} \\ &= \frac{N^2}{\left|\frac{\sin[\omega N d(\sin\varphi_{\text{b}}-\sin\varphi_{\text{s}})/2c]}{\sin[\omega d(\sin\varphi_{\text{b}}-\sin\varphi_{\text{s}})/2c]}\right|^2}.\end{aligned} \tag{10.201}$$

In such a condition, the noise reduction performance is a function of the number of microphones and the angular separation between the direction of the signal and that of the interfering noise. Figure 10.12 compares the noise suppression performance with different number of microphones as a function of incident angle φ_{b}. Generally, the bigger the angular separation is, the more the noise reduction can be achieved. For the same amount of noise cancellation, less angular separation is required for an array with more microphones.

10.9.2 Linearly Constrained Algorithms

With the same number of microphones, when the DOA of the signal of interest is known, adaptive algorithms with directional constraints can be exploited to achieve better noise reduction performance. The underlying idea is to minimize or reject signals from all other directions while maintaining a constant gain along the look direction. This can be formulated in different ways.

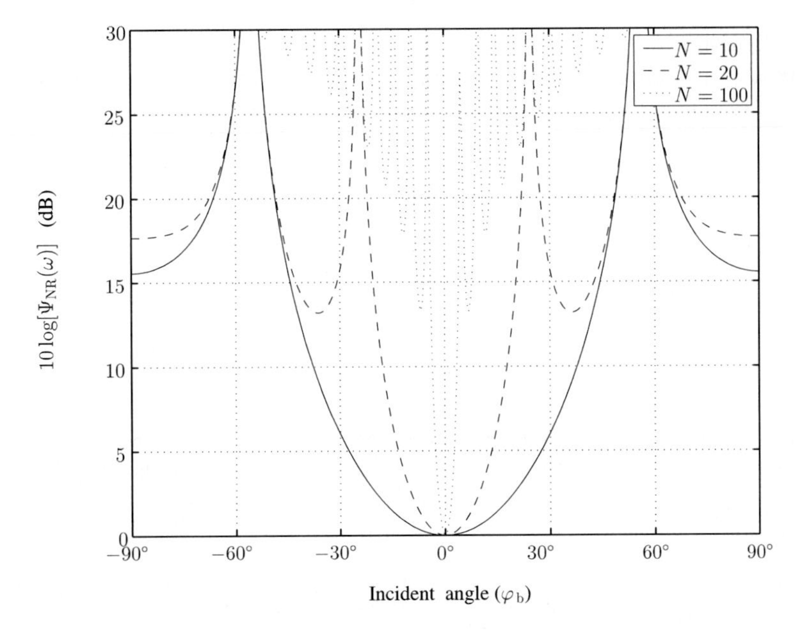

Fig. 10.12. Theoretical noise reduction performance for a uniform array under the condition that noise propagates from the far-field with an incident angle of φ_{b}. The spacing between any two neighboring microphones is 2 cm. The sampling frequency is 16 kHz. The plot is shown for a signal frequency of 4 kHz, the incident angle of the signal is $\varphi_{\mathrm{s}} = 0°$, and the SNR at the reference microphone is –10 dB.

Minimum Variance Estimator

With the signal model given in (10.190), we now consider to steer the array beam along some specific direction where the TDOA between the nth and the reference sensors is τ_n. The minimum variance algorithm is formulated as to find a weighting vector $\mathbf{w}$ such that

$$\mathbf{w} = \arg\min_{\mathbf{w}} \mathbf{w}^H \mathbf{R} \mathbf{w} \quad \text{subject to } |\mathbf{w}^H \mathbf{a}(\omega)| = 1, \tag{10.202}$$

where

$$\mathbf{a}(\omega) = \frac{1}{\sqrt{N}} [e^{j\omega\tau_1}, e^{j\omega\tau_2}, \cdots, e^{j\omega\tau_N}]^T,$$

$$\mathbf{R}_x = E\left[\mathbf{x}(k)\mathbf{x}^T(k)\right]$$

is the $N \times N$ array output covariance matrix. For a positive definite matrix $\mathbf{R}_x$, the solution to (10.202) is readily given by:

$$\mathbf{w} = \frac{\mathbf{R}_x^{-1}\mathbf{a}(\omega)}{\mathbf{a}^H(\omega)\mathbf{R}_x^{-1}\mathbf{a}(\omega)}. \tag{10.203}$$

This algorithm was first developed by Capon [60] and is often called Capon's estimator, minimum variance estimator (MV) or minimum variance distortionless response (MVDR). The output power of this array is in the form

$$E\left[|y(k)|^2\right] = \frac{1}{\mathbf{a}^H(\omega)\mathbf{R}_x^{-1}\mathbf{a}(\omega)}. \qquad (10.204)$$

By applying the Cauchy-Schwarz inequality, it can be easily shown that

$$\frac{1}{\mathbf{a}^H(\omega)\mathbf{R}_x^{-1}\mathbf{a}(\omega)} \leq \mathbf{a}^H(\omega)\mathbf{R}_x\mathbf{a}(\omega). \qquad (10.205)$$

The term in the right hand side of (10.205) is the output power of the delay-and-sum beamformer [i.e., (10.196)] in the time domain. That is, the power of the array output with adaptive directional constraints is lower than that of the delay-and-sum beamformer. Since the gain along the look direction is 1, this indicates that more noise can be attenuated with adaptive beamforming algorithms.

Spatio-Temporal Filtering with Linear Constraint

To better understand the MV algorithm, let us take a slightly different approach and reformulate it into a more general form using spatial-temporal filtering techniques. We still suppose that we have a linear array consisting of N microphones. The first sensor is selected as the reference sensor. But this time we consider a more generic signal model as given in (10.1), with both attenuation and TDOA being taken into account, i.e.,

$$x_n(k) = \alpha_n s(k - \tau_{n1}) + b_n(k), \; n = 1, 2, \cdots, N. \qquad (10.206)$$

In the following, we assume that the relative delays τ_{n1}, $n = 1, 2, \cdots, N$, are known or have been estimated. So our first step is to align the microphone signals according the TDOA information:

$$\begin{aligned} z_n(k) &= x_n(k + \tau_{n1}) = \alpha_n s(k) + b_n(k + \tau_{n1}) \\ &= y_n(k) + b_n(k + \tau_{n1}), \quad n = 1, 2, \cdots, N. \end{aligned} \qquad (10.207)$$

Let us now define the error signal, for the nth microphone, between the clean speech sample $y_n(k)$ and its estimate as,

$$e_{y_n}(k) \triangleq y_n(k) - \mathbf{h}_{:n}^T\mathbf{z}(k) = y_n(k) - \sum_{i=1}^{N}\mathbf{h}_{i:n}^T\mathbf{z}_i(k), \qquad (10.208)$$

where $\mathbf{h}_{i:n}$ are filters of length L and,

$$\begin{aligned} \mathbf{h}_{:n} &\triangleq \left[\mathbf{h}_{1:n}^T \; \mathbf{h}_{2:n}^T \; \cdots \; \mathbf{h}_{N:n}^T\right]^T, \\ \mathbf{z}(k) &\triangleq \left[\mathbf{z}_1^T(k) \; \mathbf{z}_2^T(k) \; \cdots \; \mathbf{z}_N^T(k)\right]^T. \end{aligned}$$

Since $\mathbf{z}_i(k) = \alpha_i \mathbf{s}(k) + \mathbf{b}_i(k + \tau_{i1})$, (10.208) becomes:

$$\begin{aligned} e_{y_n}(k) &= \mathbf{s}^T(k) \left[\alpha_n \mathbf{u}_1 - \sum_{i=1}^{N} \alpha_i \mathbf{h}_{i:n} \right] - \sum_{i=1}^{N} \mathbf{b}_i^T(k + \tau_{i1}) \mathbf{h}_{i:n} \\ &= \mathbf{s}^T(k) \left[\alpha_n \mathbf{u}_1 - \mathbf{D}\mathbf{h}_{:n} \right] - \mathbf{b}^T(k) \mathbf{h}_{:n} \\ &= e_{s,n}(k) - e_{b,n}(k), \end{aligned} \tag{10.209}$$

where

$$\begin{aligned} \mathbf{D} &\triangleq \left[\alpha_1 \mathbf{I} \ \alpha_2 \mathbf{I} \ \cdots \ \alpha_N \mathbf{I} \right], \\ \mathbf{b}(k) &\triangleq \left[\mathbf{b}_1^T(k + \tau_{11}) \ \mathbf{b}_2^T(k + \tau_{21}) \ \cdots \ \mathbf{b}_N^T(k + \tau_{N1}) \right]^T. \end{aligned}$$

Expression (10.209) is the difference between two error signals; $e_{s,n}(k)$ represents signal distortion and $e_{b,n}(k)$ represents the residual noise. The MSE corresponding to the residual noise with the nth microphone as the reference signal is,

$$\begin{aligned} J_{b,n}(\mathbf{h}_{:n}) &= E\left\{ e_{b,n}^2(k) \right\} \\ &= \mathbf{h}_{:n}^T E\left\{ \mathbf{b}(k)\mathbf{b}^T(k) \right\} \mathbf{h}_{:n} \\ &= \mathbf{h}_{:n}^T \mathbf{R}_b \mathbf{h}_{:n}. \end{aligned} \tag{10.210}$$

Usually, in the single-channel case, the minimization of the MSE corresponding to the residual noise is done while keeping the signal distortion below a threshold [111]. With no distortion, the optimal filter obtained from this optimization is $\mathbf{u}_1$, hence there is not any noise reduction either. The advantage of multiple microphones is that, actually, we can minimize $J_{b,n}(\mathbf{h}_{:n})$ with the constraint that $\alpha_n \mathbf{u}_1 = \mathbf{D}\mathbf{h}_{:n}$ (no speech distortion at all). Therefore, our optimization problem is,

$$\min_{\mathbf{h}_{:n}} J_{b,n}(\mathbf{h}_{:n}) \text{ subject to } \alpha_n \mathbf{u}_1 = \mathbf{D}\mathbf{h}_{:n}. \tag{10.211}$$

By using a Lagrange multiplier, we easily find the optimal solution:

$$\mathbf{h}_{\mathrm{o},:n} = \alpha_n \mathbf{R}_b^{-1} \mathbf{D}^T \left[\mathbf{D} \mathbf{R}_b^{-1} \mathbf{D}^T \right]^{-1} \mathbf{u}_1, \tag{10.212}$$

where we assumed that the noise signals $b_i(k)$ are not perfectly coherent so that $\mathbf{R}_b$ is not singular.

Given the optimal filter $\mathbf{h}_{\mathrm{o},:n}$, we can write the MMSE for the nth microphone as,

$$J_{b,n}(\mathbf{h}_{\mathrm{o},:n}) = \alpha_n^2 \mathbf{u}_1^T \left[\mathbf{D} \mathbf{R}_b^{-1} \mathbf{D}^T \right]^{-1} \mathbf{u}_1. \tag{10.213}$$

Since we have N microphones, we have N MMSEs as well. The best MMSE from a noise reduction point of view is the smallest one, which is, according to (10.213), the microphone signal with the smallest attenuation factor.

The attenuation factors α_n can be easily determined, if the power of the noise signals are known, by using the formula:

$$\alpha_n^2 = \frac{E\{z_n^2(k)\} - E\{b_n^2(k+\tau_{n1})\}}{E\{z_1^2(k)\} - E\{b_1^2(k)\}}, \quad n = 2, 3, \cdots, N. \tag{10.214}$$

For the particular case where the noise is spatio-temporally white with a power equal to σ_b^2, the MMSE and the normalized MMSE for the nth microphone are respectively,

$$J_{b,n}(\mathbf{h}_{\text{o},:n}) = \sigma_b^2 \frac{\alpha_n^2}{\sum_{i=1}^N \alpha_i^2}, \tag{10.215}$$

$$\tilde{J}_{b,n}(\mathbf{h}_{\text{o},:n}) = \frac{\alpha_n^2}{\sum_{i=1}^N \alpha_i^2}. \tag{10.216}$$

Comparing the aligned signal $z_n(k)$ with the noise-reduced signal, we can write the corresponding speech-distortion index as

$$\upsilon_{\text{sd}}(\mathbf{h}_{\text{o},:n}) = \frac{E\left\{\left[y_n(k) - \sum_{i=1}^N \mathbf{h}_{i:n}^T \mathbf{y}_i(k)\right]^2\right\}}{\sigma_{x_n}^2} \tag{10.217}$$

and the noise-reduction factor as

$$\xi_{\text{nr}}(\mathbf{h}_{\text{o},:n}) = \frac{\sigma_{b_n}^2}{E\left\{\left[\sum_{i=1}^N \mathbf{h}_{i:n}^T \mathbf{b}_i(k+\tau_{n1})\right]^2\right\}}. \tag{10.218}$$

With the optimal filter given in (10.212), for the particular case where the noise is spatio-temporally white with a power equal to σ_b^2, it can be easily shown that

$$\upsilon_{\text{sd}}(\mathbf{h}_{\text{o},:n}) = 0, \tag{10.219}$$

and

$$\xi_{\text{nr}}(\mathbf{h}_{\text{o},:n}) = \sum_{i=1}^N \alpha_i^2. \tag{10.220}$$

It can be seen that when the number of microphones goes to infinity, $\xi_{\text{nr}}(\mathbf{h}_{\text{o},:n})$ approaches to infinity, and meanwhile $\upsilon_{\text{sd}}(\mathbf{h}_{\text{o},:n}) = 0$, which indicates that the noise can be completely removed with no signal distortion at all. This clearly demonstrates the advantages of using multiple sensors over using a single sensor for noise reduction. In the former case, noise can be reduced or even eliminated with less or even no speech distortion, while in the latter situation, compromise has to be made between the conflicting requirements between noise reduction and speech distortion.

10.10 Summary

Speech signals, when propagating through acoustic channels and recorded by microphone receivers, are inevitably corrupted by noise and interference. How to estimate the speech of interest from its corrupted observations has become one of the most challenging problems in acoustic signal processing, which involves a wide variety of techniques such as source separation, channel identification, speech dereverberation, to name a few. This chapter was devoted to the speech-enhancement and noise-reduction problem, which primarily aims to recover the desired speech signal from its realizations corrupted by additive noise. It covered not only the well-recognized single-channel techniques such as Wiener filtering and spectral subtraction, but also the techniques using two or more microphone sensors. In the former case, the study demonstrated that noise reduction is possible, but at a price of distorting the clean speech, and more noise reduction corresponds to more speech degradation. Although some schemes can be arranged to restrain the speech distortion so that it will not be perceived by listeners, they in general lead to less noise reduction. This performance trade-off problem, however, becomes less serious and can even be avoided when two or more microphone sensors can be used. In this situation, both the adaptive-noise-cancellation and beamforming techniques can be employed to mitigate the unwanted noise effects with less or even no speech distortion.

11

Source Separation and Speech Dereverberation

Most environments in which we communicate with others have complicated acoustics. In addition to ambient noise, multiple competing sounds may occur simultaneously and are usually accompanied by room reverberation. Therefore, a desirable audio interface for speech communication and processing systems ought to have the ability of separating a voice from the mixtures with other interfering sounds and the ability of cutting the amount of reverberation to a minimum in the recording such that high-quality speech can still be captured from a distance.

In the previous chapter, speech enhancement via noise reduction has been studied. In this chapter, we will discuss the problem of source separation and speech dereverberation. We will begin with a review of the cocktail party effect and try to explain how humans solve the challenging source separation problem with their sophisticated binaural auditory systems. After having this knowledge, we then explore what approaches have been developed for solving this problem. Both the class of beamforming and the class of independent component analysis methods will be investigated. Finally we will present a new way to perform source separation and speech dereverberation based on blind identification of acoustic MIMO systems.

11.1 Cocktail Party Effect

It has been recognized for some time that a human has the ability of focusing on one particular voice or sound amid a cacophony of distracting conversations and background noise. This interesting psychoacoustic phenomenon is referred to as the *cocktail party effect* or *attentional selectivity* [76]. While a human hearing system and brain can effectively handle this cocktail party problem with not much effort, it is a very tricky problem in digital signal processing and human-machine interfaces. Studying and analyzing the mysterious neural mechanisms for the cocktail party effect can tell us a lot about how attention affects one's perception of ambient voice and sound. This knowledge might give us some good ideas as to whether and how we could get machines to do so. Quoting from a recent review article on the cocktail party problem by Haykin and Chen, 2005 [161]: “It does not mean that we must duplicate every aspect of the human auditory system in solving the machine cocktail party problem. Rather, it is our belief that seeking the ultimate answer to the cocktail party problem requires deep understanding of many fundamental issues that are deemed to be of theoretical and technical importance.” Therefore, in this section, we want to briefly review what has been known about the cocktail party effect and then highlight what implications we can derive for source separation algorithms that work in cocktail-party-like environments.

The cocktail party problem was first investigated by Colin Cherry in his pioneering work published in 1953 [76]. In the psychoacoustic experiments carried out by Cherry and his colleagues [76], [77], dichotic listening and shadowing were used to recreate the cocktail party effect in the laboratory and evaluate a human listener's auditory perception. In a dichotic listening experiment, two different messages are played simultaneously to an individual, one message in each ear, with the use of a pair of headphones. For evaluation, the individual is asked to shadow of the messages, which means to repeat the message that the person is instructed to attend to. The two messages differ not only semantically but also physically. The physical differences that have been examined include different spatial locations, different volume, different genders of speakers (female vs. male), different types of sound (speech vs. noise, voice vs. tone), different speaking voices (mean pitch and mean speaking speed), different accents, and so forth.

The basic finding from Cherry's seminal work and the work afterwards is that our auditory system separates acoustic signals on the basis of their physical features. The subjects of the dichotic listening and shadowing experiments can easily tell the content of the message that they were attended to. They reported that they noticed when the physical characteristics of the unattended message changed, but could not tell what the message was about. They were not aware when the meaning of the unattended message changed and when the unattended voice changed from speaking one language to another (from English to French to Spanish). Furthermore, they could not notice even if the same word had been repeated in the unattended message for a great number of times (except for such always pertinent information as the listener's name or such emotionally important messages as fire). These all suggested that signal separation/segregation is performed at a pretty early stage (before semantic and syntactic analysis) of the whole auditory perception process and depends neither on one's syntactic or lexical knowledge nor on the meaning of the messages. This is a consensus understanding among all existing influential models for the cocktail party effect, which include Broadbent's filter model [55], Treisman's attenuator model [296], Deutsch & Deutsch and Norman's late selection or response selection model [95], [241], and Johnston & Heinz's multimode or flexible filter model [193]. These models disagree only in their answers to the questions of whether the unselected auditory stream (after separation) will be passed for additional processing towards perception and whether selection is performed before or after the semantic analysis (i.e., pattern recognition). Therefore the first implication that we can derive for speech source separation algorithms is that the use of our high-level (e.g., syntactic, lexical) knowledge about speech is not necessarily beneficial.

Another important observation from the psychoacoustic experiments is that spatial hearing plays a very important role in the cocktail party effect. Our perception of speech remarkably benefits from spatial hearing. This ability is mainly attributed to the striking fact that we have two ears. This is intuitively justified by our daily experience and can be further demonstrated simply by observing the difference in understanding between using both ears and with either ear covered when listening in an enclosed space where there are multiple speakers at the same time. The advantage of spatial hearing consists of the following three elements [274]. (1) Acoustically "better" ear effect: an improvement in terms of signal-to-noise ratio (SNR) or signal-to-interference ratio (SIR) can be generally achieved at one of the two ears due to the difference in distance from the target source to the two ears and due to the filtering effect by a listener's head and body on the impinging acoustic signals. (2) Binaural processing effect: with binaural processing by correlating signals across the two ears, the SNR or SIR can be effectively improved in addition to what has been gained at the better ear. (3) Spatial attention effect: for the perceived difference in the spatial locations of the target and interfering sound sources, a listener can selectively put attention on the voice from the direction of the target source while ignoring the competing sound or noise for other directions. The difference between the binaural processing and spatial attention effects is very subtle. The former does not rely upon the difference in perceived locations, but rather on peripheral processing mechanisms [103], while the latter requires a clear perception of spatial separation and is accomplished at a central, high-level part of the auditory system. The benefit offered by the acoustically better ear and binaural processing effects is referred to as energetic masking and the benefit by the spatial attention effect as informational masking in psychoacoustics. In machine cocktail party problems, energetic masking is less useful than informational masking. On one hand, the gain in SNR and SIR due to binaural processing is relatively small (on the order of only 3-6 dB) and the underlying mechanism has not yet been well understood. On the other hand, in order to imitate the acoustically better ear effect, either a close-talk microphone or a complicated acoustic receiver like a dummy head needs to be employed, which is not very practicable from an engineering perspective. Therefore the second implication by the cocktail party effect is that the ability of spatial filtering is what source separation algorithms should possess with the use of multiple microphones.

While spatial separation is clearly an important (but not sole) cue in the cocktail party effect, temporal, spectral, and spectro-temporal cues (e.g., modulation, level, pitch, harmonics, formants, prosody, timbre, etc.) are also very intelligently used by our auditory system, for which there has been a variety of demonstrations [88], [89], [9], [182], [289], [260]. Like spatial variation, many spectro-temporal cues can produce comprehensible source separation. For example, continuity of pitch has been suggested for speech separation algorithms [249]. There was an argument that selective attention to speech is directed not just to a particular spatial location but rather to the ensemble of a particular location and the speech characteristics of a talker [325]. Recently, an investigation on the relative roles of spatial cues and a talker's prosody and vocal-track size in the cocktail party effect has been reported [90]. It was shown that natural prosodic variations and a difference in vocal-tract size that is comparable to or larger than the average male/female difference can override spatial cues. This indicates that the difference in spatial locations of the talkers in a cocktail party sometimes is not even the primary cue for selective attention. This can be somehow explained by that fact that the human auditory system has become extremely sensitive to the variation of speech signals produced by different speakers over centuries of evolution. Therefore, the third implication we can draw from the cocktail party effect for source separation algorithms is that a delicate analysis using our knowledge about the human speech production system could help identify the most evident discriminating criterion or criterion set for specific mixtures of speech signals. A data-driven strategy rather than simply assuming independent source signals in concert with spatial filtering can be expected to achieve greater successes.

The last issues that we would like to comment on are what difference a noise source can make compared to an interfering speech source and what difference is between diffuse and point noise sources in the cocktail party effect. There exist evidences showing that the cocktail party effect is more pronounced with speech sources as interference than with noise sources [158], which is clearly related to our auditory system's sensitivity to speech signals. But we are not aware of any previous psychoacoustic experiments studying the difference between diffuse and point noise sources, although empirically point noise sources might cause less problems in cocktail party problems. We believe that the presence of diffuse noise makes human auditory system find it difficult to determine the number of sound sources, which apparently impede our ability of learning the surrounding acoustic environment and separating sources in a cocktail party. Actually acoustic MIMO signal processing algorithms confront the same difficulty with diffuse noise. An ambiguous or wrong estimate of the number of sound sources could impair the (possibly implicit) effort of identifying the MIMO system for source separation. It is this similarity that inspires us to have such a discussion, which we hope can stimulate further thinking of the readers.

We have reviewed the cocktail party effect and have explained what implications we believe can be derived from the effect for developing and analyzing acoustic source separation algorithms. In the following sections, we will examine the state of the art of source separation algorithms. By comparing their performance and a human's unbeatable audition capability, we highlight what

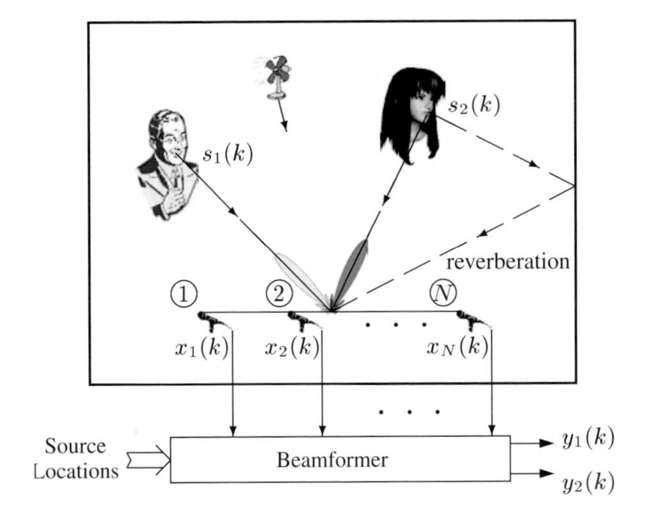

Fig. 11.1. Illustration of spatial filtering with a microphone beamformer for source separation in an enclosed environment.

more the existing source separation algorithms can take advantage of and then envision future directions of the research on this problem.

11.2 Source Separation

11.2.1 Microphone Array Beamforming

A beamformer is a spatial filter that operates on the outputs of an array of microphones in order to enhance the signal coming from one direction while suppressing noise and interference from other directions, as illustrated in Fig. 11.1. This viewpoint suggests to model the field of acoustic waves in terms of their propagating directions, which leads to a signal model that in turn shapes the formulation and affects the performance of beamforming algorithms. We will begin the review of microphone array beamforming with a discussion of the direction-oriented signal model for microphone array outputs.

Direction-Oriented Signal Model for Beamforming

The principle of superposition applies to propagating acoustic waves in the air. This allows two traveling waves to pass through each other without being disturbed and the net air pressure at any point in space and time is simply the sum of the corresponding individual wave pressures. As a result, the outputs of a microphone array can be modeled as the sum of the pressures caused by a finite number of acoustic waves traveling from different directions. In this model, signals of different directions are treated as different contributing components in the microphone outputs even though they may originate from the same source, e.g., a signal traveling from the direct path and its reverberations. This is remarkably distinctive from the MIMO model, which is based on the relationship between the sources and the microphone outputs via the concept of channels. Therefore, beamforming algorithms generally perform the same operations on the sensors' outputs regardless of the number of sources or the character of the noise present in the wavefield [192], while MIMO signal processing algorithms naturally take all these aspects into account. The dependency of the source number and noise characteristics is what we believe the most fundamental difference between microphone array beamforming and acoustic MIMO signal processing.

Beamformer Classification and Structures

Beamforming algorithms vary according to how far signal sources are located from an array relatively in comparison with the size of the array. If the sources are close to the array (in the near field), the curvature of the wave fronts must be considered and the wave propagation direction is perceptively different at different sensors. On the contrary, if the sources are far enough away from the array (in the far field), the wavefronts impinging on the array can be regarded as plane waves. In such classical array signal processing problems as radar, sonar, and digital communications, the far-field model is realistic and pragmatic, and has achieved great successes. So far most microphone array beamforming algorithms have also assumed that sound sources be in the far field in order to make the problem analytically tractable. However, the assumption may not be applicable to some acoustic applications, which leads to performance degradation for spatial filtering.

Another information that needs to be known prior to designing a beamformer is whether the target sources produce narrowband or broadband signals since this information makes a great impact on choosing the structure of the beamformer. A frequency-domain narrowband beamformer is equivalently implemented by a delay-and-sum beamformer in the time domain as shown in Fig. 11.2. For broadband signals, temporal frequency analysis is necessary for a beamformer and the narrowband decomposition can be performed by taking a discrete Fourier transform (DFT) of microphone signals using the FFT algorithm as illustrated in Fig. 11.3(a). The data across the microphone array at each frequency of interest are processed separately by their own narrowband sub-beamformer and the results are put together at the end to form the wideband beamformer outputs. The outputs can be made equivalent to that of the time-domain filter-and-sum beamformer depicted in Fig. 11.3(b) with careful data partitioning and proper selection of beamformer filter length and coefficients.

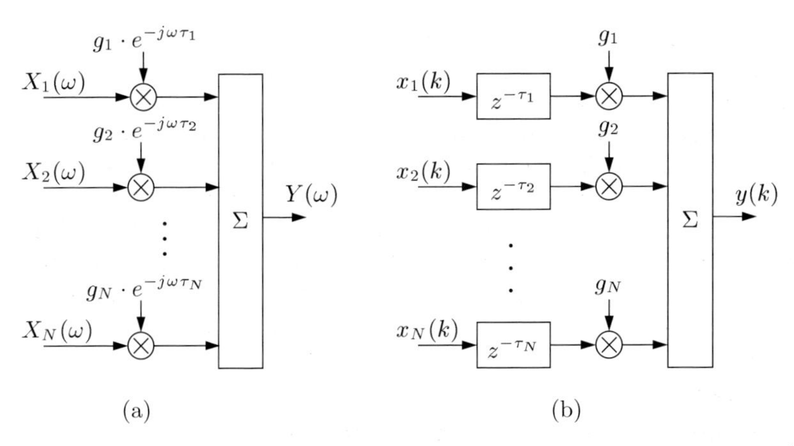

Fig. 11.2. The structure of (a) a frequency-domain narrowband beamformer, (b) a time-domain delay-and-sum beamformer.

Narrowband Beamformers

Let's consider the case of a narrowband beamformer whose incident waves are planar, i.e., sources are in the far field. Then corresponding to a propagating plane wave from the direction θ, the signal at the nth sensor is recorded as:

$$x_n(k) = s(k - t_n), \quad n = 1, 2, \cdots, N, \tag{11.1}$$

where t_n is the time delay of arrival for sensor n with respect to the plane wave. Taking the discrete-time Fourier transform of (11.1) yields:

$$X_n(\omega) = S(\omega)e^{-j\omega t_n}. \tag{11.2}$$

By applying a delay-and-sum beamformer of Fig. 11.2(b) (or equivalently a narrowband beamformer) to the array outputs, we get:

$$Y(\omega) = \sum_{n=1}^{N} X_n(\omega) g_n e^{-j\omega\tau_n}, \tag{11.3}$$

where the beamformer weights g_n $(n = 1, 2, \cdots, N)$ are scalars. Substituting (11.2) into (11.3) produces:

$$Y(\omega) = S(\omega) \cdot \sum_{n=1}^{N} g_n e^{-j\omega(\tau_n + t_n)}. \tag{11.4}$$

So the beampattern for the direction θ_{a} at which the beamformer looks (subscript a stands for array) corresponding to the set of beamformer delays $\{\tau_1, \tau_2, \cdots, \tau_N\}$ is then computed as follows:

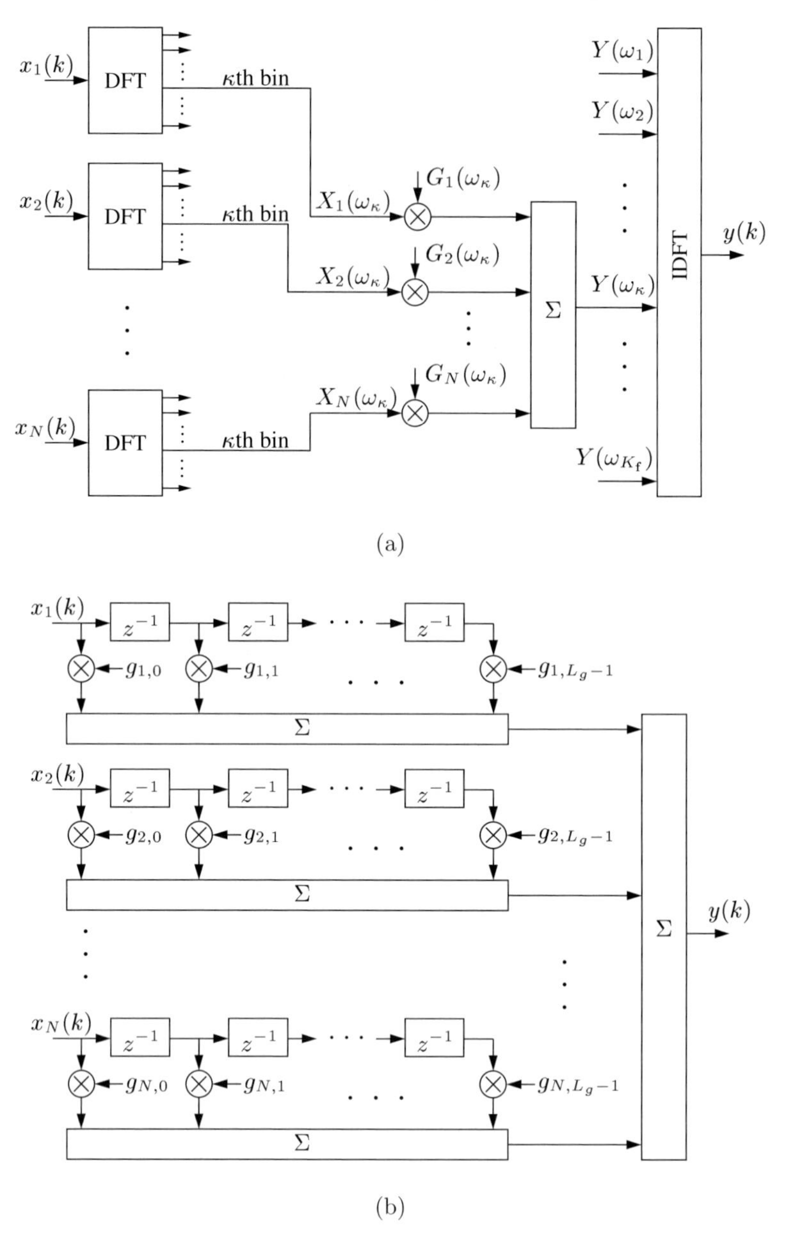

Fig. 11.3. The structure of (a) a frequency-domain broadband beamformer by narrowband decomposition, and (b) a time-domain filter-and-sum beamformer.

$$B(\omega, \theta, \theta_\mathrm{a}) \triangleq \frac{Y(\omega)}{S(\omega)} = \sum_{n=1}^{N} g_n e^{-j\omega(\tau_n + t_n)}. \tag{11.5}$$

Using the triangle inequality, we know that

$$\|B(\omega, \theta, \theta_\mathrm{a})\| = \left\| \sum_{n=1}^{N} g_n e^{-j\omega(\tau_n + t_n)} \right\| \leq \sum_{n=1}^{N} \left\| g_n e^{-j\omega(\tau_n + t_n)} \right\|, \tag{11.6}$$

and equality occurs when $\tau_n + t_n$ is equal to a constant $\forall n$.

In order to pass a particular source signal of interest from the direction θ_s corresponding to the set of time delays of arrival $\{t_{\mathrm{s},1}, t_{\mathrm{s},2}, \cdots, t_{\mathrm{s},N}\}$, we need to steer the beamformer by choosing a proper set of $\{\tau_1, \tau_2, \cdots, \tau_N\}$ such that $\|B(\omega, \theta, \theta_\mathrm{a})\|$ is maximized at the direction of $\theta = \theta_\mathrm{s}$. It is clear from (11.6) that we need to let $\tau_n + t_{\mathrm{s},n}$ be a constant. This implies that the sensor signals are aligned in phase, which is intuitively quite obvious.

Another thing that is worth noting is that in order to steer the beamformer, *only* the beamformer delays $\tau_1, \tau_2, \cdots, \tau_N$ (equivalently the phases of the complex beamformer weights) need to be altered with the scalar beamformer coefficients $g_1, g_2, \cdots, g_N$ unmodified. Therefore θ_a can be arbitrarily specified in the design of such a beamformer.

For an equally-spaced linear array as illustrated in Fig. 11.4, it is more straightforward to relate the time delays of arrival τ_n $(n = 1, 2, \cdots, N)$ to θ:

$$t_n = t_1 + \frac{(n-1)d \sin\theta}{c}, \quad n = 2, 3, \cdots, N, \tag{11.7}$$

where c is the speed of sound. Suppose that the array is steered to the broadside and let $\tau_n = 0$ $(n = 1, 2, \cdots, N)$. Then the beampattern is found as:

$$\begin{aligned} B(\omega, \theta) &= e^{-j\omega t_1} \sum_{n=1}^{N} g_n e^{\frac{-j\omega(n-1)d\sin\theta}{c}} \\ &= e^{-j\omega t_1} \sum_{n=1}^{N} g_n e^{\frac{-j2\pi(n-1)}{f_\mathrm{s}} \cdot \frac{d}{\lambda_f} \cdot \sin\theta}, \end{aligned} \tag{11.8}$$

where f_s is the sampling frequency and $\lambda_f = c/f$ is the wavelength for frequency $f = f_\mathrm{s}(\omega/2\pi)$. A constant delay t_1 in the beamformer output is unessential to the definition of beampattern and omitting it is apparently harmless. Rewrite (11.8) as follows:

$$\begin{aligned} B(\omega, \theta) &= \sum_{n=1}^{N} g_n e^{-j \cdot \left(\frac{2\pi}{f_\mathrm{s}} \cdot \frac{d}{\lambda_f} \cdot \sin\theta \right) \cdot (n-1)} \\ &= \mathrm{DTFT}_{g_n} \left(\frac{2\pi}{f_\mathrm{s}} \cdot \frac{d}{\lambda_f} \cdot \sin\theta \right), \end{aligned} \tag{11.9}$$

where $\mathrm{DTFT}_{g_n}(\omega)$ denotes the discrete-time Fourier transform (DTFT) of g_n at the radian frequency ω. Now it is clear that digital filter theory can be used to design the beamformer weights g_n for a specified beampattern.

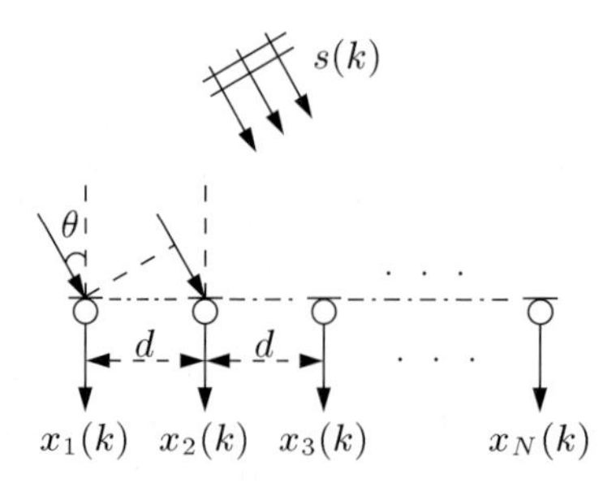

Fig. 11.4. An equally-spaced linear microphone array with sound sources being in the far field.

Broadband Beamformers for Microphone Arrays

The above discussion has been focused on a narrowband beamformer. In the development of narrowband beamforming techniques, many basic terminologies, well-known concepts, and widely-used notation were introduced. However, these techniques are not very useful for acoustic applications since speech is a typical broadband signal. The wavelength varies with frequency. As clearly shown by (11.9), the beampattern would not be the same across a broad frequency band with g_n fixed. Therefore, if we use a narrowband beamformer for broadband signals, then noise and interference signals coming from a direction different from the beamformer's look direction will not be uniformly attenuated over its entire spectrum. This "spectral tilt" results in a disturbing artifact in the array output [310].

One way to overcome this problem is to use harmonically nested subarrays [117], [201]. Every subarray is linear and equally-spaced, and is designed for operating at a single frequency. But such a solution requires a large array with a great number of microphones and the array geometry is unusual. Another way to design a broadband beamformer while still basing on classical narrowband techniques is to perform narrowband decomposition using aforementioned structure of Fig. 11.3(a), and design narrowband beamformers independently at each frequency. As explained above, a frequency-domain broadband beamformer is equivalent to a filter-and-sum beamformer if implemented in the time domain, even though developing a filter-and-sum beamforming algorithm in the time domain does not seem to have any connections to digital filter design methods at a superficial glance. We do not intent to present detailed developments of these beamforming techniques here but would like to refer the readers to [311] and references therein for more materials on this subject.

The beamformers presented above for either narrowband or broadband signals are all data independent in the sense that their weights do not depend on the array data and are chosen according to the desired beampattern for a particular signal/interference scenario [300]. The directions of noise and interference sources need to be known or prefigured prior to the design of data-independent beamformers. This knowledge is reflected in the specification of the desired beampattern with more attenuation put on those directions. Apparently the performance of a data-independent beamformer, even with a perfect design, is satisfactory only when the working signal/interference scenario matches the design one. Therefore, data-independent beamformers are not optimal in practice where any signal/interference scenarios can be encountered, not even to mention time-varying acoustic environments.

For microphone arrays, another class of beamforming algorithms is more popular. It is called statistically optimum beamforming according to [300]. In this class, the linearly constrained minimum-variance (LCMV) algorithm developed by Frost [122] is the most studied and most influential algorithm. The generalized sidelobe canceller (GSC), developed by Griffiths and Jim [150], is rather an interesting way to implement the Frost's LCMV algorithm. Using these algorithms, the beamformer's weights are chosen to minimize the variance of the array output while keeping the signals from the desired directions undistorted.

For a filter-and-sum beamformer as shown in Fig. 11.3(b), the output is computed as

$$y(k) = \sum_{n=1}^{N} \mathbf{g}_n^T \mathbf{x}_n(k) = \mathbf{g}^T \mathbf{x}(k), \tag{11.10}$$

where

$$\begin{aligned}
\mathbf{g}_n &= \begin{bmatrix} g_{n,0} & g_{n,1} & \cdots & g_{n,L_g-1} \end{bmatrix}^T, \\
\mathbf{x}_n(k) &= \begin{bmatrix} x_n(k) & x_n(k-1) & \cdots & x_n(k-L_g+1) \end{bmatrix}^T, \\
\mathbf{g} &= \begin{bmatrix} \mathbf{g}_1^T & \mathbf{g}_2^T & \cdots & \mathbf{g}_N^T \end{bmatrix}^T, \\
\mathbf{x}(k) &= \begin{bmatrix} \mathbf{x}_1^T(k) & \mathbf{x}_2^T(k) & \cdots & \mathbf{x}_N^T(k) \end{bmatrix}^T,
\end{aligned}$$

and L_g is the length of the filters in the beamformer.

For a far-field sound source from the direction θ corresponding to the time delays of arrival t_n $(n = 1, 2, \cdots, N)$, the output of the nth microphone is given by $s(k) * z^{-t_n}$, where $*$ denotes linear convolution. Let $\mathbf{u}_n$ be a vector of length L_u, whose $(t_n + 1)$th element is 1 and all other elements are 0, and where $L_u = \max\{t_1, t_2, \cdots, t_N\} + 1$. Then, the beamformer's response to such a sound source is found as

$$\begin{aligned}
\sum_{n=1}^{N} \mathbf{u}_n * \mathbf{g}_n &= \sum_{n=1}^{N} \mathbf{U}_n^{\mathrm{c}} \mathbf{g}_n, \\
&= \mathbf{U}^{\mathrm{c}} \mathbf{g},
\end{aligned} \tag{11.11}$$

where $\mathbf{U}_n^{\mathrm{c}}$ of dimension $(L_u + L_g - 1) \times L_g$ is the convolution matrix (superscript c) of $\mathbf{u}_n$

$$\mathbf{U}_n^{\mathrm{c}} \triangleq \begin{bmatrix} 0 & 0 & \cdots & 0 \\ \vdots & \vdots & \vdots & \vdots \\ 1 & 0 & \cdots & 0 \\ 0 & 1 & \cdots & 0 \\ \vdots & \vdots & \ddots & \vdots \\ 0 & 0 & \cdots & 1 \\ \vdots & \vdots & \vdots & \vdots \\ 0 & 0 & \cdots & 0 \end{bmatrix} \begin{matrix} \\ \\ \leftarrow \text{the } (t_n+1)\text{th row} \\ \\ \\ \\ \\ (L_u + L_g - 1) \times L_g \end{matrix},$$

and

$$\mathbf{U}^{\mathrm{c}} = \left[\mathbf{U}_1^{\mathrm{c}} \ \mathbf{U}_2^{\mathrm{c}} \ \cdots \ \mathbf{U}_N^{\mathrm{c}} \right].$$

The idea of the LCMV algorithm is to choose the filter $\mathbf{g}$ that minimizes the variance of the beamformer's output while allowing any signal from angle θ to be passed to the output. As a result, it is formulated as the following constrained optimization problem:

$$\begin{aligned} \min_{\mathbf{g}} E\left\{y^2(k)\right\} &= \min_{\mathbf{g}} \mathbf{g}^T \mathbf{R}_{xx} \mathbf{g}, \\ \text{subject to } \mathbf{U}^{\mathrm{c}} \mathbf{g} &= \mathbf{c}_{k_{\mathrm{d}}}, \end{aligned} \tag{11.12}$$

where $\mathbf{R}_{xx} = E\{\mathbf{x}(k)\mathbf{x}^T(k)\}$, $\mathbf{c}_{k_{\mathrm{d}}}$ is a vector of length $(L_u + L_g - 1)$ whose $(k_{\mathrm{d}}+1)$th element is the only non-zero constant, and k_{d} is the decision delay of the beamformer. By using the method of Lagrange multipliers, (11.12) can be easily solved for the LCMV filter:

$$\mathbf{g}_{\mathrm{LCMV}} = \mathbf{R}_{xx}^{-1} \mathbf{U}^{\mathrm{c}T} \left(\mathbf{U}^{\mathrm{c}} \mathbf{R}_{xx}^{-1} \mathbf{U}^{\mathrm{c}T} \right)^{-1} \mathbf{c}_{k_{\mathrm{d}}}. \tag{11.13}$$

In (11.12), there is only one single constraint corresponding to angle θ, which can be easily generalized to multiple linear constraints for a number of different angles. The signals coming from these angles can be either passed through without distortion or forced to zero, which provides us great flexibilities for better control of the beampattern.

The GSC algorithm transforms the LCMV beamforming from a constrained optimization problem into an unconstrained form. Therefore the GSC and LCMV beamformers are essentially the same while the GSC has some implementation advantages, particularly for adaptive operation in time-varying acoustic environments (see [300] for a more detailed discussion on the implementation advantages of the GSC algorithm). One main drawback of the GSC beamformer is that the cancellation of the target signals could occur in reverberant acoustic environments. Many robust variants of the GSC were proposed to combat this problem [4], [166], [134], [163].

It is clear that the LCMV or GSC algorithm does not form beams any more. However, they are still conventionally called beamformers. Now beamforming has the same connotation as spatial filtering and they are often interchangeably used in the literature. We see from the discussion above that successes have been achieved to transplant narrowband beamforming techniques to acoustic environments where speech of interest is a broadband signal. However, the problem of how to deal with near-field sound sources in reverberant acoustic environments still lacks of satisfactory solutions and even a decent framework for analysis. When the sound sources are in the near field of a microphone array and reverberation is significant, it is our belief that the MIMO model is better than the direction-oriented signal model. Formulating and solving the microphone array beamforming problem using the MIMO model seems a promising research direction in the future.

11.2.2 Independent Component Analysis and Blind Source Separation

Another class of source separation technique includes independent component analysis (ICA) [84] and blind source separation (BSS). The ICA problem was first introduced by Herault, Jutten, and Ans in 1985 [162] (a paper in French), and came into vogue in the last decade. In contrast to beamforming which separates sources exploiting the difference in their spatial locations, ICA assumes that the source signals are statistically independent. ICA and BSS algorithms process microphone signals with a *linear* de-mixing system whose outputs are the estimates of the source signals, as illustrated in Fig. 11.5 where there are M independent sound sources and N microphones. The de-mixing system is determined in a learning procedure after which the outputs become independent between each other. Existing ICA methods differ in the way the dependence of the separated speech signals is defined, which include second order statistics (SOS), higher (than second) order cumulant-based statistics, and information-theory-based measures. In this section, we will present a very brief overview of ICA and BSS algorithms but direct the reader to proper references whenever available for further exploration of the problem.

Information-Theory-Based and Cumulant-Based High-Order-Statistics ICA Approaches

The Information Maximization Principle

A number of random variables $\{y_1, y_2, \cdots, y_M\}$ are statistically independent if and only if their joint probability density function (PDF) is the product of the marginal PDFs, or equivalently if their mutual information is zero. The mutual information of the M random variables is given by:

$$I(\mathbf{y}) \triangleq E\left\{\log \frac{p_y(\mathbf{y})}{\prod_{m=1}^{M} p_m(y_m)}\right\}, \qquad (11.14)$$

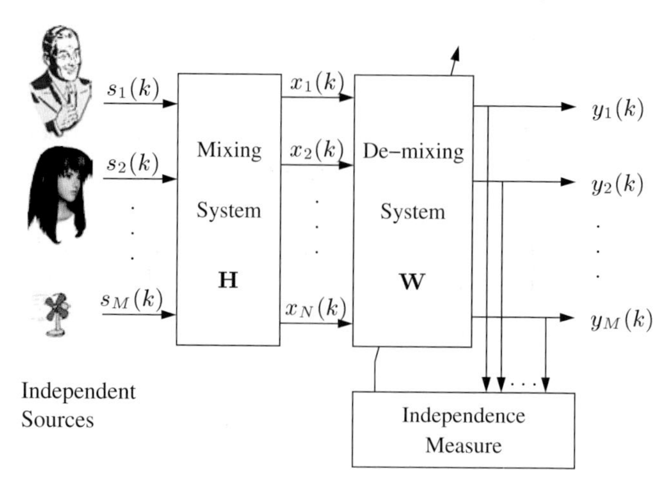

Fig. 11.5. Schematic diagram of a blind source separation system.

where

$$\mathbf{y} = \begin{bmatrix} y_1 & y_2 & \cdots & y_M \end{bmatrix}^T,$$

$p_y(\mathbf{y})$ is the joint PDF, and $p_m(y_m)$ is the marginal PDF.

Although mutual information is an ideal independence measure, its gradient with respect to the linear de-mixing system (which is essential to developing an adaptive ICA algorithm) is impossible to estimate in practice [16]. Alternatively, the joint entropy of a nonlinear transform $\boldsymbol{\psi}(\mathbf{y})$ of the de-mixing system's output $\mathbf{y}$ was proposed to serve as an independence measure [14]. If the nonlinear function

$$\boldsymbol{\psi}(\mathbf{y}) = \begin{bmatrix} \psi_1(y_1) & \psi_2(y_2) & \cdots & \psi_M(y_M) \end{bmatrix}^T$$

is properly selected, then maximizing the joint entropy

$$\begin{aligned} H(\boldsymbol{\psi}) &\triangleq -\int p_\psi(\boldsymbol{\psi}) \log p_\psi(\boldsymbol{\psi}) d\boldsymbol{\psi} \\ &= -E\left\{\log p_\psi(\boldsymbol{\psi})\right\} \end{aligned} \tag{11.15}$$

would wipe off the mutual information in $\mathbf{y}$, i.e., make y_m $(m = 1, 2, \cdots, M)$ independent between each other. This idea is now more widely known as the information maximization or *infomax* (a term coined by Bell and Sejnowski in [14]) principle. Intuitively, this makes sense since the less redundancy (mutual information) between the elements of a random vector, the more information (entropy) they can carry together. More rigorously, this relationship is explained in the following. Since

$$p_\psi(\boldsymbol{\psi}) = \frac{p_y(\mathbf{y})}{\det\left(\partial\boldsymbol{\psi}/\partial\mathbf{y}\right)} = \frac{p_y(\mathbf{y})}{\prod_{m=1}^{M} |\psi'_m(y_m)|}, \tag{11.16}$$

we can rewrite the joint entropy (11.15),

$$\begin{aligned} H(\boldsymbol{\psi}) &= -E\{\log p_{\psi}(\boldsymbol{\psi})\} \\ &= -E\{\log p_y(\mathbf{y})\} + E\left\{\log \prod_{m=1}^{M} |\psi'_m(y_m)|\right\} \\ &= -E\left\{\log \frac{p_y(\mathbf{y})}{\prod_{m=1}^{M} p_m(y_m)}\right\} + E\left\{\log \frac{\prod_{m=1}^{M} |\psi'_m(y_m)|}{\prod_{m=1}^{M} p_m(y_m)}\right\} \\ &= -I(\mathbf{y}) + \sum_{m=1}^{M} E\left\{\log \frac{|\psi'_m(y_m)|}{p_m(y_m)}\right\}. \end{aligned} \tag{11.17}$$

Clearly, if $|\psi'_m(y_m)| = p_m(y_m)$, $\forall m$, then $H(\boldsymbol{\psi}) = -I(\mathbf{y})$ and maximizing the joint entropy of $\boldsymbol{\psi}$ will make the mutual information in $\mathbf{y}$ decay to zero. The condition indicates that the desirable nonlinear transform $\psi_m(y_m)$ would be the cumulative distribution function (CDF) of the source estimate y_m. However, it is very difficult in practice to find a nonlinear transform that is exactly the CDF of the unknown sources and using a poor estimate would potentially interfere the asymptotic estimates of an infomax algorithm. For super-Gaussian distributions (like speech), a good choice for the nonlinear function is $\psi_m(y_m) = \tanh(y_m)$ [15].

Infomax-Based Adaptive ICA Algorithms

Now let's explore how to develop adaptive ICA algorithms based on the infomax principle. We consider an instantaneous mixing and de-mixing system. In the absence of noise, we have

$$\mathbf{x}(k) = \mathbf{H}\mathbf{s}(k), \tag{11.18}$$

$$\mathbf{y}(k) = \mathbf{W}\mathbf{x}(k), \tag{11.19}$$

where the signal vectors $\mathbf{s}(k)$, $\mathbf{x}(k)$, $\mathbf{y}(k)$ and the system matrices $\mathbf{H}_{N\times M}$, $\mathbf{W}_{M\times N}$ are defined similarly to what were presented in the description of a MIMO model in (2.9) with the length of channel impulse responses being 1. Then the joint entropy of $\boldsymbol{\psi}$ in terms of $\mathbf{W}$ is found as

$$H(\boldsymbol{\psi}|\mathbf{W}) = -E\{\log p_{\psi}(\boldsymbol{\psi}|\mathbf{W})\} = -E\left\{\frac{p_x(\mathbf{x})}{|J(\mathbf{x}, \boldsymbol{\psi})|}\right\}, \tag{11.20}$$

where

$$J(\mathbf{x}, \boldsymbol{\psi}) = \det(\mathbf{W}) \prod_{m=1}^{M} |\psi'_m(y_m)| \tag{11.21}$$

is the Jacobian of the transformation from $\mathbf{x}$ to $\boldsymbol{\psi}$ [247]. Substituting (11.21) into (11.20) yields

$$H(\boldsymbol{\psi}|\mathbf{W}) = H(\mathbf{x}) + \log|\det(\mathbf{W})| + \sum_{m=1}^{M} E\left\{\log|\psi'_m(y_m)|\right\}. \tag{11.22}$$

It can be easily shown that

$$\frac{\partial \log|\det(\mathbf{W})|}{\partial \mathbf{W}} = \left(\mathbf{W}^{-1}\right)^T, \tag{11.23}$$

$$\frac{\partial}{\partial \mathbf{W}}\left[\sum_{m=1}^{M} E\left\{\log|\psi'_m(y_m)|\right\}\right] = -E\left\{\boldsymbol{\phi}(\mathbf{y})\mathbf{x}^T\right\}, \tag{11.24}$$

where

$$\boldsymbol{\phi}(\mathbf{y}) = \left[\begin{array}{cccc} -\frac{\psi''_1(y_1)}{\psi'_1(y_1)} & -\frac{\psi''_2(y_2)}{\psi'_2(y_2)} & \cdots & -\frac{\psi''_M(y_M)}{\psi'_M(y_M)} \end{array}\right]^T.$$

Hence, the stochastic gradient of $H(\boldsymbol{\psi}|\mathbf{W})$ with respect to $\mathbf{W}$ is given by

$$\frac{\partial H(\boldsymbol{\psi}|\mathbf{W})}{\partial \mathbf{W}} = \left(\mathbf{W}^{-1}\right)^T - E\left\{\boldsymbol{\phi}(\mathbf{y})\mathbf{x}^T\right\}. \tag{11.25}$$

At every point in time, we can approximate $E\left\{\boldsymbol{\phi}(\mathbf{y})\mathbf{x}^T\right\}$ with an instantaneous estimate $\boldsymbol{\phi}[\mathbf{y}(k)]\mathbf{x}^T(k)$ like what the LMS algorithm does, and then deduce the infomax stochastic gradient ascent adaptive ICA algorithm [14]:

$$\mathbf{W}(k+1) = \mathbf{W}(k) + \mu_{\text{im}}\left\{\left[\mathbf{W}^{-1}(k)\right]^T - \boldsymbol{\phi}[\mathbf{y}(k)]\mathbf{x}^T(k)\right\}, \tag{11.26}$$

where μ_{im} is a positive step size and the subscript im stands for infomax.

It has been known that the infomax stochastic gradient ascent algorithm is not equivariant, i.e., its performance depends on the mixing matrix $\mathbf{H}$ [328]. As a result, the convergence of the algorithm can vary significantly from environment to environment. The natural gradient algorithm [80], [7] rescales the entropy gradient by post-multiplying it with $\mathbf{W}^T\mathbf{W}$ and update the de-mixing system matrix with

$$\mathbf{W}(k+1) = \mathbf{W}(k) + \mu_{\text{ng}}\left\{\mathbf{I} - \boldsymbol{\phi}[\mathbf{y}(k)]\mathbf{y}^T(k)\right\}\mathbf{W}(k), \tag{11.27}$$

where $\mathbf{I}$ is the identity matrix, μ_{ng} is again a positive step size, and the subscript ng stands for natural gradient. The natural gradient algorithm is not only equivariant [61] but also able to keep $\mathbf{W}(k)$ from becoming singular [328]. Therefore, by using the natural gradient, the adaptive algorithm is made more robust and more stable, while its computational complexity is reduced without the need of computing the inverse of $\mathbf{W}(k)$.

When the infomax natural gradient adaptive algorithm is used for speech signals, its convergence can be further enhanced with a normalization in the gradient for tracking the inherent non-stationarity of speech signals. An NLMS-type normalization was proposed in [61], but it falls short in practice with highly correlated signals. In [36], an RLS-type normalization was suggested and could achieve better results. We intend not to develop these modifications to the natural gradient algorithm. The interested readers can refer to those references given above for more details.

Maximum-Likelihood Principle and Its Link to the Infomax Principle

Maximum likelihood (ML) is a very useful tool in estimation theory and can be also applied to the ICA problem [132], [7], [252]. Suppose that we have K observation vectors $\mathbf{x}(0), \mathbf{x}(1), \cdots, \mathbf{x}(K-1)$ along time. Then the log-likelihood for a hypothesized source PDF $\hat{p}_s(\mathbf{s})$ is defined as

$$\mathcal{L}(\hat{p}_s) \triangleq \frac{1}{K}\sum_{k=0}^{K-1} \log p\left(\mathbf{x}(k)|\hat{p}_s\right), \tag{11.28}$$

where $p\left(\mathbf{x}(k)|\hat{p}_s\right)$ is the conditional PDF of $\mathbf{x}(k)$ given the distribution $\hat{p}_s$. When K is large, the sample average converges to its ensemble average such that

$$\lim_{K\to\infty} \mathcal{L}(\hat{p}_s) = \int p\left(\mathbf{x}|p_s\right) \log p\left(\mathbf{x}|\hat{p}_s\right) d\mathbf{x}, \tag{11.29}$$

where $p\left(\mathbf{x}|p_s\right)$ is the conditional PDF of $\mathbf{x}$ given the *true* source distribution p_s. Substituting

$$p\left(\mathbf{x}|\hat{p}_s\right) = \frac{p\left(\mathbf{x}|\hat{p}_s\right)}{p\left(\mathbf{x}|p_s\right)} \cdot p\left(\mathbf{x}|p_s\right) \tag{11.30}$$

into (11.29) produces

$$\begin{aligned}\lim_{K\to\infty} \mathcal{L}(\hat{p}_s) &= \int p\left(\mathbf{x}|p_s\right) \log \frac{p\left(\mathbf{x}|\hat{p}_s\right)}{p\left(\mathbf{x}|p_s\right)} d\mathbf{x} + \int p\left(\mathbf{x}|p_s\right) \log p\left(\mathbf{x}|p_s\right) d\mathbf{x} \\ &= -d_{\mathrm{KL}}\left[\mathbf{x}|p_s, \mathbf{x}|\hat{p}_s\right] - H(\mathbf{x}),\end{aligned} \tag{11.31}$$

where

$$\begin{aligned}d_{\mathrm{KL}}\left[\mathbf{x}|p_s, \mathbf{x}|\hat{p}_s\right] &\triangleq \int p\left(\mathbf{x}|p_s\right) \log \frac{p\left(\mathbf{x}|p_s\right)}{p\left(\mathbf{x}|\hat{p}_s\right)} d\mathbf{x} \\ &= E\left\{\log \frac{p\left(\mathbf{x}|p_s\right)}{p\left(\mathbf{x}|\hat{p}_s\right)}\right\}\end{aligned}$$

is the Kullback-Leibler divergence between the two distributions $p\left(\mathbf{x}|p_s\right)$ and $p\left(\mathbf{x}|\hat{p}_s\right)$, and $H(\mathbf{x}) = -E\left\{\log p_x(\mathbf{x})\right\}$ is the joint entropy of $\mathbf{x}$. Note that in the last step of (11.31), we have used the fact that $p(\mathbf{x}|p_s) = p_x(\mathbf{x})$.

Note that $\mathbf{x}(k) = \mathbf{H}\mathbf{s}(k)$. Hence we have

$$p(\mathbf{x}|p_s) = \frac{p_s(\mathbf{s})}{\det(\partial\mathbf{x}/\partial\mathbf{s})} = \frac{p_s(\mathbf{s})}{\det(\mathbf{H})}. \tag{11.32}$$

Using (11.32), we deduce that

$$\begin{aligned}d_{\mathrm{KL}}\left[\mathbf{x}|p_s, \mathbf{x}|\hat{p}_s\right] &= E\left\{\log \frac{p(\mathbf{x}|p_s)}{p(\mathbf{x}|\hat{p}_s)}\right\} \\ &= E\left\{\log \left[\frac{p_s}{\det(\mathbf{H})} \cdot \frac{\det(\mathbf{H})}{\hat{p}_s}\right]\right\} \\ &= E\left\{\log \frac{p_s}{\hat{p}_s}\right\} \\ &= d_{\mathrm{KL}}\left[p_s, \hat{p}_s\right].\end{aligned} \tag{11.33}$$

Consequently, (11.31) becomes

$$\lim_{K\to\infty} \mathcal{L}(\hat{p}_s) = -d_{\mathrm{KL}}[p_s, \hat{p}_s] - H(\mathbf{x}). \tag{11.34}$$

Since $H(\mathbf{x})$ is independent of $\hat{p}_s$, maximizing the likelihood with respect to $\hat{p}_s$ is equivalent to minimizing the Kullback-Leibler divergence between p_s and $\hat{p}_s$. In other words, the maximum-likelihood ICA method tries to reduce the mismatch between the true and hypothesized source distributions. In the formulation of an ICA problem, $\mathbf{y}$ is regarded as an estimate of $\mathbf{s}$. Therefore $\hat{p}_s = p_y$ and p_y is obtained from p_x through a hypothesized de-mixing matrix $\mathbf{W}$. Finally, the ML estimate of the de-mixing matrix is

$$\mathbf{W}_{\mathrm{ML}} = \arg\min_{\mathbf{W}} d_{\mathrm{KL}}[p_s, p_y(\mathbf{Wx})]. \tag{11.35}$$

However, an obstacle that prevents the use of the ML method in practice is the lack of *a priori* knowledge about the true source distribution p_s. The only thing that we suppose to know is that the M sources are independent and hence

$$p_s(\mathbf{s}) = \prod_{m=1}^{M} p_m(s_m). \tag{11.36}$$

Recall that the infomax principle confronts the same difficulty as explained in the discussion that follows (11.17). The nonlinear transforms are ideally the CDFs of the source signals. It is clear now that both the ML and infomax algorithms need to assume the form of the source distributions. If they assume the same, i.e., $|\psi'_m(y_m)| = p_m(s_m)$, then we can rewrite (11.17) as

$$\begin{aligned} H(\boldsymbol{\psi}) &= -E\left\{\log \frac{p_y(\mathbf{y})}{\prod_{m=1}^{M} p_m(s_m)}\right\} = -E\left\{\log \frac{p_y(\mathbf{y})}{p_s(\mathbf{s})}\right\} \\ &= -d_{\mathrm{KL}}[p_y(\mathbf{y}), p_s(\mathbf{s})]. \end{aligned} \tag{11.37}$$

Since the Kullback-Leibler divergence is in general not symmetric, i.e.,

$$d_{\mathrm{KL}}[p_y(\mathbf{y}), p_s(\mathbf{s})] \neq d_{\mathrm{KL}}[p_s(\mathbf{s}), p_y(\mathbf{y})],$$

the ML and infomax independence measures are not identical. However, $d_{\mathrm{KL}}[p_y(\mathbf{y}), p_s(\mathbf{s})]$ and $d_{\mathrm{KL}}[p_s(\mathbf{s}), p_y(\mathbf{y})]$ quantify, nevertheless, the same mismatch between $p_y(\mathbf{y})$ and $p_s(\mathbf{s})$ (only from different perspectives), which suggests a very close connection between the ML and infomax principles. This connection has been recognized independently in a number of previous studies [250], [62], but it is the first time (to the best of our knowledge) that the slight distinction between ML and infomax due to the asymmetry of Kullback-Leibler divergence is clearly explained.

Cumulant-Based ICA Approaches

The independence measures (also known as contrast functions) defined above using various information-theoretic principles are all centered around the PDFs of random variables. Note that the PDF of a multivariate random vector can be described with statistical moments or cumulants. This leads to another class of high-order-statistics (HOS) ICA approaches that define their independence measures using cumulants.

For a single random variable X of distribution $p_X(x)$, its pth order cumulant is defined as

$$\kappa_p(X) \triangleq \frac{d^p}{d\omega^p} \ln \left\{ \int_{-\infty}^{\infty} e^{j\omega x} p_X(x) dx \right\} \bigg|_{\omega=0} . \tag{11.38}$$

In terms of the central moments

$$\mu_p(X) = E\left\{[X - E\{X\}]^p\right\}, \tag{11.39}$$

the first several cumulants are found as

$$\begin{aligned} \kappa_1(X) &= \mu_1(X), \\ \kappa_2(X) &= \mu_2(X), \\ \kappa_3(X) &= \mu_3(X), \\ \kappa_4(X) &= \mu_4(X) - 3\mu_2^2(X), \\ \kappa_5(X) &= \mu_5(X) - 10\mu_2(X)\mu_3(X). \end{aligned} \tag{11.40}$$

The second-order cumulant is known as variance. In the ICA society, the fourth-order cumulant is also referred to as kurtosis. But this is apparently different from the classical Pearson's definition of kurtosis [251], [262] in that the fourth-order cumulant is not normalized by $\mu_2^2(X)$ [1].

For a number of random variables X_1, X_2, X_3, X_4, their most frequently used second- and fourth-order cumulants are expressed in terms of their central moments as follows:

[1] If X_1 and X_2 are two independent random variables, it holds that

$$\kappa_4(X_1 + X_2) = \kappa_4(X_1) + \kappa_4(X_2).$$

Then the kurtosis of a sum of a large number of independent random variables with positive kurtosis go to infinity (or negative infinity with negative kurtosis), which causes the conflict with the central limit theorem. According to the central limit theorem, the distribution of a sum of a large number of independent random variables is approximately Gaussian whose kurtosis is 0. However, Pearson's definition of kurtosis does not have this difficulty in explaining the central limit theorem. Therefore, we believe that, from a perspective of statistical theory, Pearson's definition of kurtosis is better than that used in the ICA society. The reason why a simplified kurtosis was employed in the ICA society is that a learning rule based on the fourth-order cumulant is easier to develop (computing its derivative is less tedious and more tractable).

$$\begin{aligned}\kappa_2(X_1, X_2) &= \mu_2(X_1, X_2), \qquad (11.41)\\ \kappa_4(X_1, X_2, X_3, X_4) &= \mu_4(X_1, X_2, X_3, X_4) - \\ &\quad \mu_2(X_1, X_2)\mu_2(X_3, X_4) - \mu_2(X_1, X_3)\mu_2(X_2, X_4) - \\ &\quad \mu_2(X_1, X_4)\mu_2(X_2, X_3), \qquad (11.42)\end{aligned}$$

where

$$\begin{aligned}\mu_2(X_1, X_2) &= E\left\{(X_1 - E\{X_1\})(X_2 - E\{X_2\})\right\}, \\ \mu_4(X_1, X_2, X_3, X_4) &= E\left\{(X_1 - E\{X_1\})(X_2 - E\{X_2\}) \times \right.\\ &\qquad \left.(X_3 - E\{X_3\})(X_4 - E\{X_4\})\right\},\end{aligned}$$

are the second- and fourth-order central moments, respectively.

Kurtosis is a classical measure of non-Gaussianity. The kurtosis of a Gaussian distribution is 0. Distributions that have a negative kurtosis are called sub-Gaussian, and those with a positive kurtosis are called super-Gaussian. Motivated by the central limit theorem which tells that a mixture of independent non-Gaussian random variables is more like a Gaussian-distributed signal, non-Gaussianity was suggested as a way to measure the independence of the separated signals in the BSS problem and kurtosis (the fourth-order cumulant) was proposed in a so-called FastICA algorithm [183]. More rigorously, the FastICA is based on the following property: assume that a separated signal y_m has a unit variance and so do the independent source signals. When y_m is identical to one of the source signals up to a sign (meaning that it has been successfully separated from the other sources), $\kappa_4(y_m)$ is locally minimized or maximized [183]. Therefore, an ICA learning rule can be developed via searching for local kurtosis extrema. While the FastICA algorithm has a nice speed of convergence, kurtosis or fourth-order cumulant alone is only a sketchy approximation of independence, which leads to inaccurate source separation results.

Actually, the idea of using a cumulant expansion to approximate an information-theory-based independence measure was first proposed by Comon in 1994 [84], prior to the FastICA algorithm. In [84], the target independence measure is mutual information. Later on, Cardoso generalized the idea and presented cumulant-based approximations to a number of information-theory-based independence measures [64]. For example, the Kullback-Leibler divergence between two multivariate distributions $p_s(\mathbf{s})$ and $p_y(\mathbf{y})$ has the following approximation:

$$\begin{aligned}& d_{\mathrm{KL}}\left[p_s(\mathbf{s}), p_y(\mathbf{y})\right] \\ &\approx \frac{1}{4}\sum_{i,j}\left[\kappa_2(s_i, s_j) - \kappa_2(y_i, y_j)\right]^2 + \\ &\quad \frac{1}{48}\sum_{i,j,p,q}\left[\kappa_4(s_i, s_j, s_p, s_q) - \kappa_4(y_i, y_j, y_p, y_q)\right]^2, \qquad (11.43)\end{aligned}$$

where only the second- and fourth-order cumulants are used.

Another thread of ideas using cumulants to develop HOS ICA algorithms was based on the observation that cumulants have multilinearity property and can be considered as tensors [85]. Consequently, for a linear transform, the cumulants of the input and the cumulants of the output signals have an algebraic structure, which can be exploited to identify the linear transform and further perform source separation. The algebraic technique will not be elaborated and the interested reader can refer to [64] for more details. We only want to point out that this class of method is very computationally intensive.

Second-Order-Statistics BSS Approaches

In general, computing HOS of a multivariate random vector is inaccurate (a biased estimate) and unreliable (with a large variance) when the observed sample space is not large enough. For example, sample kurtosis is extremely sensitive to outliers[2]. In addition, the HOS-based ICA learning rules are inevitably complex and often converge slowly. As a result, interests arose in seeking for BSS algorithms using only SOS and have been present for quite some time.

Being uncorrelated (second-order independence) is a necessary but not sufficient condition for statistical independence. Using only SOS, decorrelation can be achieved but it does not guarantee source separation [314]. For an instantaneous mixing MIMO system with white and independent inputs, it was shown in [63] that achieving mutual uncorrelation among the estimated source signals is not sufficient to solve the BSS problem. This could be more thoroughly understood if one recalls the discussion about the blind identifiability for such a system in Sect. 6.3.2. Using SOS, the mixing system $\mathbf{H}$ can be blindly identified up to an orthonormal matrix. Similarly, when blind source separation is concerned, the de-mixing system can be determined also up to an orthonormal transform. Suppose that a *whitening* matrix $\mathbf{W}$ has been computed such that the covariance matrix of $\mathbf{y} = \mathbf{W}\mathbf{x}$ is the identity matrix, i.e.,

$$\mathbf{R}_{yy} \triangleq E\left\{\mathbf{y}\mathbf{y}^T\right\} = \mathbf{W}\mathbf{R}_{xx}\mathbf{W}^T = \mathbf{I}. \tag{11.44}$$

Then, for any orthonormal transform $\mathbf{V}$, $\mathbf{y}' = \mathbf{V}\mathbf{y}$ is also a "spatially" white signal vector. Although there exists one orthonormal matrix $\mathbf{V}$ such that the product $\mathbf{VW}$ is the true solution for the BSS problem (i.e., $\mathbf{VWx}$ contains the separated source signals), we do not know how to compute $\mathbf{V}$ without supplementing other information. Of course high-order mutual independence

[2] An outlier is a sample (or measurement) that is *unusually* large or small compared to the rest of the data. Realizations of a random variable following a given distribution are expected to fall within a specific range extending from the mean to a point relative to the standard deviation. Those that do not are called outliers.

is not considered. Otherwise the developed algorithm will not be an SOS-based method.

In most acoustic applications, speech is the input signal of primary interest. Therefore, speech characteristics were explored to help solve the BSS problem in addition to SOS. Non-stationarity and non-whiteness are two most widely known features.

The idea of exploiting non-stationarity of speech signals was first proposed in [314] and was investigated in many other studies, e.g., [299], [221], [184]. Speech is inherently non-stationary, but can be regarded as a stationary signal in a short period of time. When the observed signals are grouped into frames of a proper length, short-time SOS of the outputs in each frame can be computed. Presumably the mixing system varies at a rate much slower than speech and the de-mixing system $\mathbf{W}$ needs to simultaneously diagonalize the output covariance matrices in a number of consecutive frames. When the number of frames is large enough, $\mathbf{W}$ can be uniquely identified.

Non-whiteness of speech signals can be taken advantage of in a similar manner [227]. In this case,

$$\begin{aligned}\mathbf{R}_{xx}(\tau) &\stackrel{\triangle}{=} E\left\{\mathbf{x}(k)\mathbf{x}^T(k-\tau)\right\} \\ &= \mathbf{H}\mathbf{R}_{ss}(\tau)\mathbf{H}^T, \quad \tau = 0, 1, 2, \cdots, \end{aligned} \tag{11.45}$$

does not vanish for $\tau \neq 0$. Consequently, the de-mixing system should diagonalize

$$\mathbf{R}_{yy}(\tau) = \mathbf{W}\mathbf{R}_{xx}(\tau)\mathbf{W}^T \tag{11.46}$$

for all $\tau \geq 0$. This also leads to a unique solution for $\mathbf{W}$.

Apparently, non-stationarity and nonwhiteness are not exclusive. They together can be used for solving the BSS problem with SOS [79]. Although the problems with non-stationary and colored inputs are formulated in different ways, solving them involves the same technique for joint diagonalization of covariance matrices. There is a very rich literature in joint diagonalization and the interested reader can refer to [18], [194], [195], and the references therein for a complete exposition.

Convolutive Mixtures and Frequency-Domain BSS

So far, the discussion of BSS algorithms has been concerned with only instantaneous mixing systems. However, the mixing process in most acoustic MIMO systems is not instantaneous but convolutive in nature due to reverberation. So it would be of more practical interest to investigate BSS algorithms for convolutive mixtures.

As we see above, BSS for instantaneous mixtures is by all means not an easy problem. It can be imagined that dealing with convolutive mixtures would be more challenging. But the time-domain convolutive mixture $\mathbf{x}(k)$ given in (2.9) can be transformed to several instantaneous mixtures in the frequency domain via discrete Fourier transform [104], [278], [272]. Blind source separation is then performed with respect to instantaneous mixtures independently at each frequency bin. Both HOS and SOS based techniques are ready to be applied. However, this does not indicate that the BSS problem for convolutive mixtures has been successfully solved. It must be noted that independent source signals in an instantaneous mixture can at best be blindly separated up to a scale and a permutation. This results in the possibility that the recovered signal y_m is not a consistent estimate of one of the source signals over all frequencies, which is known as the permutation inconsistency problem [272], [185]. The degradation of speech quality caused by the permutation ambiguity is only slightly noticeable when the length of the mixing channels is short. The impact becomes more evident when the channels are longer in reverberant environments. Although a number of methods were proposed to align permutations of the de-mixing filters over all the frequencies [272], [248], [238], [264], [186], [266], this is still an open problem under active research.

Fundamentally, the problem of permutation inconsistency is caused by the strategy of performing BSS independently at each frequency bin. Intuitively it can be precluded if the cost function for BSS is constructed in the time domain with each channel impulse response being treated as an unbroken polynomial as suggested by [196] and [58]. But the computational complexity for implementation and mathematical complexity for analysis are both very intensive.

11.3 A Synergistic Solution to Source Separation and Speech Dereverberation

In a realistic, reverberant acoustic environment where there are multiple sound sources, co-channel signals and their temporal interference are interwoven together in microphone outputs. Consequently, at the same time as a source separation algorithm is being developed, speech dereverberation needs to be given the same amount of attention and consideration. While a beamformer can suppress reverberation coming from a direction different from the one that it is steered to, BSS algorithms would regard independent though distorted source signals as valid solutions. Actually, neither beamforming nor BSS takes speech dereverberation into account in their problem formulation. In this section, we plan to examine the problems of source separation and speech dereverberation in a reverberant environment from a different perspective and want to develop a synergistic solution based on blind multichannel identification.

As clearly explained in Sect. 6.3, a useful solution to blind identification of acoustic MIMO systems, particularly those corresponding to reverberant environments, has not yet been successfully worked out. But this does not mean that we have been at our wit's end. It is worth noting that we already have mature tools to blindly identify a SIMO system, even with long channels.

With ingenious use of these tools, the impulse responses of a slowly-varying acoustic MIMO system can also be obtained.

In most practical scenarios like teleconferencing, people do not always speak at the same time. It can be assumed that from time to time each speaker occupies at least one exclusive interval alone and when they start talking simultaneously the room acoustics have not yet significantly changed. As a result, in each single-talk interval, a SIMO system is blindly identified and its channel impulse responses are saved for later use in source separation and speech deconvolution during double or multiple talk periods. Apparently, this requires that we are able to distinguish single and multiple talks. Speech source detection is an interesting and important technique, but is not within the scope of this book. The reader who is interested in this topic can read a recently published paper [54] and references therein. For the principle and adaptive implementation of blind SIMO identification, please see Sect. 6.2.

Having the estimated channel impulse responses and following what was developed in Sect. 7.2, we can separate co-channel interfering signals and reverberation, and convert an $M \times N$ MIMO system with $M < N$ into M SIMO systems, which are free of co-channel interference (CCI). Although source separation at this point has been achieved, the obtained multiple CCI-free speech signals would sound possibly more reverberant due to the prolonged impulse responses of the equivalent channels. Since the channel impulse responses corresponding to the same source do not share any common zeros as an prerequisite for blind SIMO identifiability in the single-talk periods, the channels of each one of these derived CCI-free SIMO systems do not share any common zeros either. In addition, the transfer functions of these channels are linear combinations of the original channel transfer functions. Finally, for these CCI-free SIMO systems, the MINT method (see Sect. 7.3) are used to eliminate reverberation. This multiple-step method of source separation and speech dereverberation for an acoustic MIMO system can be easily tailored such that it can be more efficiently implemented to solve the problem with single source [180] or multiple sources [181] of interest. Here we only strategically describe this method to avoid unnecessary repetition since the processings of each step have been all painstakingly developed in the theoretical part of this book. In the following, we will use an example to illustrate and evaluate how it works.

Example 11.1. **Extracting and Dereverberating a Speech Signal of Interest in a 3×4 System**

This example considers a 3×4 acoustic MIMO system in the varechoic chamber at Bell Labs (see Sect. 2.3.1 for a description of the varechoic chamber). A diagram of the floor plan layout is given in Fig. 11.6, which shows the positions of the four microphones and three sound sources. The first female speech source is the target for extraction. The measured channel impulse responses were collected in the database of [156]. Four panel configurations

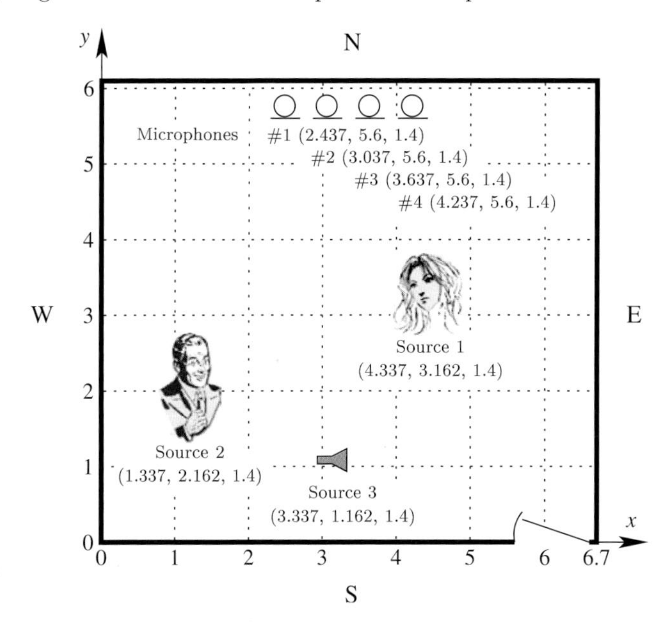

Fig. 11.6. Floor plan of the varechoic chamber at Bell Labs (coordinate values measured in meters) where a 3×4 acoustic system is constructed and its channel impulse responses were measured.

will all be employed in this example (see Table 2.1 for a list of the four panel configurations and the corresponding average reverberation time).

Figure 11.7 illustrates the whole procedure of source separation and speech dereverberation for such a system in the absence of additive noise at the microphones. Suppose that the channel impulse responses h_{nm} ($m = 1, 2, 3$, and $n = 1, 2, 3, 4$) have been blindly identified. We select three outputs at a time to construct, with the three inputs, the pth ($p = 1, 2, \cdots, P$ and $P = 4$) 3×3 subsystem. From such a subsystem, the speech of interest is extracted though distorted to get $y_{s_1,p}(k)$ by using a set of filters $\{h_{s_1,p1}, h_{s_1,p2}, h_{s_1,p3}\}$, whose transfer functions are linear combinations of $H_{nm}(z)$. Then the equivalent channel impulse responses from the three source signals to $y_{s_1,p}(k)$ are found as $f_{s_1,p}, f_{s_1,ps_2}, f_{s_1,ps_3}$, respectively, as shown in Fig. 11.7(a). It is expected that f_{s_1,ps_2} and f_{s_1,ps_3} would be equal to zero. Collecting $y_{s_1,p}(k)$ from the four subsystems, we obtain a 1×4 SIMO system with $s_1(k)$ as the sole input and then use the MINT method to carry out speech dereverberation. Figure 11.7(b) demonstrates the last step of the processing.

Experimental Setup

In this example, signals were sampled at 8 kHz and the original impulse response measurements have 4096 samples. In the cases of 89% and 75% panels open, energy in reverberation decays quickly with arrival time and we cut

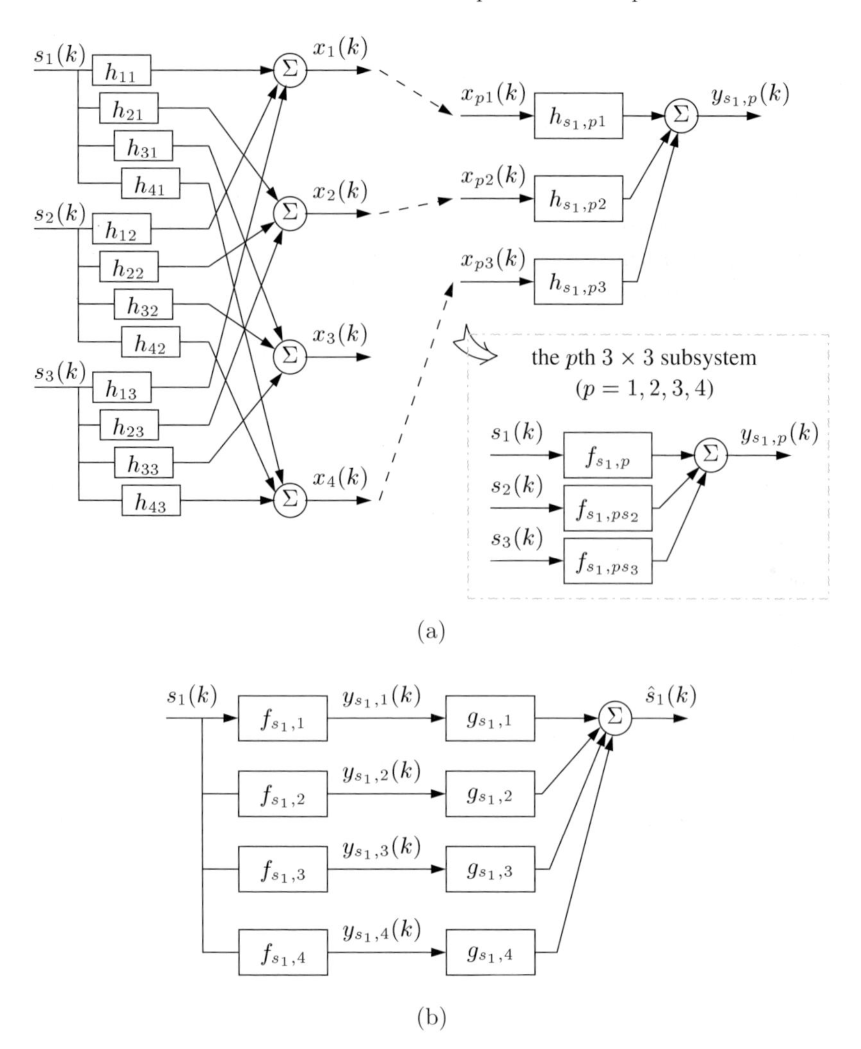

Fig. 11.7. Illustration of speech extraction and dereverberation in a 3×4 acoustic MIMO system in the absence of additive noise. (a) Extracting the speech signal of interest in the pth ($p = 1, 2, \cdots, P$ and $P = 4$) 3×3 subsystem, and (b) speech dereverberation using the MINT method for $s_1(k)$.

impulse responses at $L_h = 256$. When 30% or none of planes are open, we set $L_h = 512$. Among the three sources, the first two are speech sources, one female and the other male, and the third is babbling noise recorded in the New York Stock Exchange (NYSE). The two speech sources are equally loud in volume while the noise source is 5 dB weaker than the speech sources. The time sequence and spectrogram (30 Hz bandwidth) of these source signals for the first 1.5 seconds are shown in Fig. 11.8. From the spectrograms, we can tell that babbling noise is a quite wideband signal. Using these source signals and channel impulse responses, we calculate microphone outputs by convolution.

Performance Measures

Before presenting the experimental results, we would like to explain the performance measures to be used. To evaluate the performance of a blind SIMO identification algorithm, we will use the normalized projection misalignment (NPM) as defined in Sect. 6.2.6. To assess the performance of source separation and speech dereverberation, two measures, namely signal-to-interference ratio (SIR) and speech spectral distortion, are employed in the simulations.

For the SIR, we refer to the notion suggested in [185] but define the measure in a different manner since their definition is applicable only for an $M \times M$ MIMO system. In this example, our interest is in the more general $M \times N$ MIMO systems with $M < N$. Moreover, the M sources are equally important in [185] while here the first source is the speech source of interest and is more important than others.

Since only the first speech source is what we are interested in extracting, the SIR would be defined in a way where a component contributed by $s_1(k)$ is treated as the signal and the rest as the interference. We first define the input SIR at microphone n as:

$$\mathrm{SIR}_n^{\mathrm{in}} \triangleq \frac{E\left\{[h_{n1} * s_1(k)]^2\right\}}{E\left\{[h_{n2} * s_2(k)]^2\right\} + E\left\{[h_{n3} * s_3(k)]^2\right\}}, \quad n = 1, 2, \cdots, 4, \tag{11.47}$$

where $*$ denotes linear convolution. Then the overall average input SIR is computed as

$$\mathrm{SIR}^{\mathrm{in}} \triangleq \frac{1}{N} \sum_{n=1}^{N} \mathrm{SIR}_n^{\mathrm{in}}, \tag{11.48}$$

where $N = 4$ for this example.

The output SIR is defined in the same way. We first investigate the output SIR in the pth ($p = 1, 2, 3, 4$) subsystem:

$$\mathrm{SIR}_p^{\mathrm{out}} \triangleq \frac{E\left\{[f_{s_1,p} * s_1(k)]^2\right\}}{E\left\{[f_{s_1,ps_2} * s_2(k)]^2\right\} + E\left\{[f_{s_1,ps_3} * s_3(k)]^2\right\}}. \tag{11.49}$$

Then the overall average output SIR is found as:

$$\mathrm{SIR}^{\mathrm{out}} \triangleq \frac{1}{P} \sum_{p=1}^{P} \mathrm{SIR}_p^{\mathrm{out}}, \tag{11.50}$$

where $P = 4$ for this example.

To assess the quality of dereverberated speech signals, the Itakura-Saito (IS) distortion measure has been widely used [257]. It is the ratio of the residual energies produced by the original speech when inverse filtered using

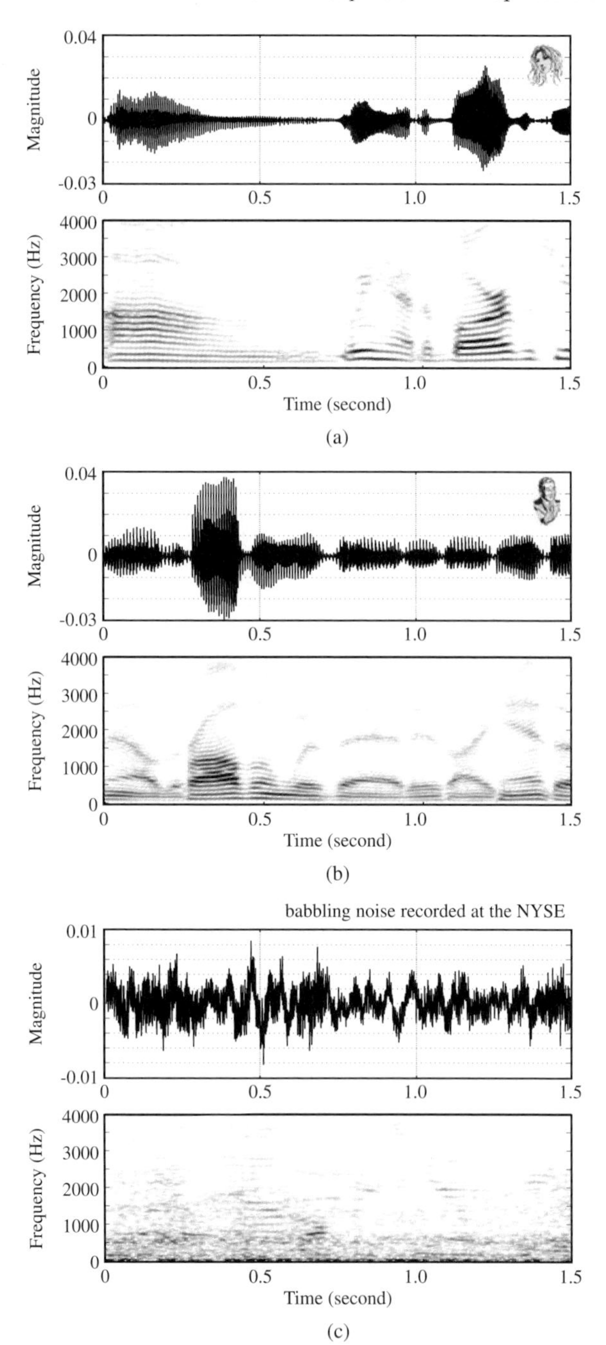

Fig. 11.8. Time sequence and spectrogram (30 Hz bandwidth) of the two speech and one noise source signals for the first 1.5 seconds. (a) $s_1(k)$ (female speaker), (b) $s_2(k)$ (male speaker), and (c) babbling noise recorded in the NYSE.

Table 11.1. Performance of the source separation and speech dereverberation algorithm for extracting a speech signal of interest from a 3×4 acoustic MIMO system in a reverberant environment simulated with data measured in the varechoic chamber at Bell Labs with different panel configurations.

Open Panels	T_{60} (ms)	L_h	NPM (dB) H_{s_1}	NPM (dB) H_{s_2}	NPM (dB) H_{s_3}	$\mathrm{SIR}^{\mathrm{in}}$ (dB)	$\mathrm{SIR}^{\mathrm{out}}$ (dB)	$d_{\mathrm{IS}}^{\mathrm{SS}}$	$d_{\mathrm{IS}}^{\mathrm{SD}}$
89%	240	256	-18.64	-16.82	-20.75	0.945	44.755	4.41	0.07
75%	310	256	-17.92	-18.93	-23.72	1.870	45.163	5.48	0.19
30%	380	512	-12.13	-13.04	-12.55	0.836	40.274	5.65	0.32
0%	580	512	-12.56	-13.51	-16.90	1.725	42.275	9.54	0.14

NOTES:
H_{s_m} represents the SIMO system corresponding to source s_m.

the LP coefficients derived from the original and processed speech. Let $\boldsymbol{\alpha}_t$ and $\boldsymbol{\alpha}'_t$ be the LP coefficient vectors of an original speech signal frame $\mathbf{s}_t$ and the corresponding processed speech signal frame $\mathbf{s}'_t$ under examination, respectively. Denote $\mathbf{R}_{tt}$ as the Toeplitz autocorrelation matrix of the original speech signal. Then the IS measure is given as:

$$d_{\mathrm{IS},t} = \frac{\boldsymbol{\alpha}_t'^T \mathbf{R}_{tt} \boldsymbol{\alpha}_t'^T}{\boldsymbol{\alpha}_t^T \mathbf{R}_{tt} \boldsymbol{\alpha}_t^T} - 1. \tag{11.51}$$

Such a measure is calculated on a frame-by-frame basis. For the whole sequence of two speech signals, the mean IS measure is obtained by averaging $d_{\mathrm{IS},t}$ over all frames. According to [255], the IS measure exhibits a high correlation (0.59) with subjective judgments, suggesting that the IS distance is a good objective measure of speech quality. It was reported in [71] that the difference in mean opinion score (MOS) between two processed speech signals would be less than 1.6 if their IS measure is less than 0.5 for various speech codecs. Many experiments in speech recognition show that if the IS measure is less than about 0.1, the two spectra that we compare are perceptually nearly identical.

In this example, IS measures are calculated at different points, after source separation and after speech dereverberation. After source separation, the IS measure is obtained by averaging the result with respect to each one of the $P = 4$ SIMO outputs $y_{s1,p}(k)$ and is denoted by $d_{\mathrm{IS}}^{\mathrm{SS}}$. After speech dereverberation, the final IS measure is denoted by $d_{\mathrm{IS}}^{\mathrm{SD}}$.

Experimental Results

Table 11.1 summarizes the results with four different panel configurations. In order to help the reader better understand the performance, we plot the results of the case with all panels being closed in Figs. 11.9 and 11.10.

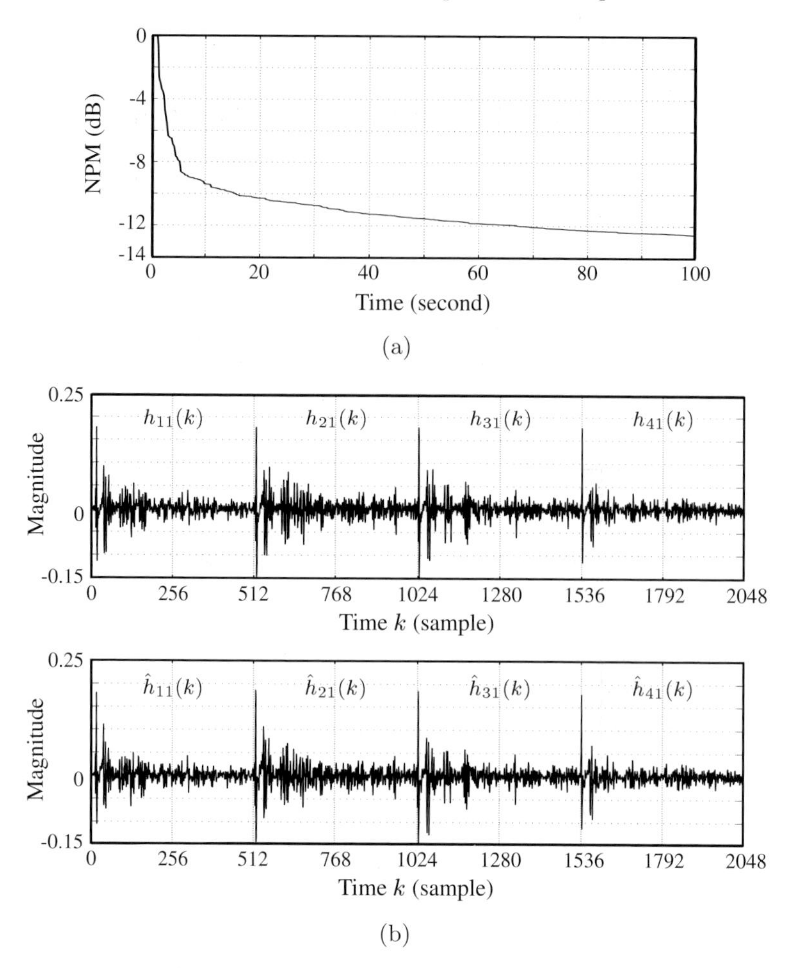

Fig. 11.9. Performance of the adaptive FNMCLMS algorithm for blind identification of the SIMO system corresponding to $s_1(k)$ with all panels being closed. (a) Running average (1000 samples) of the NPM, and (b) comparison of impulse responses between the actual channels and their estimates.

Let's first examine Fig. 11.9, which shows the accuracy of the channel impulse responses blindly estimated using the adaptive FNMCLMS algorithm (see Sect. 6.2.4 and Table 6.5). In this case with all panels being closed, the channel impulse responses are quite long ($L_h = 512$) and blindly identifying such a system is very challenging. But the FNMCLMS algorithm still performs well and converges to a point with only -12.56 dB NPM. This figure only plots the results of the SIMO system corresponding to $s_1(k)$ for a concise presentation. From Table 11.1, we can see that blind identification of the SIMO systems corresponding to $s_2(k)$ and $s_3(k)$ produces more accurate estimates.

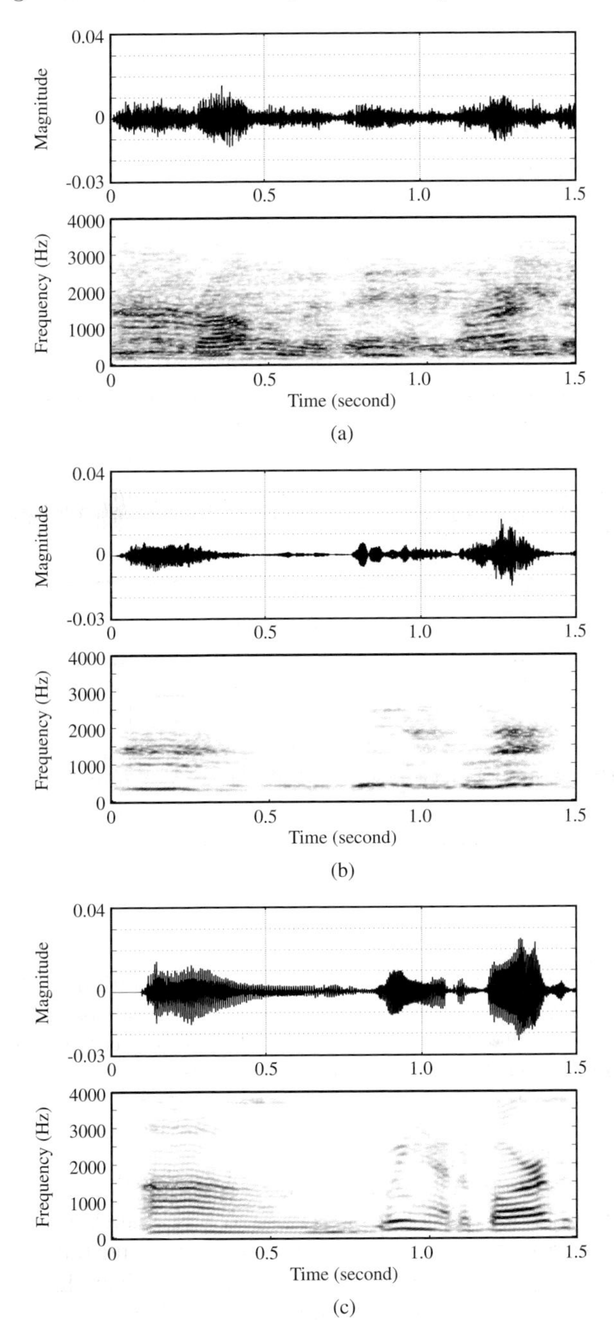

Fig. 11.10. Time sequence and spectrogram (30 Hz bandwidth) of (a) $x_1(k)$, (b) $y_{s_1,1}(k)$, and (c) $\hat{s}_1(k)$ for the experiment carried out in the varechoic chamber with all panels being closed.

Figure 11.10 visualizes how the speech signal of interest is separated from the other two competing sources and how it is dereverberated. From Table 11.1, we see that after source separation, the output SIR is 42.275 dB, which is very high. Listening tests revealed that the separated signals $y_{s_1,p}(k)$ were certainly recognizable although they sounded more echoic as we expected. This can be justified by the spectrogram plot of $y_{s_1,1}(k)$ in Fig. 11.10(b). Apparently, in periods of voiced speech on this narrow-band spectrogram, harmonics are vague, implying strong distortion which results in a large IS measure of 9.54. After dereverberation, the speech signal is satisfactorily recovered though delayed [clearly seen from the time sequence of the recovered signal $\hat{s}_1(k)$ in Fig. 11.10(c)] with a relatively low IS measure of only 0.14. From this example, we see that using only SIR to evaluate a blind source separation algorithm is inadequate, if not misleading.

Let's look at Table 11.1 again and compare the performance for different panel configurations. It is clear that the estimated channel impulse responses by the FNMCLMS algorithm become less accurate as the environment gets more reverberant and we have to model the channels with longer filters. This trend leads to a poorer performance in terms of both SIR and speech distortion. Although we should never stop looking for ways to improve our ability of blindly identifying an acoustic MIMO system, the current techniques seem already useful in source separation and speech dereverberation. Needless to say, it is appealing that the recovered speech signal can attain high perceptual quality with an IS measure lower than 0.1. But in most applications of speech processing, this is an excessive and unnecessary, if not practical, requirement. What we observed in these simulations nevertheless show some promise of successful use of the proposed algorithm in prospect speech processing systems.

11.4 Summary

Although humans can easily distinguish one voice amid a clutter of distracting speech-like interference and background noise in reverberant environments, it is not easy to decipher this ability and it is even more challenging to develop effective source separation and speech dereverberation algorithms to let machines mimic this processing in our brains. After decades of continuous research surrounding this phenomena by psychoacousticians, neural scientists, and signal processing engineers, fascinating, if not overwhelming, advancement has been achieved.

This chapter studied the state-of-the-art techniques of source separation and speech dereverberation. We began with an overview of the cocktail party effect and explained what implications we can derive for developing source separation algorithms. Then we had a survey of microphone arrays beamforming. We analyzed the direction-oriented signal model, illustrated various beamformer structures, commented on the difference between narrowband and broadband beamforming techniques, and discussed the popular linearly constrained minimum-variance (LCMV) and generalized sidelobe cancellation (GSC) algorithms. We have also reviewed the emerging independent component analysis (ICA) methods for blind source separation (BSS). Both cutting-edge high-order-statistics ICA and second-order-statistics BSS algorithms were developed. In the final section, we have presented a synergistic solution to source separation and speech dereverberation based on blind identification of acoustic MIMO systems. Its usefulness was justified by realistic experiments.

References

1. K. Abed-Meraim, E. Moulines, and P. Loubaton, "Prediction error method for second-order blind identification," *IEEE Trans. Signal Processing*, vol. 45, pp. 694–705, Mar. 1997.
2. J. S. Abel and J. O. Smith, "The spherical interpolation method for closed-form passive source localization using range difference measurements echo cancelation," in *Proc. IEEE Int. Conf. Acoust., Speech, Signal Processing*, 1987, vol. 1, pp. 471–474.
3. A. S. Abutaled, "An adaptive filter for noise canceling," *IEEE Trans. Circuits Syst.*, vol. 35, pp. 1201–1209, Oct. 1998.
4. S. Affes, S. Gazor, and Y. Grenier, "Robust adaptive beamforming via LMS-like target tracking," in *Proc. IEEE Int. Conf. Acoust., Speech, Signal Processing*, 1994, vol. 4, pp. 269–272.
5. M. Ali, "Stereophonic echo cancellation system using time-varying all-pass filtering for signal decorrelation," in *Proc. IEEE Int. Conf. Acoust., Speech, Signal Processing*, 1998, pp. 3689–3692.
6. J. B. Allen and D. A. Berkley, "Image method for efficiently simulating small-room acoustics," *J. Acoust. Soc. Am.*, vol. 65, pp. 943–950, Apr. 1979.
7. S. Amari, A. Cichocki, and H. H. Yang, "A new learning algorithm for blind signal separation," in *Advances in Neural Information Processing Systems 8*, Cambridge, MA: MIT Press, pp. 757–763, 1996.
8. S. Amari, "Natural gradient works efficiently in learning," *Neural Computation*, vol. 10, pp. 251–276, Feb. 1998.
9. P. F. Assmann and Q. Summerfield, "Modeling the perception of concurrent vowels: vowels with different fundamental frequencies," *J. Acoust. Soc. Am.*, vol. 88, pp. 680–697, Sept. 1990.
10. C. Avendano, J. Benesty, and D. R. Morgan, "A least squares component normalization approach to blind channel identification," in *Proc. IEEE Int. Conf. Acoust., Speech, Signal Processing*, 1999, vol. 4, pp. 1797–1800.
11. L. A. Baccala and S. Roy, "A new blind time-domain channel identification method based on cyclostationarity," *IEEE Signal Processing Lett.*, vol. 1, pp. 89–91, June 1994.
12. Y. Bard, *Nonlinear Parameter Estimation.* New York: Academic Press, 1974.

13. F. K. Becker and H. R. Rudin, "Application of automatic transversal filters to the problem of echo suppression," *Bell Syst. Tech. J.*, vol. 45, pp. 1847–1850, 1966.
14. A. J. Bell and T. J. Sejnowski, "An information-maxmization approach for blind separation and blind deconvolution," *Neural Comput.*, vol. 7, pp. 1129–1159, 1995.
15. A. J. Bell and T. J. Sejnowski, "The 'independent components' of natural scenes are edge filters," *Vision Research*, vol. 37, pp. 3327–3338, 1997.
16. A. J. Bell, "Information theory, independent-component analysis, and applications," in *Unsupervised Adaptive Filtering, Volume I: Blind Source Separation*, S. Haykin, Ed., New York: John Wiley & Sons, Chap. 6, pp. 237–264, 2000.
17. M. G. Bellanger, *Adaptive Digital Filters and Signal Analysis.* New York: Marcel Dekker, 1987.
18. A. Belouchrani, K. Abed-Meraim, J.-F. Cardoso, and E. Moulines, "A blind source separation technique using second-order statistics," *IEEE Trans. Signal Processing*, vol. 45, pp. 434–444, Feb. 1997.
19. J. Benesty and P. Duhamel, "Fast constant modulus adaptive algorithm," *IEE Proc.*, Pt. F, vol. 138, pp. 379–387, Aug. 1991.
20. J. Benesty and P. Duhamel, "A fast exact least mean square adaptive algorithm," *IEEE Trans. Signal Processing*, vol. 40, pp. 2904–2920, Dec. 1992.
21. J. Benesty, S. W. Li, and P. Duhamel, "A gradient-based adaptive algorithm with reduced complexity, fast convergence and good tracking characteristics," in *Proc. IEEE Int. Conf. Acoust., Speech, Signal Processing*, 1992, pp. 5–8.
22. J. Benesty, P. Duhamel, and Y. Grenier, "A multi-channel affine projection algorithm with applications to multi-channel acoustic echo cancellation," *IEEE Signal Processing Lett.*, vol. 3, pp. 35–37, Feb. 1996.
23. J. Benesty, P. Duhamel, and Y. Grenier, "Multi-channel adaptive filtering applied to multi-channel acoustic echo cancellation," in *Proc. EUSIPCO*, 1996, pp. 1405–1408.
24. J. Benesty, D. R. Morgan, and M. M. Sondhi, "A better understanding and an improved solution to the specific problems of stereophonic acoustic echo cancellation," *IEEE Trans. Speech Audio Processing*, vol. 6, pp. 156–165, Mar. 1998.
25. J. Benesty, D. R. Morgan, J. L. Hall, and M. M. Sondhi, "Synthesized stereo combined with acoustic echo cancellation for desktop conferencing," *Bell Labs Tech. J.*, vol. 3, pp. 148–158, July-Sept. 1998.
26. J. Benesty, D. R. Morgan, and M. M. Sondhi, "A hybrid mono/stereo acoustic echo canceler," *IEEE Trans. Speech Audio Processing*, vol. 6, pp. 468–475, Sept. 1998.
27. J. Benesty, D. R. Morgan, J. L. Hall, and M. M. Sondhi, "Stereophonic acoustic echo cancellation using nonlinear transformations and comb filtering," in *Proc. IEEE Int. Conf. Acoust., Speech, Signal Processing*, 1998, pp. 3673–3676.
28. J. Benesty, A. Gilloire, and Y. Grenier, "A frequency domain stereophonic acoustic echo canceler exploiting the coherence between the channels," *J. Acoust. Soc. Am.*, vol. 106, pp. L30–L35, Sept. 1999.
29. J. Benesty, D. R. Morgan, J. L. Hall, and M. M. Sondhi, "Synthesized stereo combined with acoustic echo cancellation for desktop conferencing," in *IEEE Int. Conf. Acoust., Speech, Signal Processing*, vol. 2, 1999, pp. 853–856.

30. J. Benesty, A. Gilloire, and Y. Grenier, “A frequency-domain stereophonic acoustic echo canceler exploiting the coherence between the channels and using nonlinear transformations,” in *Proc. IWAENC*, 1999, pp. 28–31.
31. J. Benesty, D. R. Morgan, and J. H. Cho, “A family of doubletalk detectors based on cross-correlation,” in *Proc. IWAENC*, Sept. 1999, pp. 108–111.
32. J. Benesty, “Adaptive eigenvalue decomposition algorithm for passive acoustic source localization,” *J. Acoust. Soc. Am.*, vol. 107, pp. 384–391, Jan. 2000.
33. J. Benesty and D. R. Morgan, “Multi-channel frequency-domain adaptive filtering,” in *Acoustic Signal Processing for Telecommunication*, S. L. Gay and J. Benesty, Eds., Boston, MA: Kluwer Academic Publishers, Chap. 7, pp. 121–133, 2000.
34. J. Benesty and D. Morgan, “Frequency-domain adaptive filtering revisited, generalization to the multi-channel case, and application to acoustic echo cancellation,” in *Proc. IEEE Int. Conf. Acoust., Speech, Signal Processing*, 2000, vol. 2, pp. 789–792.
35. J. Benesty, T. Gänsler, and P. Eneroth, “Multi-channel sound, acoustic echo cancellation, and multi-channel time-domain adaptive filtering,” in *Acoustic Signal Processing for Telecommunication*, S. L. Gay and J. Benesty, Eds., Boston, MA: Kluwer Academic Publishers, Chap. 6, pp. 101–120, 2000.
36. J. Benesty, “An introduction to blind source separation of speech signals,” in *Acoustic Signal Processing for Telecommunication*, S. L. Gay and J. Benesty, Eds., Boston, MA: Kluwer Academic, Chap. 15, pp. 321–329, 2000.
37. J. Benesty, D. R. Morgan, and J. H. Cho, “An new class of doubletalk detectors based on cross-correlation,” *IEEE Trans. Speech Audio Processing*, vol. 8, pp 168–172, Mar. 2000.
38. J. Benesty, T. Gänsler, D. R. Morgan, M. M. Sondhi, and S. L. Gay, *Advances in Network and Acoustic Echo Cancellation*. Berlin: Springer, 2001.
39. J. Benesty and S. L. Gay, “An improved PNLMS algorithm,” in *Proc. IEEE Int. Conf. Acoust., Speech, Signal Processing*, 2002, pp. 1881–1884.
40. J. Benesty and T. Gänsler, “A multichannel acoustic echo canceler double-talk detector based on a normalized cross-correlation matrix,” *Eur. Trans. Telecomm.*, vol. 13, pp. 95–101, Mar.-Apr. 2002.
41. J. Benesty, Y. Huang, and D. R. Morgan, “On a class of exponentiated adaptive algorithms for the identification of sparse impulse responses,” in *Adaptive Signal Processing: Applications to Real-World Problems*, J. Benesty and Y. Huang, Eds., Berlin: Springer, Chap. 1, pp. 1–22, 2003.
42. J. Benesty and T. Gänsler, “New insights into the RLS algorithm,” *EURASIP Journal on Applied Signal Processing*, vol. 2004, pp. 331–339, Mar. 2004.
43. J. Benesty, J. Chen, and Y. Huang, “Time-delay estimation via linear interpolation and cross correlation,” *IEEE Trans. Speech Audio Processing*, vol. 12, pp. 509–519, Sept. 2004.
44. J. Benesty, Y. Huang, and J. Chen, “An exponentiated gradient adaptive algorithm for blind identification of sparse SIMO systems,” in *Proc. IEEE Int. Conf. Acoust., Speech, Signal Processing*, 2004, vol. 2, pp. 829–832.
45. J. Benesty and Y. Huang, “The LMS, PNLMS, and exponentiated gradient algorithms,” in *Proc. EUSIPCO*, 2004, pp. 721–724.
46. J. Benesty, Y. Huang, and J. Chen, “An exponentiated gradient adaptive algorithm for blind identification of sparse SIMO systems,” in *Proc. IEEE Int. Conf. Acoust., Speech, Signal Processing*, 2004, pp. II-829–II-832.

47. J. Benesty and T. Gänsler, "Computation of the condition number of a non-singular symmetric Toeplitz matrix with the Levinson-Durbin algorithm," *IEEE Trans. Signal Processing*, vol. 54, pp. 2362–2364, June 2006.
48. A. Benveniste and M. Goursat, "Blind equalizers," *IEEE Trans. Commun.*, vol. COM-32, pp. 871–883, Aug. 1984.
49. K. Berberidis, A. Marava, P. Karaivazoglou, and J. Palicot, "Robust and fast converging decision feedback equalizer based on a new adaptive semi-blind channel estimation algorithm," in *Proc. IEEE GLOBECOM*, 2001, vol. 1, pp. 269–273.
50. N. J. Bershad, "On the optimum data non-linearity in LMS adaptation," *IEEE Trans. Acoust., Speech, Signal Processing*, vol. ASSP-34, pp. 69–76, Feb. 1986.
51. J. Blauert, *Spatial Hearing*. MIT Press, Cambridge, MA, 1983.
52. S. F. Boll, "Suppression of acoustic noise in speech using spectral subtraction," *IEEE Trans. Acoust., Speech, Signal Processing*, vol. ASSP-27, pp. 113–120, Apr. 1979.
53. D. H. Brandwood, "A complex gradient operator and its application in adaptive array theory," *Proc. IEE*, vol. 130, Pts. F and H, pp. 11–16, Feb. 1983.
54. R. F. Brcich, A. M. Zoubir, and P. Pelin, "Detection of sources using bootstrap techniques," *IEEE Trans. Signal Processing*, vol. 50, pp. 206–215, Feb. 2002.
55. D. E. Broadbent, *Perception and Communication*. London: Pergamon Press, 1958.
56. H. Buchner, J. Benesty, and W. Kellermann, "Multichannel frequency-domain adaptive filtering with application to multichannel acoustic echo cancellation," in *Adaptive Signal Processing: Applications to Real-World Problems*, J. Benesty and Y. Huang, Eds., Berlin: Springer, Chap. 4, pp. 95–128, 2003.
57. H. Buchner, J. Benesty, and W. Kellermann, "An extended multidelay filter: fast low-delay algorithms for very high-order adaptive systems," in *Proc. IEEE Int. Conf. Acoust., Speech, Signal Processing*, 2003, vol. 5, pp. 385–388.
58. H. Buchner, R. Aichner, and W. Kellermann, "A generalization of blind source separation algorithms for convolutive mixtures based on second-order statistics," *IEEE Trans. Speech Audio Processing*, vol. 13, pp. 120–134, Jan. 2005.
59. H. Buchner, J. Benesty, and W. Kellermann, "Generalized multichannel frequency-domain adaptive filtering: efficient realization and application to hands-free speech communication," *Elsevier Signal Processing*, vol. 85, pp. 549–570, Mar. 2005.
60. J. Capon, "High resolution frequency-wavenumber spectrum analysis," *Proc. IEEE*, vol. 57, pp. 1408–1418, Aug. 1969.
61. J.-F. Cardoso and B. H. Laheld, "Equivariant adaptive source separation," *IEEE Trans. Signal Processing*, vol. 44, pp. 3017–3030, Dec. 1996.
62. J.-F. Cardoso, "Infomax and maximum likelihood for source separation," *IEEE Signal Processing Lett.*, vol. 4, pp. 112-114, Apr. 1997.
63. J.-F. Cardoso, "Blind signal separation: statistical principles," *Proc. IEEE*, vol. 86, pp. 2009-2025, Oct. 1998.
64. J.-F. Cardoso, "High-order contrasts for independent component analysis," *Neural Comput.*, vol. 11, pp. 157–192, 1999.
65. G. C. Carter, A. H. Nuttall, and P. G. Cable, "The smoothed coherence transform," *Proc. IEEE*, vol. 61, pp. 1497–1498, Oct. 1973.
66. G. C. Carter, "Coherence and time delay estimation," in *Signal Processing Handbook*, C. H. Chen, Ed., New York: Marcel Dekker, 1988.

67. B. Champagne, M. Eizenman, and S. Pasupathy, "Exact maximum likelihood time delay estimation for short observation intervals," *IEEE Trans. Signal Processing*, vol. 39, pp. 1245–1257, June 1991.
68. Y. T. Chan and K. C. Ho, "A simple and efficient estimator for hyperbolic location," *IEEE Trans. Signal Processing*, vol. 42, pp. 1905–1915, Aug. 1994.
69. Y. T. Chan and K. C. Ho, "An efficient closed-form localization solution from time difference of arrival measurements," in *Proc. IEEE Int. Conf. Acoust., Speech, Signal Processing*, 1994, vol. II, pp. 393–396.
70. Y. M. Chang and D. O'Shaughnessy, "Speech enhancement based conceptually on auditory evidence," *IEEE Trans. Signal Processing*, vol. 39, pp. 1943–1954, Sept. 1991.
71. G. Chen, S. N. Koh, and I. Y. Soon, "Enhanced Itakura measure incorporating masking properties of human auditory system," *Signal Processing*, vol. 83, pp. 1445–1456, July 2003.
72. H. Chen, X. Cao, and J. Zhu, "Convergence of stochastic-approximation-based algorithms for blind channel identification," *IEEE Trans. Inform. Theory*, vol. 48, pp. 1214–1225, May 2002.
73. J. Chen, K. K. Paliwal, and S. Nakamura, "Sub-band based additive noise removal for robust speech recognition," in *Proc. EUROSPEECH*, 2001, vol. 1, pp. 571–574.
74. J. Chen, J. Benesty, and Y. Huang, "Robust time delay estimation exploiting redundancy among multiple microphones," *IEEE Trans. Speech Audio Processing*, vol. 11, pp. 549–557, Nov. 2003.
75. J. C. Chen, R. E. Hudson, and K. Yao, "Maximum-likelihood source localization and unknown sensor location estimation for wideband signals in the near-field," *IEEE Trans. Signal Processing*, vol. 50, pp. 1843–1854, Aug. 2002.
76. E. C. Cherry, "Some experiments on the recognition of speech, with one and with two ears," *J. Acoust. Soc. Am.*, vol. 25, pp. 975–979, Sept. 1953.
77. E. C. Cherry and W. L. Taylor, "Some further experiments upon the recognition of speech, with one and with two ears," *J. Acoust. Soc. Am.*, vol. 26, pp. 554–559, July 1954.
78. J. H. Cho, D. R. Morgan, and J. Benesty, "An objective technique for evaluating doubletalk detectors in acoustic cancelers," *IEEE Trans. Speech Audio Processing*, vol. 7, pp. 718–724, Nov. 1999.
79. S. Choi, A. Cichocki, and A. Belouchrani, "Blind separation of second-order nonstationary and temporally colored sources," in *Proc. IEEE Workshop on Statistical Signal Processing*, 2001, pp. 444-447.
80. A. Cichocki, R. Unbehauen, L. Moszczynski, and E. Rummert, "A new online adaptive learning algorithm for blind separation of source signals," in *Proc. ISANN*, 1994, pp. 406–411.
81. T. Claasen and W. Mecklenbrauker, "Comparison of the convergence of two algorithms for adaptive FIR digital filters," *IEEE Trans. Acoust., Speech, Signal Processing*, vol. ASSP-29, pp. 670–678, June 1981.
82. A. B. Clark and R. C. Mathes, "Echo suppressors for long telephone circuits," *Proc. AIEE*, vol. 44, pp. 481–490, Apr. 1925.
83. I. Cohen, "Modeling speech signals in the time-frequency domain using GARCH," *Signal Processing*, vol. 84, pp. 2453–2459, Dec. 2004.
84. P. Comon, "Independent component analysis: a new concept," *Signal Processing*, vol. 36, pp. 287–314, 1994.

85. P. Comon, "Tensor diagonalization, a useful tool in signal processing," in *Proc. 10th IFAC Symposimu on System Identification*, 1994, vol. 1, pp. 77–82.
86. R. M. Corless, G. H. Gonnet, D. E. G. Hare, D. J. Jeffrey, and D. E. Knuth, "On the Lambert W Function," in *Advances in Computational Mathematics*, vol. 5, pp. 329–359, 1996.
87. H. Cox, R. M. Zeskind and M. M. Owen, "Robust Adaptive Beamforming," *IEEE Trans. Acoustic., Speech, Signal Processing*, vol. ASSP-35, pp. 1365–1376, Oct. 1987.
88. C. J. Darwin, "On the dynamic use of prosody in speech perception," in *Structure and Process in Speech Perception*, A. Cohen and S. G. Nooteboom, Eds., Berlin: Springer, pp. 178–194, 1975.
89. C. J. Darwin and C. E. Bethell-Fox, "Pitch continuity and speech source attribution," *J. Exp. Psychol.: Hum. Percept. Perform.*, vol. 3, pp. 665–672, 1977.
90. C. J. Darwin and R. W. Hukin, "Effectiveness of spatial cues, prosody, and talker characteristics in selective attention," *J. Acoust. Soc. Am.*, vol. 107, pp. 970–977, Feb. 2000.
91. M. Dendrinos, S. Bakamidis, and G. Garayannis, "Speech enhancement from noise: a regenerative approach," *Speech Commun.*, vol. 10, pp. 45–57, Feb. 1991.
92. L. Deng, J. Droppo, and A. Acero, "Estimation cepstrum of speech under the presence of noise using a joint prior of static and dynamic features," *IEEE Trans. Speech Audio Processing*, vol. 12, pp. 218–233, May 2004.
93. L. Deng, J. Droppo, and A. Acero, "Enhancement of log mel power spectra of speech using a phase-sensitive model of the acoustic environment and sequential estimation of the corrupting noise," *IEEE Trans. Speech Audio Processing*, vol. 12, pp. 133–143, Mar. 2004.
94. M. Dentino, J. McCool, and B. Widrow, "Adaptive filtering in the frequency domain," *Proc. IEEE*, vol. 66, pp. 1658–1659, Dec. 1978.
95. J. A. Deutsch and D. Deutsch, "Attention: some theoretical considerations," *Psychological Review*, vol. 70, pp. 80–90, 1963.
96. J. M. B. Dias and J. M. N. Leitão, "Efficient computation of $\mathrm{tr}\{\mathbf{TR}^{-1}\}$ for Toeplitz matrices," *IEEE Signal Processing Lett.*, vol. 9, pp. 54–56, Feb. 2002.
97. J. DiBiase, H. Silverman, and M. Branstein, "Robust localization in reverberant rooms," in *Microphone Arrays: Signal Processing Techniques and Applications*, M. Branstein and D. Ward, Eds., Berlin: Springer, 2001.
98. E. J. Diethorn, "A subband noise-reduction method for enhancing speech in telephony and teleconferencing," in *Proc. IEEE Workshop on Applications of Signal Processing to Audio and Acoustics*, 1997.
99. G. Dong and R. W. Liu, "An orthogonal learning rule for null-space tracking with implementation to blind two-channel identification," *IEEE Trans. Circuits Syst. I*, vol. 45, pp. 26–33, Jan. 1998.
100. N. I. Durlach and H. S. Colburn, "Binaural phenomena," in *Handbook of Perception, Volume IV, Hearing*, E. C. Carterette and M. P. Friedman, Eds., New York: Academic Press, 1978, Chap. 10.
101. D. L. Duttweiler, "A twelve-channel digital echo canceler," *IEEE Trans. Commun.*, vol. 26, pp. 647–653, May 1978.
102. D. L. Duttweiler, "Proportionate normalized least-mean-square adaptation in echo cancelers," *IEEE Trans. Speech Audio Processing*, vol. 8, pp. 508–518, Sept. 2000.

103. B. A. Edmonds and J. F. Culling, "The spatial unmasking of speech: evidence for within-channel processing of interaural time delay," *J. Acoust. Soc. Am.*, vol. 117, pp. 3069–3078, May 2005.
104. F. Ehlers, and H. G. Schuster, "Blind separation of convolutive mixtures and an application in automatic speech recognition in a noisy environment," *IEEE Trans. Signal Processing*, vol. 45, pp. 2608–2612, Oct. 1997.
105. P. Eneroth, J. Benesty, T. Gänsler, and S. L. Gay, "Comparison of different adaptive algorithms for stereophonic acoustic echo cancellation," in *Proc. EUSIPCO*, 2000, pp. 1835–1837.
106. Y. Ephraim and D. Malah, "Speech enhancement using a minimum-mean square error short-time spectral amplitude estimator," *IEEE Trans. Acoust., Speech, Signal Processing*, vol. 32, pp. 1109–1121, Dec. 1984.
107. Y. Ephraim and D. Malah,"Speech enhancement using a minimum mean-square error log-spectral amplitude estimator," *IEEE Trans. Acoust., Speech, Signal Processing*, vol. ASSP-33, pp. 443–445, Apr. 1985.
108. Y. Ephraim, D. Malah, and B.-H. Juang, "On the application of hidden Markov models for enhancing noisy speech," *IEEE Trans. Acoust., Speech, Signal Processing*, vol. ASSP-37, pp. 1846–1856, Dec. 1989.
109. Y. Ephraim, "A Bayesian estimation approach for speech enhancement using hidden Markov models," *IEEE Trans. Signal Processing*, vol. 40, pp. 725–735, Apr. 1992.
110. Y. Ephraim, "Statistical-model-based speech enhancement systems," *Proc. IEEE*, vol. 80, pp. 1526–1554, Oct. 1992.
111. Y. Ephraim and H. L. Van Trees, "A signal subspace approach for speech enhancement," *IEEE Trans. Speech Audio Processing*, vol. 3, pp. 251–266, July 1995.
112. W. Etter and G. S. Moschytz, "Noise reduction by noise-adaptive spectral magnitude expansion," *J. Audio Eng. Soc.*, vol. 42, pp. 341–349, May 1994.
113. N. W. D. Evans and J. S. Mason, "Noise estimation without explicit speech, non-speech detection: a comparison of mean, media and modal based approaches," in *Proc. EUROSPEECH*, 2001, vol. 2, pp. 893–896.
114. E. R. Ferrara, Jr., "Fast implementation of LMS adaptive filter," *IEEE Trans. Acoust., Speech, Signal Processing*, vol. ASSP-28, pp. 474–475, Aug. 1980.
115. A. Feuer and E. Weinstein, "Convergence analysis of LMS filters with uncorrelated Gaussian data," *IEEE Trans. Acoust., Speech, Signal Processing*, vol. ASSP-33, pp. 222–230, Feb. 1985.
116. D. R. Fischell and C. H. Coker, "A speech direction finder," in *Proc. IEEE Int. Conf. Acoust., Speech, Signal Processing*, 1984, pp. 19.8.1–19.8.4.
117. J. L. Flanagan, D. A. Berkeley, G. W. Elko, J. E. West, and M. M. Sondhi, "Autodirective microphone systems," *Acustica*, vol. 73, pp. 58–71, Feb. 1991.
118. J. L. Flanagan, A. Surendran, and E. Jan, "Spatially selective sound capture for speech and audio processing," *Speech Communication*, vol. 13, pp. 207–222, Jan. 1993.
119. J. A. N. Flores and S. J. Young, "Continuous speech recognition in noise using spectral subtraction and HMM adaptation," in *Proc. IEEE Int. Conf. Acoust., Speech, Signal Processing*, 1994, vol. 1, pp. 409–412.
120. G. J. Foschini, "Layered space-time architecture for wireless communication in a fading environment using multi-element antennas," *Bell Labs Tech. J.*, vol. 1, pp. 41–59, 1996.

121. W. H. Foy, "Position-location solutions by Taylor-series estimation," *IEEE Trans. Aerosp. Electron. Syst.*, vol. AES-12, pp. 187–194, Mar. 1976.
122. O. L. Frost, III, "An algorithm for linearly constrained adaptive array processing," *Proc. IEEE*, vol. 60, pp. 926–935, Aug. 1972.
123. K. Fukunaga, *Introduction to Statistial Pattern Recognition.* San Diego, CA: Academic, 1990.
124. W. F. Gabriel, "Spectral analysis and adaptive array superresolution techniques," *Proc. IEEE*, vol. 68, pp. 654–666, June 1980.
125. T. Gänsler, M. Hansson, C.-J. Ivarsson, and G. Salomonsson, "A double-talk detector based on coherence," *IEEE Trans. Commun.*, vol. 44, pp. 1421–1427, Nov. 1996.
126. T. Gänsler and P. Eneroth, "Influence of audio coding on stereophonic acoustic echo cancellation," in *Proc. IEEE Int. Conf. Acoust., Speech, Signal Processing*, 1998, pp. 3649–3652.
127. T. Gänsler, J. Benesty, S. L. Gay, and M. M. Sondhi, "A robust proportionate affine projection algorithm for network echo cancellation," in *Proc. IEEE Int. Conf. Acoust., Speech, Signal Processing*, 2000, pp. II-793–II-796.
128. T. Gänsler and J. Benesty, "Stereophonic acoustic echo cancellation and two-channel adaptive filtering: an overview," *Int. J. Adapt. Control Signal Processing*, vol. 14, pp. 565–586, Sept. 2000.
129. T. Gänsler, S. L. Gay, M. M. Sondhi, and J. Benesty, "Double-talk robust fast converging algorithms for network echo cancellation," *IEEE Trans. Speech Audio Processing*, vol. 8, pp. 656–663, Nov. 2000.
130. T. Gänsler and J. Benesty, "A frequency-domain double-talk detector based on a normalized cross-correlation vector," *Signal Processing*, vol. 81, pp. 1783–1787, Aug. 2001.
131. T. Gänsler and J. Benesty, "Double-talk detectors for acoustic echo cancelers," in *Audio Signal Processing for Next-Generation Multimedia Communication systems*, Y. Huang and J. Benesty, Eds., Boston, MA: Kluwer Academic Publishers, Chap. 6, pp. 149–169, 2004.
132. M. Gaeta and J. L. Lacoume, "Source separation without a priori knowledge: the maximum likelihood solution," in *Proc. EUSIPCO*, 1990, pp. 621–624.
133. S. Gannot, D. Burshtein, and E. Weinstein, "Iterative and sequential Kalman filter-based speech enhancement algorithms," *IEEE Trans. Speech Audio Processing*, vol. 6, pp. 373–385, July 1998.
134. S. Gannot, D. Burshtein, and E. Weinstein, "Signal enhancement using beamforming and nonstationarity with applications to speech," *IEEE Trans. Signal Processing*, vol. 49, pp. 1614–1626, Aug. 2001.
135. M. B. Gardner, "Historical background of the Haas and/or precedence effect," *J. Acoust. Soc. Am.*, vol. 43, pp. 1243–1248, 1968.
136. S. L. Gay and S. Travathia, "The fast affine projection algorithm," in *Proc. IEEE Int. Conf. Acoust., Speech, Signal Processing*, 1995, vol. 3, pp. 3023–3027.
137. S. L. Gay, "An efficient, fast converging adaptive filter for network echo cancellation," in *Proc. of Assilomar*, vol. 1, 1998, pp. 394–398.
138. S. L. Gay and J. Benesty, Eds., *Acoustic Signal Processing for Telecommunications.* Kluwer Academic Publishers, Boston, MA, 2000.
139. S. L. Gay, "The fast affine projection algorithm," in *Acoustic Signal Processing for Telecommunication*, S. L. Gay and J. Benesty, Eds., Boston, MA: Kluwer Academic Publishers, Chap. 2, pp. 23–45, 2000.

140. S. L. Gay and S. C. Douglas, "Normalized natural gradient adaptive filtering for sparse and nonsparse systems," in *Proc. IEEE Int. Conf. Acoust., Speech, Signal Processing*, 2002, pp. 1405–1408.
141. A. Gersho, "Adaptive filtering with binary reinforcement," *IEEE Trans. Information Theory*, vol. IT-30, pp. 191–199, Mar. 1984.
142. O. Ghitza, "Auditory models and human performance in tasks related to speech coding and speech recognition," *IEEE Trans. Speech Audio Processing*, vol. 2, pp. 115–132, Jan. 1994.
143. J. D. Gibson, B. Koo, and S. D. Gray, "Filtering of colored noise for speech enhancement and coding," *IEEE Trans. Signal Processing*, vol. 39, pp. 1732–1742, Aug. 1991.
144. A. Gilloire and V. Turbin, "Using auditory properties to improve the behavior of stereophonic acoustic echo cancellers," in *Proc. IEEE Int. Conf. Acoust., Speech, Signal Processing*, 1998, pp. 3681–3684.
145. D. N. Godard, "Self-recovering equalization and carrier tracking in two-dimensional data communication systems," *IEEE Trans. Commun.*, vol. COM-28, pp. 1867–1875, Nov. 1980.
146. G. H. Golub and C. F. Van Loan, *Matrix Computations.* Baltimore, MD: The Johns Hopkins University Press, 1996.
147. M. M. Goulding and J. S. Bird, "Speech enhancement for mobile telephony," *IEEE Trans. Vehicular Technology*, vol. 39, pp. 316–326, Nov. 1990.
148. R. Gray, "On the asymptotic eigenvalue distribution of Toeplitz matrices," *IEEE Trans. Information Theory*, vol. IT-18, pp. 725–730, Nov. 1972.
149. S. M. Griebel and M. S. Brandstein, "Microphone array source localization using realizable delay vectors," in *Proc. IEEE Workshop on Applications of Signal Processing to Audio and Acoustics*, 2001, pp. 71–74.
150. L. J. Griffiths and C. W. Jim, "An alternative approach to linearly constrained adaptive beamforming," *IEEE Trans. Antennas Propagat.*, vol. AP-30, pp. 27–34, Jan. 1982.
151. M. I. Gürelli and C. L. Nikias, "EVAM: an eigenvector-based algorithm for multichannel blind deconvolution of input colored signals," *IEEE Trans. Signal Processing*, vol. 43, pp. 134–149, Jan. 1995.
152. M. I. Gürelli and C. L. Nikias, "A new eigenvector-based algorithm for multichannel blind deconvolution of input colored signals," in *Proc. IEEE Int. Conf. Acoust., Speech, Signal Processing*, 1993, vol. 4, pp. 448–451.
153. E. Hänsler and G. Schmidt, *Acoustic Echo and Noise Control: A Practical Approach.* John Wiley & Sons, Hoboken, NJ, 2004.
154. W. R. Hahn and S. A. Tretter, "Optimum processing for delay-vector estimation in passive signal arrays," *IEEE Trans. Inform. Theory*, vol. IT-19, pp. 608–614, May 1973.
155. P. S. K. Hansen, *Signal Subspace Methods for Speech Enhancement*, Ph.D. dissertation, Techn. Univ. Denmark, Lyngby, Denmark, 1997.
156. A. Härmä, "Acoustic measurement data from the varechoic chamber," Technical Memorandum, Agere Systems, Nov. 2001.
157. J. C. Hassab and R. E. Boucher, "Performance of the generalized cross correlator in the presence of a strong spectral peak in the signal," *IEEE Trans. Acoust., Speech, Signal Processing*, vol. ASSP-29, pp. 549–555, June 1981.
158. M. L. Hawley, R. Y. Litovsky, and J. F. Culling, "The benefit of binaural hearing in a cocktail party: effect of location and type of interferer," *J. Acoust. Soc. Am.*, vol. 115, pp. 833–843, Feb. 2004.

159. S. Haykin, "Radar array processing for angle of arrival estimation," in *Array Signal Processing*, S. Haykin, Ed., Englewood Cliffs, NJ: Prentice-Hall, 1985.
160. S. Haykin, *Adaptive Filter Theory.* Fourth Edition, Upper Saddle River, NJ: Prentice Hall, 2002.
161. S. Haykin and Z. Chen, "The cocktail party problem," *Neural Comput.*, vol. 17, pp. 1875–1902, Sept. 2005.
162. J. Herault, C. Jutten, and B. Ans, "Detection de grandeurs primitives dans un message composite par une architecture de calul neuromimetique un apprentissage non supervise," in *Proc. GRETSI*, 1985.
163. W. Herbordt and W. Kellermann, "Adaptive beamforming for audio signal acquisition," in *Adaptive Signal Processing: Applications to Real-World Problems*, J. Benesty and Y. Huang, Eds., Berlin: Springer, Chap. 6, pp. 155–194, 2003.
164. S. I. Hill and R. C. Williamson, "Convergence of exponentiated gradient algorithms," *IEEE Trans. Signal Processing*, vol. 49, pp. 1208–1215, June 2001.
165. H. G. Hirsch and C. Ehrlicher, "Noise estimation techniques for robust speech recognition," in *Proc. IEEE Int. Conf. Acoust., Speech, Signal Processing*, 1995, vol. 1, pp. 153–156.
166. O. Hoshuyama, A. Sugiyama, and A. Hirano, "A robust adaptive beamformer for microphone arrays with a blocking matrix using constrained adaptive filter," *IEEE Trans. Signal Processing*, vol. 47, pp. 2677–2684, Oct. 1999.
167. O. Hoshuyama, R. Goubran, and A. Sugiyama, "A generalized proportionate variable step-size algorithm for fast changing acoustic environments," in *Proc. IEEE Int. Conf. Acoust., Speech, Signal Processing*, 2004, pp. 161–164.
168. P. O. Hoyer, "Non-negative matrix factorization with sparseness constraints," *J. Machine Learning Res.*, vol. 49, pp. 1208–1215, June 2001.
169. Y. Hu and P. C. Loizou, "A generalized subspace approach for enhancing spech corrupted by colored noise," *IEEE Trans. Speech Audio Processing*, vol. 11, pp. 334–341, July 2003.
170. Y. Hua, "Fast maximum likelihood for blind identification of multiple FIR channels," *IEEE Trans. Signal Processing*, vol. 44, pp. 661–672, Mar. 1996.
171. Y. Hua and J. K. Tugnait, "Blind identifiability of FIR-MIMO systems with colored input using second order statistics," *IEEE Signal Processing Lett.*, vol. 7, pp. 348–350, Dec. 2000.
172. Y. Hua, S. An, and Y. Xiang, "Blind identification of FIR MIMO channels by decorrelating subchannels," *IEEE Trans. Signal Processing*, vol. 51, pp. 1143–1155, May 2003.
173. Y. Huang, J. Benesty, and G. W. Elko, "Adaptive eigenvalue decomposition algorithm for realtime acoustic source localization system," in *Proc. IEEE Int. Conf. Acoust., Speech, Signal Processing*, 1999, vol. 2, pp. 937–940.
174. Y. Huang, J. Benesty, and G. W. Elko, "Microphone arrays for video camera steering," in *Acoustic Signal Processing for Telecommunication*, S. L. Gay and J. Benesty, Eds., Boston, MA: Kluwer Academic, 2000.
175. Y. Huang, J. Benesty, and G. W. Elko, "Passive acoustic source localization for video camera steering," in *Proc. IEEE Int. Conf. Acoust., Speech, Signal Processing*, 2000, vol. 2, pp. 909–912.
176. Y. Huang, J. Benesty, G. W. Elko, and R. M. Mersereau, "Real-time passive source localization: an unbiased linear-correction least-squares approach," *IEEE Trans. Speech Audio Processing*, vol. 9, pp. 943–956, Nov. 2001.

177. Y. Huang and J. Benesty, "Adaptive multi-channel least mean square and Newton algorithms for blind channel identification," *Signal Processing*, vol. 82, pp. 1127–1138, Aug. 2002.
178. Y. Huang and J. Benesty, "A class of frequency-domain adaptive approaches to blind multichannel identification," *IEEE Trans. Signal Processing*, vol. 51, pp. 11–24, Jan. 2003.
179. Y. Huang, J. Benesty, and J. Chen, "Optimal step size of the adaptive multichannel LMS algorithm for blind SIMO identification," *IEEE Signal Processing Lett.*, vol. 12, pp. 173–176, Mar. 2005.
180. Y. Huang, J. Benesty, and J. Chen, "Separation and dereverberation of speech signals with multiple microphones," in *Speech Enhancement*, J. Benesty, S. Makino, and J. Chen, Eds., Berlin: Springer, Chap. 12, pp. 271–298, 2005.
181. Y. Huang, J. Benesty, and J. Chen, "A blind channel identification-based two-stage approach to separation and dereverberation of speech signals in a reverberant environment," *IEEE Trans. Speech Audio Processing*, vol. 13, pp. 882–896, Sept. 2005.
182. R. W. Hukin and C. J. Darwin, "Comparison of the effect of onset asynchrony on auditory grouping in pitch matching and vowel identification," *Perception & Psychophysics*, vol. 57, pp. 191–196, 1995.
183. A. Hyvärinen and E. Oja, "A fast fixed-point algorithm for independent component analysis," *Neural Comput.*, vol. 9, pp. 1483–1492, 1997.
184. S. Ikeda and N. Murata, "An approach to blind source separation of speech signals," in *Proc. Int. Symposium on Nonlinear Theory and Its Applications*, 1998.
185. M. Z. Ikram and D. R. Morgan, "Exploring permutation inconsistency in blind separation of speech signals in a reverberant environment," in *Proc. IEEE Int. Conf. Acoust., Speech, Signal Processing*, 2000, pp. 1041–1044.
186. M. Z. Ikram and D. R. Morgan, "Permutation inconsistency in blind speech separation: investigation and solutions," *IEEE Trans. Speech Audio Processing*, vol. 13, pp. 1–13, Jan. 2005.
187. International Organization for Standardization, Géneve, *Acoustics-Measurement of the Reverberation Time of Rooms with Reference to Other Acoustical Parameters*, 2nd Ed.
188. F. Itakura and S. Saito, "A statistical method for estimation of speech spectral density and formant frequencies," *Electron. Commun. Japan*, vol. 53A, pp. 36–43, 1970.
189. G. Jacovitti and R. Cusani,"On a fast digital method of estimating the autocorrelation of a Gaussian stationary process," *IEEE Trans. Acoust., Speech, Signal Processing*, vol. ASSP-32, pp. 968–976, Oct. 1984.
190. G. Jacovitti and R. Cusani, "An efficient technique for high correlation estimation," *IEEE Trans. Acoust., Speech, Signal Processing*, vol. ASSP-35, pp. 654–660, May 1987.
191. S. H. Jensen, P. C. Hansen, S. D. Hansen, and J. A. Sørensen, "Reduction of broad-band noise in speech by truncated QSVD", *IEEE Trans. Speech Audio Processing*, vol. 3, pp. 439–448, Nov. 1995
192. D. H. Johnson and D. E. Dudgeon, *Array Signal Processing: Concepts and Techniques*. Upper Saddle River, NJ: Prentice Hall, 1993.
193. W. A. Johnston and S. P. Heinz, "Flexibility and capacity demands of attention," *J. Exp. Psychol.: General*, vol. 107, pp. 420–435, 1978.

194. M. Joho and H. Mathis, "Joint diagonalization of correlation matrices by using gradient methods with application to blind signal separation," in *Proc. Sensor Array and Multichannel Signal Processing Workshop*, 2002, pp. 273–277.
195. M. Joho and K. Rahbar, "Joint diagonalization of correlation matrices by using Newton methods with application to blind signal separation," in *Proc. Sensor Array and Multichannel Signal Processing Workshop*, 2002, pp. 403–407.
196. M. Joho, "Blind signal separation of convolutive mixtures: a time-domain joint-diagonalization approach," in *Proc. Int. Symp. Independent Component Analysis Blind Signal Separation*, 2004, pp. 577–584.
197. Y. Joncour and A. Sugiyama, "A stereo echo canceller with pre-processing for correct echo path identification," in *Proc. IEEE Int. Conf. Acoust., Speech, Signal Processing*, 1998, pp. 3677–3680.
198. T. Kailath, "A view of three decades of linear filtering theory," *IEEE Trans. Information Theory*, vol. IT-20, pp. 146–181, Mar. 1974.
199. T. Kailath, *Linear Systems.* Englewood Cliffs, NJ: Prentice-Hall, Inc., 1980.
200. S. M. Kay, *Fundamentals of Statistical Signal Processing: Estimation Theory.* Englewood Cliffs, New Jersey: Prentice-Hall, 1993.
201. W. Kellermann, "A self-steering digital microphone array," in *Proc. IEEE Int. Conf. Acoust., Speech, Signal Processing*, 1991, vol. 5, pp. 3581–3584.
202. J. L. Kelly and B. F. Logan, "Self-adjust echo suppressor," U.S. Patent 3,500,000, Mar. 10, 1970 (filed Oct. 31, 1966).
203. R. L. Kirlin, D. F. Moore, and R. F. Kubichek, "Improvement of delay measurements from sonar arrays via sequential state estimation," *IEEE Trans. Acoust., Speech, Signal Processing*, vol. ASSP-29, pp. 514–519, June 1981.
204. J. Kivinen and M. K. Warmuth, "Exponentiated gradient versus gradient descent for linear predictors," *Inform. Comput.*, vol. 132, pp. 1–64, Jan. 1997.
205. C. H. Knapp and C. G. Carter, "The generalised correlation method for estimation of time delay," *IEEE Trans. Acoust., Speech, Signal Processing*, vol. ASSP-24, pp. 320–327, Aug. 1976.
206. H. Krim and M. Viberg, "Two decades of array signal processing research: the parametric approach," *IEEE Signal Processing Magazine*, vol. 13, pp. 67–94, July 1996.
207. S. M. Kuo and Z. Pan, "An acoustic echo canceller adaptable during double-talk periods using two microphones," *Acoustics Letters*, vol. 15, pp. 175–179, 1992.
208. H. J. Kushner, "On closed-loop adaptive noise cancellation," *IEEE Trans. Automat. Contr.*, vol. 43, pp. 1103–1107, Aug. 1998.
209. C. Kyriakakis, "Fundamental and technological limitations of immersive audio systems," *Proc. IEEE*, vol. 86, pp. 941–951, May 1998.
210. B. Lee, K. Y. Lee, and S. Ann, "An EM-based approach for parameter enhancement with an application to speech signals," *Signal Processing*, vol. 46, pp. 1–14, Sept. 1995.
211. J. C. Lee and C. K. Un, "Performance analysis of frequency-domain block LMS adaptive digital filters," *IEEE Trans. Circuits Syst.*, vol. 36, pp. 173–189, Feb. 1989.
212. H. Lev-Ari and Y. Ephraim, "Extension of the signal subspace speech enhancement approach to colored noise," *IEEE Trans. Speech Audio Processing*, vol. 10, pp. 104–106, Apr. 2003.
213. N. Levinson, "The Wiener rms (root-mean-square) error criterion in filter design and prediction," *J. Math. Phy.*, vol. 25, pp. 261–278, Jan. 1947.

214. J. S. Lim and A. V. Oppenheim, "Enhancement and bandwidth compression of noisy speech," *Proc. IEEE*, vol. 67, pp. 1586–1604, Dec. 1979.
215. H. Liu, G. Xu, and L. Tong, "A deterministic approach to blind equalization," in *Proc. 27th Asilomar Conf. on Signals, Systems, and Computers*, 1993, vol. 1, pp. 751–755.
216. T. Lotter, *Single and Multichannel Speech Enhancement for Hearing Aids.* Ph.D. Dissertation, Aachener Beitrge zu Digitalen Nachrichtensystemen, P. Vary, Ed., RWTH Aachen University, 2004.
217. R. E. Mahony and R. C. Williamson, "Prior knowledge and preferential structures in gradient descent learning algorithms," *Journal of Machine Learning Research*, vol. 1, pp. 311–355, Sept. 2001.
218. D. Mansour and A. H. Gray, Jr., "Unconstrained frequency-domain adaptive filter," *IEEE Trans. Acoust., Speech, Signal Processing*, vol. 30, pp. 726–734, Oct. 1982.
219. R. Martin, "Noise power spectral density estimation based on optimal smoothing and minimum statistics," *IEEE Trans. Speech Audio Processing*, vol. 9, pp. 504–512, July 2001.
220. R. K. Martin, W. A. Sethares, R. C. Williamson, and C. R. Johnson, Jr., "Exploiting sparsity in adaptive filters," *IEEE Trans. Signal Processing*, vol. 50, pp. 1883–1894, Aug. 2002.
221. K. Matsuoka, M. Ohya, and M. Kawamoto, "A neural net for blind separation of nonstationary signals," *Neural Networks*, vol. 8, pp. 411–419, 1995.
222. R. J. McAulay and M. L. Malpass, "Speech enhancement using a soft-decision noise suppression filter," *IEEE Trans. Acoust., Speech, Signal Processing*, vol. ASSP-28, pp. 137–145, Apr. 1980.
223. J. M. Mendel, "Tutorial on higher-order statistics (spectra) in signal processing and system theory: theoretical results and some applications," *Proc. IEEE*, vol. 79, pp. 278–305, Mar. 1991.
224. C. D. Meyer, *Matrix Analysis and Applied Linear Algebra.* Philadelphia, PA: SIAM, 2000.
225. U. Mittal and N. Phamdo, "Signal/noise KLT based approach for enhancing speech degraded by colored noise," *IEEE Trans. Speech Audio Processing*, vol. 8, pp. 159–167, Mar. 2000.
226. M. Miyoshi and Y. Kaneda, "Inverse filtering of room acoustics," *IEEE Trans. Acoust., Speech, Signal Processing*, vol. 36, pp. 145–152, Feb. 1988.
227. L. Molgedey and H. G. Schuster, "Separation of a mixture of independent signals using time delayed correlations," *Physical Review Lett.*, vol. 72, pp. 3634–3636, 1994.
228. M. Montazeri and P. Duhamel, "A set of algorithms linking NLMS and block RLS algorithms," *IEEE Trans. Signal Processing*, vol. 43, pp. 444–453, Feb. 1995.
229. T. K. Moon and W. C. Stirling, *Mathematical Methods and Algorithms.* Upper Saddle River, NJ: Prentice-Hall, 1999.
230. B. C. J. Moore, *An Introduction to the Psychology of Hearing.* London: Academic Press, 1989, Chap. 3.
231. D. R. Morgan and S. G. Kratzer, "On a class of computationally efficient, rapidly converging, generalized NLMS algorithms," *IEEE Signal Processing Lett.*, vol. 3, pp. 245–247, Aug. 1996.

232. D. R. Morgan, "A parametric error analysis of the backward integration method for reverberation time estimation," *J. Acoust. Soc. Am.*, vol. 101, pp. 2686–2693, May 1997.
233. D. R. Morgan, V. N. Parikh, and C. H. Coker, "Automated evaluation of acoustic talker direction finder algorithms in the varechoic chamber," *J. Acoust. Soc. Am.*, vol. 102, pp. 2786–2792, Nov. 1997.
234. D. R. Morgan, J. Benesty, and M. M. Sondhi, "On the evaluation of estimated impulse responses," *IEEE Signal Processing Lett.*, vol. 5, pp. 174–176, July 1998.
235. D. R. Morgan, J. L. Hall, and J. Benesty, "Investigation of several types of nonlinearities for use in stereo acoustic echo cancellation," *IEEE Trans. Speech Audio Processing*, vol. 9, pp. 686–696, Sept. 2001.
236. E. Moulines, O. Ait Amrane , and Y. Grenier, "The generalized multidelay adaptive filter: structure and convergence analysis," *IEEE Trans. Signal Processing*, vol. 43, pp. 14–28, Jan. 1995.
237. E. Moulines, P. Duhamel, J. F. Cardoso, and S. Mayrargue, "Subspace methods for the blind identification of multichannel FIR filters," *IEEE Trans. Signal Processing*, vol. 43, pp. 516–525, Feb. 1995.
238. N. Murata, S. Ikeda, and A. Ziehe, "An approach to blind source separation based on temporal structure of speech signals," *Neurocomputing*, vol. 41, pp. 1–24, Oct. 2001.
239. S. T. Neely and J. B. Allen, "Invertibility of a room impulse response," *J. Acoust. Soc. Am.*, vol. 68, pp. 165–169, July 1979.
240. T. Nishiura, T. Yamada, S. Nakamura, and K. Shikano, "Localization of multiple sound sources based on a CSP analysis with a microphone array," in *Proc. IEEE Int. Conf. Acoust., Speech, Signal Processing*, 2000, pp. 1053–1055.
241. D. A. Norman, "Toward a theory of memory and attention," *Psychological Review*, vol. 75, pp. 522–536, 1968.
242. K. Ochiai, T. Araseki, and T. Ogihara, "Echo canceler with two echo path models," *IEEE Trans. Commun.*, vol. COM-25, pp. 589–595, June 1977.
243. A. V. Oppenheim and R. W. Schafer, *Discrete-Time Signal Processing.* Englewood Cliffs, NJ: Prentice-Hall, 1989.
244. A. V. Oppenheim, E. Weistein, K. C. Zangi, M. Feder, and D. Gauger, "Single-sensor active noise cancellation," *IEEE Trans. Speech Audio Processing*, vol. 2, pp. 285–290, Apr. 1994.
245. K. Ozeki an T. Umeda, "An adaptive filtering algorithm using an orthogonal projection to an affine subspace and its properties," *Electron Commun. Japan*, vol. 67-A, pp. 19–27, 1984.
246. K. K. Paliwal and A. Basu, "A speech enhancement method based on Kalman filtering," in *Proc. IEEE Int. Conf. Acoust., Speech, Signal Processing*, 1987, pp. 177–180.
247. A. Papoulis, *Probability, Random Variables, and Stochastic Pocesses.* New York: McGraw-Hill, 1991.
248. L. Parra and C. Spence, "Convolutive blind separation of non-stationary sources," *IEEE Trans. Speech Audio Processing*, vol. 8, pp. 320–327, May 2000.
249. T. W. Parsons, "Separation of speech from interfering speech by means of harmonic selection," *J. Acoust. Soc. Am.*, vol. 60, pp. 911–918, Oct. 1976.
250. B. Pearlmutter and L. Parra, "Maximum likelihood blind source separation: a context-sensitive generalization of ICA," in *Advances in Neural Information Processing Systems 9*, Cambridge, MA: MIT Press, 1997, pp. 613–619.

251. K. Pearson, "Skew variation, a rejoinder," *Biometrika*, vol. 4, pp. 169–212, 1905.
252. D.-T. Pham and P. Garat, "Blind separation of mixture of independent sources through a quasi-maximum likelihood approach," *IEEE Trans. Signal Processing*, vol. 45, pp. 1712–1725, July 1997.
253. J. Prado and E. Moulines, "Frequency-domain adaptive filtering with applications to acoustic echo cancellation," *Ann. Télécommun.*, vol. 49, pp. 414–428, 1994.
254. W. H. Press, B. P. Flannery, S. A. Teukolsky, and W. T. Vetterling, *Numerical Recipes in C: The Art of Scientific Computing.* Cambridge: Cambridge University Press, 1988.
255. S. R. Quackenbush, T. P. Barnwell, and M. A. Clements, *Objective Measures of Speech Quality.* Englewood Cliffs, NJ: Prentice-Hall, 1988.
256. T. F. Quatieri and R. B. Dunn, "Speech enhancement based on auditory spectral change," in *Proc. IEEE Int. Conf. Acoust., Speech, Signal Processing*, May 2002, vol. 1, pp. 257–260.
257. L. R. Rabiner and B. H. Juang, *Fundamentals of Speech Recognition.* Englewood Cliffs, NJ: Prentice-Hall, 1993.
258. D. V. Rabinkin, R. J. Ranomeron, J. C. French, and J. L. Flanagan, "A DSP implementation of source location using microphone arrays," in *Proc. SPIE*, vol. 2846, pp. 88–99, 1996.
259. A. Rezayee and S. Gazor, "An adpative KLT approach for speech enhancement," *IEEE Trans. Speech Audio Processing*, vol. 9, pp. 87–95, Feb. 2001.
260. M. M. Rose and B. C. J. Moore, "Effects of frequency and level on auditory stream segregation," *J. Acoust. Soc. Am.*, vol. 108, pp. 1209–1214, Sept. 2000.
261. P. R. Roth, "Effective measurements using digital signal analysis," *IEEE Spectrum*, vol. 8, pp. 62–70, Apr. 1971.
262. D. Rupert, "What is kurtosis? an influence function approach," *Amer. Statist.*, vol. 41, pp. 1–5, 1987.
263. W. C. Sabine, *Collected Papers on Acoustics.* Cambridge, MA: Harvard University Press, 1922.
264. H. Saruwatari, S. Kurita, K. Takeda, F. Itakura, T. Nishikawa, and K. Shikano, "Blind source separation combining independent component analysis and beamforming," *EURASIP Journal on Applied Signal Processing*, vol. 2003, pp. 1135–1146, Nov. 2003.
265. Y. Sato, "A method of self-recovering equalization for multilevel amplitude-modulation," *IEEE Trans. Commun.*, vol. COM-23, pp. 679–682, June 1975.
266. H. Sawada, R. Mukai, S. Araki, and S. Makino, "A robust and precise method for solving the permutation problem of fequency-domain blind source separation," *IEEE Trans. Speech Audio Processing*, vol. 12, pp. 530–538, Sept. 2004.
267. H. C. Schau and A. Z. Robinson, "Passive source localization employing intersecting spherical surfaces from time-of-arrival differences," *IEEE Trans. Acoust., Speech, Signal Processing*, vol. ASSP-35, pp. 1223–1225, Aug. 1987.
268. R. O. Schmidt, "A new approach to geometry of range difference location," *IEEE Trans. Aerosp. Electron.*, vol. AES-8, pp. 821–835, Nov. 1972.
269. M. R. Schroeder, "New method of measuring reverberation time," *J. Acoust. Soc. Am.*, vol. 37, pp. 409–412, Mar. 1965.
270. M. R. Schroeder, U.S. Patent No. 3,180,936, filed Dec. 1, 1960, issued Apr. 27, 1965.

271. M. R. Schroeder, U.S. Patent No. 3,403,224, filed May 28, 1965, issued Sept. 24, 1968.
272. C. Servière, "Feasibility of source separation in frequency domain," in *Proc. IEEE Int. Conf. Acoust., Speech, Signal Processing*, 1998, vol. 4, pp. 2085–2088.
273. S. Shimauchi, Y. Haneda, S. Makino, and Y. Kaneda, "New configuration for a stereo echo canceller with nonlinear pre-processing," in *Proc. IEEE Int. Conf. Acoust., Speech, Signal Processing*, 1998, pp. 3685–3688.
274. B. G. Shinn-Cunningham, "Influences of spatial cues on grouping and understanding sound," in *Proc. Forum Acusticum*, 2005.
275. D. V. Sidorovich and A. B. Gershman, "Two-dimensional wideband interpolated root-MUSIC applied to measured seismic data," *IEEE Trans. Signal Processing*, vol. 46, pp. 2263–2267, Aug. 1998.
276. H. F. Silverman, "Some analysis of microphone arrays for speech data analysis," *IEEE Trans. Acoust., Speech, Signal Processing*, vol. ASSP-35, pp. 1699–1712, Dec. 1987.
277. D. Slock, "Blind fractionally-spaced equalization, prefect reconstruction filerbanks, and multilinear prediction," in *Proc. IEEE Int. Conf. Acoust., Speech, Signal Processing*, 1994, vol. 4, pp. 585–588.
278. P. Smaragdis, "Efficient blind separation of convolved sound mixtures," in *Proc. IEEE Workshop on Applications of Signal Processing to Audio and Acoustics*, 1997.
279. J. O. Smith and J. S. Abel, "Closed-form least-squares source location estimation from range-difference measurements," *IEEE Trans. Acoust., Speech, Signal Processing*, vol. ASSP-35, pp. 1661–1669, Dec. 1987.
280. M. M. Sondhi and A. J. Presti, "A self-adaptive echo canceler," *Bell Syst. Tech. J.*, vol. 45, pp. 1851–1854, 1966.
281. M. M. Sondhi, "An adaptive echo canceler," *Bell Syst. Tech. J.*, vol. 46, pp. 497–511, Mar. 1967.
282. M. M. Sondhi and D. A. Berkley, "Silencing echoes on the telephone network," *Proc. of the IEEE*, vol. 68, pp. 948–963, Aug. 1980.
283. M. M. Sondhi, D. R. Morgan, and J. L. Hall, "Stereophonic acoustic echo cancellation – an overview of the fundamental problem," *IEEE Signal Processing Lett.*, vol. 2, pp. 148–151, Aug. 1995.
284. J.-S. Soo and K. K. Pang, "Multidelay block frequency domain adaptive filter," *IEEE Trans. Acoust., Speech, Signal Processing*, vol. 38, pp. 373–376, Feb. 1990.
285. V. Stahl, A. Fischer, and R. Bippus, "Quantile based noise estimation for spectral subtraction and Wiener filtering," in *Proc. IEEE Int. Conf. Acoust., Speech, Signal Processing*, 2000, vol. 3, pp. 1875–1878.
286. P. E. Stoica and A. Nehorai, "MUSIC, maximum likelihood and Cramèr-Rao bound," *IEEE Trans. Acoust., Speech, Signal Processing*, vol. ASSP-37, pp. 720–740, May 1989.
287. J. A. Stuller and N. Hubing, "New perspctives for maximum likelihood timedelay estimation," *IEEE Trans. Signal Processing*, vol. 45, pp. 513–525, Mar. 1997.
288. A. Sugiyama, J. Berclaz, and M. Sato, "Noise-robust double-talk detection based on normalized cross correlation and a noise offset," in *Proc. IEEE Int. Conf. Acoust., Speech, Signal Processing*, vol. 3, 2005, pp. 153–156.

289. V. Summers and M. R. Leek, "F0 processing and the separation of competing speech signals by listeners with normal hearing and with hearing loss,", *J. Sp. Lang. Hear. Res.*, vol. 41, pp. 1294–1306, Dec. 1998.
290. M. Tanaka, Y. Kaneda, S. Makino, and J. Kojima, "Fast projection algorithm and its step-size control," in *Proc. IEEE Int. Conf. Acoust., Speech, Signal Processing*, vol. 2, 1995, pp. 945–948.
291. I. E. Telatar, "Capacity of multi-antenna Gaussian channels," Tech. Rep., Bell Labs, 1995.
292. L. Tong, V. C. Soon, R. Liu, and Y. Huang, "AMUSE: a new blind identification algorithm," in *Proc. IEEE ISCAS*, 1990, vol. 3, pp. 1784–1787.
293. L. Tong, G. Xu, and T. Kailath, "A new approach to blind identification and equalization of multipath channels," in *Proc. 25th Asilomar Conf. on Signals, Systems, and Computers*, 1991, vol. 2, pp. 856–860.
294. L. Tong and S. Perreau, "Multichannel blind identification: from subspace to maximum likelihood methods," *Proc. IEEE*, vol. 86, pp. 1951–1968, Oct. 1998.
295. J. R. Treichler and B. G. Agee, "A new approach to multipath correction of constant modulus signals," *IEEE Trans. Acoust., Speech, Signal Processing*, vol. ASSP-31, pp. 459–272, Apr. 1983.
296. A. Treisman, "Contextual cues in selective listening," *Quarterly J. Exp. Psychol.*, vol. 12, pp. 242–248, 1960.
297. R. J. Vaccaro, "The past, present, and future of underwater acoustic signal processing," *IEEE Signal Processing Magazine*, vol. 15, pp. 21–51, July 1998.
298. P. P. Vaidyanathan, *Multirate Systems and Filter Bank.* Englewood Cliffs, NJ: Prentice-Hall, 1993.
299. S. Van Gerven and D. Van Compernolle, "Signal separation by symmetric adaptive decorrelation: stability, convergence, and uniqueness," *IEEE Trans. Signal Processing*, vol. 43, pp. 1602–1612, July 1995.
300. B. D. Van Veen and K. M. Buckley, "Beamforming: a versatile approach to spatial filtering," *IEEE ASSP Magazine*, vol. 5, pp. 4–24, Apr. 1988.
301. P. Vary, "Noise suppression by spectral magnitude estimation–mechanism and theoretical limits," *Signal Processing*, vol. 8, pp. 387–400, July 1985.
302. S. V. Vaseghi, *Advanced Digital Signal Processing and Noise Reduction.* West Sussex, England: John Wiley & Sons, 2000.
303. J. Vermaak, C. Andrieu, A. Doucet, and S. J. Godsill, "Particle methods for Bayesian modeling and enhancement of speech signals," *IEEE Trans. Speech Audio Processing*, vol. 10, pp. 173–185, Mar. 2002.
304. N. Virag, "Single channel speech enhancement basd on masking properties of human auditory system," *IEEE Trans. Speech Audio Processing*, vol. 7, pp. 126–137, Mar. 1999.
305. S. A. Vorobyov and A. Cichocki, "Adaptive noise cancellation for multi-sensory signals," *Fluctuation Noise Lett.*, vol. 1, pp. 13–23, 2001.
306. C. Wang and M. S. Brandstein, "A hybrid real-time face tracking system," in *Proc. IEEE Int. Conf. Acoust., Speech, Signal Processing*, 1998, vol. 6, pp. 3737–3741.
307. D. L. Wang and J. S. Lim, "The unimportance of phase in speech enhancement," *IEEE Trans. Acoustic., Speech, Signal Processing*, vol. ASSP-30, pp. 679–681, Aug. 1982.
308. H. Wang and P. Chu, "Voice source localization for automatic camera pointing system in videocomferencing," in *Proc. IEEE Int. Conf. Acoust., Speech, Signal Processing*, 1997, pp. 187–190.

309. D. B. Ward and G. W. Elko, "Mixed nearfield/farfield beamforming: a new technique for speech acquisition in a reverberant environment," in *Proc. IEEE ASSP Workshop Appls. Signal Processing Audio Acoustics*, 1997.
310. D. B. Ward, R. C. Williamson, and R. A. Kennedy, "Broadband microphone arrays for speech acquisition," *Acoustics Australia*, vol. 26, pp. 17–20, Apr. 1998.
311. D. B. Ward, R. A. Kennedy, and R. C. Williamson, "Constant directivity beamforming," in *Microphone Arrays*, M. Brandstein and D. Ward, Eds., Berlin: Springer, 2001.
312. W. C. Ward, G. W. Elko, R. A. Kubli, and W. C. McDougald, "The new varechoic chamber at AT&T Bell Labs," in *Proc. Wallance Clement Sabine Centennial Symposium*, 1994, pp. 343–346.
313. M. Wax and T. Kailath, "Optimum localization of multiple sources by passive arrays," *IEEE Trans. Acoust., Speech, Signal Processing*, vol. ASSP-31, pp. 1210–1218, Oct. 1983.
314. E. Weinstein, M. Feder, and A. V. Oppenheim, "Multi-channel signal separation by decorrelation," *IEEE Trans. Speech Audio Processing*, vol. 1, pp. 405–413, Oct 1993.
315. R. D. Wesel, "Cross-correlation vectors and double-talk control for echo cancellation," Unpublished work, 1994.
316. B. Widrow and M. E. Hoff, Jr., "Adaptive switching circuits," *IRE WESCON Conv. Rec.*, Pt. 4, 1960, pp. 96–104.
317. B. Widrow, P. Mantey, L. Griffiths, and B. Goode, "Adaptive antenna systems," *Proc. IEEE*, vol. 55, pp. 2143–2159, Dec. 1967.
318. B. Widrow, "Adaptive filters," in *Aspects of Network and System Theory*, R. E. Kalman and N. DeClaris, Eds., Holt, Rinehart and Winston, NY, 1970.
319. B. Widrow, J. R. Glover, J. M. McCool, J. Kaunitz, C. S. Williams, R. H. Hearn, J. R. Zeidler, E. Dong, and R. C. Goodlin, "Adaptive noise canceling: principles and applications," *Proc. IEEE*, vol. 63, pp. 1692–1975, Dec. 1975.
320. B. Widrow and S. D. Stearns, *Adaptive Signal Processing*, Englewood Cliffs, NJ: Prentice Hall, 1985.
321. N. Wiener and E. Hopf, "On a class of singular integral equations," *Proc. Prussian Acad., Math.-Phys. Ser.*, p. 696, 1931.
322. N. Wiener, *Extrapolation, Interpolation, and Smoothing of Stationary Time Series.* New York: John Wiley & Sons, 1949.
323. F. L. Wightman and D. J. Kistler, "The dominant role of low-frequency interaural time differences in sound localization," *J. Acoust. Soc. Am.*, vol. 91, pp. 1648–1661, Mar. 1992.
324. F. L. Wightman and D. J. Kistler, "Factors affecting the relative salience of sound localization cues," in *Binaural and Spatial Hearing in Real and Virtual Environments*, R. H. Gilkey and T. R. Anderson, Eds., New Jersey: LEA Publishers, 1997, Chap. 1.
325. D. L. Woods, S. A. Hillyard, and J. C. Hansen, "Event-related brain potentials reveal similar attentional mechanisms during selective listening and shadowing," *J. Exp. Phychol.: Hum. Percept. Perform.*, vol. 10, pp. 761–777, 1984.
326. G. Xu, H. Liu, L. Tong, and T. Kailath, "A least-squares approach to blind channel identification," *IEEE Trans. Signal Processing*, vol. 43, pp. 2982–2993, Dec. 1995.
327. M. Xu and Y. Grenier, "Time-frequency domain adaptive filter," in *Proc. IEEE Int. Conf. Acoust., Speech, Signal Processing*, 1989, pp. 1154–1157.

328. H. H. Yang and S. Amari, "Adaptive online learning algorithms for blind separation: maximum entropy and minimum mutual information," *Neural Comput.*, vol. 9, pp. 1457–1482, 1997.
329. H. Ye and B. X. Wu, "A new double-talk detection algorithm based on the orthogonality theorem," *IEEE Trans. Commun.*, vol. 39, pp. 1542–1545, Nov. 1991.
330. J. R. Zeidler, "Performance analysis of LMS adaptive prediction filters," *Proc. of the IEEE*, vol. 78, pp. 1781–1806, Dec. 1990.
331. Q. Zhao and L. Tong, "Adaptive blind channel estimation by least squares smoothing," *IEEE Trans. Signal Processing*, vol. 47, pp. 3000–3012, Nov. 1999.

Index

a posteriori error signal, 45, 61, 99, 137
a posteriori Kalman gain, 66
a posteriori SNR, 277
a priori error signal, 61, 99, 137
a priori Kalman gain, 66
a priori SNR, 277
acoustic echo, 185
acoustic echo canceler, 185, 190
acoustic echo path, 189
acoustic impulse response, 16, 28, 54, 78, 104
acoustically better ear effect, 321
adaptive eigenvalue decomposition algorithm, 227
adaptive infomax algorithm for independent component analysis, 333
adaptive noise cancellation, 262, 302
affine projection algorithm (APA), 65, 99
all-pass filter, 179
angular frequency, 218
attentional selectivity, 319
audio bridge, 206
auditory perception, 320

babbling noise, 344
backward predictor, 36
backward spatial linear prediction, 230
beamforming, 263, 309, 323
beampattern, 325
Bezout theorem, 18, 180
binaural processing effect, 321
blind channel identification, 109
blind equalizer, 177
blind identification via decorrelating subchannels algorithm, 157
blind MIMO identification, 147
blind SIMO identifiability, 111
blind SIMO identification, 111
blind source separation, 331
blindly identifiable, 111
block error signal, 87
block Toeplitz, 49
body language, 1
body-worn microphone, 2
broadband beamformer, 328
Broadbent's filter model, 320

central limit theorem, 337
central moment, 337
channel invertibility, 16
circulant matrix, 52, 88
co-channel interference, 169, 342
co-prime channel impulse responses, 180
co-prime polynomials, 18
cocktail party effect, 207, 319
cocktail party problem, 320
coherence function, 53, 304
coherence method, 204
colorization effect, 15
comb filter, 197
common-zero problem, 18
component-normalization constraint, 113
condition number, 35
constrained multichannel LMS algorithm, 113

constrained multichannel Newton algorithm, 117
constrained optimization problem, 330
continuous-time Dirac Delta function, 28
convergence analysis, 93
convergence in mean, 93
convergence in mean square, 94
convolution matrix, 329
convolutive MIMO system, 154
convolutive mixture, 340
corollary to the principle of orthogonality, 33
correlation matrix, 33
covariance matrix, 193
Cramèr-Rao lower bound, 220, 239, 241
cross relations, 112
cross-correlation coefficient, 233, 234
cross-correlation function, 217, 219
cross-correlation method, 202, 219
cross-correlation vector, 33, 193
cross-relation method, 111
cumulant, 337
cumulative distribution function, 333

de-mixing system, 331
decorrelation, 195
delay-and-sum beamformer, 310, 324
delta function, 150
desktop conferencing, 206
detection, 206
detection statistic, 201
deterministic algorithm, 41
Deutsch & Deutsch and Norman's late selection or response selection model, 320
dichotic listening and shadowing, 320
Dirac filter, 20
direct inverse equalizer, 178
direction of arrival (DOA), 310
direction-oriented signal model, 323
distinct power spectra, 153
distortion measure, 196
double-talk, 187, 200
double-talk detector, 200

echo canceler, 187
echo suppressor, 187
EG± algorithm, 70
eigendecomposition, 42
eigenvalue decomposition, 151
energetic masking, 321
ensemble interval histogram (EIH), 196
error signal, 32, 86, 192
Euclidean distance, 28, 138
excess mean-square error, 45
exponentially decaying filter, 20
exponentiated gradient algorithm, 68
exponentiated RLS algorithm, 72
extended multidelay filter (EMDF), 98

face-to-face communication, 2
far-end, 187, 190
far-field, 309, 324
FastICA algorithm, 338
filter-and-sum beamformer, 324
Fisher information matrix, 241
forgetting factor, 192
forward predictor, 37
forward spatial linear prediction, 228
Fourier matrix, 88, 125
frequency-domain blind identification of convolutive MIMO systems, 157
frequency-domain blind source separation, 340
frequency-domain block error signal, 126
frequency-domain constrained multichannel LMS algorithm, 127
frequency-domain criterion, 89
frequency-domain error signal, 88
frequency-domain LMS (FLMS) algorithm, 85, 98
frequency-domain Newton's algorithm, 132
frequency-domain unconstrained multichannel LMS algorithm, 127
frequency-domain Wiener filter, 287
frequency-selective fading, 15
Frobenius norm, 37
full column rank, 18, 151
full rank, 111
full-duplex, 187
full-duplex exchange, 2
fully excited, 111

Geigel algorithm, 202

generalized cross-correlation function, 224, 225
generalized cross-correlation method, 225
generalized multidelay filter (GMDF), 86
generalized Rayleigh quotient, 280
generalized sidelobe canceller, 329
geometric progression, 97
greatest common divisor, 18, 173

half-duplex, 187
half-wave rectifier, 195
hands free, 2
hybrid, 186
hyperbolic least-squares error, 245

ill-conditioned SIMO system, 142
image analysis technique, 26
image method, 29
image model, 24
immersive acoustic interface, 3
immersive experience, 2
improved PNLMS algorithm, 63
impulse response, 14, 189, 216
independence measure, 332
independent component analysis, 331
infomax, 332
infomax natural gradient adaptive independent component analysis algorithm, 334
infomax stochastic gradient ascent adaptive independent component analysis algorithm, 334
information maximization principle, 331
information-theoretic principle, 337
informational masking, 321
inter-symbol interference, 15, 169
interchannel intensity difference, 209
interchannel time difference, 210
inverse-square law, 15, 28
irreducible, 18
irreducible MIMO system, 151
Itakura distance, 266
Itakura-Saito (IS) measure, 196
Itakura-Saito distance, 268
Itakura-Saito distortion, 345

Jacobian, 333
Jacobian matrix, 241
Johnston & Heinz's multimode or flexible filter model, 320
joint diagonalization, 340
joint entropy, 332
joint probability density function, 331

Kullback-Leibler divergence, 68, 138, 335, 336
kurtosis, 337

Lagrange multiplier, 248
Lambert W function, 73
lateralization, 209
learning curve, 44
least-mean-square (LMS) algorithm, 44
least-squares estimator, 244
Levinson-Durbin algorithm, 40
linear predictive coding (LPC), 196
linear predictor, 282
linear shift-invariant system, 9, 14
linear system, 190, 192
linear update, 60
linear-correction least-squares estimator, 248
linearly constrained algorithm, 312
linearly constrained minimum-variance algorithm, 329
log-likelihood function, 222
log-spectral distance, 266

magnitude-difference function, 219
magnitude-difference method, 220
marginal probability density function, 331
masking, 195
maximum likelihood, 220, 242, 243, 335
maximum likelihood estimator, 243
maximum likelihood method, 224
maximum-likelihood independent component analysis method, 336
mean absolute error, 46
mean opinion score (MOS), 196, 269, 347
mean-square error, 32
memoryless MIMO system, 151
microphone array, 309
MIMO system, 12
MIMO Wiener filter, 48
MIMO Wiener-Hopf equations, 49

minimum mean-square error, 34
minimum statistics, 295
minimum variance estimator, 313
minimum-phase filter, 16, 179
MINT equalizer, 179
MINT method, 342
MINT theory, 180
misalignment correlation matrix, 95
misalignment vector, 41, 93
MISO frequency-domain algorithms, 101
MISO NLMS algorithm, 49
MISO normal equations, 102
MISO system, 12
MISO Wiener-Hopf equations, 49
MMSE, 273, 274
MMSE equalizer, 179
model filter, 86, 193
modified frequency-domain normalized multichannel LMS algorithm, 135
MSE, 271, 273
multi-party teleconferencing, 1
multichannel, 190
multichannel audio bridge, 207
multichannel cross correlation, 227, 228
multichannel cross-correlation coefficient, 235, 237
multichannel cross-correlation method, 237
multichannel diversity, 18
multichannel EG± algorithm, 139
multichannel identification, 190, 192
multidelay filter (MDF), 86, 98
multipath, 216
multipath delay spread, 15
multipath propagation, 169, 177
multiple talks, 342
mutual information, 331
MVDR, 314

narrowband beamformer, 324, 325
natural gradient, 67
natural modes, 41
natural prosodic variation, 322
natural RLS algorithm, 67
near-end, 187, 190
near-field, 309, 324
negative half-wave rectifier, 195
network echo, 186
Neumann series, 253
noise reduction, 261, 275
noise spectrum, 295
noise-reduction factor, 264
non-Gaussianity, 338
non-linear update, 60
non-stationarity, 340
non-whiteness, 340
nonlinearity, 195
nonuniqueness problem, 192
normal equations, 90, 193
normal rank, 175
normalized projection misalignment, 345
normalized cross-correlation method, 203
normalized frequency-domain multichannel LMS algorithm, 133
normalized LMS algorithm, 46, 62
normalized minimum mean-square error, 34
normalized misalignment, 41, 53, 78, 194
normalized MMSE, 273, 274
normalized projection misalignment, 141
normalized step-size parameter, 46
null space, 51, 163, 193

online conferencing, 1
optimum step size, 120
outlier, 339
overlap save, 124

parametric Wiener filter, 296
parametric-model-based method, 270
Pearson's kurtosis, 337
permutation ambiguity, 150
permutation inconsistency problem, 158, 341
pitch continuity, 322
positive half-wave rectifier, 195
precedence effect, 209
principle of orthogonality, 33
principle of superposition, 14
probability of detection, 206
probability of false alarm, 206
probability of miss, 206

probability vector, 68
projection matrix, 65
proportionate NLMS algorithm, 63
prosody, 322

quadratic function, 34
quasi-stationarity, 158

rank, 51
rank deficient, 163
real-time system, 254
receiver operating characteristic, 206
receiving room, 190
recursive least-squares (RLS) algorithm, 66
recursive least-squares error, 192
reflection coefficient, 26
regularization, 36, 96
relative entropy, 68, 138
reverberation, 15, 169, 177, 185
reverberation time, 16, 24, 343
Riemannian space, 67
robust statistics, 205

sample kurtosis, 339
scale ambiguity, 113, 150
Schroeder's backward integration method, 16
Schur complement, 36
second-order independence, 339
second-order-statistics blind source separation approach, 339
Separating Co-Channel and Temporal Interference, 170
sign-data algorithm, 47
sign-error algorithm, 47
sign-sign algorithm, 47
signal cancellation problem, 305
signal subspace, 270
signal-to-interference ratio, 321, 345
signal-to-noise ratio, 265, 321
SIMO system, 11
single-path, 216
single-talk interval, 342
singular value decomposition, 151
SISO system, 9
sound rendering, 2
source localization, 215, 238, 239, 255, 256
source separation, 323
spaciousness, 15
sparse, 19, 59
sparseness measure, 19
spatial attention effect, 321
spatial correlation matrix, 229, 230, 232, 233, 237
spatial correlation vector, 230
spatial filtering, 169
spatial hearing, 321
spatial linear interpolation, 231
spatial realism, 2
spatio-temporal equalization, 169
spectral magnitude restoration, 292
spectral restoration, 269
spectral subtraction, 293
spectral tilt, 328
spectrogram, 344
speech dereverberation, 347
speech detector, 188
speech enhancement, 261
speech intelligibility, 15
speech spectral distortion, 345
speech-distortion index, 265
speech-distortion measure, 263
spherical interpolation, 247
spherical interpolation estimator, 247
spherical intersection estimator, 247
spherical least-squares error, 245
statistical-model-based method, 270
steepest descent, 303
steepest-descent algorithm, 41
step-size parameter, 41
stereo audio bridge, 208
stochastic gradient algorithm, 44
sub-Gaussian, 338
suboptimal filter, 283
super-Gaussian, 338
super-Gaussian distribution, 333
supervised equalizer, 177
Sylvester matrix, 18, 163
synthetic sound, 209
system identification, 86

tail effect, 34, 194
tele-collaboration, 1
teleconferencing, 190, 254
temporal filtering, 262
temporal interference, 169

tethered microphone, 2
time constant, 43
time delay estimation, 215, 235
time difference of arrival (TDOA), 215
time-domain Wiener filter, 270
time-varying channel impulse responses, 14
time-varying filter, 198
Toeplitz, 33
Toeplitz matrix, 87
tracking, 81, 191
transmission room, 190
Treisman's attenuator model, 320
two-channel FLMS algorithm, 104
two-channel MDF, 104
two-path model, 205

unconstrained frequency-domain LMS (UFLMS) algorithm, 85, 98
unconstrained multichannel LMS algorithm, 121
uniform filter, 20
unit-norm constraint, 113

varechoic chamber, 16, 22, 342
variable step-size unconstrained multichannel LMS algorithm, 122
video camera steering, 254
vocal-track size, 322
voice activity detector, 295
voice communication, 2

well-conditioned SIMO system, 142
wideband beamformer, 324
Wiener filter, 32, 34, 269, 271, 274, 296
Wiener gain filter, 299
Wiener-Hopf equations, 33, 272, 274

zero-forcing equalizer, 178

Signals and Communication Technology

(continued from page ii)

Multimedia Communication Technology
Representation, Transmission
and Identification of Multimedia Signals
J.R. Ohm
ISBN 3-540-01249-4

Information Measures
Information and its Description in Science
and Engineering
C. Arndt
ISBN 3-540-40855-X

Processing of SAR Data
Fundamentals, Signal Processing,
Interferometry
A. Hein
ISBN 3-540-05043-4

Chaos-Based Digital Communication Systems
Operating Principles, Analysis Methods,
and Performance Evaluation
F.C.M. Lau and C.K. Tse
ISBN 3-540-00602-8

Adaptive Signal Processing
Applications to Real-World Problems
J. Benesty and Y. Huang (Eds.)
ISBN 3-540-00051-8

**Multimedia Information Retrieval
and Management**
Technological Fundamentals
and Applications
D. Feng, W.C. Siu, and H.J. Zhang (Eds.)
ISBN 3-540-00244-8

Structured Cable Systems
A.B. Semenov, S.K. Strizhakov,
and I.R. Suncheley
ISBN 3-540-43000-8

UMTS
The Physical Layer of the Universal Mobile
Telecommunications System
A. Springer and R. Weigel
ISBN 3-540-42162-9

Advanced Theory of Signal Detection
Weak Signal Detection
in Generalized Observations
I. Song, J. Bae, and S.Y. Kim
ISBN 3-540-43064-4

Wireless Internet Access over GSM and UMTS
M. Taferner and E. Bonek
ISBN 3-540-42551-9

Printing: Krips bv, Meppel
Binding: Stürtz, Würzburg